Sensors & their Applications XVI

Journal of Physics: Conference Series Volume 307

Cork, Ireland
12-14 September 2011

ISBN: 978-1-61839-127-8
ISSN: 1742-6588

Printed from e-media with permission by:

Curran Associates, Inc.
57 Morehouse Lane
Red Hook, NY 12571

Printed by Curran Associates, Inc. (2011)

For permission requests, please contact the Institute of Physics
at the address below.

Institute of Physics
Dirac House, Temple Back
Bristol BS1 6BE UK

Phone: 44 1 17 929 7481
Fax: 44 1 17 920 0979

techtracking@iop.org

Additional copies of this publication are available from:

Curran Associates, Inc.
57 Morehouse Lane
Red Hook, NY 12571 USA
Phone: 845-758-0400
Fax: 845-758-2634
Email: curran@proceedings.com
Web: www.proceedings.com

TABLE OF CONTENTS

INVITED

SESSION 1: NANOSENSING, MODELLING AND SIMULATION

SESSION 2: OPTICAL SENSING

SESSION 3: BIOLOGICAL AND BIOMEDICAL SENSING AND BIOENGINEERING

EXHIBITED PAPERS

PREFACE

This volume records the Proceedings of the sixteenth conference in the biennial Sensors and Their Applications series which took place at the Clarion Hotel, Cork, Ireland between 12–14 September 2011. The conference is organized by the Instrument Science and Technology Group of the Institute of Physics. On this occasion, the conference was hosted by Tyndall National Institute at University College Cork. This year the conference returns to Ireland, having last been held in Limerick in 2003.

The conference proceedings record the continuing growth of the sensors community nationally and internationally. The conferences bring together contributions from scientists and engineers from academia, research institutes and industrial establishments, and therefore provide an excellent opportunity for these communities to present and discuss the latest results in the field of sensors, instrumentation and measurement.

Amongst the more traditional themes, such as optical sensing, there is growth in new areas such as biomedical sensing and instrumentation, and nanosensing, which is reflected in this volume. Similarly the contribution of modelling and simulation techniques in sensor and instrumentation design and their applications is acknowledged by a session in this area. The sessions across the conference are supported by notable contributions from invited speakers.

We would like to thank all of our colleagues in the sensor and instrumentation community who have supported this event by contributing manuscripts. Our thanks also go to Tyndall National Institute for hosting this conference and all the sponsors who, with their generous financial and in-kind contributions, enabled the better organization of this conference. We would also like to thank all the members of the Instrument Science and Technology Group for their support, and in particular for refereeing the submitted manuscripts. We are also pleased to express our thanks to the Conference Department of the Institute of Physics for their invaluable support in organising this event. We are especially grateful to Dawn Stewart for her responsive and efficient day-to-day handling of this event, as well as to Claire Garland for her planning and management of this event.

We hope that the conference authors, participants and a wider audience will find these proceedings to be of interest and to serve as a useful reference text.

Panicos Kyriacou
Conference Chairman
Alan O'Riordan
Conference Local Chairman

All papers published in this volume of *Journal of Physics: Conference Series* have been peer reviewed through processes administered by the proceedings Editors. Reviews were conducted by expert referees to the professional and scientific standards expected of a proceedings journal published by IOP Publishing.

Detecting Disease Biomarkers Using Nanocavities and Nanoparticle Composites

Robert J. Forster, Colm Mallon, Anitha Devadoss and Tia E. Keyes

National Centre for Sensor Research, School of Chemical Sciences, Dublin City University, Dublin 9, Ireland

Email: robert.forster@DCU.ie

Abstract. The convergence of electrochemistry, materials, photonics and biomedical science at the nanoscale opens up significant opportunities for developing advanced sensors. In this contribution, we present examples of our use of nanometer dimensioned electrodes, nanocavities and nanoparticle-metallopolymer composites to create high sensitivity detection platforms and materials for detecting proteins and nucleic acids. The application of these approaches in the diagnosis and prognosis of cancers such as neuroblastoma, as well as point-of-care detection of infectious disease, will be discussed.

1. Introduction

The detection of disease biomarkers well in advance of clinical symptoms offers the possibility of improved patient outcomes and lower overall healthcare costs. However, the concentrations of these protein or nucleic acid biomarkers is typically very low, nanomolar to picomolar and for point-of-care applications using sample volumes of the order of a microlitre is desirable. Thus, the analytical challenge is often to detect a few thousand biomolecules in a complex sample. Success demands highly sensitive detection strategies and sensors that can address small sample volumes. Reducing the critical dimension of electrodes to the micron and smaller length scale offers unique analytical advantages in terms of the sample volume required, the speed of analysis and their high current density which can dramatically enhance the detection sensitivity.[1] Recessed micro-electrodes, offer further advantages and have the potential for single molecule detection limits. Moreover, the cavity can be functionalized, e.g., packed with an electrochemiluminescent nanocomposite,[2-4] to create sensitive sensors that are mechanically stable under real world conditions. However, typical recessed electrodes either have a rather simple geometry, e.g., a recessed planar disk, or the geometry is ill-defined, e.g., when microdisk electrodes are etched with acid, the resulting electrode shape depends strongly on the etch rate and irregular structures may be formed. Therefore, there is a significant need for new approaches that yield reproducible recessed ultramicroelectrodes of well defined geometry without using costly micro-fabrication techniques.

This contribution describes the use of nano-sphere lithography,[5] to create single conducting nanocavities, or arrays of cavities, and significantly shows how their shape can be controlled from spheres to cuboids. Moreover, a strategy to selectively modify their top and interior surfaces using self-assembled monolayers is described. Finally, the direct, electrochemiluminescent detection of biomolecules using a $[Ru(bpy)_2PVP_{10}](ClO_4)_2$ metallopolymer: gold nanoparticle nanocomposite is

Published under licence by IOP Publishing Ltd

demonstrated; bpy is 2,2′-bipyridyl and PVP is poly(4-vinyl pyridine). The limit of detection achieved using this combination of micro- or nano-electrodes and nanocomposites is several orders of magnitude lower than conventional systems.

2. Results and Discussion

2.1 Single and Arrays of Nanocavity Electrodes.
Figure 1 illustrates the approach used to construct single nanocavity electrodes.

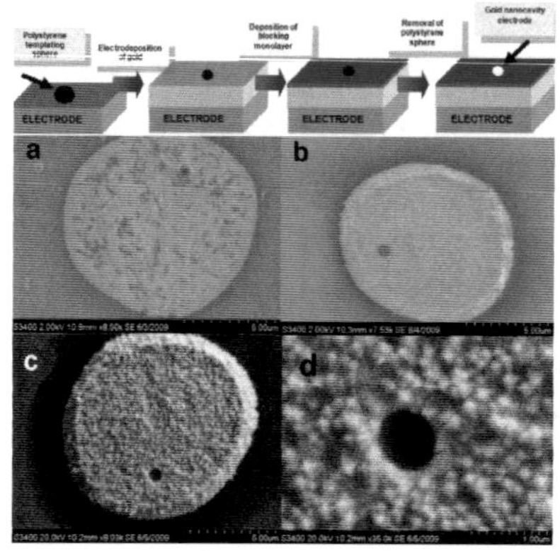

Figure 1. Scheme showing the formation of a nano-cavity electrode and SEM images of (a) a single polystyrene sphere on a 10 μm gold electrode (b) gold which has been electrodeposited around the sphere to a thickness of 600 nm (c) electrode surface after sphere removal, showing a circular pore where the sphere was present and (d) higher magnification image of the pore.

First, a single polystyrene sphere, or an array of spheres, is deposited on an electrode and gold is electrodeposited around the sphere. A blocking layer of 1-hexadecanethiol is then self-assembled at the upper planar gold surface before sphere removal. This blocking layer confines the electrochemical reactions to the inside of the nanocavity. After sphere removal, a truncated spherical cavity is obtained, the dimensions of which is controlled by the templating sphere size and the thickness of electrodeposited gold. The choice of templating sphere limits the cavity size, while the thickness of gold deposited around the sphere controls the actual depth and pore opening produced. Also, by using an intermediate step in which a polydimethylsiloxane, PDMS, mould of the array is cast, the shape of the metal nanocavities can be altered by mechanical stretching. Figure 1 shows an example of a single sphere (r=410 nm) deposited on a 5 μm radius gold microdisk. Figure 1b shows the formation of the nanocavity following gold deposition to a depth of 600 nm. SEM reveals that the cavity opening is approximately 750 nm wide, which is consistent with the charge passed and a gold deposit of approximately 600 nm thickness.

The inset of Figure 2 shows the voltammetric response for a 5 μm radius microelectrode in contact with a 2 mM $[Fe(CN)_6]^{4-}$ dissolved in aqueous 0.2 M Na_2SO_4 as the supporting electrolyte. The electrode corresponds to that shown in Figure 1(b) whereby gold has been electrodeposited around a single templating sphere but no blocking of the surface with a monolayer has taken place and the sphere remains in place. Consistent with a response under radial diffusion control, a well defined steady state current, i_{ss}, is observed.[6]

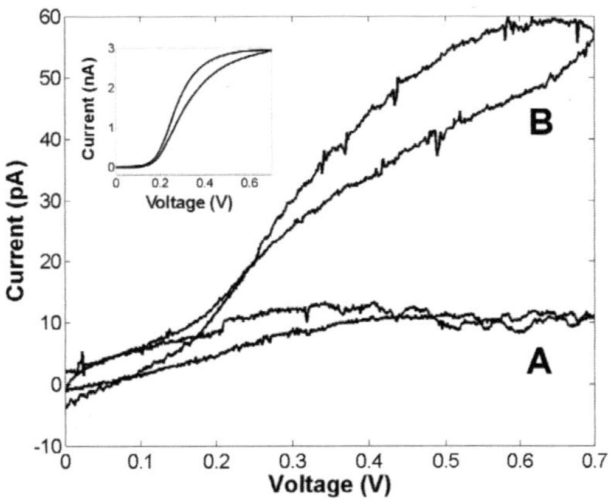

Figure 2. CVs obtained at the nano-cavity electrode in 2 mM $[Fe(CN)_6]^{4-}$ with 0.2 M Na_2SO_4 as the supporting electrolyte. Curves A and B were obtained before and after sphere removal. The exposed gold surface was blocked with a $C_{16}SH$ layer before sphere removal and the inset shows the voltammetric response of the unblocked electrode.

Figure 2 (curve A) shows the response of this electrode after deposition of the C16SH self assembled monolayer, but before polystyrene sphere removal, corresponding to the electrode shown in Figure 1(c). In sharp contrast to the response illustrated in the inset of Figure 2, no well defined steady state current is observed and the current is approximately 350 times lower than that observed before monolayer formation. The blocking response seen here is typical behavior for an alkanethiol layer and a hydrophilic probe such as $[Fe(CN)_6]^{4-}$. When the sphere is removed, but the alkanethiol remains assembled on the upper gold surface, corresponding to Figure 1(d), and the solution phase probe can access the cavity. Figure 2 (curve B) shows the voltammetric response obtained in 2 mM $[Fe(CN)_6]^{4-}$ with 0.2 M Na_2SO_4 as the supporting electrolyte after the sphere has been removed by sonication in THF for 30 minutes. Here, steady state behavior, reminiscent of that seen at the unmodified microelectrode (inset), is clearly observed. However, the current magnitude is significantly smaller than observed at unblocked surface, which reflects the small electrochemically active area of the nanocavity electrode.

2.2 Nanocomposites for High Sensitivity Electrochemiluminescence Detection of Biomolecules.

As well as developing physically small sensor platforms capable of addressing small sample volumes, highly sensitive detection of biomolecules also requires the development of advanced sensor materials. Electrochemiluminescence (ECL), an electrochemically-induced process of generating light at the electrode surface and produces a high signal to noise ratio because an excitation source is not required. Specifically, metallopolymers of the form $[M(bpy)_2PVP_{10}](ClO_4)_2$, where M may be Ru or Os, bpy is 2,2′-bipyridyl, PVP is poly (4-vinylpyridine); are attractive due to their relatively long emission wavelength and stability in both oxidized and reduced forms.[7] Polymers in which the metal complex is coordinated are attractive because they eliminate the need for the luminophore to diffuse to the electrode surface and thin films can be easily formed. Enhancing the electrochemiluminescence intensity generated by these metallopolymer films relative to the background is a significant approach to improving the limits of detection. We recently demonstrated that incorporating 4-(N, N′-dimethylamino) pyridine protected gold nanoparticles (DMAP-AuNPs) enhances the ECL emission intensity of the $[Ru(bpy)_2PVP_{10}]^{2+}$ metallopolymer by up to 7 fold depending on the nanoparticle diameter where the co-reactant is oxalate.[8] This enhancement arises predominantly from a greater film conductivity rather than plasmonic effects. However, the ability of nanocomposites to improve the signal-to-noise ratio and provide lower limits of detection remains under explored.

Figure 3 shows the ECL response for electrodes modified with the nanocomposites at different concentration of β-NADH. It is observed that the ECL emission intensity increases with an increasing β-NADH concentration. The nanocomposite of the metallopolymer and 12.5 nm DMAP protected gold nanoparticles dramatically lowers the detection limit of β-NADH due to the enhanced ECL emission intensity and fM concentrations can be detected.

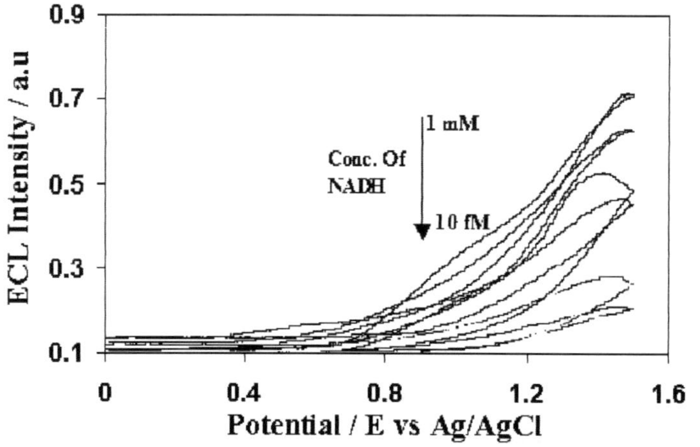

Figure 3. Dependence of ECL emission intensity of ruthenium metallopolymer-gold nanoparticle composite with various concentrations of β-NADH ranging from (top to bottom) 1 mM, 750 μM, 500 μM, 1 μM, 1 nM, 1 pM, 100 fM and 10 fM, respectively.

3. Conclusions
Nanocavity electrodes are attractive for biosensing including electrochemical detection of molecules in flow systems and offer greater sensitivity through metal enhanced emission. The small volume of the cavities (less than 300 picolitres) is also useful for self-metering of sample volumes. The ability to fill the cavity with fluid or to selectively functionalize its interior with reagent whose release can be electrochemically triggered, may open up new applications in sensing and localized drug release. Combinations of these novel platforms with metallopolymer nanocomposites ought to enable the detection of ultralow concentrations (femto molar) of biomolecules.

4. Acknowledgement
This material is based upon work supported by the Science Foundation Ireland under Grant No. [10/IN.1/B3021].

References
[1] Forster R J and Keyes T E 2006 *Handbook of Electrochemistry*, Ed. Zoski, C.
[2] Forster R J, Bertoncello P and Keyes T E 2009 *Ann. Rev. Anal. Chem.* **2** 359
[3] Bertoncello P and Forster R J 2010 *Biosens. and Bioelect.* **24** 3191
[4] Devadoss A, Dickinson C, Keyes T E and Forster R J 2011 Anal. Chem. **83** 2383
[5] Mallon C T, Zuliani C, Keyes T E and Forster R J 2010 *Chem. Comm.* **46** 7109
[6] Forster R J 2003 *Encyclopedia of Electrochemistry*, Eds. P. A. Unwin and A. J. Bard, p. 160–

195, Wiley, New York
[7] Hogan C F and Forster R J 1999 *Anal. Chim. Acta* **396** 13
[8] Devadoss A, Spehar-Deleze A, Bertoncello P, Marthi R, Keyes T E and Forster R J 2010 *Langmuir*, **26** 2130

Presenter: Christopher Whalen
Organization: Sierra Sensors GmbH

Title:

A New form of Microfluidic Sample Delivery for High Throughput Biosensor Analysis

Abstract:

As the biosensor industry continues to mature many of the big breakthroughs in analysis performance will not come from huge advances in detection, but from the tangential technologies. One of these technologies is sample delivery. It is now clearly understood that the way samples are presented to the sensing surface(s) significantly impacts the performance and applicability of analytical biosensors. There is also the ever present desire to simultaneously analyze as many samples as possible. A sample delivery system that provides high throughput, high performance, and flexibility, could dramatically change the use of analytical biosensors. This presentation will focus on a new form of microfluidic sample delivery called Hydrodynamic Isolation (HI). Through a combination of hydrodynamic focusing and location specific sample delivery and evacuation, HI can simultaneously deliver different sample solutions to each sensing location on any two-dimensional detection array. Even in an open array the different solution streams are completely independent, are delivered as very discrete volumes, and a single solution can be addressed to one or more sensor locations. HI eliminates the need for mechanical micro-valves close to the detection chamber, making it simple to build and multiplex. By enabling full flexibility in sample delivery across 2-D arrays, HI has the potential to greatly improve analysis throughput but it will also have a big impact on assay design and development times. The use of HI in an array based real-time label-free analysis platform will be presented.

Journal of Physics: Conference Series **307** (2011) 012003 doi:10.1088/1742-6596/307/1/012003

Closing the Loop with Sensors in Commercial Building Systems: Applying Lessons from Automotive Vehicles

Automotive systems have evolved extensively over the past 50 years, providing a fully integrated system of sub-systems that work in concert for optimal vehicle level closed loop control. In this talk we look at several automotive sub-systems: stability and control, safety and security, emissions and comfort, diagnostics and maintenance, infotainment and communications; with an eye toward understanding their technology drivers and associated value propositions. Conversely, we examine how commercial building systems currently are represented as a collection of sub-systems that often work independently of each other for local optimization, often relying upon open loop control systems developed and installed decades ago. Reasoning primarily by analogy we explore opportunities for energy and efficiency, comfort and environment, and safety/security; asking whether there is sufficient value associated with a new class of building sensors and how those technologies might be brought to bear in improving performance. Finally, we examine the fundamental architecture of detection systems built upon sensing elements, with the aim of understanding trade-offs between: detection, false alarm rate, power, and cost.

Journal of Physics: Conference Series **307** (2011) 012003 doi:10.1088/1742-6596/307/1/012003

Biographical Sketch

Dr. Joseph Mantese is a Research Fellow at United Technologies Corporation's Research Center; specializing in electronic materials, components, sensors, and packaging. Prior to joining UTRC, Dr. Mantese was Department Head of Delphi Research Laboratories (Materials, Components, and Packaging) the central research laboratory responsible for developing advanced technologies for automotive systems, including systems for: safety, entertainment, HVAC, connection, and emission. Before joining Delphi in 1999, Dr. Mantese was a member of General Motors Research and Development Laboratories where he was Section Leader of sensor development. Dr. Mantese is the recipient of an R&D 100 Award (1997) for the development of industrial scale plasma ion implantation, winner of UTRC's 2010 Outstanding Achievement Award for his work related to multi-species chemical sensing, is twice winner of General Motors' Campbell Award (1990 and 1995) for scientific breakthroughs in materials science, an inductee and subsequent honoree of Delphi Corporation's Hall of Fame (2000, 2004) for scientific research and creation of corporate intellectual property, and the winner of Wayne State University's Socius Collegii Award (2004) for collaborative research with the school of engineering. Dr. Mantese is the holder of 32 patents pertaining to electronic materials, sensors, MEMS, and components. He is also the author of over 90 peer reviewed papers, a book on the fundamentals of graded ferroic materials, and three book chapters.

Towards Nanowire (Bio) Sensors

K Dawson and A O'Riordan[1]
Tyndall National Institute, University College Cork, Lee Maltings, Cork, Ireland

E-mail: alan.oriordan@tyndall.ie

Abstract. Nanowire based electrochemical devices represent a new and emerging class of sensors. In this work, we fabricate gold nanowires with well controlled critical dimensions using hybrid-electron beam lithography. Individual nanowire electrodes are routinely fabricated and are easily electrically contacted using overlay electrodes. Following fabrication, nanowire device performance is assessed using both electrical and electrochemical characterization techniques. We observe low electrical resistances (~ 900 Ω) with typical linear Ohmic responses from fully packaged nanowire devices. Steady-state cyclic voltammograms in ferrocenemonocarboxylic acid demonstrate scan rate independence up to 1000 mV s^{-1}, with low interfacial capacitance. The applicability of these nanowire electrodes for sensing was explored using square wave voltammetry.

1. Introduction

In recent years, the application area of biosensors has experienced substantial growth with the global biosensor market estimated to exceed $10.6 billion by 2013. [1] The emerging trend for point-of-care instrumentation is driving the development of miniaturized systems capable of reliable and rapid quantification of a wide range of biomolecules. [2-5] To address this opportunity, nanosensors based on one-dimensional nanostructures such as silicon nanowires and carbon nanotubes FET devices are being explored since these nanostructures offer the potential to achieve highly sensitive integrated sensors with direct electrical readout. [6-8] In these sensor systems, the perturbation in the local electrostatic environment of a nanowire-based FET by a charged analyte alters the conductance, which is employed as the sensor signal. [6] While elegant, sensor technologies such as this require complex heterogeneous integration approaches following fabrication and certain device architectures, may require the application of back gate voltages as high as 40 V to be applied. [9]

By contrast, electrochemical sensors typically have much lower voltages (-1.5 – 1.5 V) requirements. Nanoscale electrodes have tremendous potential as electrochemical sensors exhibiting enhanced performance, compared to traditional macroelectrodes. As critical dimensions of the electrodes enter the nano regime, radial analyte diffusion profiles dominate with a corresponding increase in mass transport rate, higher current densities giving rise to an increased I_F/I_C ratio (due to smaller surface area), higher S/N ratios and steady-state sigmoidal voltammograms. [10]

Nanowires represent a new and emerging powerful class of sensors and to date have been utilized for direct, sensitive and rapid detection of biological and chemical species. [2, 3, 11-14] Noble metal nanowires such as gold, platinum and silver, are of particular importance due to their excellent

[1] To whom any correspondence should be addressed.

electrical and optical properties. Nanowire sensors based on surface enhanced Plasmon resonance, electrochemical, and surface enhanced Raman scattering detection mechanisms have been reported. [15-17] Gold nanowires are of particular interest due to their high chemical and thermal stability and biocompatibility. However, widespread take-up of nanoscale electrodes has been limited by the lack of practical and effective methodologies for fabrication of robust nanoelectrodes.

A great variety of nanowire fabrication techniques for have been reported to date, [14, 17-23] a versatile approach for fabrication of robust and reproducible nanowires is electron beam lithography (EBL), metal deposition and lift-off. This approach delivers on-chip nanowires with reproducible well defined geometries that are easily contactable with overlaid interconnecting electrodes (using standard micron-scale photolithography techniques). In this work, standard electroanalytical techniques such as cyclic voltammetry and square wave voltammetry were employed to investigate the electrochemical characteristics and the potential applications of nanowire electrodes fabricated by hybrid-EBL as future sensors.

2. Experimental

2.1. Nanowire Fabrication

Nanowire electrodes were fabricated using a hybrid EBL-photolithography process on silicon substrates with a ~300 nm layer of silicon dioxide thermally grown. Nanowire structures were patterned by direct electron beam writing in resist (ZEP 520 Nippon Zeon, beam voltage 50k V, beam current 100 pA and beam dose 120 $\mu C\ cm^{-2}$), followed by metal evaporation (Ti/Pt/Au 5/3/30 nm) and lift-off. Optical lithography was employed to overlay contact electrodes and metal interconnection tracks were onto pre-fabricated nanowires using optical lithography, with metal evaporation (Ti/Au 10/200 nm) and liftoff. A passivation layer of photoresist (~1 μm) was spin coated onto a chip and a trench opened in this passivation layer directly over nanowire electrode to allow contact the electrolyte solution and the nanowire exclusively. To complete device packaging, chips were assembled onto specifically designed printed circuit boards, electrically contacted using wedge wire bonding (25 μm gold wires) and the bond wires protected using epoxy.

2.2. Nanowire characterisation:

Scanning electron microscopy (SEM) images of single Au nanowires were acquired using a field emission SEM (JSM-6700F, JEOL UK Ltd.) operating at beam voltages between 5 and 10 kV. Optical micrographs were acquired using a calibrated microscope (Axioskop II, Carl Zeiss Ltd.) equipped with a charge-coupled detector camera (CCD; DEI-750, Optronics). Two point electrical measurements were performed on a probe station (Model 6200, Micromanipulator Probe Station) in combination with a source meter (Keithly 2400) using a dedicated LabVIEW™ V8.0 program. For these current-voltage (I-V) measurements, the source electrode was grounded, a bias sweep up to ± 10 mV was applied to the drain electrode, and the current through the nanowire was measured.

2.3. Electroanalysis at Nanowire Electrodes:

All electrochemical studies were performed using a CHI660a Electrochemical Analyzer and Faraday Cage CHI200b (CH Instruments). Experiments were carried out in a three-electrode cell employing the single nanowires as working electrodes, with platinum coil (BAS Inc) as counter electrode and Ag/AgCl (KCl sat) reference electrode (CH Instruments). Voltammetric experiments on ferrocene monocarboxylic acid, FcCOOH (1 mM) (Alfa Aesar) were carried out in phosphate buffered saline, PBS (10 mM) (Sigma Aldrich) solutions at pH 7.4, (0.0 to 0.6 vs Ag/AgCl). Square wave voltammetry was performed in across a potential range of 0.0-0.5 V, applying an increment potential of 1 mV, pulse amplitude of 25 mV and a frequency of 25 Hz. All solutions were prepared with deionized water; 18.2 MΩ cm (ELGA Pure Lab Ultra).

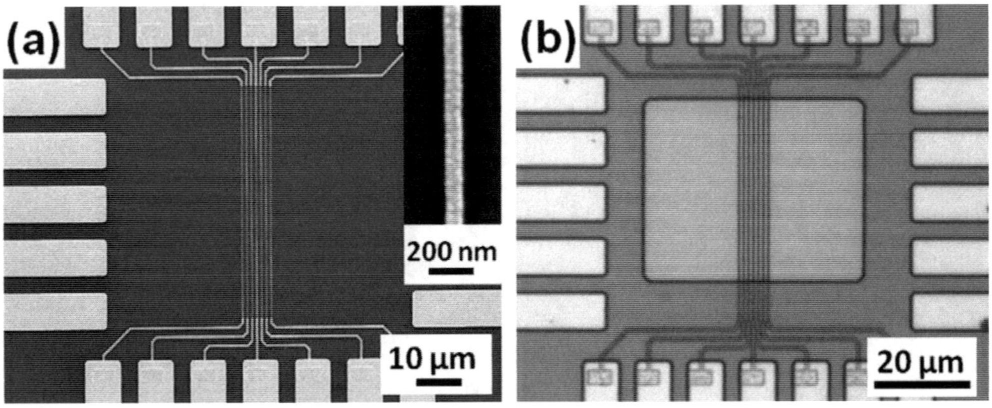

Figure 1: (a) SEM micrograph of individual nanowire electrodes with overlaid metal contacts (1,000x mag). Inset: A high resolution SEM micrograph of single gold nanowire electrode (70,000x mag) fabricated by EBL. (b) An optical micrograph of fully integrated and passivated nanowire electrodes.

3. Results and Discussion:

3.1. Nanowire Structural characterization:

Individual gold nanowire electrodes were fabricated using a hybrid EBL and photolithography process at silicon wafer substrates bearing a 300 nm layer of thermally grown silicon dioxide, see experimental section. Following device fabrication, nanowire electrodes were characterized using a combination of, SEM and optical microscopy. Individual single nanowire electrodes of width ~ 100 ± 6 nm were routinely fabricated by EBL, as depicted at 70,000x magnification in figure 1a inset. Electrical contacts and interconnection tracks were overlaid onto nanowire electrodes by standard optical lithography, followed by metal deposition and liftoff. Each nanowire electrode was individually contacted as depicted in figure 1a. Figure 1b shows an optical micrograph of a fully packaged array of individual nanowire devices, electrically contacted and passivated with an overlaying insulating photoresist layer. The insulating overlayer facilitated prevention of unwanted electrochemical reactions occurring between metal interconnection tracks and electrochemically active species. To allow direct interaction exclusively between a nanowire and electrolytic solution, photolithography was employed to pattern a trench (40 x 50 µm), which was selectively opened in the photoresist layer, directly above the nanowire array.

3.2. Nanowire Electrical Characterization:

Electrical characterization using standard two-point *I-V* measurements was performed to confirm the functionality of packaged nanowire devices in the voltage range of -10 mV to +10 mV. Several nanowire electrodes were characterized in this fashion and each displayed Ohmic linear responses confirming electrical contact to the nanowires by the interconnection tracks. Fully packaged nanowire devices exhibited low resistance (890-911 Ω); shown in figure 2a. Standard open- and short- circuit control experiments were also performed on the packaged chips to confirm the observed electrical characteristics were exclusively due to individual nanowires and without stray contributions from the substrate.

3.3. Nanowire Electrochemical Characterization:

Cyclic Voltammetry at Nanowire Electrodes: Electrochemical investigations were undertaken employing a potentiostat with a Faraday cage to apply an appropriate potential sweep range to a

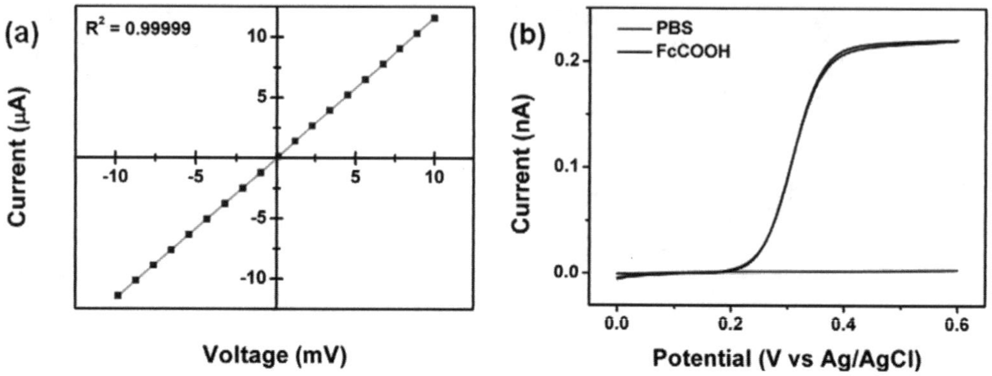

Figure 2: (a) Two point I-V measured across the voltage range; -10 mV to 10 mV, for a typical fully integrated nanowire displaying Ohmic behavior with low resistance, 890-911 Ω. (b) Cyclic voltammograms measured at a single gold nanowire electrode in 10 mM PBS, pH 7.4, (N_2 Sat) in the absence and presence of 1 mM FcCOOH at 5 mV/s.

nanowire working electrode with versus Ag/AgCl (saturated KCl) reference with Pt coil as a counter electrode. Ferrocenemonocarboxylic acid was selected as a suitable analyte for this study due to its good chemical stability and simple electrochemical redox behavior.[24] Figure 2b depicts a typical cyclic voltammogram (CV) acquired using a gold nanowire device in 1 mM FcCOOH in 10 mM PBS (N_2 Sat), pH 7.4 and 25 °C. The CVs measured displayed steady-state voltammograms, typical of nanoelectrodes for the single-electron oxidation of FcCOOH. At nanowire electrodes mass transport is greatly enhanced, due to the small critical dimensions of the nanowires and becomes comparable to or larger than the rate of electron transfer from the FcCOOH to the electrode surface. [25] Control CV experiments were also undertaken at nanowire electrodes in deaereated 10 mM PBS buffer (in the absence of FcCOOH), pH 7.4, 5 mV s^{-1} and 25 °C. Under these conditions no Faradaic current was measured, with very low charging currents ~ 4 pA indicating that the interconnection tracks were sufficiently insulated by the resist layer from the electrolytic solution.

3.3.1. Capacitance: The steady-state nature of the current response at nanowire electrodes in 1 mM FcCOOH was preserved with increasing scan rate up to 1000 mV/s, as shown in figure 3a. This permits rapid signal acquisition, which is critical for sensing applications. Slight increases in the magnitude of the limiting current may be attributed to the presence of capacitive current which increases linearly with respect to increasing scan rate as described by equation 1 [26];

$$I_C = 2AC_d\upsilon \qquad (1)$$

where I_C is the difference in capacitance current from the forward and reverse scans of a CV, A is the active electrode area, C_d is the capacitance and v denotes scan rate. To confirm this, the capacitance of individual nanowires was thus estimated from the non Faradaic region of CVs captured in deaereated 1 mM FcCOOH in 10 mM PBS at scan rates from 5 mV/s to 1000 mV/s. The difference in capacitive current (I_C) (measured at 0.1 V vs. Ag/AgCl) was found to have a strong linear dependence on scan rate, (υ) as depicted in figure 3b. From the slope of this relationship ($2AC_d$) the interfacial capacitance (C_d) of nanowire electrodes was estimated to be 15.6 ± 2 nF/cm^2 which is consistent with literature for nanoscale electrodes. [24] In this case, the capacitance has been normalized to the geometric area of the passivated interconnection tracks and the exposed nanowire electrodes. The observed low capacitive contributions to CVs measured at nanowire electrodes

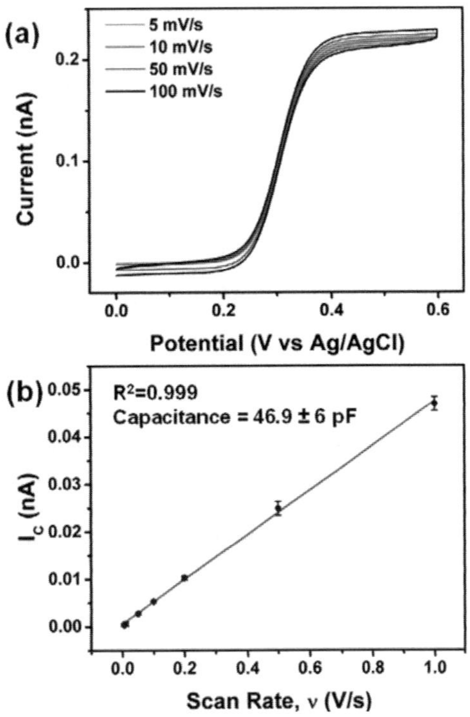

Figure 3: (a) Cyclic voltammograms at a single nanowire electrode in 1 mM FcCOOH in 10 mM PBS (N_2 Sat) at a range of scan rates, (b) relationship between the capacitive current (I_C) versus scan rate (v). I_c values measured at the non-Faradaic region (0.1 V vs. Ag/AgCl) of CVs in 1 mM FcCOOH at single gold nanowire.

enable their use over fast time intervals with greater accuracy than larger electrodes. Reducing the contributions of capacitance to the overall current response at an electrode can greatly enhance the signal to noise ratio of electrode for sensing, allowing lower limits of detection to be achieved. .

3.3.2. Square Wave Voltammetry: Pulse voltammetric techniques; such as square wave voltammetry (SWV), are routinely employed as methods of measurement for electrochemical sensors, as they are known to amplify electrochemical signals, enabling lower limits of detection. To demonstrate the applicability of nanowire based electrodes to future sensing square wave voltammetry was carried out in a range of concentrations of FcCOOH in 10 mM PBS (N_2 Sat). SWVs were achieved at amplitude of 25 mV, frequency of 25 Hz, with a 1 mV step increment across an applied potential range of 0.0 V to 0.5 V. The resultant voltammograms, seen in figure 4, displayed typical current peaks at ~ 220 mV vs Ag/AgCl which is in good agreement with previous reports.[24] While capacitive current observed in SWVs appears to become more pronounced at increasingly positive potentials, the ratio of Faradaic current to capacitive current (I_F/I_C) increased linearly with concentration. This strongly suggests that pulse voltammetric current sampling at single nanowire electrodes further broadens the potential sensing applications that can be developed based on these devices.

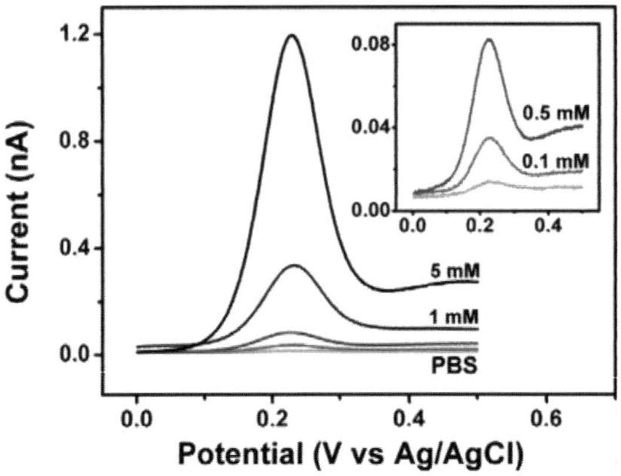

Figure 4: Square wave voltammetry at single gold nanowire in deaerated 10 mM PBS and different concentrations of FcCOOH. Inset: Close up of the square wave voltammograms of lower concentrations (0.1 mM and 0.5 mM) of FcCOOH and the blank PBS.

4. Conclusions:

Individual gold nanowire electrodes, with highly reproducible critical dimensions, were fabricated a hybrid EBL-photolithography process on silicon substrates. Two point electrical experiments of fully packaged nanowire electrodes revealed low resistances. Investigations of the electrochemical properties of individual nanowires via cyclic voltammetry demonstrated reproducible steady-state behavior typical of electrodes with nanoscale critical dimensions in the presence of a standard redox mediator, ferrocenemonocarboxylic acid. This implies very fast mass transport of the analyte to electrode, which could increase the sensitivity of future sensing devices based on this platform. Preservation of the steady-state nature of CVs over a wide range of scans, coupled with low capacitive contributions demonstrates the potential suitability for such electrodes to be employed in rapid analysis at low concentrations. As a result of these findings the devices described in this work appear to be very amenable to future applications as electrochemical based sensors. To further explore this potential, nanowire electrodes have been investigated using square wave voltammetry in the presence of a redox mediator. The observed results indicated future applicability of these devices towards sensing applications in low concentrations.

5. Acknowledgments:

This work was supported by Science Foundation Ireland under the Research Frontiers Programme (SFI/09/RFP/CAP2455).

6. References:

[1] Frost & Sullivan, Biosensor Market Report, 2007.
[2] J. Hahm and C. M. Lieber, *Nano Lett.*, 2004, 4, 51-54.
[3] X. Wang and C. S. Ozkan, *Nano Lett.*, 2008, 8, 398-404.
[4] M. Manning and G. Redmond, *Langmuir*, 2005, 21, 395-402.
[5] C. O. Parker, Y. H. Lanyon, M. Manning, D. W. M. Arrigan and I. E. Tothill, *Analytical Chemistry*, 2009, 81, 5291-5298.
[6] F. Patolsky, G. F. Zheng and C. M. Lieber, *Nature Protocols*, 2006, 1, 1711-1724.

[7] M. Curreli, C. Li, Y. H. Sun, B. Lei, M. A. Gundersen, M. E. Thompson and C. W. Zhou, *Journal of the American Chemical Society*, 2005, 127, 6922-6923.
[8] K. Maehashi, T. Katsura, K. Kerman, Y. Takamura, K. Matsumoto and E. Tamiya, *Analytical Chemistry*, 2007, 79, 782-787.
[9] E. Stern, J. F. Klemic, D. A. Routenberg, P. N. Wyrembak, D. B. Turner-Evans, A. D. Hamilton, D. A. LaVan, T. M. Fahmy and M. A. Reed, *Nature*, 2007, 445, 519-522.
[10] D. W. M. Arrigan, *Analyst*, 2004, 129, 1157-1165.
[11] K. Ramanathan, M. A. Bangar, M. Yun, W. Chen, N. V. Myung and A. Mulchandani, *Journal of the American Chemical Society*, 2005, 127, 496-497.
[12] M. A. Bangar, C. M. Hangarter, B. Yoo, Y. Rheem, W. Chen, A. Mulchandani and N. V. Myung, *Electroanalysis*, 2009, 21, 61-67.
[13] R. H. Wang, C. M. Ruan, D. Kanayeva, K. Lassiter and Y. B. Li, *Nano Lett.*, 2008, 8, 2625-2631.
[14] K. Dawson, J. Strutwolf, G. Herzog, D. W. M. Arrigan, A. J. Quinn and A. O'Riordan, *ECS Transactions*, 2010, 28.
[15] M. Moskovits, *Reviews of Modern Physics*, 1985, 57, 783.
[16] T. Neumann, M. L. Johansson, D. Kambhampati and W. Knoll, *Advanced Functional Materials*, 2002, 12, 575-586.
[17] X. Zhang, D. Li, L. Bourgeois, H. Wang and P. A. Webley, *Chemphyschem*, 2009, 10, 436-441.
[18] S. Keebaugh, A. K. Kalkan, W. J. Nam and S. J. Fonash, *Electrochemical and Solid State Letters*, 2006, 9, H88-H91.
[19] E. J. Menke, M. A. Thompson, C. Xiang, L. C. Yang and R. M. Penner, *Nat. Mater.*, 2006, 5, 914-919.
[20] R. Ji, W. Lee, R. Scholz, U. Gosele and K. Nielsch, *Advanced Materials*, 2006, 18, 2593-+.
[21] J. Jorritsma, M. A. M. Gijs, J. M. Kerkhof and J. G. H. Stienen, *Nanotechnology*, 1996, 7, 263-265.
[22] E. C. Walter, B. J. Murray, F. Favier, G. Kaltenpoth, M. Grunze and R. M. Penner, *Journal of Physical Chemistry B*, 2002, 106, 11407-11411.
[23] J. Chai, D. Wang, X. N. Fan and J. M. Buriak, *Nature Nanotechnology*, 2007, 2, 500-506.
[24] Y. H. Lanyon, G. De Marzi, Y. E. Watson, A. J. Quinn, J. P. Gleeson, G. Redmond and D. W. M. Arrigan, *Analytical Chemistry*, 2007, 79, 3048-3055.
[25] Heller, J. Kong, H. A. Heering, K. A. Williams, S. G. Lemay and C. Dekker, *Nano Lett.*, 2005, 5, 137-142.
[26] A.J. Bard, Faulkner, L.R.,, *Electrochemical Methods, Fundamentals & Applications*, Wiley & Sons., New York, 2001.

Surface Plasmon assisted extraordinary transmission in metallic nanohole arrays and its suitability as a bio-sensor

D Jones, N. Liu, B. Corbett, P. Lovera, A.J. Quinn, A O'Riordan
Nanotechnology Group, Tyndall National Institute, University College Cork, Lee Maltings, Cork, Ireland.

E-mail: alan.oriordan@tyndall.ie

Abstract. In this work plasmonic nanostructures in the form of free standing films of sub-wavelength holes are fabricated and characterised using a variety of methods. It is shown how the light transmitted through a nanohole array is altered by the dielectric constant of the medium above the array facilitating the development of new types of plasmonic and photonic devices for future bio-sensing applications.

1. Introduction

There has been considerable interest in the emerging field of nanoplasmonics enhanced extraordinary transmission through sub wavelength nanohole arrays since the phenomena was first reported [1]. Surface plasmon polaritons (SPPs) exist as propagating waves on planar metal films with amplitudes that extend further into the dielectric region compared with the metal region [2]. Thus, the electromagnetic field is confined to the vicinity of the interface between the metal and dielectric. There is an enhancement of the field in this area which results in a high sensitivity of SPPs to the surface conditions enabling a wide variety of application areas that can use this sensitivity [3]. These include: surface adsorbate characterization, determination of surface roughness, in addition to chemical and biological sensors [4]. To excite SPPs on a surface using free-space light, additional momentum additional must be provided. One method to provide this extra momentum is by the patterning of an optical grating structure on the film [5]. The SPP waves interact with periodic structures on the metal surface leading to enhanced or extraordinary transmission [6]. This enhanced transmission leads to much smaller devices areas. In this paper we develop an approach for development of a nano SPP device and discuss its potential applications to a range of different areas including emitters and detectors.

Extraordinary transmission through sub-wavelength periodic nanopore arrays, fabricated in ultrathin metal films, may be attributed to the excitation of surface plasmon polaritons of specific resonant orders. The approximate equation for the SPP Block wave modes (λ_{SPP}) in a gold film when the incident light is normal to a square array of nanoholes with a periodicity P in a gold film is given by

$$\lambda_{SPP} = \frac{P}{\left(i^2 + j^2\right)^{1/2}} \left(\frac{\varepsilon_{Au}\varepsilon_d}{\varepsilon_{Au} + \varepsilon_d} \right)^{1/2}$$

where i and j define the order of the Block mode, ε_{Au} is the permittivity of Au and ε_d the permittivity of the adjacent dielectric material [7]. This is used to calculate the wavelengths an array of a certain period will have enhanced transmission and whether it is caused by the metal-substrate interface or the metal-dielectric interface as is required. The location of the peaks exhibited in extraordinary transmission spectra are dependant not only on the periodicity of the optical grating but also on the optical characteristics, i.e., the refractive index of the dielectric. Increasing the dielectric constant results in red-shifting of resonant SPP block wave modes. This well known phenomenon has been exploited by many research groups as a transducer signal for sensing devices [8-10].

Nanoplasmonic optical sensors are typically fabricated by focused ion beam milling nanohole arrays into thin gold films. A key limitation with this approach is that the depth of milling is extremely difficult to control. In this way, the underlying transparent substrate can be damaged, significantly decreasing the amount of transmitted light. Recently, the Odom group has demonstrated fabrication of nanohole arrays on gold films deposited on sacrificial chrome layers. Following fabrication, the underlying chrome layer is wet etched to release the nano-patterned gold. A self-assembly approach is then employed to deposit the gold layer on a transparent substrate for subsequent characterization. In this paper we apply this approach for fabrication of gold films bearing nanohole arrays and explore their positional as future nanoplasmonic sensors platforms [11-12].

2. Experimental section

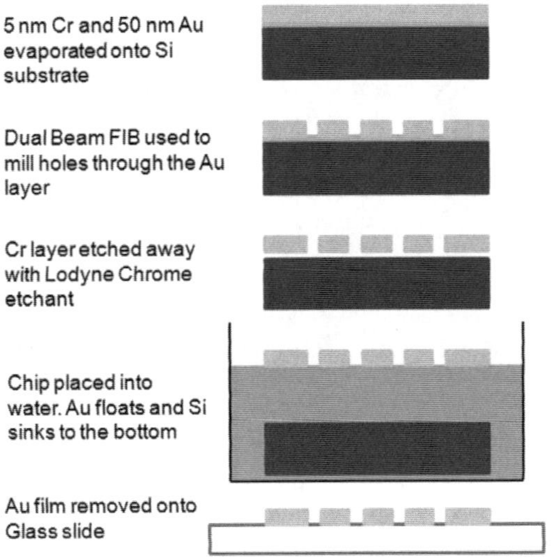

Figure 1. Schematic showing fabrication process for nanohole array in gold film.

2.1. Fabrication of nanohole arrays

To fabricate nanohole arrays, a 5 nm Cr film was first deposited on a clean Si substrate by e-beam evaporation, followed by evaporation of 50 nm thick Au layer. The sample was then loaded into the Quanta 3D 200i dual-beam Focused Ion Beam (FIB). Individual 60 μm × 60 μm periodic nanohole arrays were fabricated on the sample by focused ion beam milling; 30 kV beam voltage, 100 pA beam current, 1 pass, a defocus of 14 μm and dwell time of 5.5 ms. Individual nanoholes had a diameter of 110 nm and a pitch of 450 nm were routinely fabricated. The gold film was then freed from the underlying Si substrate by wet etching the chromium layer (Chrome Etchant Lodyne, Grower Chemicals Ltd.) for 10 minutes. The chip was then removed and immersed into deionised water where the Au film floated on the surface due to surface tension of the water meniscus while the Si substrate sank to the bottom. Using this approach the highly fragile gold films were easily freed from the silicon substrate in a contactless manner and prevented any damage occurring to the film. Finally, a glass microscope slide was immersed under the gold nanostructured film and used to lift the film from the water and dried for 24 hours. After transfer, the film adhered strongly to the glass substrate *via* Van Der Waals interaction, which permitted its optical characterization.

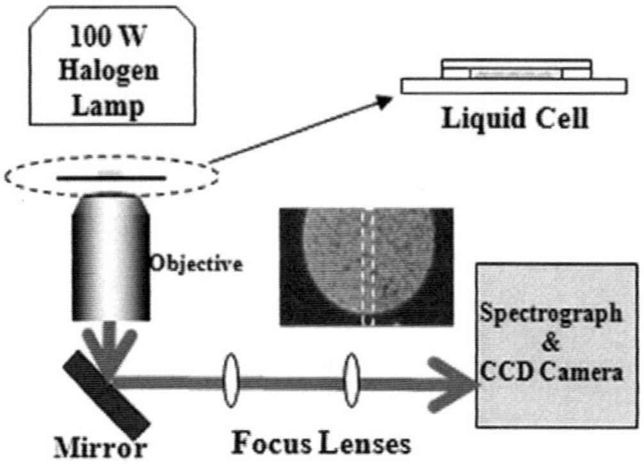

Figure 2. Schematic diagram of the setup used to characterise the samples.

2.2. Optical characterisation system

The normal transmission spectra of the sample were collected with a combination of an Olympus IX71 inverted microscope and an Acton SP2300i spectrograph. A 100W Halogen white light source was used to illuminate the sample, with the incident angle of light within 10 degrees with respect to the normal of sample. A crude microfluidic cell was constructed by gluing two pieces of 0.15 mm thick cover slip on either side of the adsorbed film (using UV curable glue) on a microscope slide. A third cover slip was bridged over these pieces of cover to form a lid of the cell. Spectra were acquired in air and in a range of saline solutions with refractive index varying from 1.33 to 1.37 by varying the concentration of NaCl in deionised water.

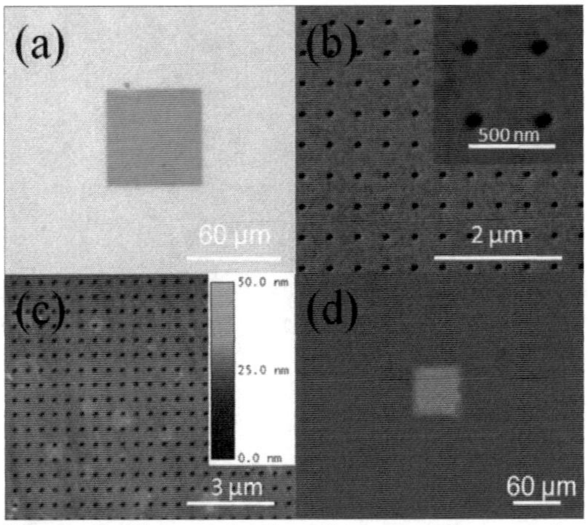

Figure 3. Characterisation of sample. The array is 60 µm by 60 µm in size; the period is 450 nm and the nanohole size 110 nm (a) Optical microscope image of an array. (b) SEM image of array. (c) AFM image and (d) Optical microscopy image with the light source underneath the sample.

3. Results and Discussion

Figure 3(a) shows an opitical image of a 60 µm by 60 µm array of nanopores using a calibrated microscope (Axioskop II, Carl Zeiss Ltd.) equipped with a charge-coupled detector camera (CCD; DEI-750, Optronics). Optical images gave a good first indication of whether the transfer process had damaged the film or not. In a few cases small wrinkles were visibleFigure 3(b) shows a scanning electron microscope image of a portion of the nanohole array on the Si substrate prior to lift-off. Scanning electron microscopy (SEM) images of nanohole arrays were acquired using a field emission SEM (JSM-6700F, JEOL UK Ltd.) operating at beam voltages between 3 and 5 kV. SEM analysis was undertaken for visual characterization of array quality and to determine the pitch and size of the nanoholes. Nanoholes with diameters of 110 nm and a pitch of 450 nm are routinely fabricated. If the Ion bean in the FIB was not perfectly focused the size of the nanoholes would be altered. Similarly the stigmatism of the FIB affects the shape of the hole and consequently the spectrum of transmission light. If the stigmatism is slightly off it results in nanoholes which not circular. Each array was checked after fabrication to ensure the nanoholes were as required.

The topography of the films was characterized using a calibrated atomic force microscope (AFM; Dimension 3100, Veeco Instruments Inc.) in tapping mode with commercial tapping mode probes (MP-11100, Veeco Instruments Inc; typical radius of curvature ~ 10 nm and front/side cone angles of 15°/ 17.5°, respectively). Figure 3 (c) shows a typical AFM image of a portion of an array. The depth of the holes was measured alone with the roughness of the Au surface.

Figure 3(d) show an optical micrograph of light transmitted through a nanohole array patterned in a gold film. The skin depth of gold is ~ 30 nm above which light will not be transmitted through the gold film. It is clear from the image that light is only transmitted through the array as expected.

3.1. Spectroscopic characterisation

Figure 4 shows the transmission spectra measured for a nanohole array in the presence of solution of varying refractive index (n=1.333 – n=1.37). Arrays with a pitch of 450 nm and a nanohole diameter of 110 nm, exhibited a broad characteristic peak of the bulk gold plasmon resonance at 500 nm and spectral peaks characteristic of SPPs at 680 nm and 630 nm, respectively. For a periodicity of 450 nm, one expects a (1,0) Au, NaCl solution SPP peak between 620 nm and 700 nm. However there is a Rayleigh-Wood anomaly at approximately 660 nm which interacts with the peak to form two separate peaks at either side. Rayleigh-Wood anomaly's occurs when light is diffracted at an angle parallel to the surface of the gold film [13]. It occurs for wavelengths satisfying the following equation

$$\lambda_{RA} = \frac{P}{\left(i^2 + j^2\right)^{1/2}} \sqrt{\varepsilon}$$

where P is the periodicity of the hole structure, i and j the integer pair that corresponds to the particular order of the RA and ε the permittivity of the substrate. The main peak at 680 nm has the narrowest spectral features but a smaller shift in nm compared to the broader peak at 630 nm or to the trough at 660 nm. When the refractive index of the top surface is changed from that of water (n=1.333) to that of a 21% NaCl solution the peaks shift to longer wavelengths as expected. This is the property of the nanohole array suitable for bio-sensing applications. When a molecule is attached to the film the peak will be at a specific wavelength. Any change to the surface of the Au (be it addition of another molecule or removal of the first) would produce a shift in the SPP peak which would be used to detect said change.

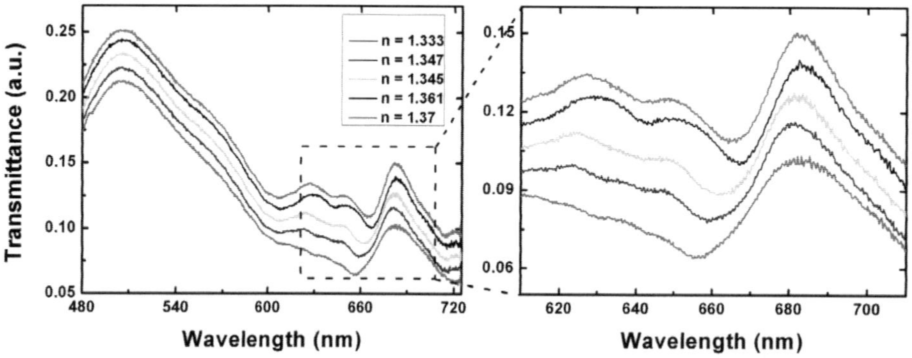

Figure 4. Transmission Spectra through Nanohole array in the presence of solution of varying refractive index (n=1.333 – n=1.37). The spectra are offset for clarity. The peaks shift to the right as the refractive index increases showing how the spectral features from the Nanohole arrays can be exploited for highly sensitive refractive sensing.

To determine the optimal size of the array a number of different size arrays were fabricated varying in size from 10 μm by 10 μm, to 100 μm by 100 μm. Spectra were acquired through each of these arrays and the relative peak intensity measured. As is shown in figure 5(a), it was found that the optimal size of an array for our optical setup was about 60 μm x 60 μm as intensity of transimited light reached a maximum at this size.

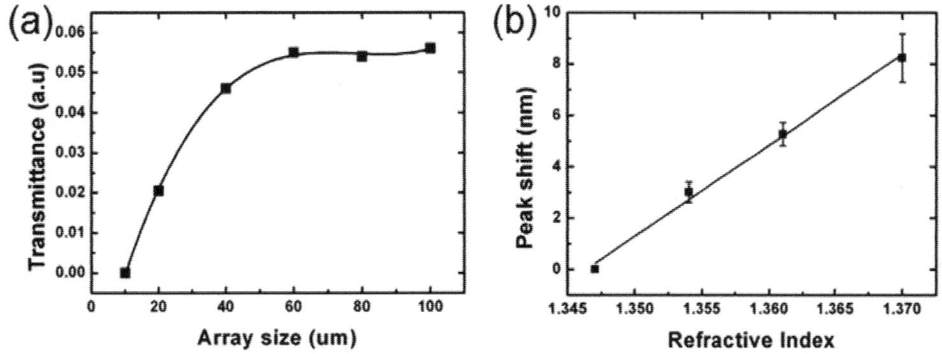

Figure 5. (a) is the intensity through arrays which vary in size from 10 x 10 µm to 100 x 100 µm (black line is a guide to the eye) and (b) is the peak shift in nm plotted against the refractive index of the solution.

The peak shift versus refractive index is plotted in figure 5(b). For a 0.023 RIU (refractive index unit) change there is a 8 nm shift in the peak which shows a sensitivity of 347 nm/RIU which is in line with similar research [14]. Therefore it should be possible to detect when a monolayer is attached to the gold surface. When molecules are attached to the Au surface, RI of superstrate deviates from 1.333, the value for water. The change of RI varies with the type of molecules as well as the thickness of the attached layer. For example, a 7 ~ 15 nm thick molecular layer, with RI of 1.45 will give a roughly equivalent effect of a superstrate with RI of 1.342, if we assume that the light extends 300 nm into the superstrate. Work in now ongoing to apply this approach for the sensitive detection of a variety important of chemical and biological monolayers.

References

1. Ebbesen, T.W., et al., *Extraordinary optical transmission through sub-wavelength hole arrays.* Nature, 1998. **391**(6668): p. 667-669.

2. Raether, H., *Surface plasmons.* 1988, Springer, Berlin.

3. Zayats, A.V., I.I. Smolyaninov, and A.A. Maradudin, *Nano-optics of surface plasmon polaritons.* Physics reports, 2005. **408**(3-4): p. 131-314.

4. Wong, D., Chemical and Biochemical Sensing with Optical Fibers and Waveguides. Measurement Science and Technology, 1997. **8**.

5. Henzie, J., M.H. Lee, and T.W. Odom, *Multiscale patterning of plasmonic metamaterials.* Nature nanotechnology, 2007. **2**(9): p. 549-554.

6. Degiron, A. and T. Ebbesen, The role of localized surface plasmon modes in the enhanced transmission of periodic subwavelength apertures. Journal of Optics A: Pure and Applied Optics, 2005. **7**: p. S90.

7. McMahon, J.M., et al., Tailoring the sensing capabilities of nanohole arrays in gold films with Rayleigh anomaly-surface plasmon polaritons. Optics Express, 2007. **15**(26): p. 18119-18129.

8. Liang, W., et al., *Highly sensitive fiber Bragg grating refractive index sensors.* Applied Physics Letters, 2005. **86**: p. 151122.

9. Polynkin, P., et al., Evanescent field-based optical fiber sensing device for measuring the refractive index of liquids in microfluidic channels. Optics letters, 2005. **30**(11): p. 1273-1275.

10. Krioukov, E., et al., *Sensor based on an integrated optical microcavity.* Optics letters, 2002. **27**(7): p. 512-514.

11. Gordon, R., et al., *A new generation of sensors based on extraordinary optical transmission.* Accounts of chemical research, 2008. **41**(8): p. 1049-1057.

12. Brolo, A.G., et al., Surface plasmon sensor based on the enhanced light transmission through arrays of nanoholes in gold films. Langmuir, 2004. **20**(12): p. 4813-4815.

13. Chang, S.H., S. Gray, and G. Schatz, Surface plasmon generation and light transmission by isolated nanoholes and arrays of nanoholes in thin metal films. Optics Express, 2005. **13**(8): p. 3150-3165.

14. Henzie, J., et al., *Nanofabrication of Plasmonic Structures*.* Annual review of physical chemistry, 2009. **60**: p. 147-165.

Elliptical nanohole array in thin gold film as micrometer sized optical filter set for fluorescent-labelled assays.

P Lovera, D Jones and A O'Riordan[1]
[1] Tyndall National Institute, Nanotechnology Group, Lee Malting, Dyke Parade, Cork, Ireland

E-mail: alan.oriordan@tyndall.ie

Abstract. In this paper, we study the polarisation properties of light transmitted through an array of elliptical nanoholes patterned in a thin gold film. The light transmitted through the sample strongly depends on the polarization of the excitation light and shows two distinct peaks, separated by 65 nm. These peaks are related to Rayleigh-Woods anomaly on the Au/glass interface. This type of devices has tremendous potential as micrometer size optical filter set for fluorescent-labelled assays.

1. Introduction

Since Ebbesen [1] observed a strong and unexpected enhancement of light transmission through arrays of subwavelength holes, numerous experiments have been performed to characterize and model this abnormal transmission. This effect, known as extraordinary optical transmission, is generally admitted to be due to excitation of surface plasmons on the upper and lower surfaces of the metallic array [2]. A typical transmission spectrum shows one or multiple peak whose wavelength could be simply tailored by changing the periodicity of the array or the dimension of the nanoholes. While arrays of circular nanohole have been thoroughly studied, less work has been done on other shape of nanoholes. Degiron [3] showed that, for an equivalent surface area, transmission through rectangular nanohole is ten times the one of circular nanohole. Also, it was demonstrated that the shape anisotropy of elliptical or rectangular nanohole induced polarisation sensitive transmission spectra in both single nanohole and arrays [4-9]

In this paper, we study the polarisation properties of light transmitted through an array of elliptical nanoholes in a thin gold film. The light transmitted through the sample strongly depends on the polarization of the excitation light and shows two distinct peaks in the far visible. These peaks can be related to Rayleigh-Woods anomaly on the Au/glass interface. This type of samples may find application as miniaturised spectral filters for on chip fluorescent-labelled assays.

[1] To whom any correspondence should be addressed.

2. Experimental section

2.1. Fabrication of the array of elliptical nanohole in thin gold film

10 nm Cr film was first deposited on the clean Si substrate by e-beam evaporation, followed by 50 nm thick Au layer deposition. The sample was then loaded into the Quanta 3D 200i dual-beam Focused Ion Beam (FIB). Patches of 60 µm × 60 µm periodic nanohole arrays were fabricated on the sample by focused ion beam. The elliptical nanoholes had diameters of 150 nm and 350 nm and were spaced by 450 nm in each direction. The long axis of the nanohole had a 45° angle with respect to the array. The perforated Au layer was transferred to a microscope glass slide using a lift-off procedure. The gold film was freed from the Si substrate by wet etching the Chromium layer in Chrome Etchant Lodyne (Grower Chemicals Ltd.). The chip was lowered into water where the Au film floated on the surface and the Si substrate sank. A glass microscope slide was used to lift the film out of the water and dried for 24 hours.

2.2. Optical characterisation of the sample

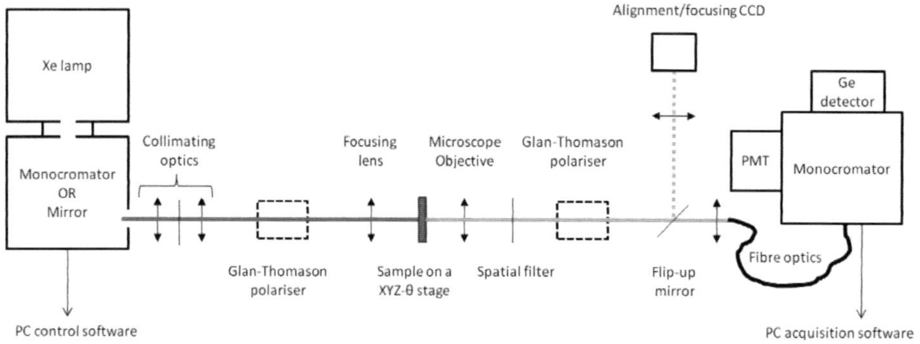

Scheme S1. Schematic of the optical set-up used for measurements of transmission spectra of nanohole array in thin Au film.

Transmission spectra of the nanohole arrays in Au thin film were recorded using a home-built optical system, see Scheme S1. The light from a Xe arc lamp was collimated by a combination of lenses, pinhole and slightly focused onto the sample. The angle of incidence of the beam was chosen to be 90°. The transmitted light (in zeroth-order diffraction) through the sample was collimated by a 50X objective (LMPlanFL, Olympus, n.a. 0.5)) and focused onto a collection fiber bundle connected to a Triax 190 0.19 metre monochromator (1200 lines / mm grating). Visible spectra were recorded at an integration time of 0.1 second using a R928 PMT (Hamamatsu Photonics). Infrared spectra were recorded at an integration time of 0.1 second using a nitrogen cooled germanium detector (EI-L, Edinburgh Instruments). All acquired spectra were divided the one of a glass substrate. Analysis was performed using Jobin-Yvon's SpectRad software. A CMOS camera (DCC1645C, Thorlabs) was used to assure proper focusing of the nanohole array. Polarised measurements were performed by adding a Glan-Thomson polarizer (NT47-045, Edmund Optics Ltd.) in the incident and/or collection path. Polarisation bias of the optical setup was determined by taking spectra of a blank substrate with light polarized every 15°. No polarization bias was observed.

2.3. Finite Difference Time Domain (FDTD) Simulations:

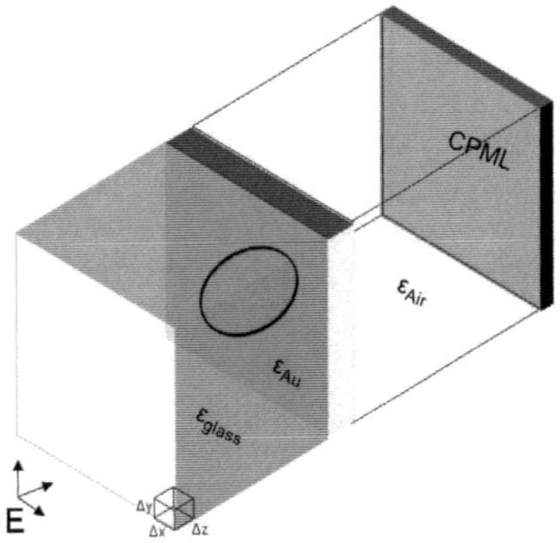

Scheme S2. computational domain used for the FDTD simulation

2.3.1. Computational volume. Finite Difference Time Domain (FDTD) simulations were performed using JFDTD3D (version 2.0) program available under the General Public License (GPL) (www.thecomputationalphysicist.com. The program was based on a 3-D, parallel FDTD implementation using the Message Passing Interface library. The code was run on 8 nodes, each node having 2 Opteron CPUs (2.6GHz, 4GB RAM). The grid spacing in each spatial dimension was 5 nm. The computational grid was chosen to be 450x450x1600 nm^3, see Scheme S2. The extra lengths in the z directions allowed the resolution of Wood's anomalies, if present. Periodic boundary conditions (450 nm) in the x and y directions were applied to simulate an infinite square array. Perfectly Matching Layers (PML) boundary conditions were imposed in the z direction in order to avoid reflections from the edges of the computational window. The setup was modelled (see Scheme S2) by a 50 nm Au film with a 650 nm thick glass substrate ($\varepsilon_{Glass} = 2.31\varepsilon_0$) and a 900 nm air ($\varepsilon_{Air} = \varepsilon_0$) superstrate. In the plane of the film, the nanohole was centered in the middle of the square ($x_0 = y_0 = 225.10^{-9}\,m$) and the equations used to define the ellipse were:

$$\frac{\sqrt{2}*\left[(x_{pos}-x_0)+(y_{pos}-y_0)\right]^2}{(150*10^{-9})^2}+\frac{\sqrt{2}*\left[(x_{pos}-x_0)-(y_{pos}-y_0)\right]^2}{(350*10^{-9})^2}\leq 1 \quad (1),$$

and

$$\frac{\sqrt{2}*\left[(x_{pos}-x_0)-(y_{pos}-y_0)\right]^2}{(150*10^{-9})^2}+\frac{\sqrt{2}*\left[(x_{pos}-x_0)+(y_{pos}-y_0)\right]^2}{(350*10^{-9})^2}\leq 1 \quad (2).$$

2.3.2. Permittivity of Gold. The permittivity of gold was modelled using a Drude plus two-pole Lorentz model:

$$\varepsilon_{Au}(\omega)=\varepsilon_\infty+\sum_{1,2}\frac{\omega_{Ln}^2\Delta\varepsilon_{Ln}}{\omega(\omega+i2\delta_{Ln})-\omega_{Ln}^2}-\frac{\omega_D^2}{\omega^2+i\gamma_D\omega} \quad (3)$$

With Drude parameters
$\varepsilon_\infty = 5.40,\ \gamma_D = 0.103*10^{15}\,Hz\ \ and\ \ \omega_D = 0.140*10^{17}\,Hz$

And Lorentz parameters:

$\omega_{L1} = 0.42 * 10^{16}\, Hz,\ \omega_{L2} = 0.523 * 10^{16}\, Hz,\ \Delta\varepsilon_{L1} = 2.542 * 0.279,\ \Delta\varepsilon_{L1} = 2.542 * 0.721,$
$\delta_{L1} = 0.435 * 10^{15}\, Hz \text{ and } \delta_{L1} = 0.661 * 10^{15}\, Hz$

2.3.3. Transmission spectrum. In order to simulate the excitation source, a Gaussian damped sinusoidal pulse (having a frequency content of 1 to 6 eV) with a 45° polarisation was launched from the glass side and simulation run (for 200.10^{-15} sec) in the time domain. Total transmission spectra were obtained by Fourier transforming the simulated electric and magnetic fields on a surface above the holes and constructing the surface integral of the outward Poynting vector.

3. Results and Discussion

3.1. Optical imaging of the sample

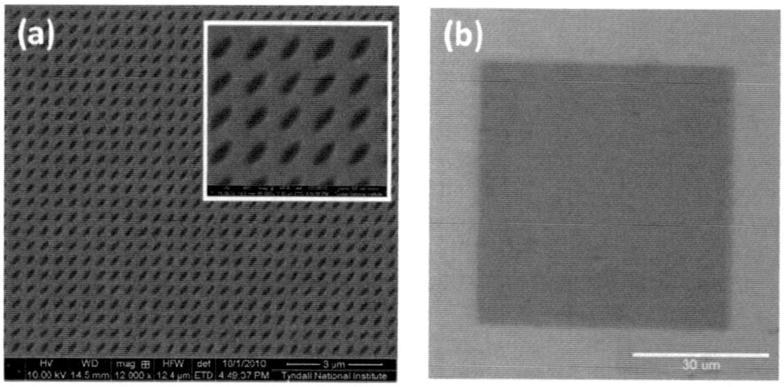

Figure 1. (a) scanning electron microscope and (b) optical image of the elliptical nanohole array after the transfer. The periodicity of the nanohole structure is 450 nm in each direction. The ellipses are oriented at 45° and have a 350 nm long axis and a 150 nm short axis. The sample is a 50 nm thick Au film. The array is 60x60μm^2

Figure 1a shows a scanning electron microscope image of the gold thin film on the Si substrate. High quality array of elliptical nanohole have been successfully made. The inset of figure 1a shows the surface of the film appears smooth and the edges of the nanohole sharp and well defined. Figure 1b shows an optical micrograph (50X objective) of the sample after the transfer onto a microscope slide. Despite being only 50 nm in thickness, the film was strong enough to survive the transfer. However, in some samples, SEM images performed on thin film after the lift off process showed localized wrinkles. After the transfer, the film adhered to the glass substrate *via* Van Der Waals interaction, which permitted its optical characterization.

3.2.Polarised transmission through the sample

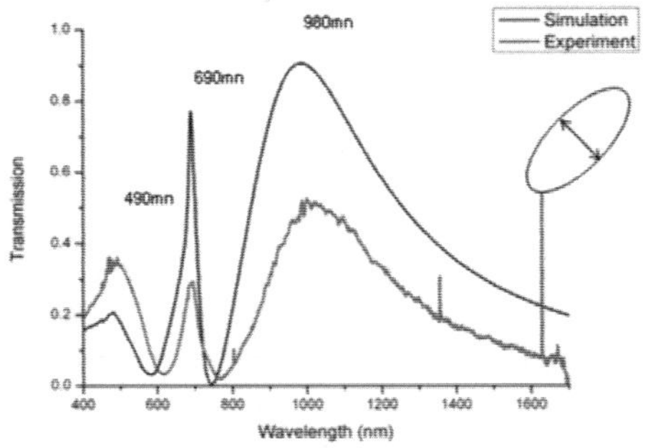

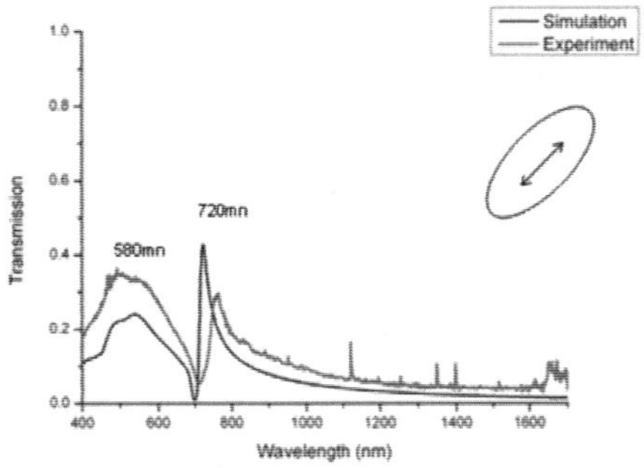

Figure 2. FDTD-simulated (black) and measured (red) zero-order transmission through the elliptical nanohole array on a glass substrate. Excitation polarisation along (a) the short axis and (b) the long axis of the ellipses.

3.2.1. Eigenmodes of the system. Polarised transmission spectra of the sample were acquired using the system and method described in section 2.2. For any polarisation state of the excitation, the spectrum could always be described as a combination of the two spectra shown in figure 2a and 2b. These two spectra thus describe the two eigenmodes of the sample. They correspond to excitation light polarized along the short and the long axis of the nanohole structure.

3.2.2. Comparison with FDTD simulation. FDTD simulations (see section 2.3 for details) were carried out for the two polarizations states. As can be seen in figure 2, there is a very good agreement between the experimental data and the simulated one, in terms of peak position and overall trend. The discrepancy in the relative intensity can be attributed to minor imperfections in the experimental hole structure or the simulated one (errors in the holes dimensions, imperfection in the Drude-Lorentz model).

3.2.3. Rayleigh-Woods anomaly and Surface Plasmon polariton

Both spectra exhibited a broad peak around λ=490 nm, with a transmission of 20-35% (simulation and experiment, respectively). This corresponds to the intraband transition for gold (direct transmission), and occurs irrespective of the incident polarization [10]. However, the rest of a spectrum differs significantly for the two polarisation states.

When light is polarized along the short axis of the ellipse (figure 2a), two peaks can be observed, a sharp, asymmetric, one at λ=690 nm and a broad one in the near infrared at λ=980 nm. In between these two peaks, transmission reaches zero at λ=750 nm. This dip in the transmission spectrum can be assigned to the (1,1) Rayleigh-Wood anomaly on the glass-gold interface. Rayleigh-Wood anomaly corresponds to light diffracted at an angle parallel to the surface of the gold film. It occurs for wavelengths satisfying equation [11] (4)

$$\lambda_{RA} = \frac{P}{\left(n_x^2 + n_y^2\right)^{1/2}} \sqrt{\varepsilon} \tag{4}$$

where P is the periodicity of the hole structure, n_x and n_y integer numbers indicating the exciting diffraction order and ε the permittivity of the external dielectric. In the present case of a grating with a periodicity of 450 nm, one expects a (1,1) Rayleigh-Wood anomaly at 735 nm for the glass-Au interface, close to the observed minimum at 750 nm..

When the light is polarised along the long axis of the ellipse, one can observe that above 900 nm, the transmission is very low and this is due to the cut-off wavelength of the nanohole waveguide along this direction. Also noticeable on this spectrum is a shoulder at λ=580 nm in the intraband transition peak. The shoulder can be assigned to the (1,1) SPP peak on the glass-gold interface (i.e., coupling of light to propagating plasmon mediated by the grating). For normal incidence of light onto a square-symmetry periodic lattice, the transmission maxima occur at the wavelengths satisfying a two-dimensional grating coupling condition for surface Plasmon [12] (this equation is intrinsic as ε_{Au} is dependant on λ):

$$\lambda_{SPP} = \frac{P}{\left(n_x^2 + n_y^2\right)^{1/2}} \left(\frac{\varepsilon_{Au}\varepsilon}{\varepsilon_{Au} + \varepsilon}\right)^{1/2} \tag{5}$$

For a periodicity of 450 nm, one expects a (1,1) SPP peak at 572 nm, close to the observed shoulder at 580 nm. Lastly, the most interesting feature in figure 2a is a very sharp, asymmetric peak at λ=720 nm with an associated minimum at 700 nm. The special shape of this peak suggests a Fano resonance.

3.2.4. Fano resonances. Fano resonances are a general characteristic of systems where two transmission pathways interfere, a resonant and a non-resonant one, or, expressed differently, between a discrete resonance and a continuum. The general equation describing a Fano resonance can be written as [13]:

$$I \sim \frac{\left(q\gamma + \omega - \omega_0\right)^2}{\left(\omega - \omega_0\right)^2 + \gamma^2} \tag{6}$$

Where ω_0 and γ are the position and the width of the resonance, respectively, and q the Fano factor, describing the ratio of probability of taking the first path (resonant) over the second path (non resonant). In the present case, the continuum could be represented by scattered light transmitted through the hole, and the resonant process by the (1,1) Rayleigh-Wood anomaly (which is the same as in section 3.2.3, i.e., at 735 nm). Fitting the experimental peak with equation (6) gave a resonance position at 714 nm, a width of 9.5 nm a q factor of 1.4.

3.2.5. Arrays of elliptical nanohole as microscale filter for fluorescent label assay

It can be seen on figure 2a and 2b that in the 650-750 nm range, the maximum transmission for one polarisation corresponds to the minimum transmission for the other transmission. This means that two of these filters in a crossed configuration could be used as micrometer sized optical filter set for on-chip transmission fluorescence assay [14]. For examples, iFluor™ 700 dyes from BIOMOL GmbH have fluorescence excitation and emission maxima close to 690 nm and 710 nm respectively and would be a suitable label for the current sample. Also, equation (4) and (5) show that changing the periodicity of the array allows the tuning of the wavelength at which the zero transmission or maximum transmission occurs. This suggests that changing the periodicity would allow different filter sets to be made and so different labels to be probed. Work is ongoing to explore this effect.

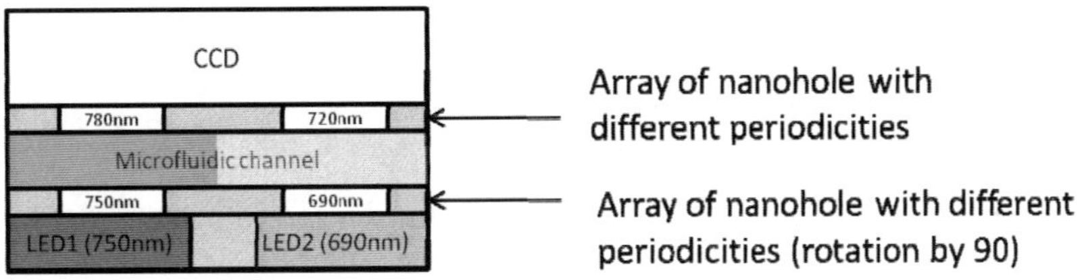

Figure 3. Schematics showing how the elliptical nanohole array could be integrated into an on-chip fluorescence assay.

4. Conclusion

Array of elliptical nanohole have been successfully patterned in thin gold film by focused ion beam. The thin film could be lifted off and transferred onto a glass substrate. Transmission of polarized light through the sample showed two eigenmodes for the system. The two modes corresponded to light polarized along the short and the long axis of the ellipses. When the light was polarized along the short axis, a broad peak could be seen at 980 nm, and a sharper one at 690 nm. The latter was assigned to Rayleigh-Wood anomaly at the glass-Au interface. When the light was polarized along the long axis, a sharp, asymmetric peak was observed at 720 nm. This peak was described as a Fano resonance between the Rayleigh-Wood anomaly and direct transmission. In the 650-750 nm rang, the maximum transmission in one polarisation corresponds to the minimum transmission in the other one. This means two of these filters in a crossed configuration could be used as miniaturised optical filter set for on-chip fluorescence assay.

Acknowledgements
This work was supported by the EU-funded project PhastID (FP7-ICT-2009-5- 258238)

References

[1] T. W. Ebbesen, *et al.*, "Extraordinary optical transmission through sub-wavelength hole arrays," *Nature,* vol. 391, pp. 667-669, Feb 1998.
[2] H. Gao, *et al.*, "Direct Evidence for Surface Plasmon-Mediated Enhanced Light Transmission through Metallic Nanohole Arrays," *Nano Letters,* vol. 6, pp. 2104-2108, 2006.

[3] A. Degiron and T. W. Ebbesen, "The role of localized surface plasmon modes in the enhanced transmission of periodic subwavelength apertures," *Journal of Optics a-Pure and Applied Optics,* vol. 7, pp. S90-S96, Feb 2005.
[4] B. Sepulveda, *et al.*, "Shape effects in the localized surface plasmon resonance of single nanoholes in thin metal films," *Optics Express,* vol. 16, pp. 5609-5616, Apr 2008.
[5] J. Elliott, *et al.*, "Polarization control of optical transmission of a periodic array of elliptical nanoholes in a metal film," *Optics Letters,* vol. 29, pp. 1414-1416, Jun 2004.
[6] C. Genet and T. W. Ebbesen, "Light in tiny holes," *Nature,* vol. 445, pp. 39-46, Jan 2007.
[7] R. Gordon, *et al.*, "Strong polarization in the optical transmission through elliptical nanohole arrays," *Physical Review Letters,* vol. 92, Jan 2004.
[8] K. J. K. Koerkamp, *et al.*, "Strong influence of hole shape on extraordinary transmission through periodic arrays of subwavelength holes," *Physical Review Letters,* vol. 92, May 2004.
[9] X. F. Ren, *et al.*, "Polarization properties of subwavelength hole arrays consisting of rectangular holes," *Applied Physics B-Lasers and Optics,* vol. 91, pp. 601-604, Jun 2008.
[10] J. M. McMahon, *et al.*, "Tailoring the sensing capabilities of nanohole arrays in gold films with Rayleigh anomaly-surface plasmon polaritons," *Optics Express,* vol. 15, pp. 18119-18129, 2007.
[11] R. W. Wood, "Anomalous Diffraction Gratings," *Physical Review,* vol. 48, p. 928, 1935.
[12] H. F. Ghaemi, *et al.*, "Surface plasmons enhance optical transmission through subwavelength holes," *Physical Review B,* vol. 58, pp. 6779-6782, Sep 1998.
[13] B. Luk'yanchuk, *et al.*, "The Fano resonance in plasmonic nanostructures and metamaterials," *Nat Mater,* vol. 9, pp. 707-715, 2010.
[14] W. L. Barnes, *et al.*, "Surface plasmon subwavelength optics," *Nature,* vol. 424, pp. 824-830, 2003.

Sensors & their Applications XVI IOP Publishing
Journal of Physics: Conference Series **307** (2011) 012007 doi:10.1088/1742-6596/307/1/012007

RAST Model: simulation of tensiotraces to facilitate drophad engineering

Philippe Pringuet[1], N D McMillan[2], G Doyle[2], S R P Smith[3], M ONeill[4], S Riedel[5]

[1] Bahrain Polytechnic, BP33349, Isa Town, Kingdom of Bahrain
[2] Institute of Technology Carlow, Kilkenny Road, Carlow, Ireland
[3] Physics Department, University of Essex, Wivenhoe Park, Colchester, Essex CO4 3SQ, UK
[4] Carl Stuart Ltd, Whitestown Business Park, Tallaght, Dublin 24, Ireland
[5] Dublin Institute of Technology, Dublin, Ireland

E-mail: `philippe.pringuet@polytechnic.bh`

Abstract.
Tensiography is a technique that determines the physical and chemical properties of a liquid by illuminating a growing pendant drop from within using a source fibre. Light reflected internally at the surface of the drop is recieved by a collector fibre and is converted into an electric signal called a tensiotrace, which is a graph of reflected light as a function of drop volume. The instrument obtaining this signal is called multianalyser. A numerical model that simulates tensiotraces through a raytracing analysis (RAST - Raytracing Analysis for the Simulation of Tensiotraces) of the multianalyser as been developed to define theoretically how the tensiotrace describes the physical and chemical properties of a liquid. The purpose of this study is to investigate the model as an engineering/design assistant leading to discoveries and improvements to the multianalyser.

1. Introduction

1.1. Tensiography

Tensiography [1] is a technique that determines physical and chemical properties of a liquid [2] from illuminating a growing pendant drop from within by a source fibre. Light reflected at the surface of the drop is received by a collector fibre and is converted into an electric signal called a tensiotrace, the graph of reflected light intensity versus drop volume. The tensiotrace records the entire drop history from remnant drop initiations until the mature drop separates. The optical tensiograph [3] (also known as drop analyser or multi analyser) is an Amplitude Modulated Fibre Optic Sensor (AMFOS) instrument that records tensiotraces.

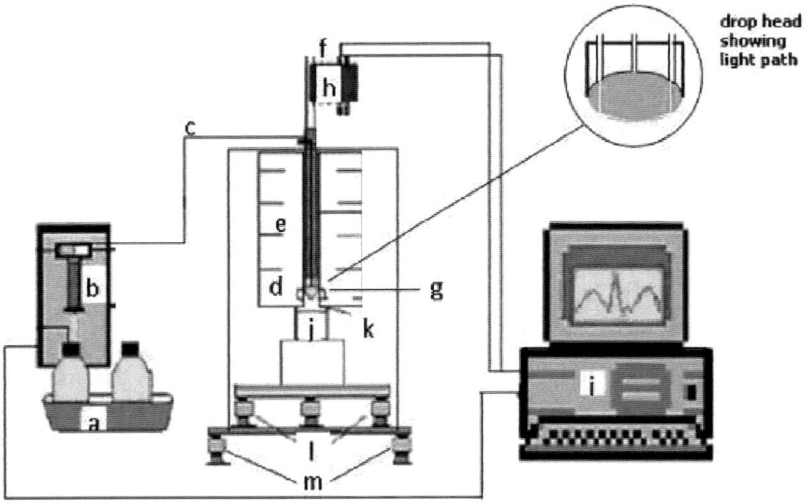

	Description		Description
a	Sample for analysis	h	Detector
b	Stepper-motor-pump	i	Computer
c	Delivery tube	j	Chamber to ensure saturation of atmosphere
d	Drophead	k	Temperature sensors
e	Temperature control	l	Auto levelling drives
f	Light Source	m	Anti vibration balancing feet
g	Optical Eyes		

Figure 1: Schematic diagram of the tensiograph instrument

The scheme of the instrument arrangement is shown in figure 1. The liquid a, which is to be analysed is pumped via the liquid delivery tube c to the drophead d using a motor stepper pump b and a pendant drop grows at a uniform volume rate (isochoric) in size until it separates from the drophead. Throughout this process, light from a light emitting diode (LED) f is injected into the drop through the source-fibre and the reflected signal, picked up by the collector-fibre is transmitted to a photodiode h. The amplification of the low photocurrent into an adequate voltage signal is necessary before the signal is passed to an acquisition board (an electronic system based on a programmable interface controller (PIC) and other suitable electronics hardware) incorporated in a personal computer (PC) i. The Peltier temperature control block e maintains the temperature of the LUT at usually $20°C$ or $25°C 0.1°C$. The optical eyes g enable the signal to be scissored - the initial trigger occurs when the initiation drop falls through the optical eye and triggers the start of the data acquisition, indicating when the measurement drop has arrived on the drophead and then the data acquisition continues until this drop falls off the drophead and the second trigger locates precisely the ending of recording of the tensiotrace (a data set recorded over the life cycle of the drop). Two-stages anti-vibration mounts l and m protect the drop from vibrations. The analogue to digital converter (ADC) of the acquisition board converts the analogue voltage (amplified optoelectronic signal from the detector) into a digital signal, which is then subsequently transferred to the memory of the PC where it is stored for later analysis. There is a source-detector employing a CCD (charged couple device) or CMOS (complementary metal oxide semi-conductor) detector in conjunction with a spectral source such as fibre deuterium, xenon, halogen or tungsten sources. Such a system provides a spectral array of tensiotraces; one for every measurement wavelength.

The instrument has been engineered to deliver measurements of the physical properties of the LUT, namely surface tension/density ratio, refractive index (η), colour at a specific wavelength ($absorption = A_\lambda$) and turbidity at a specific wavelength. The instrument can also be used for quality assurance for solid products such as pharmaceutical products if these are dissolved in an appropriate solvent for this purity/fingerprinting assay. It can also be used to sensitively monitor chemical and biochemical kinematics processes.

Figure 2 shows a typical concave drophead. The practical problem with such a design is fundamental in that physical damage can be caused due to exposure of the fibres. As a result of this study, a patented quartz drophead has been designed that gives the same universal fingerprint capability as this traditional drophead. This new design of drophead is one of the key innovations that facilitates accurate, reproducible and sensitive tensiography. Tensiography has potential to contribute to quality assurance in the wine industry in terms of chemical analysis and fingerprinting.

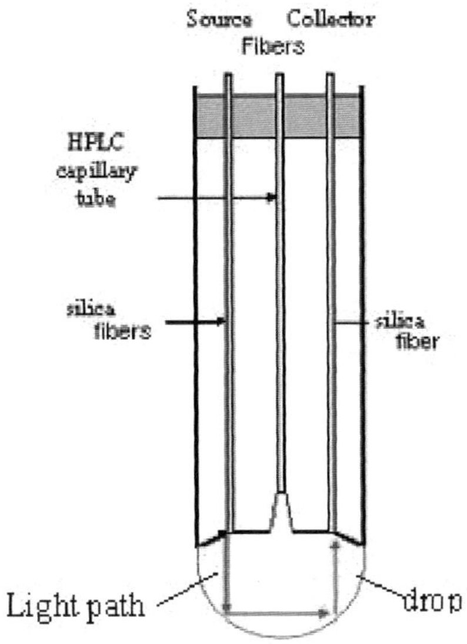

Figure 2: Drophead configuration

1.2. Tensiotraces

The key concept of the tensiograph is the fact that all the various physical and chemical processes that modulate the coupled light in the drop produce a signal, which is unique in

Journal of Physics: Conference Series **307** (2011) 012007 doi:10.1088/1742-6596/307/1/012007

a theoretical sense for every specific combination of properties of the LUT. These processes include (i) partial reflection of the light beam inside the pendant drop, (ii) absorption of the light from chromophores in the liquid, (iii) scattering of the light from turbid particulate matter, (iv) changes in the emission angle from the source fibre, and indeed perhaps other processes. The detected light intensity is amplitude modulated during the life cycle of a drop i.e. the reflecting surface of the inner drop changes continuously. Figure 3 shows the characteristic features of a tensiotrace. The physical features of tensiotraces are defined by the physical and chemical properties of the LUT such as density, colour, refractive index and adbsorbance. Although each of these can be measured from tensiotraces, the measurands are not statistically independant.

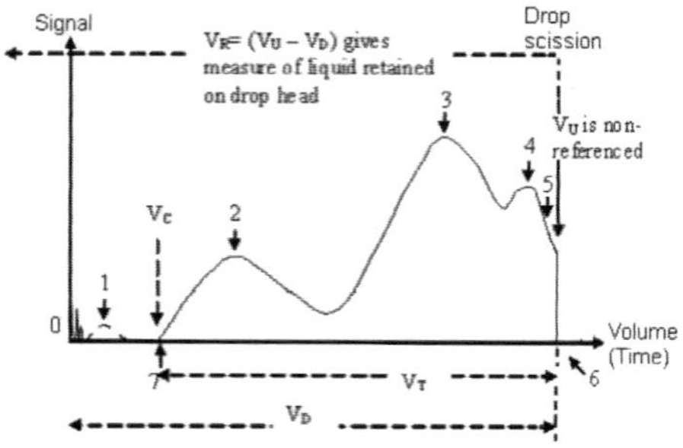

Label	Description
0	Separation vibration
1	First-order peak (protopeak)
2	Rainbow (second-order deuteropeak)
3	Tensiopeak (third-order tritopeak)
4	Shoulder peak (fourth-order tetertopeak)
5	Separation peak (fifth-order pemptopeak)
6	Drop period
7	Rainbow peak commencement

Figure 3: Tensiotrace showing the important trace features associated with drop mechanics, labeled 0 to 7

The first-order peak (proto-peak) is not seen in the current drophead design where the first order coupling is geometrically impossible because the measurement drop is too flat with the widely spaced fibres. The horizontal axis can be either volume or time as both are equivalent for constant volume delivery from stepper pumps. Here only volume is shown but this quantity is derived from a time measurement.

The positioning of the fibres in the drophead has been engineered to optimise the measurement of various physical/chemical properties. The rainbow peak height and position in the tensiotrace is a good measure of refractive index. It gets its name from the fact that in rainbows a measurement of the angle of the bow to that of the observer provides a measurement of the refractive index of the liquid. The height of the major tensiograph peak (3-tensiopeak) provides a good measurand for determining the colour and/or turbidity of the solution in that this decreases with respectively the increasing absorptive or scattering power of the liquids. The drop period can be used to give a good measure of the surface tension/density ratio. Other physical measurements have been shown to be possible with the tensiograph, but the principal use of the tensiograph is as a fingerprinting technique. The combination of the physical and chemical properties of the drop defines the form of the tensiotrace and thus the tensiotrace provides a very sensitive fingerprint technique for liquids, or indeed solid samples dissolved in a solvent.

The computer model described below reveals that the coupling of the light from source fibre to collector fibre can be represented as a 2-D process as the drop grows and with this volume growth higher reflection orders develop. The light must enter the collector fibre within the acceptance angle. Therefore the first order 'single' reflection off a position close to the centre of the drop base in the smallest drop does not couple to the detector. An important measurement position is known as the commencement or RPC (7 in figure 1) at which the light from source fibre is received by the collector fibre and can be transmitted in the multimode fibre to the detector. In the drophead design used in this study no light couples to the detector until the second order reflection begins to develop in a larger drop with reflections off either side of the drop.

1.3. The RAST model

A computer program that simulates tensiotraces through a raytracing analysis (RAST model) has been developed using Matlab to define the formation of tensiotraces. Modeled tensiotraces from this model were compared qualitatively and quantitatively to experimental tensiotraces of different liquids in order to evaluate the model.

A representation of the drophead is coded in the program. All dimensions including indent of the quartz disc, position and size of the fibres are normalised on the radius of the drophead. This is the basis of a dimensionless invariant model. Figure 4 shows a 2D cross section of the drophead model; notice the scale is normalised to -1 ... 1. The drop head dimensions are extremely important for the model and are later used when simulating the rays of light.

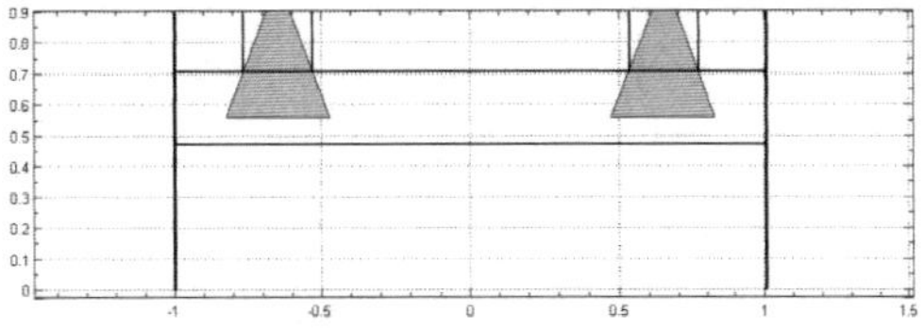

Figure 4: Representation of a drophead

In this schematic representation of the quartz drop head (figure 4), the quartz disc protecting the fibres is clearly visible. The two vertical lines on the outside of the disc represent the edges of the lip and the two small rectangles at the top are the fibers. The emitting fiber cone of light and the detector fiber acceptance angle are represented by the triangles. This representation is only an approximation since these dimensions can only be fabricated by Starna with a tolerance of 0.1 mm. In particular the edge of the drophead cannot be fabricated as represented.

The RAST model is best described by figure 5. Each drop shape is grown incrementally from the apex using Runge-Kunta 4 algorithms and following the Laplace-Young equation until the radius of the drop shape is 1 (normalised drophead diameter = 2).

The life cycle of the drop is created by modelling the drop shapes from the value of X_0 that corresponds to the solution with the smallest permissible volume. Solutions for smaller drop shapes could be obtained but the corresponding drop would not attach to the drophead edge (angle of attachment too small) and as such would be smaller than the remnant drop. X_0 is then increased incrementally and more drop shapes are calculated. A point is eventually reached when a further increase in X_0 actually results in a decrease in volume. This point in the drop cycle is considered to be the theoretical drop separation. Indeed, as more liquid is pumped in the drop, the drop has to increase in size or fall. Since there is no more solution to the equation forming a larger drop, the drop must have fallen. This is a theoretical maximum and it is believed that the drop separation can be slightly earlier as vibrations and liquid fluid mechanics are not taken in consideration here.

To form a tensiotrace, light emitted from the fibre goes through the drop and is coupled to the detector fiber. The model follows the same principle, modelling the light as a collection of rays. The emission from the multimode fibre is modelled by a cone of rays from the source fiber

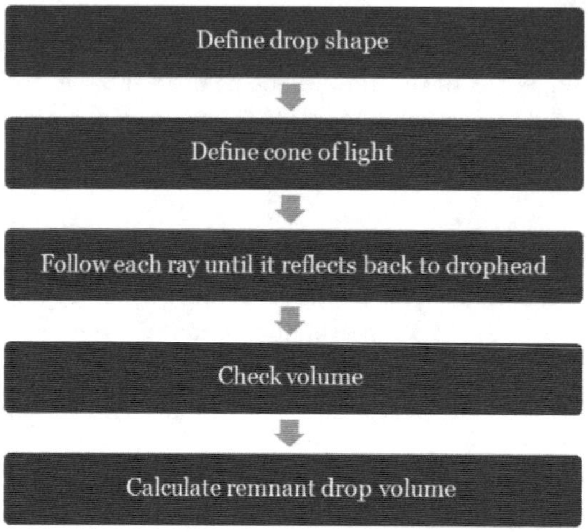

Figure 5: RAST algorithm

comprising at least 1000 rays (100,000 is more typical). Each ray starts from the surface of the fiber and has a direction limited by the numerical aperture of the fiber. This collection of rays forms the cone of light previously mentioned. Figure 4 shows the cone of light given by the numerical aperture of the fibers. In order to cover the fibre end uniformly, a parameter n_{th} is defined, corresponding to the number of possible values of theta (angle of ray with vertical norm) between 0 and the numerical aperture NA. $n_{th} = 200$ corresponds to 125830 rays calculated per dropshape and is the value used for data collection. The emission from a fibre is non-uniform and the intensity is accurately described by a $cos\theta$ function [4] based on the Warn's illumination model [5].

Rays first have to go through the quartz disc protecting the fibre from the liquid. Light is refracted at each interface and the new direction is calculated using the Fresnel equation [6]. Similarly to the emission end, rays when reaching the quartz disc are refracted on to the fibres level. The position is calculated as well as the angle of incidence. Rays are coupled (transmitted along the fiber and reach the detector) only if they hit the collector fiber with an incidence angle less or equal to the numerical aperture of the fiber. If the ray is coupled to the detector fiber, its intensity is added to the tensiotrace intensity for that volume.

2. Experimental

The testing involves the engineering process to create the new patented quartz drop head (see figure 6) made by Starna and the repercussions on the measurement capabilities of the instrument. The main objective is to show how the model can be used to improve the drophead design for a particular application. Each set of result gives a better understanding of tensiography and the tensiotraces. Also testing a new drophead design in order to obtain the desired tensiotrace can be very costly both financially and in man-hours. Indeed it requires changing the design and building a new prototype but also retesting the new drophead and analysing the results. The model facilitates this analysis at a fraction of the cost.

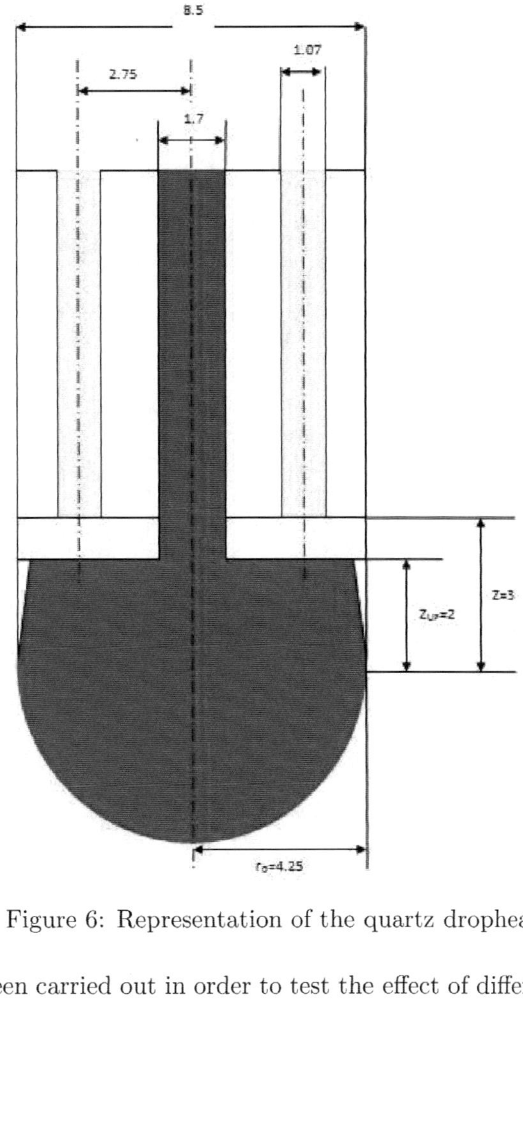

Figure 6: Representation of the quartz drophead

Simulations have been carried out in order to test the effect of different measurements of the new quartz drophead.

- the disc thickness
- the fibre diameter
- the fibre position
- the edge height (Z_{UP})

A cross section analysis was also carried out in order to study the impact position of rays and evaluate the possibility of simultaneous multi-wavelength analysis.

3. Results

3.1. Analysis of the disc thickness

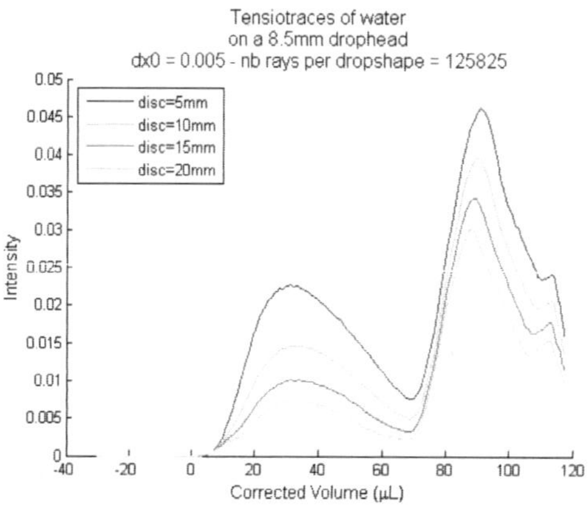

Figure 7: Modeled Tensiotraces of water for various disc thickness

Figure 7 shows the effect of the size of the quartz disc on tensiotraces. Although quartz is a transparent medium, the thicker the disc, the less light goes through; this results in lowering the intensity of the tensiotrace. It should be noted that the quartz disc is protecting the fibres from the LUT and the possible damages from proteins ??. Another effect of the quartz disc is to narrow the numerical aperture of the fibres as the refractive index of the disc is different from the refractive index of the fibres. This is shown by a larger reduction of intensity on the rainbow peak than on the tensio peak. This is shown in figure 8 where the downward slope for RPH is greater than for TPH. Ideally the smaller thickness is preferred as optical coupling is optimised but fabrication issues force the minimum size to 1mm.

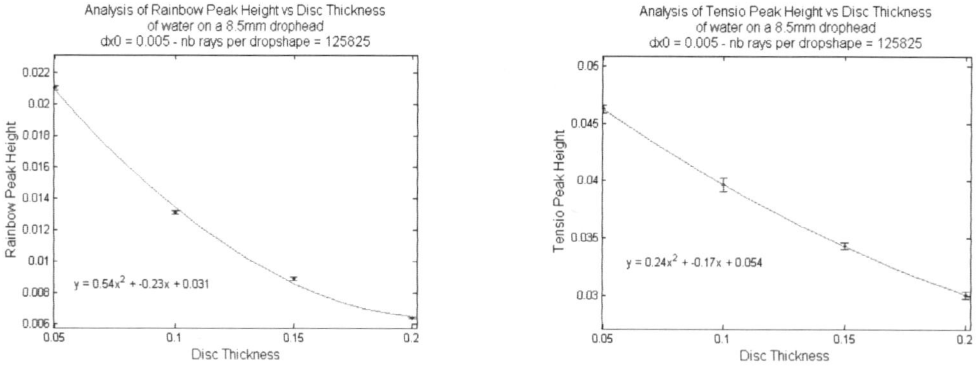

(a) Rainbow Peak Height (b) Tensio Peak Height

Figure 8: Analysis of effect of disc thickness on RPH and TPH

3.2. Analysis of the fibre diameter

Figure 9 shows the effect on tensiotraces when changing the diameter of both fibres. Even though the simulations could have been carried out with input and output fibres of different diameter,

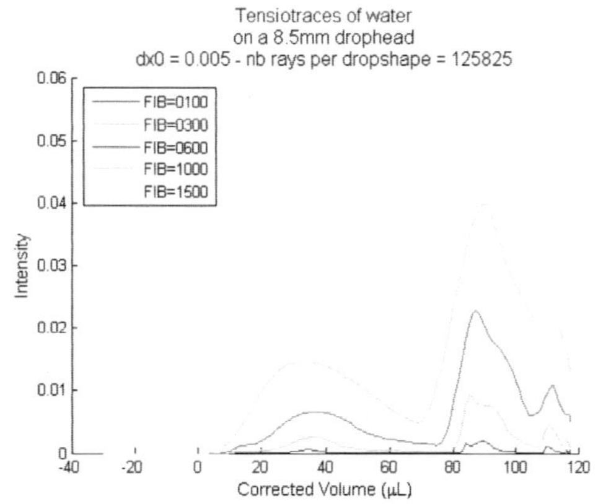

Figure 9: Modeled tensiotraces of water for various fibre diameters

this would cause problems when fabricating the drop heads as drilling quartz is a challenging process. The different fibres simulated were taken from the artphotonics catalogue [7].

Increasing the diameter of the fibres effectively increases the light being transmitted to the detector and as such probably increases the detection limit. In drop spectroscopy good quantitative capability is required. It is achieved by maximising the optical throughput (i.e. increasing the fibre diameter). However, the drophead was designed in order to maximize the resolution of each peak. The larger fibre simulated here ($FIB = 1500\mu m$) affects the quality of the tensio peak to such a degree that the shoulder peak disappears. It can be argued that 600 micron fibres give a better resolution of separation peak and noticeably better symmetry in the rainbow peak. In practice, a compromise is found using 1000 micron fibres but this does not seem to be the best option in all cases. It is important that the drop head is designed for purpose. The RAST model can help achieve this. Figure 10 shows that the fibre diameter has a linear relationship with the tensiopeak height.

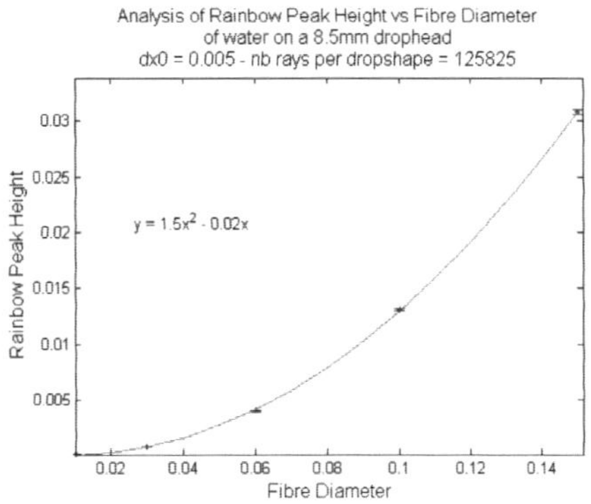

(a) Rainbow Peak Height

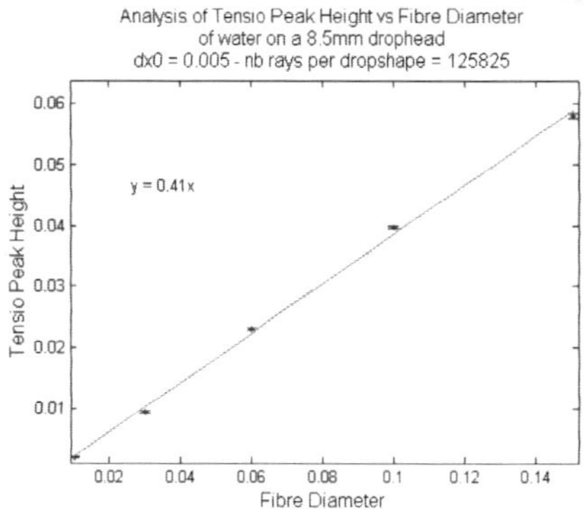

(b) Tensio Peak Height

Figure 10: Analysis of effect of fibre diameter on RPH and TPH

3.3. Analysis of fibre position

Tensiotraces of water were modelled on a quartz drophead with different positions for centre of the input and output fibres between 2.5 and 3.5mm from the centre of the drophead.

The relation between Rainbow Peak Commencement and Remnant Drop Volume was first analysed to make sure that the full tensiotrace could be recorded on the multianalyser ($RPC < 0$ result in the start of the rainbow peak not being recorded). The definition of the peak was also rated from poor to good in order to discard poorly defined tensiotraces. It was concluded that symmetric position of the fibres often give better defined peaks. Also fibres cannot be placed too close to the edge of the drophead (3.25mm and 3.5mm from centre) to avoid problems with the start of the rainbow peak or fabrication. A symmetric design of the drophead is also an advantage in the construction stage.

Peak heights were then analysed (see figure 11) in order to find the tensiotrace with the best throughput. It should be noted that the tensiotraces from drophead B and C and from drophead E and F are identical because it does not matter which fibre is connected to the source and which is connected to the detector.

Test	Position of fibres		Peaks analysis			
Nb	source	detector	RPH	TPH	RP symmetry	SP definition
A	2.50mm	2.50mm	0.0063	0.0365	good	good
B	2.50mm	2.75mm	0.0080	0.0345	good	poor
C	2.75mm	2.50mm	0.0080	0.0345	good	poor
D	2.75mm	2.75mm	0.0146	0.0397	average	good
E	2.75mm	3.00mm	0.0189	0.0383	good	poor
F	3.00mm	2.75mm	0.0189	0.0383	good	poor
G	3.00mm	3.00mm	0.0340	0.0440	poor	average

Table 1: Analysis of water tensiotraces peaks for various fibre positions

Test A gives a tensiotrace of lesser quality with a very flat (although very symmetrical) rainbow peak which may cause accuracy problems when calculating apex coordinates. Visual inspection reveals that the best position for the fibres seems to be symmetrical 3mm from the centre although the rainbow peak is not well balanced. Current version of the quartz drophead (Test D - symmetrical 2.75mm from the centre) gives a well defined tensiotrace with arguably a better balanced and more symmetric rainbow peak but with a smaller throughput. However water is at the bottom of the RI range. A low RPH for water gives a greater dynamic range for RI. Hence this choice of fibre position is best overall.

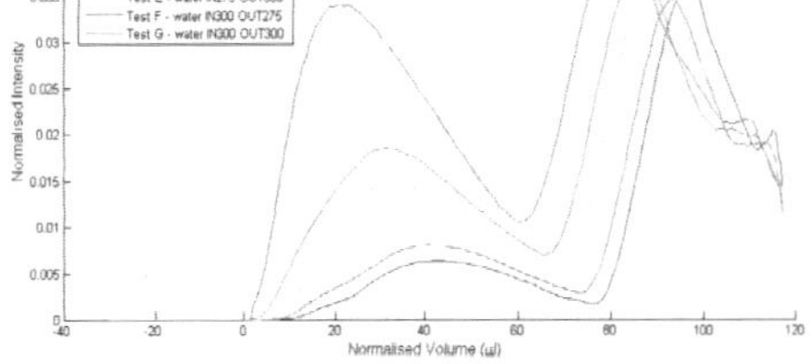

Figure 11: Analysis of water tensiotraces for various fibre positions

3.4. Analysis of Z_{UP}

The Z_{UP} dimension corresponds to the width of the protruding edges on the drophead (see figure 6). $Z_{UP} = 0$ corresponds to a flat drophead. Increasing Z_{UP} has a similar effect to increasing the disc thickness with the overall intensity of the drophead decreasing as well as the numerical aperture. It also has a strong effect on the tensiopeak, especially in the case of coloured LUT. Increasing Z_{UP} also increases the path length of the rays in the liquid, decreasing the tensiopeak height in a similar way to absorbance.

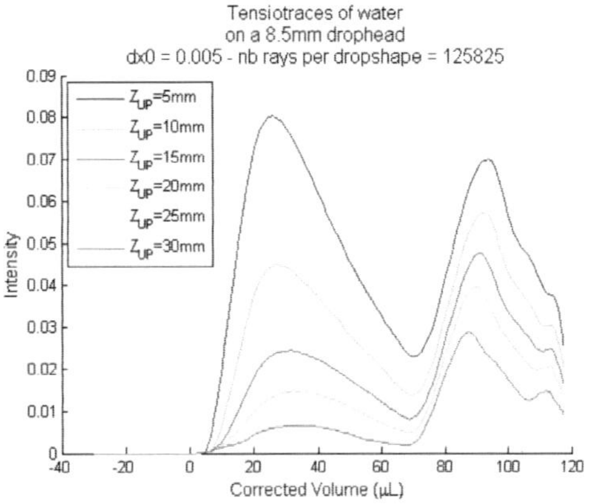

Figure 12: Modeled tensiotraces of water for various lip height

The modelling (see figure 12) shows the result of using a 2mm design is compromised as this is about 10 times poorer than the 0mm head. However, these protruding edges are necessary. It was designed to counter the wetting problem encountered with the first design of the quartz drophead and stop liquids with high surface tension to leak onto the upper edge of the drophead. This practical problem is one that stops proper measurement. Although $Z_{UP} = 0$ gives the best result it is not a practical option as there is no lip. Starna's engineering capability is pushed to the limit with small values and they cannot make them reproducibly with values of less than 2mm. Figure 13 shows the lip height has a greater effect on RPH because it increasing the lip height has a similar effect as changing the numerical aperture of the fibres.

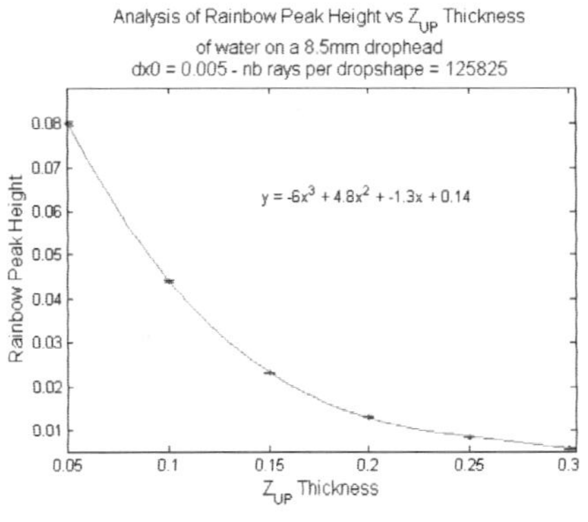

(a) Rainbow Peak Height

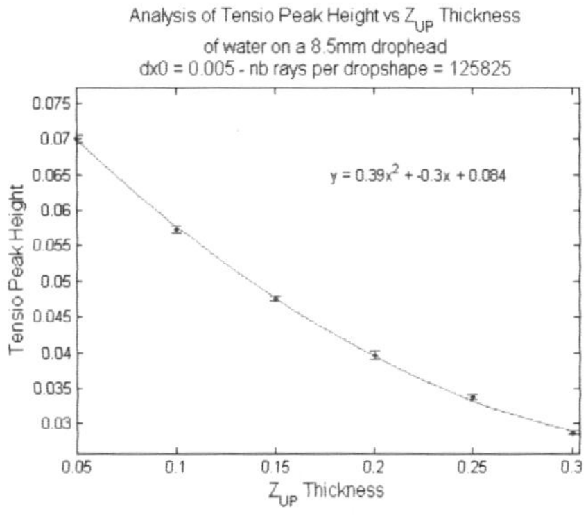

(b) Tensio Peak Height

Figure 13: Analysis of effect of lip height on RPH and TPH

3.5. Cross section analysis

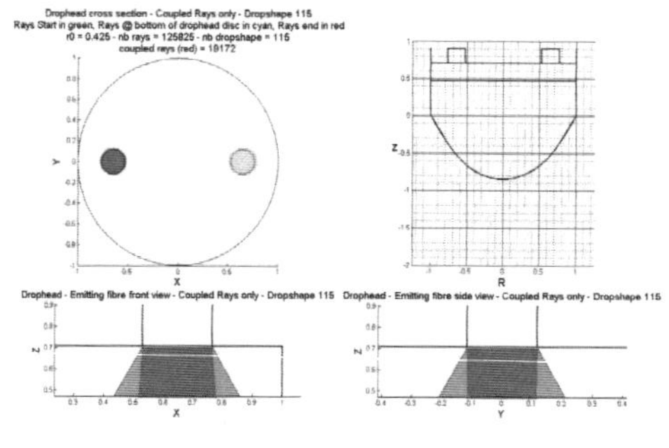

Figure 14: Drop head cross section analysis of coupled rays at Rainbow Peak

3.5.1. Analysis of coupled rays Top left corner of figure 14 shows a cross section view of the drophead and the coupled rays for the rainbow peak. As described above input rays cover uniformly the input fibre (right fibre). The detector fibre (left fibre) also makes full use of fibre end to collect light. The insert graphs at bottom respectively show a front view (bottom left insert) and side view (bottom right insert) of the input fibre. Only coupled rays are shown. Left insert reveals that although light is shining in every direction (within the aperture), only rays shining approximately vertically find their way to the detector fibre. Also, only **75%** of the cone of light is coupled, confirming the choice of fibre to be adequate. This is even more true in the side view of the emitting fibre.

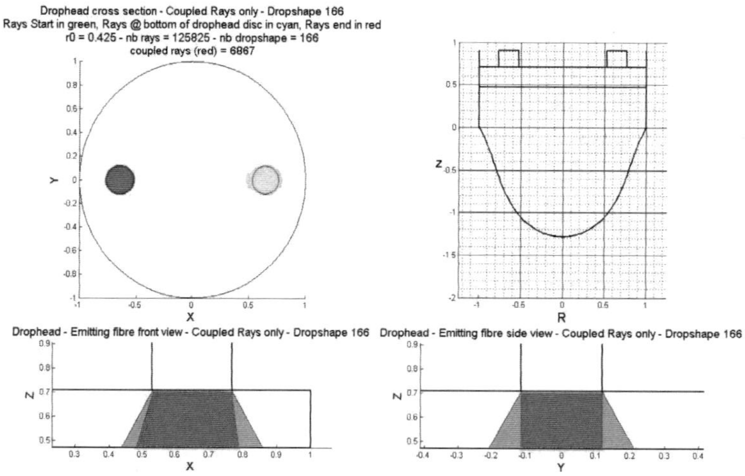

Figure 15: Drop head cross section analysis of coupled rays at Tensio Peak

Analysis of the coupled rays for the tensiopeak (see figure 15) and at the drop period (see figure 16) revealed the same conclusions as for the rainbow peak even though drop shapes are radically different. This confirms that positioning of the fibres, disc thickness and edge width are essential design features and are here optimised with regards to fabrication capabilities.

Analysis of the rainbow peak commencement (see figure 17) confirms experimental results that RPC is bound to the geometry of the drop and the drophead design, i.e. spacing between fibres.

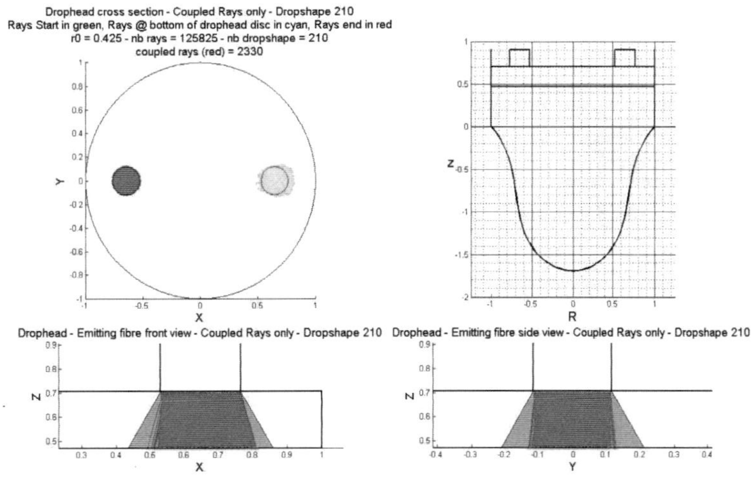

Figure 16: Drop head cross section analysis of coupled rays at Drop Period

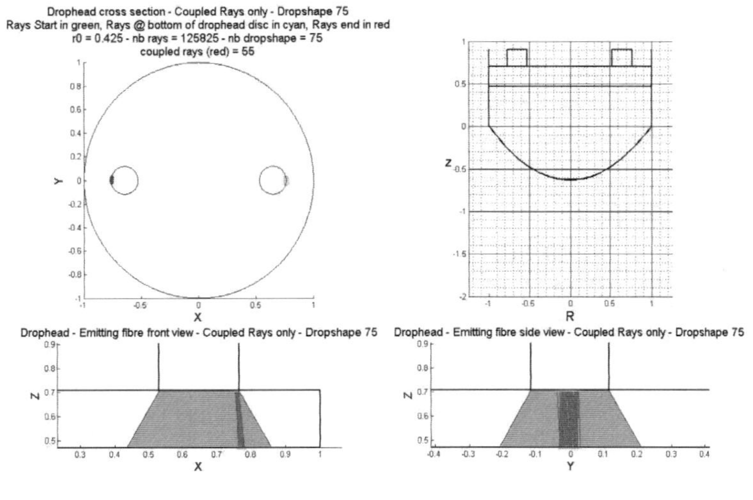

Figure 17: Drop head cross section analysis of coupled rays at Rainbow Peak Commencement

3.5.2. Analysis of lost rays Analysis of non coupled rays (lost rays) can help predict the potential of a design change, especially in regards to fibres diameter and position.

Figure 18 shows impacts of rays that did not reach the detector for the rainbow peak. They are not coming from a particular area of the input fibre nor a particular angle. However, most (**over 50%**) seem to hit the bottom of the drophead within the vicinity (within 2mm) of the detector fibre. This leads to the conclusion that a wider detector fibre (=2mm) would increase the overall intensity of the tensiotrace thus increasing the detection limit. However it might have a negative effect on the definition of peaks. Also, with it being a difficult process to drill accurately into quartz, in order to keep costs down it is a prerequisite to have both fibres of the same diameter. As with coupled rays, it is shown here that rays converge towards the detector fibre due the concave shape of the drop. It is then possible to add one or two pairs of fibres inside

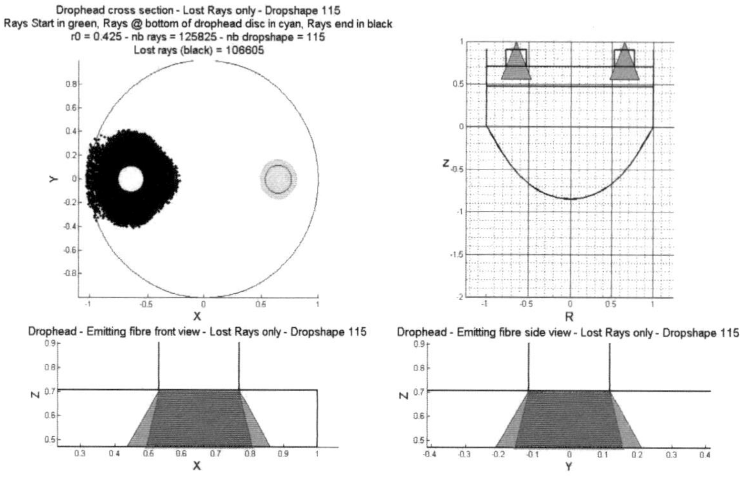

Figure 18: Drop head cross section analysis of non coupled rays at Rainbow Peak

the drophead in order to simultaneously analyse the LUT using different wavelength. Rays of different fibres would not affect each other.

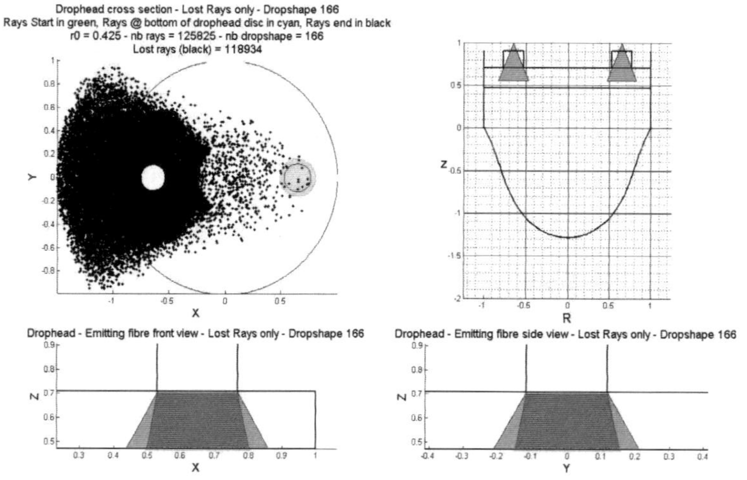

Figure 19: Drop head cross section analysis of non coupled rays at Tensio Peak

Rays that did not couple at the tensiopeak however do not converge toward the detector fibre but seem to cover a wider area thus reducing the possibility of simultaneous analysis using multiple fibre pairs (see figure 19).

4. Conclusion

This paper shows that theoretical testing was crucial to create the new patented quartz drop head and explains design choices and their repercussions on the measurement capabilities of the instrument. The main objective (to show how the model can be used to improve the drophead design for a particular application) was achieved at a fraction of the normal cost. The RAST model allows to design the drophead to fit the purpose but in practice there has to be some compromises with Starna's engineering capability.

References

[1] McMillan N D 1988 Method and apparatus for analysing liquid properties.
[2] McMillan N 1993 *Opto-Laser Europe* 42–43
[3] McMilllan N, Lawlor V, Baker M and Smith S 1998 *Drops and Bubbles in Interfacial research* **6** 593–705
[4] Adams M 2000 Fibre emission cone itensity
[5] Wernert E A 2002 Interactive computer graphics lectures - lighting and shading models
[6] Hecht E 2002 *Optics* 4th ed (San Francisco ; London : Addison-Wesley , c2002.)
[7] Artphotonics 2008 Uv - sillica fibers `http://www.artphotonics.de/SIF/002/index.php`

Journal of Physics: Conference Series **307** (2011) 012008 doi:10.1088/1742-6596/307/1/012008

New Chemically Functionalized Nanomaterials for Electrical Nerve Agents Sensors

Jean-Pierre Simonato*, Simon Clavaguera*, Alexandre Carella*, Michael Delalande*, Nicolas Raoul*, Stephane Lenfant, Dominique Vuillaume** and Emmanuel Dubois****

* CEA Grenoble / LITEN, 17 rue des Martyrs, 38054 Grenoble Cedex 9, France

** IEMN – CNRS, Avenue Poincaré, 59652 Villeneuve d'Ascq Cedex, France

E-mail: jean-pierre.simonato@cea.fr

Abstract. A chemical receptor specific to traces of organophosphorus nerve agents (OPs) has been synthesized and grafted to carbon nanotubes and silicon nanowires in order to make electrical sensors. Our results show that it is possible to detect efficiently sub-ppm traces of OPs with excellent selectivity notably with the use of silicon nanowires by monitoring the Drain-Source current of the SiNW-FET at an optimum back Gate voltage as a function of time. First developments of a prototype have also been realized.

1. Introduction

The threat of a chemical attack on homeland and military forces continues to grow and recent examples such as Iraqi gas attacks or Tokyo terrorist act have clearly demonstrate that organophosphorous agents are powerful neurotoxic molecules that can actually be used as chemical warfare agents and weapons of chemical terrorism. [1]

Due to the structural diversity of toxic chemicals, there is no single ideal sensor for all potential chemical agents, and many sensors with widely different sensitivities are needed for operational processes. Presently, the methods used to detect organophosphorus nerve agents are mostly inadequate for today's expending requirements. New sensors are highly expected, with improved performances such as low weight, very limited false positives, low power consumption, high sensibility, high selectivity, etc.

The ease of production and extreme toxicity of OP nerve agents underscores the need to detect these odorless and colorless chemicals. Today, there is an important need for wandering applications of selective sensors for organophosphorus (OP) based toxic gases for instance for individual soldier protection or homeland security. In the same context we have to take into account the urgent need of passive and low consumption multipoint sensors with extremely low false alarm rate for preventing toxic gas attacks that could occured in public areas such as subway stations, stadiums, malls...

Journal of Physics: Conference Series **307** (2011) 012008 doi:10.1088/1742-6596/307/1/012008

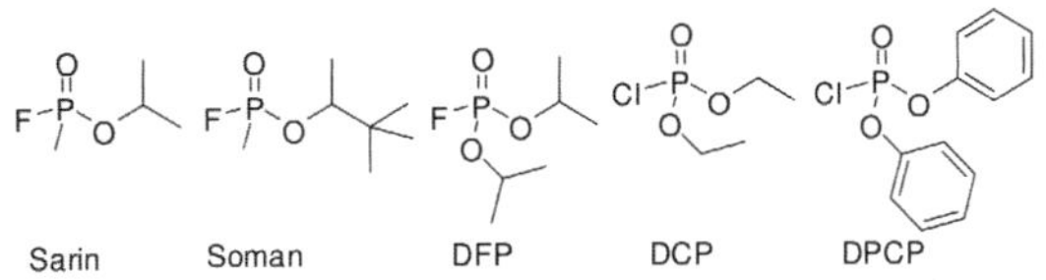

Scheme 1. Chemical structures of some OPs

In this communication we will present new sensing techniques developed for the detection of organophosphorus nerve agents based on the recent developments of nanosciences and nanotechnologies. The miniaturized detectors devices based on electrical detection, i.e. resistors or transistors, using semiconducting parts made either of functionalized carbon nanotubes or silicon nanowires.

2. Carbon nanotube based sensors

The first kind of sensors we realized where made of single walled carbon nanotubes (SWCNTs) in a resistor configuration. First, the SWCNTs where dispersed in NMP (N-methylpyrrolidone) by ultrasonication for 30 minutes. Centrifugation at 13 000 g afforded a grey solution of SWCNTs suitable for spray coating technique. The substrates are made of gold electrodes, obtained by standard lithography on silicon dioxide. The density of the network of SWCNTs was controlled by spray duration and controlled by SEM analysis.

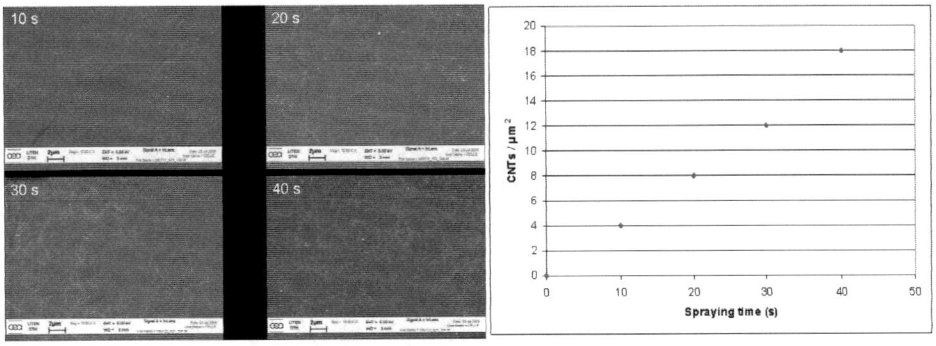

Figure 1. SWCNT network density as a function of spray duration

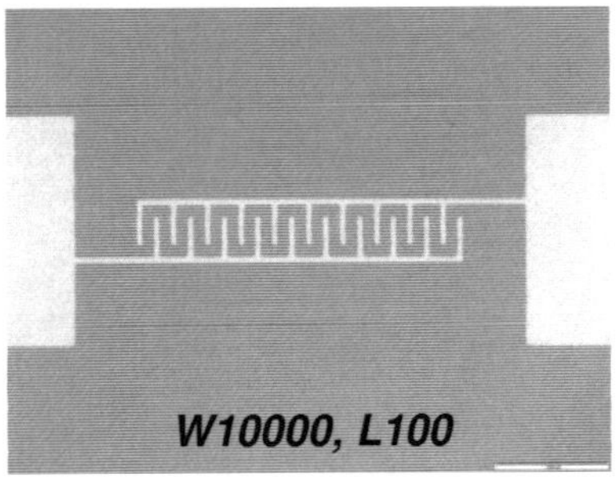

Figure 2. Electrodes of SWCNT based sensors

The as made sensors where then characterized in air and with DiPhenylChloroPhosphate (DPCP, see Scheme 1), a simulant of Sarin-like molecules. The measured vapour pressure of DPCP was around 600-800 ppb in our experimental conditions.

When resistive devices made of pristine nanotubes were exposed to DPCP vapour pressure, an immediate change in resistivity was observed, the intensity increased by about 50 % in 10 minutes. This indicates that SWCNTs are intrinsically sensitive to DPCP, but they are generally sensitive to many analytes. In order to improve the selectivity towards OPs, we modified the carbon nanotubes with the synthesised molecule **1**, originally developed by Rebek et al. [2]

The pyrene part ensures good stabilizing interactions with SWCNTs whereas the rest of the molecule allows good reactivity towards OPs. Using this functionalization route, the response of the sensor was significantly improved, with a 800 % increase of intensity under identical operating conditions (Figure 3).

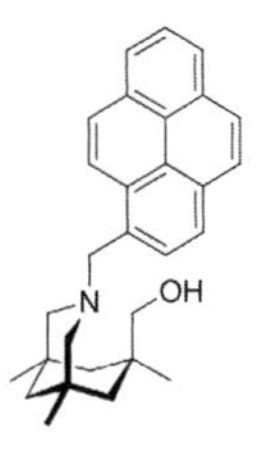

1

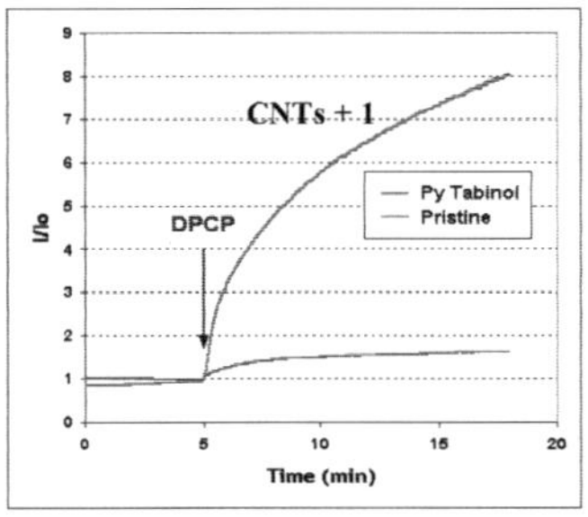

Figure 3. Response of vapour pressure of DPCP by SWCNT based sensors

Other chemical functionalizations gave also satisfactory results and will be presented during the oral communication.[3]

3. Silicon nanowire based sensors

Silicon nanowires are also interesting semiconducting nanomaterials useful for making very small sensors. We used silicon nanowires fabricated from silicon-on-insulator wafers. [4]

Silicon surface can be easily functionalized by different techniques, including thermal hydrosilylation of alkenes or alkynes.

The same OP reactive moiety was used for functionalization of silicon nanowires, but with an alkyne group as the anchoring species. The synthesis was realized from Kemp's triacid in only four steps with an overall yield exceeding 40%.

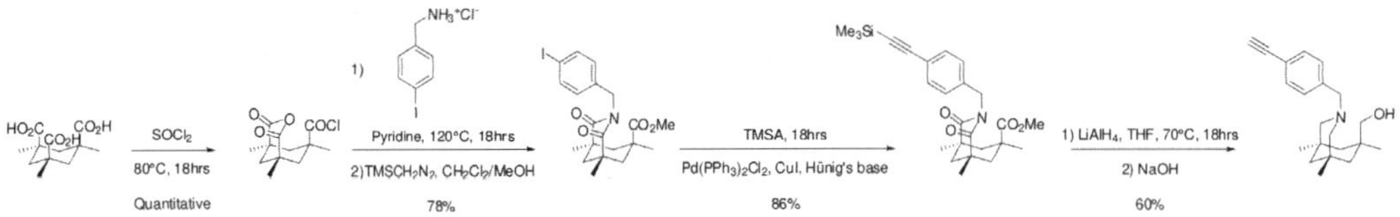

Figure 4. Synthesis of the OP chemical receptor

The molecule was grafted onto silicon nanowires by thermal hydrosilylation in aromatic solvents, and the as-made molecular monolayer was characterized by contact angle, MIR and XPS. A scheme of the functionalized device is presented in figure 5.[5]

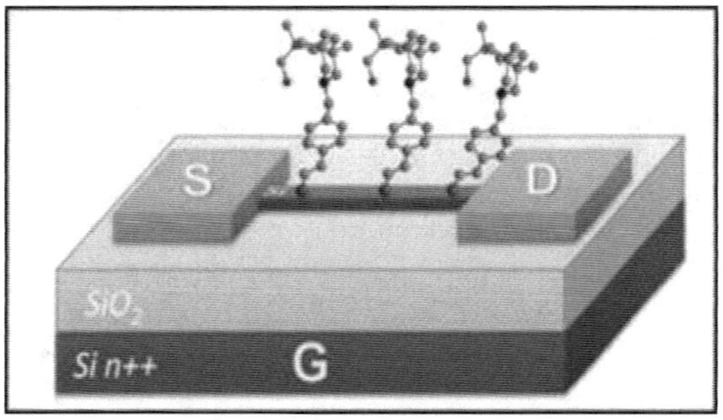

Figure 5. Scheme of the silicon nanowire base sensor (transistor configuration)

The sensitivity of this device in resistive configuration (Vg = 0) was quite good since a significant ten-fold decrease of the resistance RSD was observed upon exposure to vapours of DPCP. When the device was used in transistor configuration, two to five orders of magnitude were measured before and after exposure to sub-ppm vapours of OP simulant (Figure 6).[6]

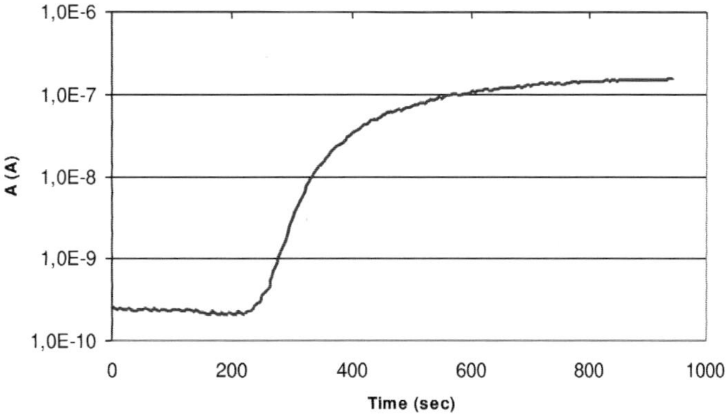

Figure 6. Typical response of a 4x1 μm^2 functionalized SiNW-FET(VDS=–1 V, VGS=V0=-2 V). DPCP vapours were introduced at t= 240 s.

Furthermore, our first attempts to evaluate selectivity of the SiNW-FETs was rather successful since no significant variation of intensity was observed upon exposure to various organic compounds (cyclohexane, dichloromethane, acetone, acetonitrile, triethylamine, acetic acid, exhaust gas, fragrances…).

A first prototype was developed (figure 7).[7] This device bearing the adequate functionality on the SiNW-FET, a power source and a microcontroller for data processing was fabricated in order to investigate its relevance for the detection of nerve agents in complex environments. We found that this sensor exhibits exceptional performances for the detection of nerve agent simulants with extremely high sensitivity and excellent selectivity.

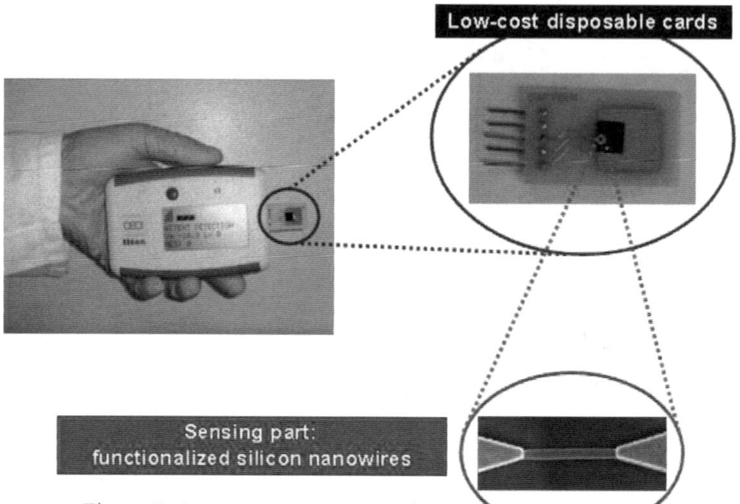

Figure 7. Autonomous prototype for OP detection

4. Conclusion

In summary, we have developed two kinds of nerve agent sensors based on electrical transduction of a chemical reaction occurring on the surface of a functionalized nanomaterial. Up to now, the best results were obtained with silicon nanowires which showed a very fast and marked response at sub-ppm level of OPs simulant. Development were carried out for the transfer of nanoscience knowledge up to effective functional devices. Further developments are still in progress. We hope this technology of hybrid sensors will help to develop sensitive, compact, low-cost and low-consumption nomade devices for widespread applications in the fields of defense and homeland security.

References

[1] R. J. Brennan, J. F. Waeckerle, T. Sharp, S. R. Lillibridge, Ann. Emerg. Med. 1999, 191-204
[2] T. J. Dale, J. Jr. Rebek, J. Am. Chem. Soc. 2006, 128, 4500-4501
[3] M. Delalande, S. Clavaguera, M. Toure, A. Carella, S. Lenfant, D. Vuillaume, and J.-P. Simonato, submitted
[4] V. Passi, F. Ravaux, E. Dubois, J.-P. Raskin, Proceeding of IEEE Int. MEMS Conference, pp. 464-467, 2010
[5] S. Clavaguera, A. Carella, L. Caillier, C. Celle, J. Pécaut, S. Lenfant, D. Vuillaume, and J.-P. Simonato, Angewandte Chemie Int. Ed. 2010, 19, 4063-66
[6] V. Passi, F. Ravaux, E. Dubois, S. Clavaguera, A. Carella, C. Celle, J.-P. Simonato, L. Silvestri, S. Reggiani, J.-P. Raskin, D. Vuillaume, submitted
[7] S. Clavaguera, N. Raoul, A. Carella, and J.-P. Simonato, submitted

This work was funded in part by the French program ANR Camigaz.

Comparison of thermistor linearization techniques for accurate temperature measurement in phase change materials

S B Stankovic[1,3] and P A Kyriacou[1,2,3]

[1] School of Engineering and Mathematical Sciences, City University London, Northampton Square, London EC1V 0HB, UK

[2] E-mail: p.kyriacou@city.ac.uk

Abstract. Alternate energy technologies are developing rapidly in the recent years. A significant part of this trend is the development of different phase change materials (PCMs). Proper utilization of PCMs requires accurate thermal characterization. There are several methodologies used in this field. This paper stresses the importance of accurate temperature measurements during the implementation of T-history method. Since the temperature sensor size is also important thermistors have been selected as the sensing modality. Two thermistor linearization techniques, one based on Wheatstone bridge and the other based on simple serial-parallel resistor connection, are compared in terms of achievable temperature accuracy through consideration of both, nonlinearity and self-heating errors. Proper calibration was performed before T-history measurement of RT21 (RUBITHERM® GmbH) PCM. Measurement results suggest that the utilization of serial-parallel resistor connection gives better accuracy (less than ±0.1°C) in comparison with the Wheatstone bridge based configuration (up to ±1.5°C).

1. Introduction

Recent years have witnessed an increasing development of alternative energy technologies. One of the main objectives in this area can be achieved through application of phase change materials (PCMs) for reduction of energy consumption.

Thermal characterization of PCMs is essential prior to any profound utilization of these materials. Current methods for the investigation of thermo-physical properties of PCMs (e.g. differential scanning calorimeter – DSC) show significant drawbacks especially in terms of the limited sample size [1] [3]. Therefore, the measurements in this study were performed through implementation of the method specifically designed for the investigation of PCMs, the temperature history (T-history) method [3].

Since PCMs are able to store/release large amounts of heat in a narrow temperature range of few degrees, the accurate thermal investigation of these materials is of key importance. However, the majority of the T-history studies in the literature have not emphasized the accuracy of temperature measurements [3]-[8]. Also, it must be taken into an account that sensors for measurements inside the PCM samples need to be small in order to reduce the interference with the phase change process. Consequently small thermistors (model Newark MA100GG103A) were selected for this investigation [9].

[3] The research funding was provided by Engineering and Physical Sciences Research Council (EPSRC).

Thermistor's nonlinearity and self-heating errors are the biggest problems in any temperature application. Several linearization techniques have been developed over the years based on voltage divider or bridge circuits, discrete active elements, and different software solutions [10] [11]. In this paper two techniques based on the Wheatstone bridge design and simple serial-parallel resistor circuit are compared in terms of achievable temperature accuracy.

T-history based characterization of RT21 PCM (RUBITHERM® GmbH) specimen with the expected phase change temperature range between 18°C and 23 °C requires measurements in the range from 11°C to 30 °C [12]. Hence, the thermistor characteristic was linearized in the range from 10°C to 39 °C in order to meet application requirements. Firstly, the linearization methods were theoretically evaluated through calculations of temperature-output voltage dependencies and absolute errors introduced by linearization circuits. Subsequently calibration was performed in 1 °C step temperature program from 10 °C to 39 °C and calibration curves were determined along with the absolute errors. Finally, T-history measurement on RT21 PCM and distilled water as a reference material was performed.

2. Materials and methods

2.1. Temperature sensors

Since the sensors used for temperature measurements in PCMs need to be small, a negative temperature coefficient (NTC) MA100GG103A thermistor model with a 2 mm diameter probe and a 28 wire gauge was used [9]. The temperature dependence of the thermistor resistance R_t is given by the table of resistances with the nominal resistance of 10kΩ at 25 °C. Its operating range is from 0 to 50 °C. Therefore, linearization was performed for the range between 10°C and 39 °C due to indicated application requirements.

2.2. Linearization circuits

The principle of thermistor linearization circuit is to provide relatively linear dependency between temperature and the circuit's output voltage. In addition to linearization the main goal of this study were accurate temperature measurements. The first design goal was to minimize the thermistor self-heating error ΔT to 0.05 °C. This error defines the value of thermistor maximum permissible current I_{max} as indicated by equation (1):

$$I_{max} = \left(\frac{\Delta T C}{R_{t,min}}\right)^{-\frac{1}{2}} \quad (1)$$

where C represents thermistor dissipation constant. Critical value (in air) of this constant for the selected thermistor is $2.5 \cdot 10^{-3}$ $mW/°C$. $R_{t,min}$ represents the minimal thermistor resistance in the operating temperature range. The second design goal was to keep the supply voltage V_{CC} equal to the standard value of 5V in order to minimize the circuit's power dissipation. Nonlinearity error is determined by appropriate temperature-output voltage curve fitting. The first (linear) and the third (cubic) order polynomial fitting equations were determined.

The first linearization circuit is more common and based on Wheatstone bridge (WB) (see figure 1). Since the Wheatstone bridge is just a form of a voltage divider the maximum permissible thermistor current defines the value of resistor R_3 and automatically determines the shape of temperature-output voltage curve and nonlinearity error. The values of the two other resistors determine the output voltage range, but do not affect the shape of the temperature-output voltage linearization curve.

On the other hand, the second linearization circuit, based on simple series-parallel resistor connection (Ser_Par) in figure 2, is more flexible since the self-heating error condition only determines the value of a single resistor (here R_1) in the connection. Additional resistor (R_2) in this configuration allows shaping of temperature-output voltage curve in order to minimize the nonlinearity error and achieve better accuracy in the particular temperature range.

The value of resistor R_2 (in figure 2) was determined by equating the second temperature derivative of temperature-output voltage transform function with zero in order to determine the linear region of the transform function.

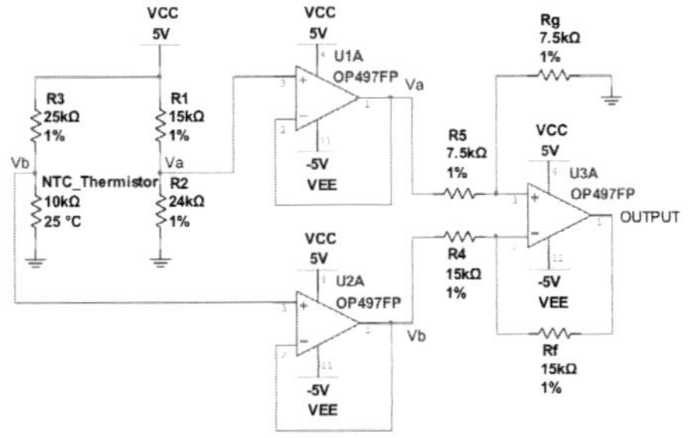

Figure 1. Wheatstone bridge thermistor linearization circuit.

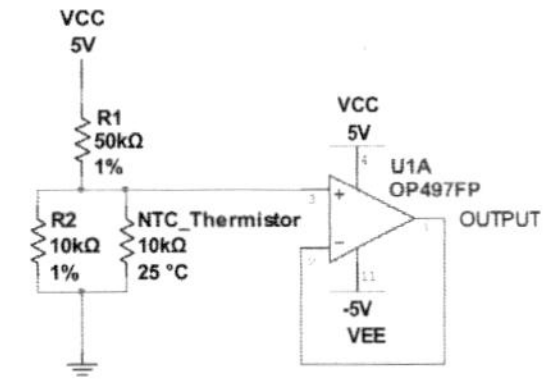

Figure 2. Serial-parallel resistor thermistor linearization circuit.

2.2.1. Theoretical evaluation of temperature-output voltage dependencies

The temperature-output voltage dependency was calculated for both linearization circuits based on the temperature-thermistor resistance characteristic. Calculations were performed for a temperature range between 10°C and 39 °C. MATLAB® code was developed to determine linear and cubic order polynomial fitting equations of calculated curves [13]. The absolute errors between real and fitted temperature values were calculated (see figure 3 and figure 4).

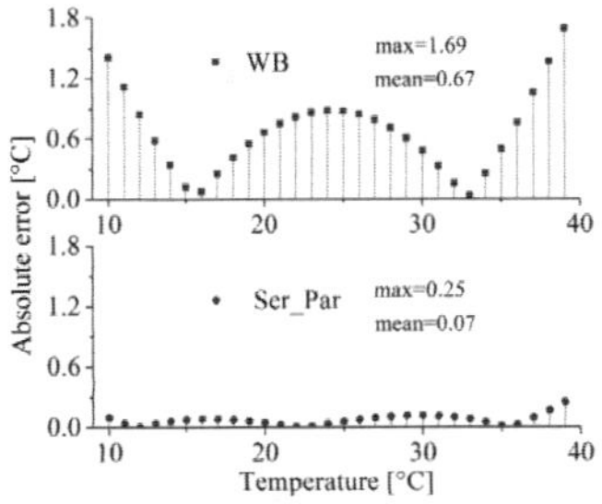

Figure 3. Theoretical absolute error values when using linear fitting for WB and Ser_Par linearization circuits.

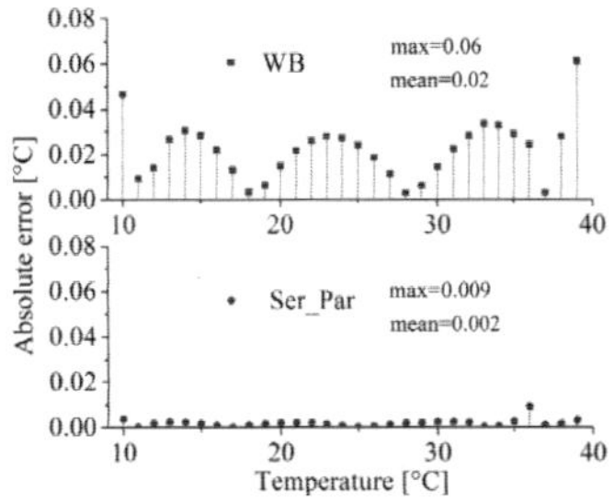

Figure 4. Theoretical absolute error values when using cubic fitting for WB and Ser_Par linearization circuits.

2.2.2. Practical evaluation of temperature-output voltage dependencies

Prior to PCM T-history measurements proper sensor calibration for the two linearization circuits was performed. Each sensor was subjected to 1 °C step temperature program from 10 °C to 39 °C. The calibration measurement was performed in a thermally controlled environment. Data acquisition performed utilising a NI 6212 USB 14-bit DAQ card at a sampling rate of 10Hz [14]. Recorded output voltage data were evaluated at known temperatures (10 °C to 39 °C in 1°C step) and the calibration curves were determined. Once again, the calculation of absolute errors between expected and fitted measured temperature data was evaluated based on MATLAB® linear and cubic calibration curve fitting equations [13].

2.3. PCM T-history experimental protocol

T-history measurement of RT21 PCM was performed in accurate temperature controlled chamber. PCM sample and distilled water (H_2O) reference were subjected to the sharp temperature step between 30°C and 11 °C and their temperature history recorded along with the environmental temperature. Measurements were performed using both linearization circuits. Data acquisition parameters were the same as those used in the calibration measurements. Temperature values from the measured voltage data were calculated through previously determined calibration curve fitting equations.

3. Results

3.1. Result from theoretical evaluation of thermistor linearization circuits

As indicated in figure 3 linear fitting equations for the theoretical temperature-output voltage relationship for the Wheatstone bridge configuration suggest an unacceptable maximum and mean errors of 1.69 °C and 0.67 °C respectively. On the other hand the errors for the serial-parallel resistor configuration are much smaller, with maximum and mean errors of 0.25 °C 0.07 °C respectively.

Cubic curve fitting gives much better and acceptable results for both, WB and Ser_Par circuits. Again as seen in figure 4, the serial-parallel resistor connection gives one order of magnitude better results (maximum and mean errors of 0.009 °C and 0.002 °C respectively) in comparison with the WB results (maximum and mean errors of 0.06 °C and 0.02 °C respectively).

3.2. Results from the calibration measurements

As indicated in figure 5 and figure 6 the linear fitting results for calibration based measurement for the two linearization circuits show good agreement with the theoretical predictions; therefore utilization of Ser_Par configuration seems acceptable even for the case of linear curve fitting (maximum and mean errors being not greater than 0.26 °C and 0.07 °C, respectively for all three channels and maximum error occurring high above 30 °C – highest temperature in T-history measurement).

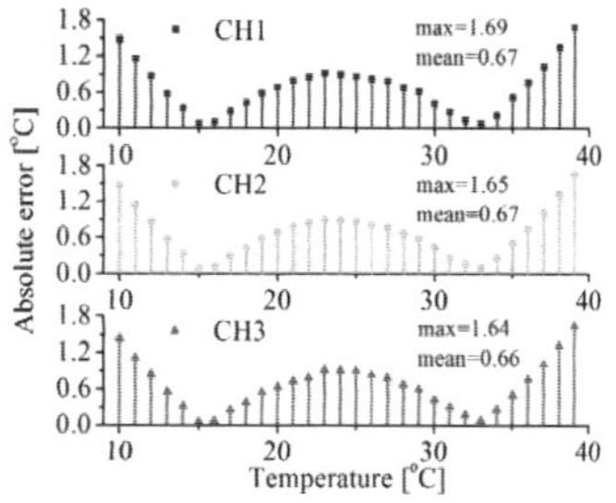

Figure 5. Calibration measurement based absolute error values when using linear fitting for WB linearization circuit; CH1 – later used for environment (ENV) measurement; CH2 – PCM measurement; CH3 – H_2O measurement.

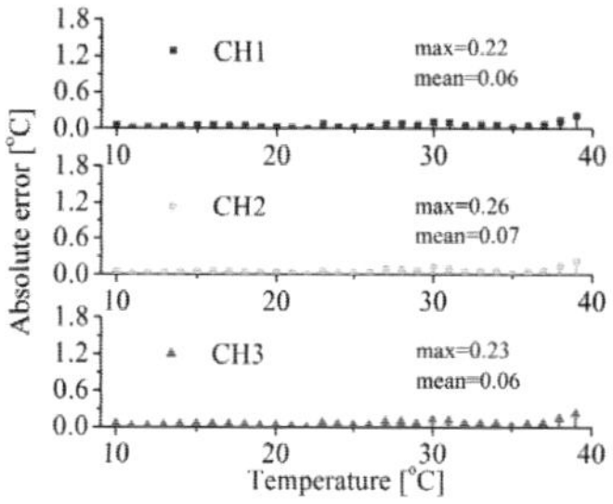

Figure 6. Calibration measurement based absolute error values when using linear fitting for Ser_Par linearization circuit; CH1 – later used for environment (ENV) measurement; CH2 – PCM measurement; CH3 – H_2O measurement.

Meanwhile, figure 7 and figure 8 show the cubic fitting results for the calibration based measurement in both linearization cases. These results show an order of magnitude higher absolute error in comparison with the theoretical predictions. However, these errors are still acceptable to some degree for the WB configuration (maximum and mean errors being not greater than 0.9 °C and 0.25 °C, respectively for all three channels) and entirely for the Ser_Par circuit (maximum and mean errors being not greater than 0.05 °C and 0.01 °C, respectively for all three channels). It is important to emphasize that the absolute error result obtained by measurements are better for the Ser_Par configuration even for the case of linear fitting, in comparison with the WB configuration (either linear or cubic fitting).

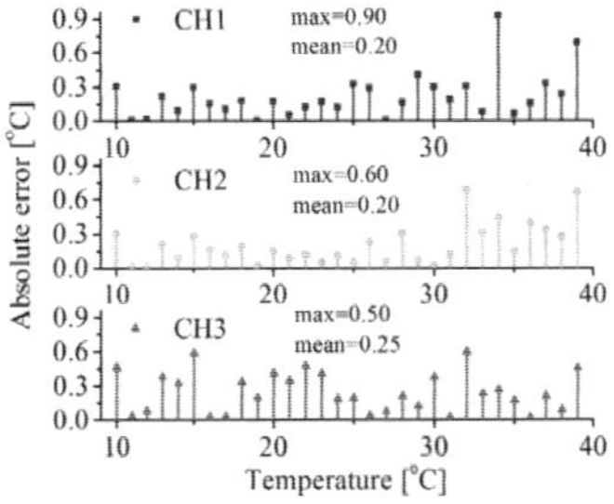

Figure 7. Calibration measurement based absolute error values when using cubic fitting for the WB linearization circuit; CH1 – later used for environment (ENV) measurement; CH2 – PCM measurement; CH3 – H_2O measurement.

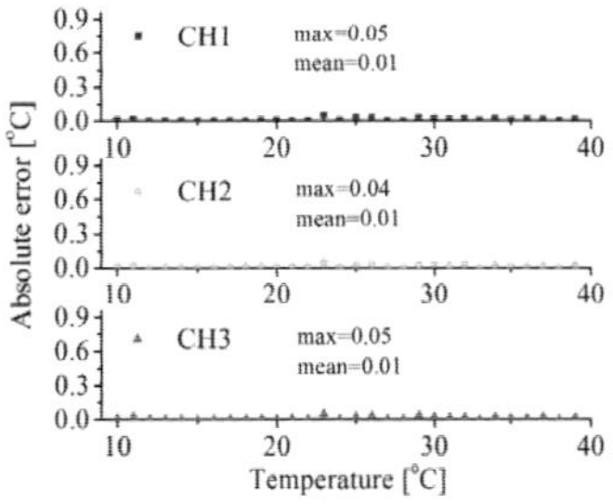

Figure 8. Calibration measurement based absolute error values when using cubic fitting for the Ser_Par linearization circuit; CH1 – later used for environment (ENV) measurement; CH2 – PCM measurement; CH3 – H_2O measurement.

3.3. Results from PCM T-history measurement

PCM T-history temperature measurement was performed between 30 °C and 11 °C. Results from the WB configuration are shown in figure 9 and figure 10 for the linear and cubic fit respectively. Parts of the graphs are *zoomed in* at the expected temperature of 11 °C. It can be seen that the linear fit shows a large deviation of around 1.5 °C, while the cubic fit gives a deviation of 0.2 °C. As seen in figure 11 and figure 12 the deviation from 11 °C in the case of the Ser_Par circuit does not exceed 0.1 °C for both, linear and cubic fittings.

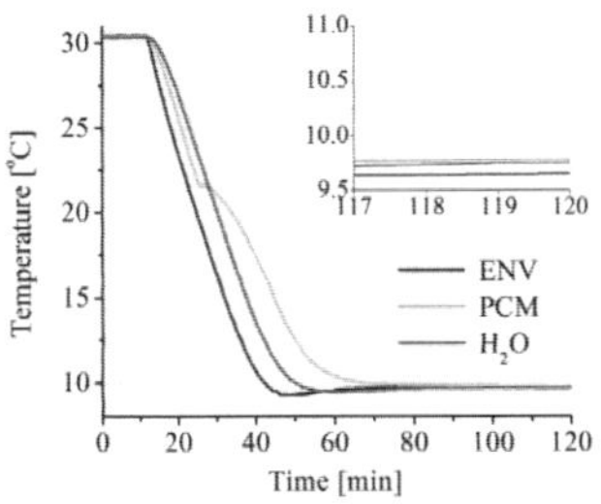

Figure 9. The WB configuration - PCM T-history curves when using linear fitting.

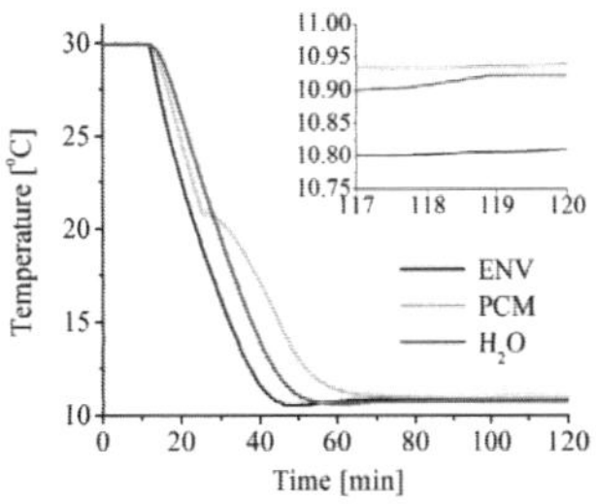

Figure 10. The WB configuration - PCM T-history curves when using cubic fitting.

4. Conclusions

Since PCMs store/release large amounts of heat in a narrow temperature range of a few degrees, accurate T-history measurements are of great importance. If the temperature sensor size is taken into an account thermistors seem like a good solution due to the availability of miniaturized probes.

The overall accuracy of the thermistor based measurement systems depend on self-heating and nonlinearity errors. These errors can be reduced by selecting the most appropriate temperature linearization method.

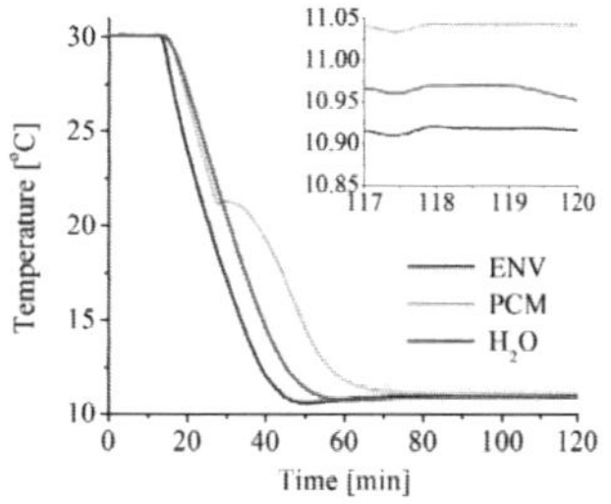

Figure 11. The Ser_Par configuration - PCM T-history curves when using linear fitting.

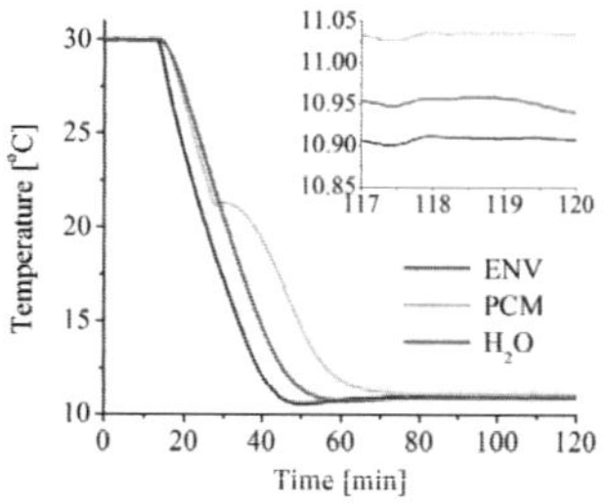

Figure 12. The Ser_Par configuration - PCM T-history curves when using cubic fitting.

Experimental results in this study showed that contrary to theoretical prediction, a lower absolute error was obtained by measurements performed when using the serial-parallel resistor connection in the case of linear fitting than with those performed with the Wheatstone bridge configuration in the case of cubic fitting. Therefore, it is concluded that the linearization techniques based on voltage divider and bridge circuits do not provide sufficient accuracy. Circuits with additional series or parallel resistors compared to voltage divider circuits enable more flexible linearization and therefore better accuracy in certain temperature ranges of interest.

References

[1] Mehling H and L F Cabeza 2008 *Heat and Cold Storage with PCM* (Berlin: Springer)
[2] Castellon C, Günther E, Mehling H, Hieble S and Cabeza L F 2008 *Int. J. Energy Res.* **32** 1258-65
[3] Zhang Y, Jiang Y and Jiang Y 1999 *Meas. Sc. Technol.* **10** 201-5
[4] Marin J M, Zalba B, Cabeza L F and Mehling H 2003 *Meas. Sc. Technol.* **14** 184-9
[5] Lázaro A, Günther E, Mehling H, Hiebler S, Marin J M and Zalba B 2006 *Meas. Sci. Technol.* **17** 2168–74
[6] Lázaro A, Zalba ., Bobi M, Castellon C, Cabeza L F 2006 *Environ. Energy Eng.* **52** 804-8
[7] Kravvaritis E D, Antonopoulos K A and Tzivanidis C 2010 *Meas. Sci. Technol.* **21** 5103-12
[8] Moreno-Alvarez L, Herrera J N and Meneses-Fabian C 2010 *Meas. Sci. Technol.* **21** 7001-5
[9] http://www.newark.com/
[10] Tsai C-F, Li L-T, Li C-H and Young M-S 2009 *Proc. 5th Int. Conf. on Intelligent Infromation Hiding and Multimedia Signal Processing (Kyoto, Japan, 12-14 September 2009)* (Washington: IEEE Computer Society) p 530
[11] Nenova Z P and Nenov T G 2009 *IEEE Trans. Instr. Meas.* **58** 441-9
[12] http://www.rubitherm.com/english/index.htm
[13] http://www.mathworks.com/products/matlab/
[14] http://www.ni.com/

Journal of Physics: Conference Series **307** (2011) 012010 doi:10.1088/1742-6596/307/1/012010

Using EMCCD's to improve the photometric precision of ground-based astronomical observations

Alan Giltinan, Dylan Loughnan, Adrian Collins, Niall Smith

Cork Institute of Technology Blackrock Castle Observatory, Castle Road, Cork, Ireland

alan.giltinan@cit.ie

Abstract. We report on a novel data reduction technique that uses the fast, low-noise operation of EMCCD's to provide improvements in ground-based astronomical photometry of up to 48% by removing much of the incoherent wavelength-dependent absorption of the atmosphere. The technique has significant application potential in multicolour photometry where inter-colour short-timescale variability are particularly prone to atmospheric effects. We comment on a two-colour photometer we have developed to take advantage of the technique.

1. The importance of precision photometry in astronomy

Variability is a defining characteristic of many astrophysical phenomena, and timescales can range from milliseconds to years. Measuring variability enables competing theoretical models to be assessed and compared. In some cases the variability is periodic in nature; in others it is stochastic. Variability which is stochastic in nature is the most difficult to reliably quantify and characterize because the observer has only one chance to measure it. The data cannot be epoch-folded nor analysed using periodicity techniques. Having precise, reliable photometry is most crucial in such circumstances. In other instances, such as exoplanet transit-hunting, the variability amplitude gets smaller as the transiting planet gets smaller. Since the holy grail of exoplanet hunting is to find an earth-sized planet, the motivation to reach ever-increasing photometric precision is obvious.

In optical astronomy flux from a source is now typically measured using a Charge Coupled Device (CCD) which provides both *intensity* and *spatial* information. A well documented advantage of using CCDs for photometry is that, along with the target, the flux from a number of reference sources can be measured simultaneously which allows for the technique of *differential photometry* to be performed. Differential photometry measures the difference between the target source and a number of nearby reference sources (usually stars) to provide a potentially high precision differential measurement. Inherent in this technique is the assumption that objects that are spatially close to one another are similarly affected by variations in the atmosphere. The approach has long been recognized as the most effective technique for achieving the highest precision photometry [1],[2].

2. EMCCD's as the recording mechanism in photometry

Characterising low-amplitude variability which occurs on timescales of minutes or less necessitates the use of integrations which are substantially shorter than that. Electron Multiplying CCD's (EMCCDs) have the potential to provide such a capability because their onboard gain effectively reduces the otherwise catastrophic readnoise that accompanies high-speed readouts in conventional CCD's. EMCCD's offer the opportunity to increase both the S/N in individual frames and the frame rates themselves.

2.1. Photometric Precision of EMCCD's

We performed laboratory tests to determine the limiting photometric precision that is achievable with EMCCDs to ascertain whether they are suitable detectors for our work. Tests were carried out on Andor Technology's iXon DV887-BI (512x512 pixels) and DU88E (1kx1k pixels). The tests employed an artificial starfield and consisted of taking approximately 100,000 images of 0.1s duration. We rebinned the data into successively longer time bins, equivalent to longer integrations per data-bin, and calculated the associated photometric error using differential photometry. The corresponding graphs (an example of which is shown in figure 1) indicate the limiting precision achievable and are consistent with an *rms* value of ~0.04%.

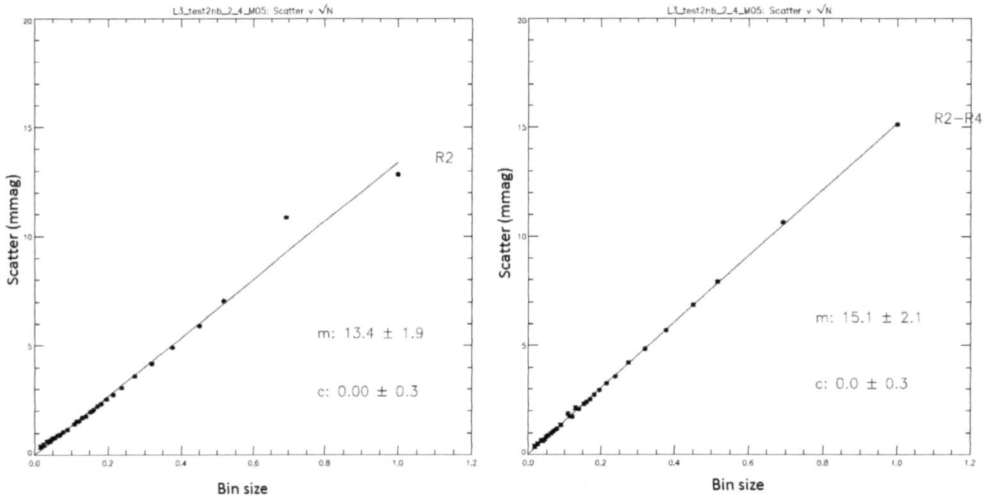

Fig 1. Scatter plots showing the instrument limiting precision of 0.4mmag (0.04%)

2.1.1. Gain, linearity and CIC. While this limiting photometric precision is acceptable for our work, it is worth noting that the use of EMCCD's is not as straightforward as that of conventional CCD's. We performed gain measurements on the DV887-BI using the technique described by Newberry [3] and found the gain to range from 1 to 2600 which requires careful choosing of the exposure/gain combination so as not to enter the non-linear region of the device. Non-linearity starts occurring at about 85% of full-well capacity. Using a technique described by Tubbs [4] we also measured the clock induced charge (CIC), which is actually the dominant source of instrument noise in EMCCD's. The result of 0.4% noise per frame suggests that CIC is not generally problematical for our application.

3. Using EMCCD's to remove atmospheric extinction – Selective Photometry

While the photometric precision achievable is ultimately limited by photon statistics, the observed precision from real data often deviates largely from laboratory experiments [5]. This suggests that factors other than the imaging instrumentation are responsible for limiting the photometric precision observed.

Many factors are known to effect photometry [1], however, short-timescale turbulent atmospheric conditions that occur on timescales of 0.01 – 0.1s are considered to be a dominant factor that are rarely, if ever, corrected for. Pockets of heterogeneous air temperatures combine to distort incoming light (variously referred to as *scintillation* and *seeing*). The effects are incoherent, wavelength-dependent, and cannot be completely corrected for by differential techniques. Indeed the photometric limitations imposed by atmospheric effects are a large driver in the development of spaceborne telescopes such as the Kepler planet-search spacecraft [6]. The cost associated with such systems, however, limits the accessibility to many leaving terrestrial based observations the only option.

SelPhot (Selective Photometry) is a new technique which we have developed that utilises the low-noise, high frame rates achievable with EMCCD technology to improve the photometric precision. The *SelPhot* technique is an automated analysis routine which is deployed on reduced (flux extracted) photometric data. Using stars as reference targets, the *SelPhot* technique searches for segments within the data where the effect of the atmosphere on the photometric data is 'most similar'. The underlying principal of the technique is that for short time intervals, differentially, the atmosphere has little effect on photometric measurements. Finding these intervals through an informed selection process allows the effect of the varying atmosphere to be removed.

An example of the application of the technique to real astronomical data is presented in Fig 2. The data was acquired during an engineering observing session at Calar Alto Observatory (altitude of 2168m) in February 2003 using an Andor iXon back-illuminated CCD97 (512 x 512 pixels) under what would be classed as good/excellent photometric conditions. Fig 2a shows the results of the standard analysis technique and the *SelPhot* analysis technique. It is evident that the error on each data point is significant reduced (average 32% reduction in error) using the latter. Fig 2b highlights the effect of the technique through plotting only the errors of the original technique and *SelPhot* technique. The optimised selection process is seen to have the effect of both homogenising and reducing the errors. Importantly, some of the suggestive short-timescale structure in the original data has been removed, suggesting its origin is extrinsic to the source.
Analysis of each reference star after the *SelPhot* technique has been applied shows an average reduction in *rms* error of 48% confirming that the technique reduces scatter in the data. A detailed discussion of the *SelPhot* technique is reported in [7].

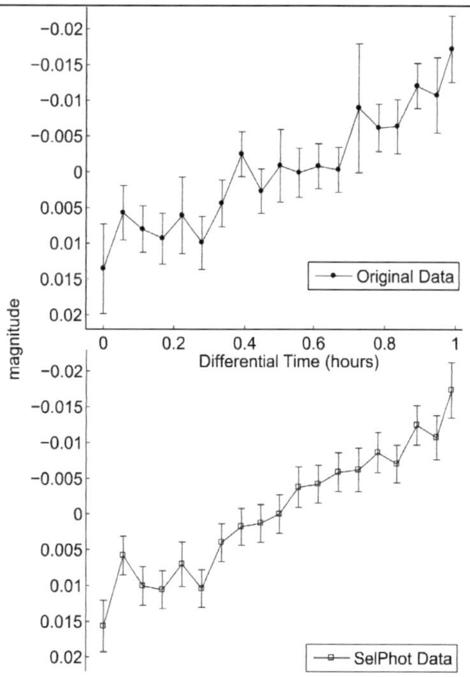

Fig 2a. Photometric light curve from the active galaxy S5 0716+714. The upper plot shows the light curve obtained using the standard approach and all the data. The lower plot shows the light curve obtained when the data has been processed using the SelPhot technique. The post-SelPhot data shows a significant reduction in error across the entire light curve validating the technique.

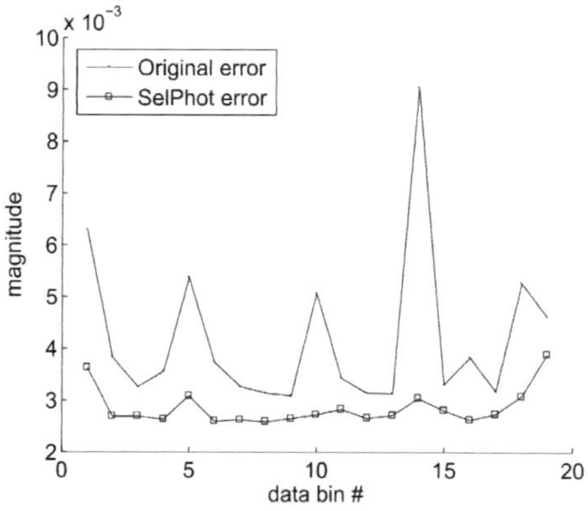

Fig 2b. Error-only plot. This figure shows a direct comparison between the original error and the error following the SelPhot technique.

4. Simultaneous Multi-Colour Photometry

The ability to discriminate between competing models of time-variable phenomena is significantly increased if photometry is performed in more than one colour simultaneously. The major problem with two-channel photometers to date has been the wavelength dependent incoherent effects of the atmosphere which can introduce apparent variability *between* colours that are not intrinsic to the source. Using the *SelPhot* technique we are able to significantly reduce such effects as long as the photometer records data sufficiently fast. TOΦCAM (Two Channel Optical Photometric Imaging Camera, pronounced *toffee-cam*) is a Science Foundation Ireland funded instrument that we have developed at Blackrock Castle Observatory to do just that. It utilizes two Andor Technologies iXon+ DU88E 1K EMCCDs which sample two colour bands simultaneously and is briefly described here.

4.1 Opto-Mechanical Layout of TOΦCAM

The system has been designed in a modular fashion to allow the instrument to be used on telescopes of different types and focal lengths whilst maintaining optimal pixel sampling. The front section of the instrument consists of a 3" Crayford focuser which is used for primary focusing of the instrument and is independent of the telescope focusing mechanism. The focuser can also house focal reducers, tele-compressors or field flatteners to optimise the system on different telescopes.

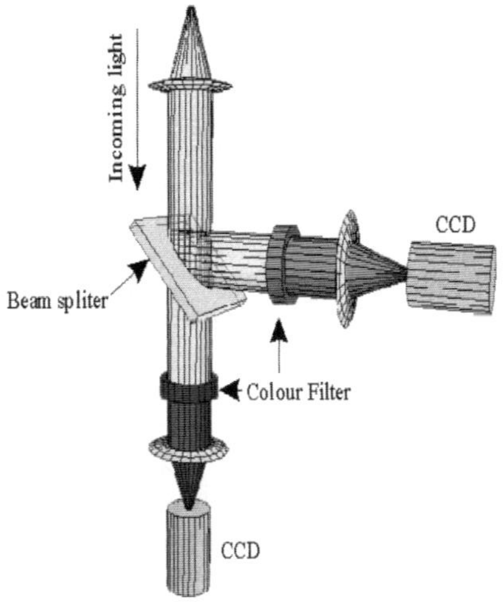

Fig 3. Graphical representation of the TOΦCAM system

The main body of the instrument houses a changeable beam splitter. This beam splitter can be chosen to suit different applications. For example, for non-polarising sources dichroic beam splitters can be used to split the incident light into two colour paths. Alternatively, for sources which show variations in polarisation angle, an optically inactive dielectric beam splitter can be used. This negates the effects on the photometry of the optically active dichroics. A

4.2 Camera Control and Data Acquisition

To support the short-timescale integrations necessary to apply the *SelPhot* technique, TOΦCAM has been designed to be capable of recording up to 1TB of data per night. Our data acquisition and control system can simultaneously acquire, timestamp and stream data to disk storage, whilst running both EMCCD's at their maximum frame full rates.

The system has been designed to be highly fault tolerant through buffering of data, teamed network interfaces and redundant arrays of disks. The data acquisition system has been successfully tested and is capable of running for extended periods (days) of time at data rates of up to 64MB per second.

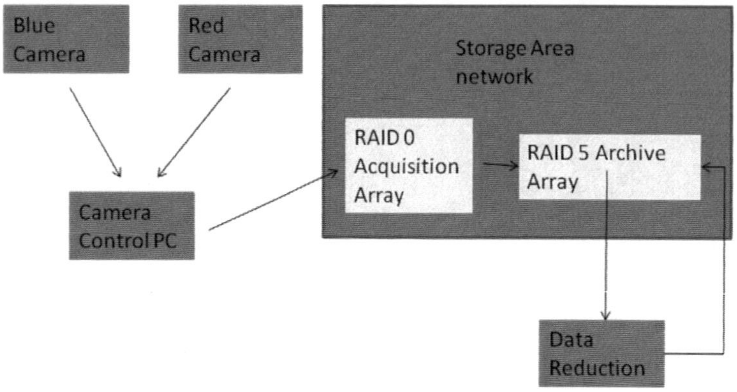

Fig 4: Architecture of TOΦCAM.

Fig 4 shows the general layout of the data acquisition system developed for TOΦCAM. Both cameras are controlled by a single PC running a customised version of Windows XP. Each science frame is time stamped via GPS to millisecond accuracy and then buffered in 4GB of onboard RAM. The data is then asynchronously streamed to a 4 Disk RAID 0 array via ISCSI to the Storage Area Network. The SAN is attached via a teamed Gigabit Ethernet connection. Once the acquisition is completed the data is moved to a more fault tolerant RAID 5 array on the same SAN where it is stored before data reduction.

TOΦCAM is currently undergoing observational tests at Blackrock Castle Observatory, Cork, Ireland.

5 Conclusion

The use of low-noise EMCCD's operated with short integrations allows much of the wavelength-dependent atmospheric extinction to be removed, improving the quality of single- and multi-colour ground-based photometry by substantial amounts. The tradeoff is the generation of large quantities of data. Future possibilities include improving the photometry further by employing faster imagers, approaching precisions heretofore requiring spaceborne instrumentation.

References

[1] A. T. Young et al., "Precise automatic differential stellar photometry," *PASP*, vol. 103, pp. 221-242, Feb. 1991.
[2] S. Howell, *Handbook of CCD Astronomy*. Cambridge University Press, 2006.
[3] AXIOM TECH NOTE: 'Measuring The Gain of a CCD camera'. www.axres.com/technote1.html
[4] R.N. Tubbs. PhD thesis, University of Cambridge, 2003
[5] S. B. Howell and M. E. Everett, "Ultra-High Precision CCD Photometry," in *Third Workshop on Photometry, p. 1*, 2001, pp. 1-23
[6] Website: http://kepler.nasa.gov
[7] A. Collins and N. Smith, "A new Technique for Improving High Precision Photometry in Turbulent Atmospheric Conditions," *in prep.*, 2011.

Fundamental limits of measurement in telecommunications: Experimental and modeling studies in a test optical network on proposal for the reform of telecommunication quantitations

James Egan, Normal McMillan, David Denieffe

IT Carlow, Ireland

eganj@itcarlow.ie

Abstract. Proposals for a review of the limits of measurement for telecommunications are made. The measures are based on adapting work from the area of chemical metrology for the field of telecommunications. Currie has introduced recommendations for defining the limits of measurement in chemical metrology and has identified three key fundamental limits of measurement. These are the critical level, the detection limit and the determination limit. Measurements on an optical system are used to illustrate the utility of these measures and discussion is given into the advantages of using these fundamental quantitations over existing methods.

1. Introduction

When describing a substance, sample or signal the most common approach is to measure the area under investigation using a quantitative approach. The questions arise as to how does a measurement approach recognize that a substance, sample or signal is detected or not? These questions of detection are fundamental in all fields of measurement.

The limits of quantification in instrumentation have been extremely controversial and has evoked much debate over decades [1, 2]. Currie [3] has introduced recommendations for defining the limits of measurement for chemical metrology instrumentation. The introduction of Curries recommendations for defining the limits of measurement has led to the establishment of international recommendations for this field. This paper proposes the application of these methods into the field of telecommunications.

2. Evaluation of the limits of measurement in telecommunications and comparing these to other scientific fields

For many years an international effort has been undertaken by metrological scientists to establish a series of objective measurements for detection and measurement purposes. A common triple question arises when trying to quantitatively analyze a substance

1. Is the signal present?
2. What are the chances that the signal can be detected?
3. What magnitude is the signal?

Before the establishment of an accepted approach for detection limit calculation a wide range of terms, conventions and concepts that were contradictory have been used [2]. In analytical science the limits of measurement have long been a controversial [3]. Not until the introduction of Curries landmark paper did anyone truly understand the level of confusion to the terms detection-limit, detection threshold, minimum detectable signal and detection. It was not until 1995 that recommendations were published to establish the limits of measurement for these scientific fields. Instead of talking about vague detection limits of a system three separate limiting levels were defined.

1. The critical level: The signal level above which an observed instrument response may be reliable recognized as "detected". Here the signal emerges from the background noise. A signal has been detected. The signal is greater than the noise.

Published under licence by IOP Publishing Ltd

2. The detection limit: The true net signal level that may be expected to lead to detection. The magnitude of the signal is 3 times the standard deviation of the noise. The 3 sigma limit is a term associated with the detection limit [4].
3. The determination limit: The signal level above which a quantitative measurement can be stated with a relative uncertainty.

These recommendations were extremely important and the aim here is to introduce Curries terminology in telecommunications for the analysis of signals.

In telecommunications a variety of detection, measurement and analysis approaches have been defined as seen in the previous section. In telecommunications the calculation that one signal is above a certain confidence level is not good enough. The idea of the detection and measurement limits is different therefore for telecommunications and Chemistry.

When applying the terminology used by Currie to telecommunications to describe the quality of signals the usefulness becomes clear. A unified scientific approach has evolved.

Critical level: If above the critical level but below the detection limit, a signal response has been detected but signal is not of high enough quality to make an accurate prediction of the signal state.
Detection limit: If the signal is between the determination level and the detection limit, the signal may be detected accurately but not with enough accuracy for the telecommunications system to receive a BER certificate.
Determination level: Finally if the signal is above the determination level. The signal is of sufficient enough quality for reliable communication to occur and to receive a BER certificate.

From Currie the three levels may be defined as;

$$L_C = k_\alpha \sigma_0 \qquad (1)$$

Critical level, k= point on the x axis where there would be an error of deciding that a substance is present when it is not, The signal is present and is misinterpreted (false negative)= k_α , σ is the standard deviation of the blank signal.

$$L_D = L_C + k_\beta \sigma_0 \qquad (2)$$

Detection limit. k= point on the x axis where there would be an error of deciding that a substance is not present when it is present, σ is the standard deviation of the detected signal. The signal is not present but is misinterpreted as a signal (false positive)= k_β

$$L_Q = k_Q \sigma_Q \qquad (3)$$

Determination limit: k= point on the x axis where the signal is a specific level of quality. σ is the standard deviation of the detected signal.

4. Examples of practical applications in telecommunications

In this section the validity of the application of this approach is shown. A signal will fall into one a number of different categories. The critical level, detection limit and determination level.

Example 1:

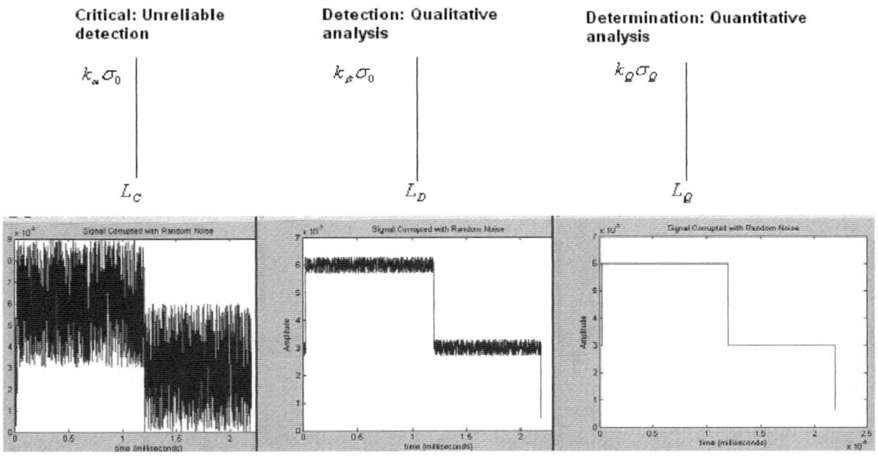

Figure 1: Simple illustration of Curries 3 levels.

In figure 1 a simple illustration of Curries 3 levels is shown. It can be seen when the signal is at the critical level as signal has been detected but it is very unreliable to measure the signal at this stage. At the detection limit the signal can be qualitative determined. At the determination level the signal is at a very high quality and it meets a particular quality of service.

It is proposed in this paper to introduce Curries three levels to telecommunications. Each technique used in the receiving side of a telecommunications system may be itself categorized as belonging to one of Curries three levels. Some of the techniques may belong to more than one field.

Example 2:
In telecommunications it is preferable to have as many signals as possible multiplexed onto the link. This is providing that the multiplexing does not affect the quality of the signals on the link. In the following 3 figures it can be seen that signals have been multiplexed together. A question arises into how many signals can be multiplexed and how close together then can be spaced before they interact with each other and cause an error.

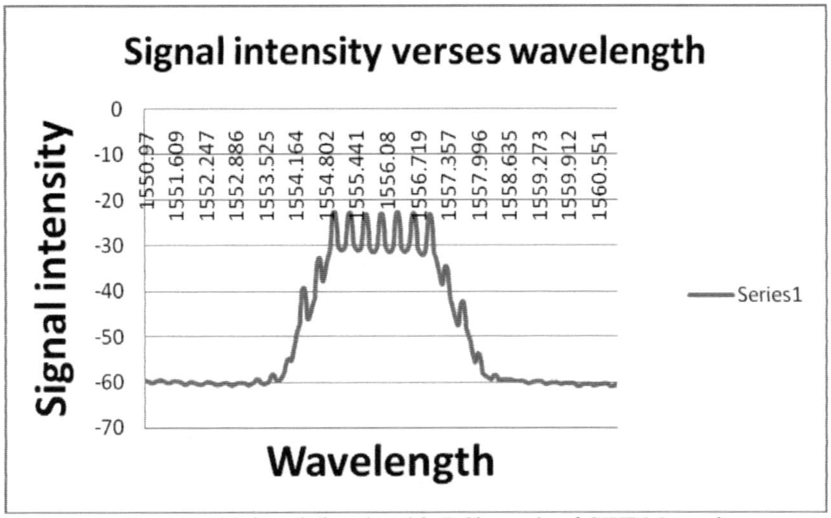

Figure 2: CWDM multiplexed signals with 7 Channels of CWDM spacing.

Techniques such as CWDM and DWDM are constantly being improved and developed. With DWDM channel spacing is less that 1000Ghz. Maximum channel counts currently exceed 200. There are many limitations associated with increasing channel capacity. The higher the level of line width the greater the crosstalk with adjacent channels. The introduction of a universal language for specifying the limits of detection in this area via Curries methodology is proposed.

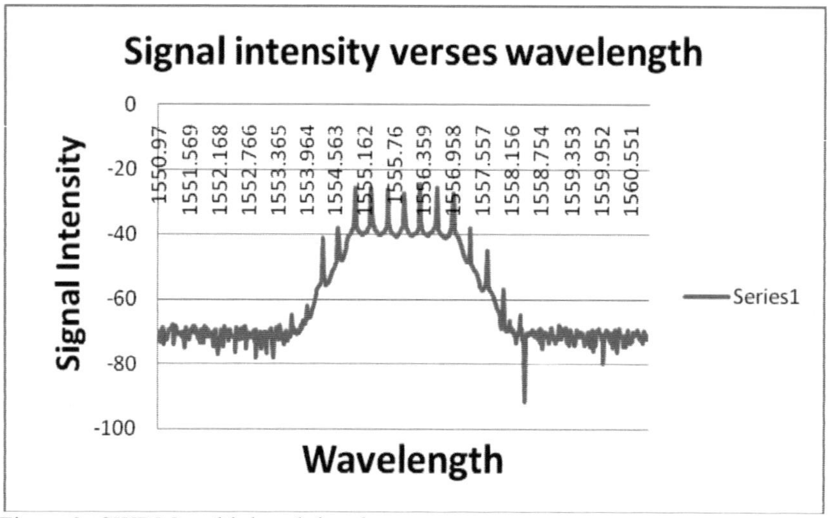

Figure 3: CWDM multiplexed signals

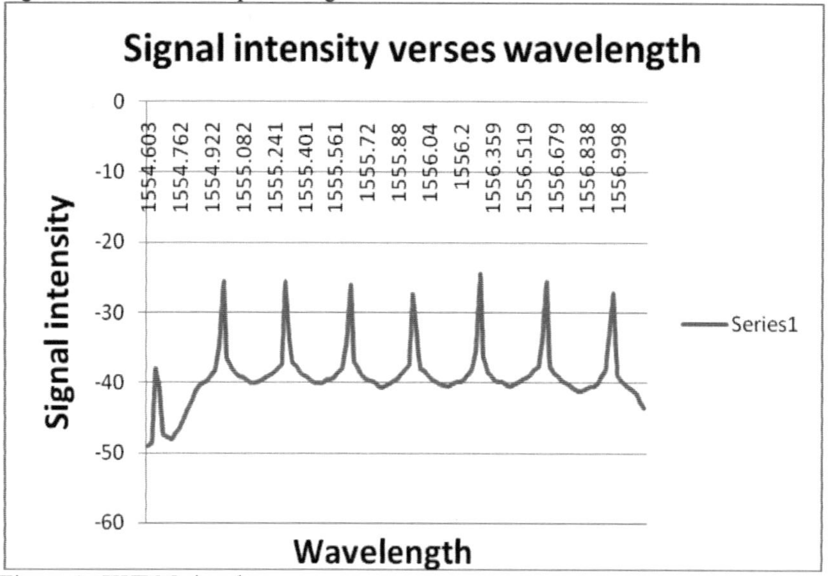

Figure 4: CWDM signals

In Figure 4 if we highlight in on the area of interest we see that the CWDM signals are spaced very close together. A question arises as to what is the maximum spacing the signals can be spaced together. Currently this is defined by the Reilleigh Limit telecommunications has been all about improving and finding new limits. To describe any improvements on this technique it would be very useful to have a standard way across all areas of science to show how the measurements have been improved. By using Curries 3 methods this can be achieved. The terminology can be reused time and time again.

5. Application of approach for telecommunications

Practical issues and implications of adopting these new structured set of quantitations in telecommunications are listed.

1. There are little practical problems associated with integrating these measurements for use in telecommunications.
2. The advantage of using this approach is that they can be used in conjunction with and are not intended to replace current methods of detection in telecommunications. Some examples of the usage of this approach: Using the eye-diagram it can be seen that the detection limit has been achieved but not the determinations level: The BER test has confirmed that the determination level has been reached.
3. There is the possibility to use this approach and integrate into the data sheets of any hardware product.
4. The advantage of using such an approach is to have a unity of terms across all scientific disciplines. Across many scientific domain terms there a wide verity of terms used to describe detection and the use of standard set of terms for telecommunications would be very useful.
    - Detection limit
    - Signal Detection Theory
    - Minimum detectable signal,
    - Detection Threshold
    - Limit of detection (LD)
    - Limit of quality of detection: LQ
    - Detection Threshold
    - Critical level
    - Determination level
    - Decision theory

6. Conclusion

Proposals for a fundamental review of the limits of measurement for telecommunications are made. The synthesis of this paper is that the current methods of detection and classification used in telecommunications can be integrated into one of three areas of measurement. These are the critical level, the detection limit and the determination limit. Practical measurements on an optical signal are used to illustrate these measures and some discussion is then given on the advantages of using what are fundamental quantitations over those of the existing methods.

It is the conclusion of this study that there is an opportunity to group the detection techniques used in telecommunications into three distinct areas. By using one of these three terms when speaking about detection it leaves the reader with a more comprehensive understanding about what is being discussed.

There is a significant challenge to get industry to apply such an approach. This would be a difficult as each section of science has developed its own detection limit terminology.

References

[1] R. M. Lindstrom, "Limits for Qualitative Detection and Quantitative Determination,," NIST Special Publication 958: A Century of Excellence in Measurements, Standards, and Technology, vol. 1901-2000, pp. 164-167, 2000.

[2] L. Currie, "Detection and quantification limits: origins and historical overview," Analytica Chimica Acta, vol. 391, pp. 127-134, 1999.

[3] L. Currie, "Limits for qualitative detection and quantitative determination. Application to radiochemistry," Analytical Chemistry, vol. 40, pp. 586-593, 1968.

[4] P. Loconto, "Instrument Detection Limits: Can We Ever Let 3s Go?" American laboratory (Fairfield), vol. 37, pp. 36-40, 2005.

Extending Murty interferometry to the Terahertz part of the spectrum

S. Hadjiloucas, G.C. Walker and J.W. Bowen

Cybernetics, School of Systems Engineering, The University of Reading, RG6 6AY, UK

Correspondance email: s.hadjiloucas@reading.ac.uk

Abstract. We discuss some novel technologies that enable the implementation of shearing interferometry at the terahertz part of the spectrum. Possible applications include the direct measurement of lens parameters, the measurement of refractive index of materials that are transparent to terahertz frequencies, determination of homogeneity of samples, measurement of optical distortions and the non-contact evaluation of thermal expansion coefficient of materials buried inside media that are opaque to optical or infrared frequencies but transparent to THz frequencies. The introduction of a shear to a Gaussian free-space propagating terahertz beam in a controlled manner also makes possible a range of new encoding and optical signal processing modalities.

1. Introduction

Shearing interferometry is a well established measurement modality which has been extensively implemented using various optical topologies at the optical or infrared parts of the spectrum [1-21]. Advances in monochromatic terahertz sources as well as electro-optic detector technologies now enable us to directly translate some of these systems to longer wavelengths. Advantages from the proposed approach are the longer wavelength implementation which results in better fringe stability and visibility due to low phase noise, the ability to accurately control the source wavelength through heterodyne frequency locking techniques, the use of very efficient broadband polarizing beam splitting components (wire grids), a naturally diffractive spreading of the beam, reduced scattering, greater penetration length of terahertz radiation through fog or ash clouds, etc. The parallel or slightly wedged plate that is used in most shearing interferometers can be made out of high density polyethylene, TPX or other polymer material that happens to be transparent at the terahertz part of the spectrum. Of particular interest is the possibility to use pyrolytic Boron Nitride which is transparent at THz frequencies [22], and has a high refractive index. Detection of the sheared wavefronts should be possible either directly with bolometer arrays which are currently under development [23-27] or indirectly through an electro-optic imaging detection system, where the THz radiation is spatially diplexed with a polarized infrared beam incident on an electro-optic crystal so as to induce a change of infrared polarization which is then monitored using a CCD camera. Alternatively, a Wollaston prism can be used to spatially separate the two polarization states so that a differential imaging photodetection imaging scheme (pixel by pixel) may provide the required fringe visibility. In the following section we discuss some interesting measurement modalities that can be explored.

In any shear interferometer, the original wavefront $W(x, y)$ interferes with a copy of itself laterally displaced by an amount S in the x direction so the new wavefront is $W(x-S, y)$. By expanding in a taylor series, the optical path difference between the two wavefronts may be written as:

Published under licence by IOP Publishing Ltd

$$W(x,y) - W(x-S,y) = \left(\frac{\partial W}{\partial x}\right)S - \left(\frac{\partial^2 W}{\partial x^2}\right)\frac{S^2}{2} + \ldots \tag{1}$$

For very small displacement S, only the first terms needs to be considered i.e., $S << 2\left(\frac{\partial W}{\partial x}\right) / \left(\frac{\partial^2 W}{\partial x^2}\right)$.

To a first order approximation, the shape of the fringes is given by: $(\partial W / \partial x)S = n\lambda$. A lateral shear interferometer, therefore, measures the transverse aberration in the direction of the shear. The lateral shear for a plate of thickness T and refractive index n assuming an incidence angle θ, is given by:

$$S = \frac{T \sin 2\theta}{\sqrt{n^2 - \sin^2\theta}} \tag{2}$$

2. Direct measurement of refractive index and birefringence of materials,

In Figure 1, a THz shearing interferometer for the measurement of the refractive index of a liquid is presented.

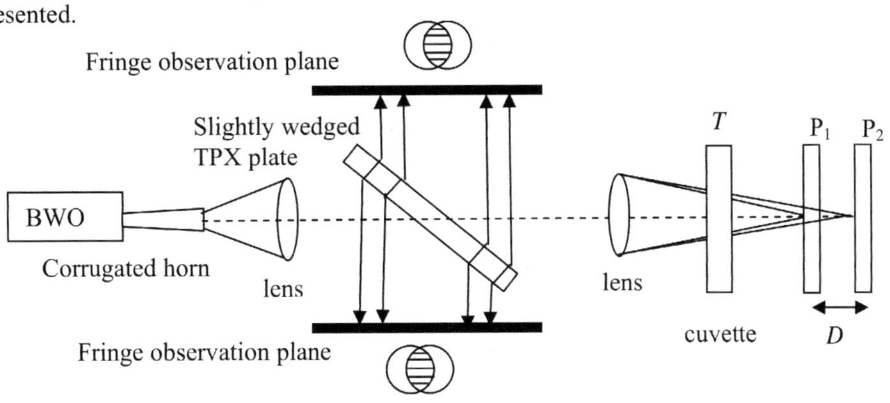

Figure 1. parallel plate shearing interferometer for the measurement of refractive index of liquids.

Initially a measurement is performed with the empty cuvette in place and horizontal fringes are observed for a plane mirror positiononed at P_1. The introduction of liquid in the cuvette, leads to a new position P_2 for which the fringes are parallel in the observation plane. The refractive index of the liquid introduced in the cuvette is given from:

$$n_l = \frac{T}{T - D} \tag{3}$$

For the measurement of birefringence, three measurements need to be performed, by moving the lens in three consecutive positions and the cuvette is replaced by a parallel plate sample and similar expressions for the refractive index of the ordinary and extraordinary rays as shown in equation 3 are performed. Changing the polarization of the source from horizontal to vertical is necessary in this arrangement, this can be conveniently accomplished by introducing a Martin-Puplett interferometer between the source and the wedged plate. The birefringence can then be obtained from:

$$n_e - n_o = \frac{1}{2}\left[\left(n_1^* - n_1\right) + \left(n_2^* - n_2\right) + \left(n_3^* - n_3\right)\right] \tag{4}$$

where subscripts indicate the refractive index of the ordinary and extraordinary rays measurements and that measurements are performed along the three directions of the sample. Similar arrangements can be used to measure the refractive index of materials that are transparent to terahertz frequencies, determine the homogeneity of samples, measure optical distortions and to perform the non-contact evaluation of thermal expansion coefficient of materials buried inside media that are opaque to optical or infrared frequencies but transparent to THz frequencies.

3. Shearing interferometry with phase-conjugate surfaces

It is well known that phase-conjugate surfaces developed for the terahertz part of the spectrum, when placed at regularly spaced intervals, can be used to efficiently eliminate wavefront distortion (Figure 2). For an incident wave-front $e^{i(\omega t+\phi)}$ on a non-linear surface with a phase correction $e^{i2\omega t}$ the reflected waves will have a phase $e^{i(\omega t+\phi)}e^{i2\omega t} = e^{i(2\omega t-\omega t-\phi)} = e^{i(\omega t-\phi)}$. The concept has already been demonstrated at mm-wave frequencies using an electronic approach to conjugate the signal at specific antenna elements [28]. Sampling at sufficiently dense intervals it is possible to conjugate an entire wave-front thus eliminating any distortions. This information can be used for improving the coupling between THz waveforms.

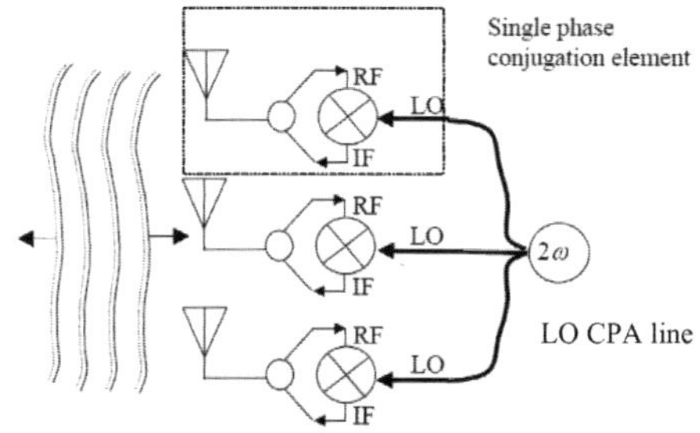

Figure 2. Array of phase conjugation elements sampling an entire THz wavefront with a local oscillator (LO) signal distributed by a network of CPA lines.

The proposed system can be used to introduce a programmable shear to a free space propagating beam or used in tandem with a shearing interferometer to perform accurate measurements by nulling the observed change in the fringes.

4. The application of shearing interferometry for information processing

It is well known that a thin lens can produce a 2-dimensional Fourier transform of an image on the back of its plane. Introduction of a shear in a controlled manner to a free space propagating Gaussian beam, provides a new degree of freedom for encoding and decoding information. The shear performs a transformation of fractional order, which can be used for phase retrieval, signal analysis, and filtering [29-54]. The relevance of fractional order transforms to understand the free space propagation of Gaussian beams described by Gauss-Laguerre or Gauss Hermite polynomials will be discussed in more detail at the conference.

5. Conclusion

A range of new wavefront shearing measurement modalities that can be implemented in the THz part of the spectrum have been suggested. Terahertz shearing interferometry is still at its infancy but because of the possible applications, it is likely to become an emergent research area once fringe imaging technologies become more widely available.

References

[1] Bates W.J., A wavefront shearing interferometer *Proc. Phys. Soc.* (*Lond.*), **59**, (1947), 940.
[2] Drew R.L., A simplified shearing interferometer *Proc. Phys. Soc.* (*Lond.*), **864**, (1951), 1005.
[3] Francon M., *Progress in microscopy.* Pergamon Press, New York, 1951 p. 150.
[4] Brown D. S. The application of shearing interferometry to routine optical testing *J. Sci. Instrum.* **32** (1955) 137.

[5] Baker J., An interferometer for measuring the spatial frequency response of a lens system. *Proc. Phys. Soc. (Lond.)*, **868**, (1955), 871.
[6] Gates J.W., Reverse shearing interferometry *Nature* **176** (1955) 259.
[7] Smith F.H., In *Modern Methods of Microscopy*. Butterworths, London, 1956, p. 76.
[8] Hariharan, P., and Sen, D., Cyclic shering interferometer. *J. Sci. Instrum.* **37** (1960) 340.
[9] Saunders, J.B., Wavefront shearing prism interferometer. *J. Res. Nat. Bur. Stand.* **68C** (1964) 155.
[10] Murty, M.V.R.K., Some modifications of Jamin interferometer useful in optical testing. *Appl. Optics*, **3** (1964) 535.
[11] Francon, M., Polarization apparatus for interference microscopy of isotropic transparent objects. *J. Opt.Soc. Am.,* **47** (1967) 528.
[12] Saunders, J.B., A simple inexpensive wavefront shearing interferometer. *Appl. Optics*, **6** (1967) 1581.
[13] Murty, M.V.R.K., Interferometry applied to the testing of optics. *Bull. Opt. Soc. India*, **1** (1967) 33.
[14] Lohmann, A., and Bryndahl O.,A., A lateral wavefront shearing interferometer with variable shear. *Appl. Optics*, **6** (1967) 1934.
[15] Van Rooyen, E., Design for a variable shear prism interferometer *Appl. Optics*, **7** (1968) 2423.
[16] Briers J.D., Prism shearing interferometer. *Opt. Technol.*, **1** (1969) 196.
[17] Van Rooyen, E., and Van Houten A.G., Design of a wavefront shearing interferometer useful for testing large aperture optical systems. *Appl. Optics*, **8** (1969) 19.
[18] Murty, M.V.R.K.,A compact lateral shearing interferometer based on the Michelson interferometer. *Appl. Optics*, **9** (1970) 1146.
[19] Kelly J.C. and Hargreaves R.A., A rugged inexpensive shearing interferometer. *Appl. Optics*, **9** (1970) 948.
[20] Nyssonen, D. and Jerke J.M., Lens testing with a simple wavefront shearing interferometer. *Appl. Optics*, **12** (1973) 2061.
[21] Hariharan P. Simple laser interferometer with variable shear and tilt. *Appl. Optics*, **14** (1976) 1056.
[22] M. Naftaly, R.Dudley and J. Leist, 'Terahertz spectroscopy of Boron Nitride,' *J. Phys. Conf. Ser.,* Dielectrics 2011 (*in press*).
[23] R Qi Li, Sheng-Hui Ding, Rui Yao, and Qi Wang Real-time terahertz scanning imaging by use of a pyroelectric array camera and image denoising, *J. Opt Soc. Am. A*, **27**, (2010) 2381-2386.
[24] Miller A.J., Luukanen A., Grossman E.N., 'Micromachined antenna-coupled uncooled microbolometers for terahertz imaging arrays, terahertz for military and security applications II,' ed. R.J. Hwu, D.L. Woolard, *Proc. SPIE* **5411**, (Bellingham WA, 2004) 18-24
[25] Wei Lee A. and Hu Q., 'Real-time, continuous-wave terahertz imaging by use of a microbolometer focal-plane array,' *Optics Lett.*, **30**, (2005) 2563-2565.
[26] Wu, Q. Hewitt, T. D. and Zhang, X.-C., 'Two-dimensional electro-optic imaging of THz beams,' *Appl. Phys. Lett.* **69** (1996) 1026-1028.
[27] Mickan S., Abbott, D. Munch, J. Zhang X.-C., van Doorn T., Analysis of system trade-offs for terahertz imaging,' *Microelectron. Journal* **31** (2000) 503–514.
[28] Chang, Y., Fetterman, H., Newberg, I. and Panaretos, S. "Millimeter-wave phase conjugation using artificial nonlinear surfaces," *Appl. Phys. Lett.* **72**, (1998) 745-747.
[29] Flannery D.L. and Horner J.L., Fourier optical signal processors, *Proc. IEEE,* **77**, (1989), 1511-1527.
[30] Walther A., Radiometry and Coherence, *J. Opt Soc. Am.*, **58** (1968), 1256-1259
[31] Walther A., Propagation of the generalized radiance through lenses, *J. Opt Soc. Am.*, **68** (1978), 1606-1610.
[32] Mendlovic, D. and Ozaktas, H.M., Fractional Fourier transforms and their optical implementation I, *J. Opt Soc. Am. A.*, **10** (1993), 1875-1881.
[33] Mendlovic, D. and Ozaktas, H.M., Fractional Fourier transforms and their optical implementation II, *J. Opt. Soc. Am. A.*, **10** (1993), 2324-2329.
[34] Ozaktas, H.M., and Mendlovic, D., Fourier transforms of fractional order and their optical interpretation, *Opt. Commun.*, **101**, (1993), 163-169.

[35] Lohmann A.W., Image rotation, Wigner rotation, and the fractional Fourier transform, *J. Opt. Soc. Am. A.*,10, (1993) 2181-2186.
[36] Lohmann A.W., Mendlovic, D., Zalevsky Z., and Dorch R.G., Some important fractional transformations for signal processing, *Opt. Commun.,* **125**, (1996), 18-20.
[37] Ozaktas, H.M., Zalevsky Z. and Kutay M.A., *The fractional Fourier transform with applications in Optics and Signal Processing,* Wiley New York, 2001.
[38] Torre A., The fractional Fourier transform and some of its applications in optics, in *Progress in Optics,* **43**, ed. E. Wolf, Nonrth-Holland, Amsterdam, (2002), 531-596.
[39] Almeida L.B., The fractional Fourier transform and time-frequency representations, *IEEE Trans. Signal Process.,* **42**, (1994), 3084-3091.
[40] Alieva, T. and Bastiaans, M.J., On fractional Fourier transform moments, *IEEE Signal Process. Lett.*, **7**, (2000), 320-323.
[41] Alieva, T. and Bastiaans, M.J., Phase-space distributions in quasi-polar coordinates and the fractinal Fourier transform *J. Opt Soc. Am. A.*, **17** (2000), 2324-2329.
[42] Teague R.M., Deterministic phase retrieval: a Green function solution, *J. Opt Soc. Am.*, **73** (1983), 1434-1441.
[43] Alieva, T. Stancović, L.J. and Bastiaans, M.J., Signal reconstructin from from two close fractional Fourier power spectra, *IEEE Trans. Signal Process.,* **51**, (2003), 112-123.
[44] Bastiaans, M.J., and Wolf K.B., 'Phase reconstruction from intensity measurements in linear systems,' *J. Opt. Soc. Am., A*, **20** (2003), 1046-1049.
[45] Dong B.Z., Zhang Y. Gu B.Y. and Yang G.Z., Numerical investigation of phase retrieval in a fractional Fourier transform, *J. Opt. Soc. Am., A*, **14** (1997), 2709-2714.
[46] Alieva T. and Agullo-Lopez F., Diffraction analysis of random fractal fields, *J. Opt. Soc. Am., A*, **15** (1998), 669-674.
[47] Merlo D.R., Romo J.A.RM., Alieva T. and Calvo M.L., Fresnel diffraction by deterministic fractal gratings, an experimental study, *Opt. Spectrosc.,* **95**, (2003) 131-133.
[48] Alieva T. and Calvo M.L., Paraxial diffraction on structures generated by multiplicative iterative procedures, *J. Opt. A., Pure Appl. Opt.,* **5**, (2003) S324-S328.
[49] Ozaktas, H.M., Barshan, B., Mendlovic, D. and Onural, L., 'Convolution, filtering and multiplexing in fractional Fourier domainsand their relationships to chirp and wavelet transforms,' *J. Opt. Soc. Am., A*, **11** (1994), 547-559.
[50] Almeida L.B., 'Product and convolution theorems for the fractional Fourier transform,' *IEEE Signal Process. Lett.,* **4**, (1997) 15-17.
[51] Mendlovic, D., Ozaktas, H.M. and Lohmann A.W., Fractional correlation, *Appl. Opt.,* **34**, (1995) 303-309.
[52] Lohmann A.W., Zalevsky Z. and Mendlovic, D., 'Synthesis of pattern recognition filters for fractional Fourier processing,' *Opt. Commun.,* **128**, (1996) 199-204.
[53] Alieva, T. and Calvo M.L., 'Generalized fractional convolution,' in *Perspectives in Modern Optics and Optical Instrumentation*, eds. J. Joseph, A. Sharma, and V.K. Rastogi, Anita Publications New Delhi, (2002), 163-173.
[54] Akay O., and Boudreaux-Bartels G.F., 'Fractional convolution and correlationvia operator methods and an application to detection of linear FM signals,' *IEEE Trans. Signal Process.,* **49** (2001) 979-993.

Testing by photometric measurement and camera study of theoretical prediction of microvolume universal sessile dropshape

S.R.P. Smith[1], M. O'Neill[2], N.D. McMillan[2], K. Arthure[2] , S. Smith [2] and S.Riedel[2]

[1] Physics Centre, Essex University, Wivenhoe Park, Colchester, Essex, CO4 3SQ, UK
[2] Drop Technology Ltd., Tallaght Business Park, Tallaght, Dublin 24, Ireland

E-mail: stephen@smith99.co.uk

Abstract. The approach to the theory of sessile dropshapes held on a cylindrical drophead is discussed. It reveals an 'undifferentiable' universal micro-dropshape for volumes below 3μL. Camera studies demonstrate the veracity of this prediction exploited in the design of a new microvolume spectrometer. The mean pathlength of light injected through a microvolume sessile drop has been determined both from the model and from experiment. Drop volumes determine accurately the mean pathlength and with this Beer's law relationship is experimentally confirmed. The Transmitted Light Drop Analyser uses this universal 'natural cuvette' to deliver both high-performance UV spectra and absorbance measurements at discrete wavelengths.

1. Introduction

The Transmitted Light Drop Analyser (TLDA) has been developed by the authors based on serious attention to drop physics including developing a model of the optics to optimize the performance of the instrument[1]. This theoretical study was the extension of a major piece of work developing drop spectroscopy for pendant drops[2] that extended from the first 1992 patents. These studies place drop spectroscopy on a firm theoretical foundation. This present study is undertaken to prove theoretical facts that are of some practical importance in it shows that all microvolume drops supported on a cylindrical drophead have the same shape irrespective of the properties of the liquid. Given this fact, it is an obvious step to suggest these universal liquid entities could be a 'natural cuvette' for spectroscopy.

2. Experimental

2.1 The Laplace Dropshape Model and Camera Testing

The modelling of drop shapes is done using a numerical integration of the Laplace-Young equation, which describes the shape of a sessile liquid drop in equilibrium under surface tension and gravitational forces. The drop has radius r_0 at its base, where it sits on the drophead. At a point Q on its surface (radius r, height z above the bottom of the drop), there is a pressure difference Δp between the inside and the outside of the drop.

The pressure forces are balanced by surface tension forces given by the classic Young-Laplace equation:

$$\Delta p = p(\text{inside}) - p(\text{outside}) = \gamma\left[\frac{1}{R_\perp} + \frac{1}{R_\parallel}\right] \qquad (1)$$

where $R_\perp$ and $R_\parallel$ are respectively the radii of curvature out of plane and in plane at Q, and γ is the surface tension.

The derivation of the equations used in the drop shape model is well known. The equations that are important here as they are the ones used in the computer model using the two characteristic dimensionless parameters β and X_0, defined by

$$\beta = \frac{\rho g r_0^2}{2\gamma} \quad X_0 = \frac{r_o}{R_o} \qquad (2)$$

β is characteristic of the liquid that constitutes the drop, and is the same for all drops formed with the same liquid and the same orifice, whilst the parameter X_0 (the inverse relative radius of curvature at the apex of the drop) changes with drop size. R_0 is the radius of curvature at the apex of the drop.

The computer model for dropshape is implemented using MatLAB and Runge-Kutta methods that are well known. Figure 1 shows the result for a 3μL drop of radius r_0 = 1 mm. It is obvious from (2) that β will be controlled by the r_0^2 term. For small drops it is found that despite density and surface tension variations, the value of β becomes small regardless of the value of the physical quantities and all drops have consequently indistinguishable shapes. A simple physical way of viewing this matter is the surface tension forces become much larger than the gravitational forces and constrain all drop types to have measurably indistinguishable shapes. The drops here are attached to a drophead but if they were free of this surface they would all be spherical.

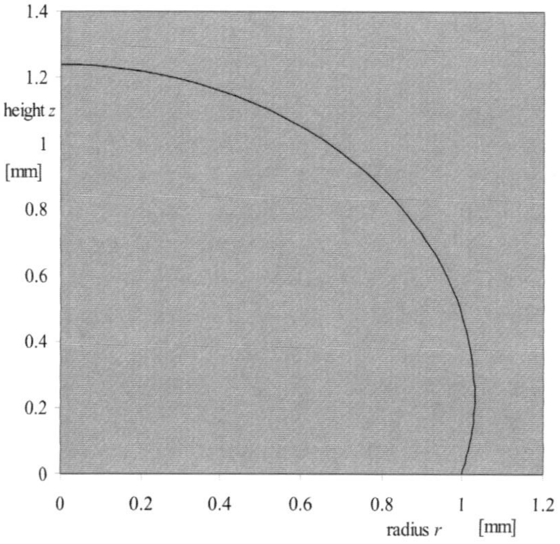

Figure 1. Semi-drop shape for a 3 μℓ drop of radius r_0 = 1 mm.

2.2 Pathlength Studies and Beer's Law Relationship of Drop Spectroscopy

The ray-tracing model developed for the TLDA is quantitatively accurate and can be used to predict the pathlength of rays refracted through the drop sample. The most important application for this type of technology is probably DNA analysis and the experimental work has as a consequence been done here on this type of fluid.

It does not follow directly that if drop shapes are indistinguishable the mean pathlength through a drop is the same, because the refractive index variations introduce differences in ray paths even if drop shapes are identical. The model suggests this is a perturbation and indeed mean pathlengths are almost indistinguishable. Experimentally this hypothesis was tested using a range of potassium dichromate solutions which were measured on the TLDA and a conventional UV/Visible Spectrophotometer (1cm cell). Good agreement is shown between the experimental pathlength results from the TLDA and the physical model.

Camera images of water, the TE buffer and three DNA solutions of different concentrations were recorded. The four repeat drop measurements of each solution were shown to overlap in drop profile. Therefore, one sample drop from each solution was taken and overlayed using the Layers function in Photoshop to confirm the profiles were the same for small volume drops. A summary of the results are shown visually in Figure 2.

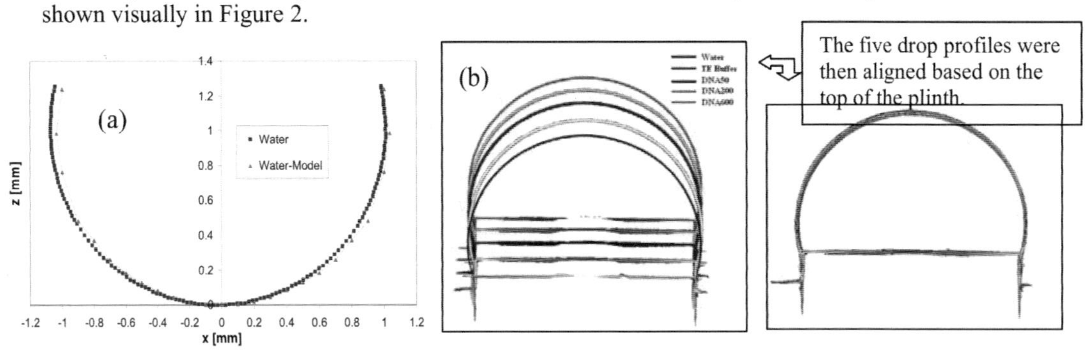

Figure 2(a). Comparison of the drop profile of a 3μL water drop from a camera image and the physical model.
Figure 2(b). Shows the profiles obtained from camera images of water, buffer and DNA drops. This was to determine whether the different small volume liquids had the same shape and hence the same pathlength.
Figure 2(c). Overlaying of the profile images for the different samples in (b).

The drop profiles of water, TE buffer and 3 DNA drops of different concentration are shown in Figure 2(b). These drop images show that the drop shape and size are indistinguishable for water and DNA samples of such small volume drops (3μL). This is not much of a theoretical challenge in tacking extremes of physical constants, but indeed water and ethanol drops would also be indistinguishable. Obviously, the choice of sample in these studies have been undertaken specifically to support the DNA assay work.

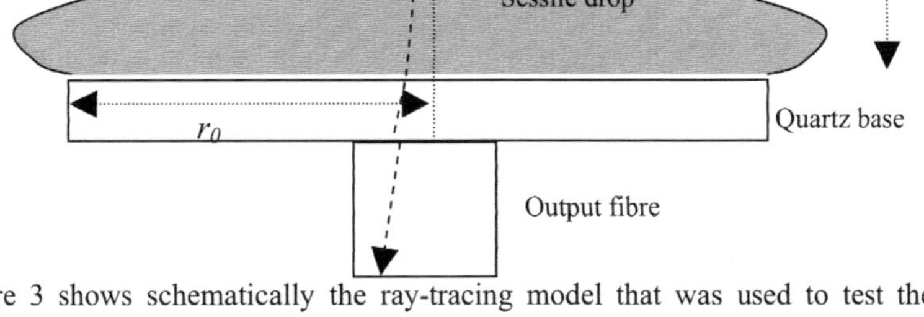

Figure 3 shows schematically the ray-tracing model that was used to test theoretically the issues involved with the mean pathlength. This was fully described in a paper by McMillan et al [2]. The two most important findings of this study were that for drop spectroscopy there is a simple analytical Beer's Law relationship confirmed experimentally by potassium dichromate standards, the results of this study are presented in Figure 4.

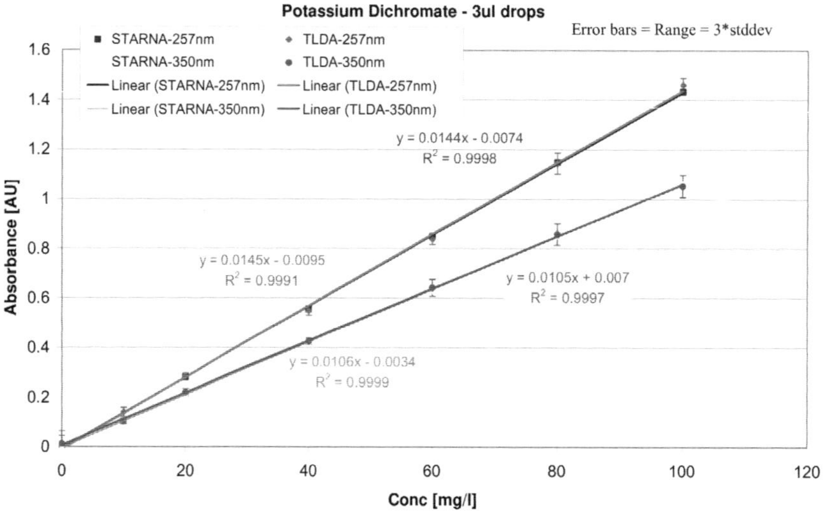

Figure 4. Results showing Beer's law relationship for TLDA with Starna standards and results from replicate readings showing 3σ–error bars.

Secondly, the pathlength variation seen in Figure 5 is known theoretically and has been confirmed experimentally. The scope of this study is obviously too large here to present fully. A large number of measurable quantities beyond those seen in this figure have however been predicted and successfully confirmed.

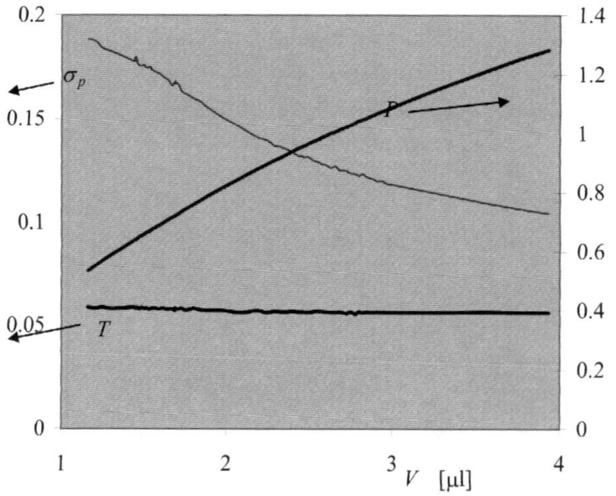

Figure 5. Dependence of transmission, mean pathlength and variance in pathlength (T, P and σ_p) on V for a large radius (2 mm) input fibre.

The mean pathlength shows a very small variation with position of the input fibre. If the fibre has a diameter of greater than 2mm (modeling results shown here in Figure 5) then it has been discovered that no variability is found, although in practice it is very difficult to obtain such multimode fibers. Using 1mm or 600 micron fibres still however gives a very acceptable performance for the TLDA.

3. Conclusions

The cuvettes for the TLDA are designed by nature and microdrops are of a precise reproducible shape for any given volume. The pathlengths are known from modelling and experimentally determined. Errors nevertheless occur in the TLDA pathlength derived from errors in pipette volumes. The study has provided insights into important physical aspects of the TLDA design and given some insights into the thinking that led to its invention. The study has combined modelling, photonic testing and camera studies to demonstrate the universal (liquid independent) sample shape of a chosen sample volume. The TLDA uses microvolume drops that allow mean pathlength to be determined by pipetted drop volume. The dynamic range of the TLDA is consequently much greater than on a traditional UV-visible spectrophotometer perhaps 0-50 A-units. Positional invariance of fibre position can be obtained if a large diameter source fibre (above 2mm) is used but very good instrument performance can be obtained with fibres above 500 microns.

References

[1] McMillan N.D, O'Neill M, Smith S.R.P, Hammond J, Riedel S, Arthure K and Smith S 2009 *Reliable TLDA-microvolume UV spectroscopy with applications in chemistry and biosciences for microlitre analysis and rapid pipette calibration*, J. Opt. A: Pure Appl. Opt. pp 11-22 doi:10.1088/1464-4258/11/5/054016

[2]] McMillan N.D, O'Neill M, Smith S.R.P, Pringuet P, O'Mongain E, O'Rourke B, Riedel S, Neill M, Augousti A, Wüustneck N, Wüstneck R, McMillan D.D.G, Colin F, Hennerbert P, Pottecher G and Kennedy D 2008 *Quantitative drop spectroscopy using the drop analyser: theoretical and experimental approach for microvolume applications of non-turbid solutions,* Meas. Sci. Technol. 19 pp19-36

A novel approach to the assessment of vascular endothelial function

S Sathasivam[1], S Phababpha[2], P Sengmeuan[2], P Detchaporn[2], Z Siddiqui[1], U Kukongviriyapan[2] and S Greenwald[1]

1. Pathology Group, Blizard Institute of Cell & Molecular Science, Barts & The London School of Medicine & Dentistry, Queen Mary University of London, UK

2. Department of Physiology, Khon Kaen University, Khon Kaen, Thailand

Email: s.e.greenwald@qmul.ac.uk

Abstract. Impaired endothelial function (EF) is associated with atherogenesis, and its quantitative assessment has prognostic value. Currently, methods based on assessing flow-mediated dilation (FMD) are technically difficult and expensive. We tested a novel way of assessing EF by measuring the time difference between pulses arriving at the middle fingers of each hand (f-fΔT), whilst FMD is induced in one arm. We compared f-fΔT with standard methods in healthy and diseased subjects. Our findings suggest that the proposed simple and inexpensive technique gives comparable results and has the potential to qualitatively assess EF in the clinical setting, although further work is required.

1. Introduction

Impaired endothelial function (EF) is an important early step in the atherosclerotic disease process and quantitative assessment of EF has prognostic value; hence, a readily available, non-invasive, method of assessment can provide a powerful screening tool for vascular disease [1].

The current gold-standard method of assessment uses duplex ultrasound to measure brachial artery flow-mediated dilation (FMD), which is dependent on the availability of nitric oxide released from endothelial cells. It follows that the greater the FMD, the better the EF. However, using duplex ultrasound is technically challenging, requires highly skilled operators, is expensive and therefore difficult to use as a readily available screening tool [2].

In an attempt to find an appropriate alternative, several novel biomechanical approaches have been explored, including measuring pulse wave velocity (PWV) to assess FMD and hence EF. PWV is the speed at which the pulse of blood propagates along vessels [3]. When a vessel dilates following vascular smooth muscle relaxation, its wall becomes acutely more compliant and the PWV decreases [4]. Therefore, measuring the PWV in the forearm (brachial-finger PWV, b-f PWV) whilst FMD is induced in the brachial artery (as a result of reactive hyperaemia), and comparing this to the b-f PWV when the arm was at rest, a quantitative measure of the brachial artery dilation can be derived. Studies have suggested that the assessment of FMD using this approach gives comparable results to that of the gold-standard duplex ultrasound method [5,6].

Yet another approach involves monitoring the pulses arriving at the middle finger (MF) of each hand, whilst the right arm has FMD induced in it and the left arm is at rest [7]. In this case, the time delay (f-f ΔT) between the arrival of the pulses at the right MF and the left MF will reflect a change in

Published under licence by IOP Publishing Ltd

the PWV and therefore dilation in the right brachial artery. Again, comparing this to the f-f ΔT of both arms at rest gives a quantitative measure of the dilation in the brachial artery. To our knowledge, use of this novel approach has not been reported yet.

The overall objective of this study is to validate the f-f ΔT method for the assessment of FMD. To do this we carried out five experiments, each of which, provided a partial validation. In Experiment-1 we tested the hypothesis that f-f ΔT will be greatest at the start of FMD and will decrease to resting levels as FMD diminishes with time; and that therefore, f-f ΔT will follow the same time course and magnitude as b-f PWV. In Experiment-2 we compared the assessment of FMD in healthy controls and thalassaemic adolescents using the b-f PWV approach and the f-f ΔT approach to see if these techniques could discriminate between healthy and diseased individuals. In Experiment-3 we tested the FMD response in healthy subjects and patients at risk of CVD using the f-f ΔT approach in order to see if it could discriminate between healthy subjects and subjects at risk of CVD. In Experiment-4 we assessed FMD in subjects with varying severity of coronary artery disease (CAD) using the established venous occlusion plethysmography (VOP) method and finally, in Experiment-5 we compared the assessment of FMD in normal, overweight and obese adolescents using the b-f PWV approach and the VOP approach. Therefore, in summary, our experiments were comparing the assessment of FMD using the three different techniques (f-f ΔT, b-f PWV and VOP) and also determining whether each technique could discriminate between different groups (normal and diseased subjects) on the basis of FMD assessment.

2. Method

2.1. Subjects

- Experiment-1: 10 healthy subjects with no CVD (mean [±SD] age 35±18.6 years) were recruited.
- Experiment-2: 30 controls (mean [±SD] age 14.4±1.2 years) and 29 patients with β-thalassaemia/Haemoglobin E (mean [±SD] age 13.8±1.5 years).
- Experiment-3: 28 healthy volunteers with low CVD risk (mean [±SD] age 21±1.29 years) and 12 patients with high CVD risk (mean [±SD] age 62±7.57 years) were recruited. "High CVD risk" implies having CVD or having CVD risk factors (diabetes mellitus, hypercholesterolaemia or smoking).
- Experiment-4: 33 CAD patients with risk factors (RF) ≤1, 46 CAD patients with RFs =2 and 71 CAD patients with RFs ≥3 were recruited. Mean [±SD] age for RFs ≤1 group was 63±1.4 years, for RFs =2 group it was 60±1.2 years and for RFs ≥3 group it was 62±1.0 years.
- Experiment-5: 54 male and 64 female adolescents were recruited and divided on the basis of BMI into 3 groups: 44 controls (mean [±SD] age 14.7±1.3 years) with a normal weight (mean BMI [±SD] 20.6±1.7); 35 overweight subjects (mean [±SD] age 14.5±1.4 years, mean BMI [±SD] 25.6±1.4) and 39 obese subjects (mean [±SD] age 14.5±1.6 years, BMI [±SD] 31.9±3.9).

2.2. Protocol

The standard method of inducing FMD was employed: a cuff placed around the upper arm is inflated to suprasystolic pressure to induce ischaemia for 5 minutes and, upon deflation of the cuff, hyperaemic flow induces endothelial-dependent dilation of the brachial artery [8].

For f-f ΔT measurements standard photoplethysmography (PPG) finger probes (Nelcor pulse oximeter probes) were used to monitor pulses at each middle finger (MF). For b-f PWV measurements a PPG finger probe and custom made flat PPG probes were used to monitor pulses at the MF and brachial artery. Alternatively, pulses at the brachial artery were monitored using 4 or 8Mhz (as appropriate) ultrasound probes driven by a continuous wave Doppler flow velocimeter (Dopplex II, Huntleigh Healthcare, Cardiff, UK) in measurements for Experiments 2 and 5. The b-f distance was measured with a flexible tape to the nearest 5mm for PWV calculation. One minute baseline f-f ΔT and b-f PWV were recorded before inflating the cuff so that post-occlusion values could later be compared to these baseline values to assess the magnitude and time course of any changes.

The PPG probes and Dopplex outputs were connected to a custom built arterial compliance monitor configured to accept inputs from a variety of probe types. The signals were captured at a rate of 1kHz, amplified and passed via a USB interface to a lap top computer running custom written software (PMD 3.0) which processed the signals and gave a real-time display of the pulses. The data were analysed offline using the PMD software.

Forearm blood flow (FBF) was measured by VOP according to previously described procedures [9,10] with some modifications. Briefly, a mercury-filled Silastic tube (D. E. Hokanson, Inc., Issaquah, WA, USA) was attached to the widest part of the forearm, and an occlusion cuff was placed just proximal to the elbow. Blood flow was measured following rapid inflation (<1 second) of the upper arm occlusion cuff to 40mmHg for 10 seconds to occlude venous outflow from the arm. The cuff was then allowed to deflate rapidly and remain deflated for 10 seconds. Each of these 20 second cycles was repeated 10 times. The FBF output signal was recorded and analysed with a BIOPAC data acquisition system (BIOPAC Systems, Goleta, CA, USA). Basal FBF values from the 10 inflation/deflation cycles were averaged. Reactive hyperaemia was induced by a brief period of ischemia achieved by inflating the upper arm cuff to a pressure of 200mmHg for 4 minutes. Following release of the high pressure occlusion, the FBF following reactive hyperaemia and its return to baseline was tracked by repeating the 20 second inflation/deflation cycles for a period of 3 minutes. The subjects were then allowed to rest for 10 min and the entire procedure was repeated.

2.3. *Analysis*

In experiments involving f-f ΔT or b-f PWV, two sets of pulses were obtained: pulses at right MF and pulses at left MF (f-f ΔT); and pulses at the brachial artery and pulses at the MF (b-f PWV). In both cases, the time delay between corresponding pulses in each set was determined by the foot-to-foot method (Figure 1). A mean f-f ΔT or b-f PWV for a time interval was calculated, and the absolute difference from baseline f-f ΔT or percentage change from b-f PWV was determined. In experiments involving VOP, the maximum increase in FBF during FMD (as a percentage change from baseline FBF) was calculated.

In Experiments 1 and 3, we used the time taken for the f-f ΔT or b-f PWV to return to baseline as the quantitative measure of brachial artery dilation. In Experiments 2, 4 and 5, the maximum change in f-f ΔT, b-f PWV or FBF was used for the quantitative measure of brachial artery dilation. In Experiment 4 we also used the area under the FBF-time curve as a quantitative measure of brachial artery dilation.

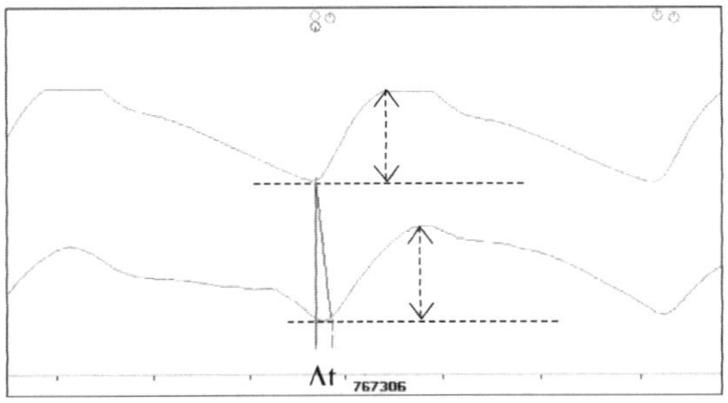

Figure 1. Images of pulses as seen on the PMD 3.0 software. The operator enters the height of each wave into the software and the software identifies the rising edge of the systolic wave (the 'foot') from the smoothed second differential of each signal. The foot-to-foot time (Δt) delay between the two signals is calculated for each heart beat.

3. Results

3.1. Experiment-1

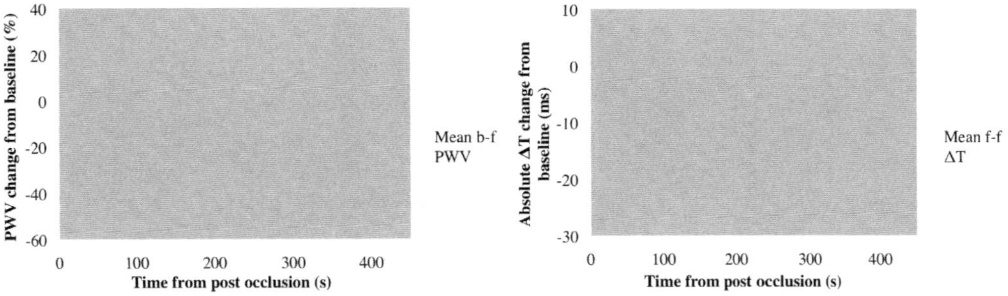

Figure 2. The change in b-f PWV (left) and f-f ΔT (right) over the course of FMD. Error bars show standard error of the mean (SEM). Zero on each plot represents baseline values of b-f PWV and f-f ΔT. Between 0–30s there was a significant difference in b-f PWV from baseline and between 0–40s there was a significant difference in f-f ΔT from baseline f-f ΔT ($p<0.05$ on paired t-test).

3.2. Experiment-2

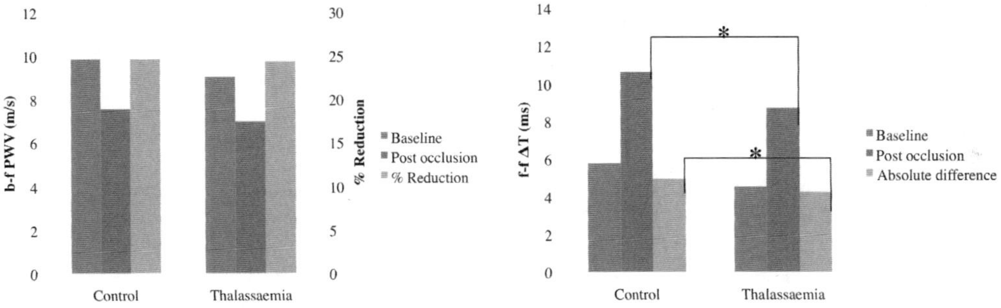

Figure 3. Maximum change from baseline in b-f PWV (left) and f-f ΔT (right) during FMD in controls and thalassaemia patients. Error bars show standard deviation (SD). * indicates significant difference ($p<0.05$ on t-test) between the groups.

3.3. Experiment-3

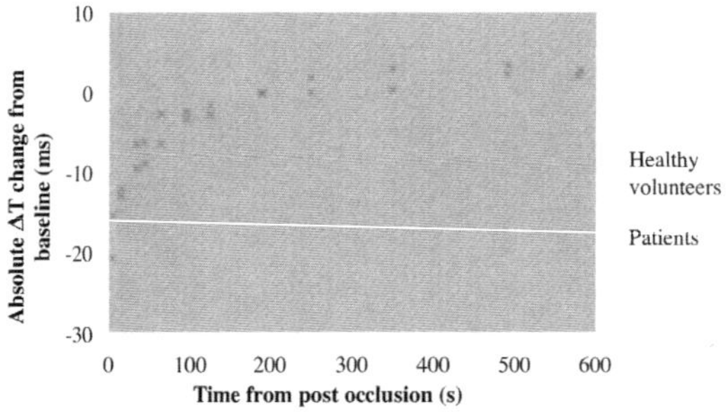

Figure 4. The change in f-f ΔT over the course of FMD in healthy volunteers and patients. Error bars show SEM. Zero on plot represents baseline f-f ΔT. Between 0–120s in healthy volunteers and 0–30s in patients there was a significant difference in f-f ΔT from baseline ($p<0.05$ on paired t-test).

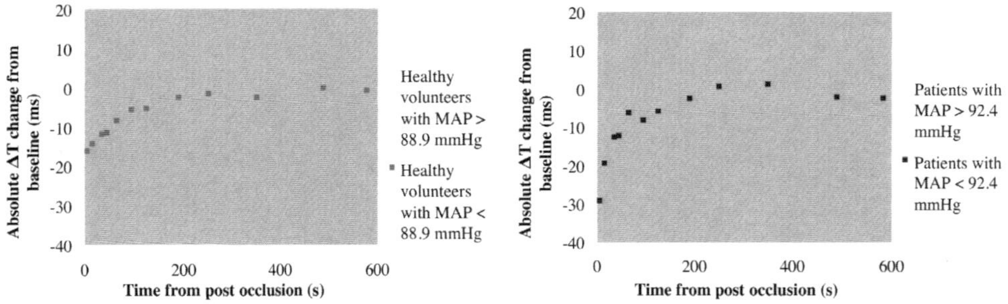

Figure 5. The change in f-f ΔT over the course of FMD in healthy volunteers (left) and patients (right) after stratifying each group into sub-groups of those with mean arterial pressure (MAP) greater than and less than the group's median MAP. Error bars show SEM. Zero on plot represents baseline f-f ΔT. There is significant difference ($p<0.05$ on t-test) between the healthy volunteer sub-groups at all times except 0s, 60s and 240s. There is significant difference ($p<0.05$ on t-test) between the patient sub-groups at all times except 0s, 60s and 180–600s.

3.4. Experiment-4

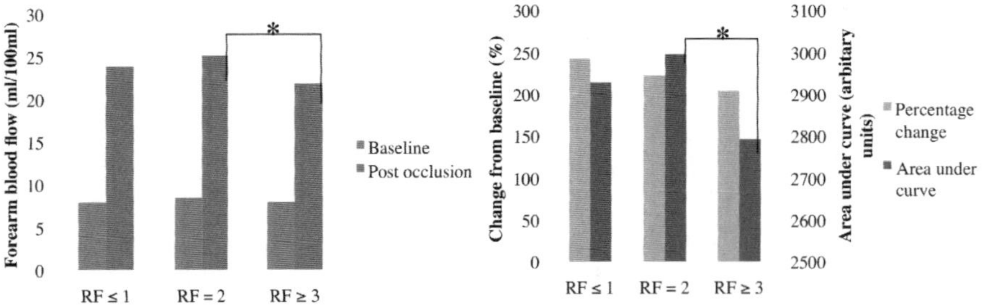

Figure 6. Maximum change from baseline in FBF during FMD in CAD patients with varying numbers of risk factors (left). Maximum percentage change from baseline in FBF and the area under the FBF-time curve during FMD in the same groups (right). Error bars show SEM. * indicates significant difference ($p<0.05$ on t-test) between the groups.

3.5. Experiment-5

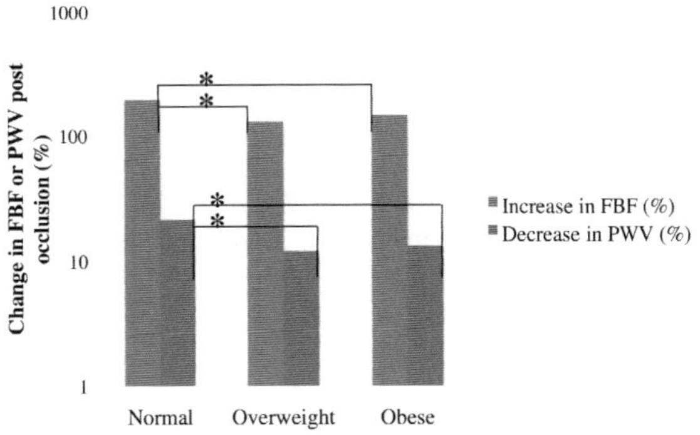

Figure 7. Maximum change from baseline in FBF and b-f PWV during FMD for normal controls, overweight and obese subjects. Error bars show standard deviation (SD). * indicates significant difference ($p<0.05$) between the groups.

4. Discussion

As predicted, the change in f-f ΔT from the baseline f-f ΔT was greatest at the start of FMD and it decreased to baseline as FMD diminishes with time. This pattern was complemented by the b-f PWV curve which showed a drop in PWV before returning to baseline. Furthermore, the return to baseline in f-f ΔT and b-f PWV was at similar times (Figure 2). However, the b-f PWV data are highly variable when compared to the f-f ΔT data, which may reflect the relative ease of the f-f ΔT technique in obtaining stable signals using only PPG finger probes. Naka et al. showed that a significant drop in PWV is observed the upper limb and lower limb during FMD [5], whilst Torrado et al. showed that a significant drop in carotid-radial PWV is observed during FMD [6].

Comparison of the b-f PWV and f-f ΔT methods during the assessment of FMD in thalassaemia patients also gave comparable results but the f-f ΔT method shows the responses in controls and thalassaemia patients are significantly different whilst this is not the case using the b-f PWV method (Figure 3). This may suggest that the f-f ΔT method is more sensitive than the b-f PWV method. However, since baseline f-f ΔT will tend to be near 0ms it is not numerically useful to calculate the percentage changes from baseline when using the f-f ΔT method, which is a limitation.

Using f-f ΔT to compare FMD in healthy volunteers (with low CVD risk) and patients (with high CVD risk) allowed us to test whether this novel approach is sensitive enough to detect the differences in EF, which we reasonably assumed existed, between the two groups. It took 120s for f-f ΔT (2.6±1.11ms [±SEM], $p>0.05$) to return to baseline in controls, whilst it took 30s for f-f ΔT (-6.2±2.61ms (±SEM), $p>0.05$) to return to baseline in patients (Figure 4). Therefore, using f-f ΔT, we were able to detect an attenuated FMD response in patients compared to the healthy volunteers. However it should be emphasised the mean age of the control group was very much less than that of the patients so much of the difference in response between the two groups could be due to age differences. However in Experiment 4, the groups were aged matched and attenuated response was seen in subjects with a greater disease risk, although measurements were confined to FBF. For both healthy volunteers and patients, sub-cohort analysis suggests that individuals with a lower MAP have a greater f-f ΔT change from baseline throughout FMD. This finding is statistically significant at most times throughout FMD, particularly in healthy volunteers and, to a lesser extent, in patients (Figure 5). Although, this suggest that the f-f ΔT approach is sensitive enough to spot such differences further work needs to be done to confirm this.

Using VOP to assess FMD in CAD patients with varying numbers of risk factors suggests that FBF can discriminate between high and low risk groups. There was a significant difference between the

RFs =2 vs. RFs ≥3 groups when comparing the FBF or the FBF-time curve area (but not when comparing the percentage change from baseline FBF). There was no significant difference between the RFs ≤1 vs. RFs =2 groups, and interestingly, none existed between RFs ≤1 vs. RFs ≥3 groups (the group we expected to have the largest difference) (Figure 6). When comparing FMD in normal, overweight and obese adolescents using the VOP and b-f PWV technique, significant differences between the groups in both FBF and b-f PWV were achieved. The maximum response (decrease from baseline) in b-f PWV during FMD is significantly greater in controls with a normal BMI than in overweight or obese adolescents. This is expected as obesity is a risk factor for CVD and therefore individuals would have impaired EF. The same is suggested by the maximum response in FBF using the VOP technique; normal controls have a significantly greater FBF than overweight and obese adolescents (Figure 7). However, one might expect obese individuals to have more impairment in their response than overweight individuals, however ANOVA followed by post hoc testing would be needed to investigate this further.

In this study we compared three techniques of assessing FMD in healthy and diseased subjects. The results suggest that the novel f-f ΔT method of assessing FMD, and hence EF, is simple and sensitive enough to be developed into a readily available screening tool. However, further work is required to elucidate some issues discussed here. Furthermore, the f-f ΔT changes observed during FMD still needs to be compared to the FBF measured using VOP, and it also needs to be compared to direct measurements of diameter changes seen in FMD (using duplex ultrasound or MRI, the current gold-standard methods for assessing EF). Other simple approaches to assessing EF have been reported, including the measurement, during FMD, of the time delay between the ECG R wave and the arrival of the pulse wave at a peripheral site [11].

References

[1] Deanfield JE, Halcox JP, Rabelink TJ (2007). Endothelial function and dysfunction: testing and clinical relevance. *Circulation* **115** 1285-95A reference

[2] Corretti MC, Anderson TJ, Benjamin EJ, et al. (2002). Guidelines for the ultrasound assessment of endothelial-dependent flow-mediated vasodilation of the brachial artery: a report of the International Brachial Artery Reactivity Task Force. *J Am Coll Cardiol* **39** 257-65

[3] Mackenzie IS, Wilkinson IB, Cockcroft JR (2002). Assessment of arterial stiffness in clinical practice. *QJM* **95** 67-74

[4] Kinlay S, Creager MA, Fukumoto M, et al. (2001). Endothelium-derived nitric oxide regulates arterial elasticity in human arteries in vivo. *Hypertension* **38** 1049-53

[5] Naka KK, Tweddel AC, Doshi SN, et al. (2006). Flow-mediated changes in pulse wave velocity: a new clinical measure of endothelial function. *Eur Heart J* **27** 302-9

[6] Torrado J, Bia D, Zocalo Y, et al. (2009). Reactive hyperemia-related changes in carotid-radial pulse wave velocity as a potential tool to characterize the endothelial dynamics. *Conf Proc IEEE Eng Med Biol Soc* **2009** 1800-3

[7] Chowienczyk PJ, Lawson CP, Millasseau SC (2003). Method of measuring endothelial function in a person. USA. US6908436B2

[8] Moens AL, Goovaerts I, Claeys MJ et al. (2005). Flow-mediated vasodilation: a diagnostic instrument, or an experimental tool? *Chest* **127** 2254-63

[9] Higashi Y, Sasaki S, Nakagawa K et al. (2001). Effect of obesity on endothelium-dependent, nitric oxide-mediated vasodilation in normotensive individuals and patients with essential hypertension. *Am J Hypertens* **14** 1038-45

[10] Kukongviriyapan V, Somparn N, Senggunprai L et al. (2008). Endothelial dysfunction and oxidant status in pediatric patients with hemoglobin E-beta thalassemia. *Pediatr Cardiol* **29** 130-5

[11] Selvaraj N, Jaryal AK, Santhosh J, et al. (2009). Monitoring of reactive hyperemia using photoplethysmographic pulse amplitude and transit time. *J Clin Monit Comput* **23** 315-22

Sensors & their Applications XVI IOP Publishing
Journal of Physics: Conference Series **307** (2011) 012015 doi:10.1088/1742-6596/307/1/012015

Investigation of optimal parameters for finite element solution of the forward problem in magnetic field tomography based on magnetoencephalography

K Y Aristovich[1], S H Khan[1], and A I Borovkov[2]

[1]School of Engineering and Mathematical Sciences, City University London, Northampton Square, London EC1V 0HB, UK

[2]St Petersburg State Polytechnic University, Polytechnicheskaya Street 29, St Petersburg, 195251, Russia

kirill.aristovich.1@city.ac.uk

Abstract. This paper presents an investigation of optimal parameters for finite element (FE) solution of the forward problem in magnetic field tomography (MFT) brain imaging based on magnetoencephalography (MEG). It highlights detailed analyses of the main parameters involved and evaluates their optimal values for various cases of FE model solutions (e.g., steady-state, transient, etc.). In each case, a detail study of some of the main parameters and their effects on FE solution and its accuracy are carefully tested and evaluated. These parameters include: total number and size of 3D FE elements used, number and size of elements used in surface discretisation (of both white and grey matters of the brain), number and size of elements used for approximation of current sources, number of anisotropic properties used in steady-state and transient solutions, and the time steps used in transient analyses. The optimal values of these parameters in relation to solution accuracy and mesh convergence criteria have been found and presented.

1. Introduction

Magnetic field tomography (MFT) is a relatively new imaging modality which involves localisation and subsequent imaging of active areas in the brain by measuring the extremely weak neuromagnetic fields (10–100 fT) produced by neuronal currents in these areas associated with cognitive processing (magnetoencephalogram). This approach, called the magnetoencephalography (MEG) technique (recording of magnetic fields produced by electrical activity in the brain) is the only truly noninvasive method which could provide information about functional brain activity. Compared to other imaging modalities it is the only imaging modality that combines high temporal with high spatial resolution.

The forward problem in MFT involves the computation of magnetic field distribution from known neuronal current generators (sources) in the brain [1-3]. The inverse problem localizes and images these generators using MEG data measured around the head and the data obtained from the forward solution. An accurate solution of the forward problem is also needed for design, configuration and placement of SQUID sensors, used to measure the neuromagnetic fields around the head, and which constitute the sensing subsystem of the MFT system.

The FE solution of the forward problem involves very accurate modelling of the human brain together with methodologies for accurate evaluation of various solutions. However, these aspects are rarely covered in the current literature, especially with respect to solution errors and accuracy of FE parameters used for simulations. Human brain is an extremely complex structure for FE modelling both in terms of geometric and material parameters [4, 5]. Although accurate incorporation of these has obvious effects on FE solution accuracy, it needs to be balanced against computational time and complexity involved. Based on various FE analyses, this paper describes optimisation of various parameters highlighting optimal requirements for FE models of the brain and the current sources used.

2. Finite element models and strategy for parameter optimisation

2.1. Model description

The FE models used in this study are based on the methodologies and tools developed by the authors for building realistic models of the human brain [6, 2]. Parametrical solid CAD spline models are used for further FE discretisation which incorporates geometrical accuracy limited only by existing MRI equipment. A typical outline of the solid model used is presented in the Figure 1.

The FE mesh is created, firstly by standard discretisation method involving surface meshing of brain features used and then by an optimised procedure for building 3D volume elements [7-9]. Here the surface mesh, made up of triangular elements is fully closed and the volume is divided into 3D tetrahedral elements. Although the size of these 3D elements can be defined independent of the surface mesh, to ensure the connectivity and transition of regions, the edge-lengths of 3D elements are matched with those of 2D surface elements. Also the 2D mesh is defined to be uniform within 50% variation of the average element size apart from small regions to be occupied by source currents. Therefore in optimisation analyses the overall FE discretisation is almost fully dependent upon initial size of the surface mesh. It is convenient, however to define FE mesh density by the total number of 3D elements in the model [8, 9] which is also the case in this paper.

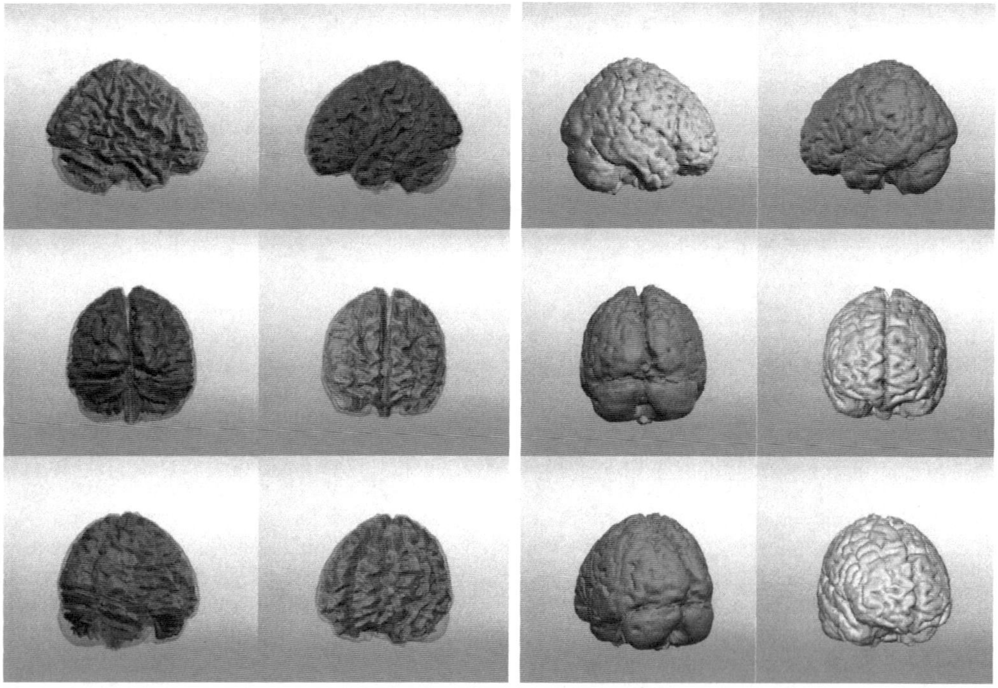

Figure 1. Initial solid model of the human brain: outline of white (left) and grey (right) matters.

2.2. Description of the test problem

As mentioned earlier, the forward problem in MFT based on MEG involves modelling and computation of magnetic fields produced by known neuronal currents on the brain. These currents are modelled as current sources within the FE model domain and can be approximated in a number of different ways [3]. However, they are all based on placing the current source inside the geometric model domain and calculating the resulting magnetic field inside as well as outside the actual brain geometry. This magnetic field is measured on an appropriate simulated detection surface placed outside the head that represents the real measurement surface for sensor arrays (e.g., SQUID sensors) in MFT brain imaging [10].

Because of the linearity of magnetic field problem involved in the forward solution in MFT, superposition principle can be applied to evaluate the effects of multiple current sources based on multiple solutions obtained for a single current source appropriately placed inside the brain geometry. Hence for parameter optimisation of a single current source is used for the test problem (Figure 2).

2.3. Method for the evaluation of optimal parameters

The criterion for mesh convergence was used as a general method for FE optimisation and the refinement of solution accuracy [8]. The method involves successive refinement of FE parameters until the solution achieves a required, predefined accuracy (Figure 3). Thus the outcome of such modelling tests will evaluate the behaviour of model solutions and, hence practical model optimisation can be performed based on required data analyses.

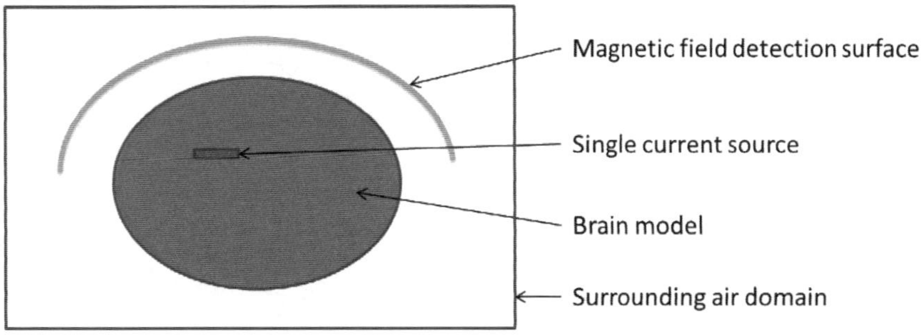

Figure 2. Schematic for the test problem solution

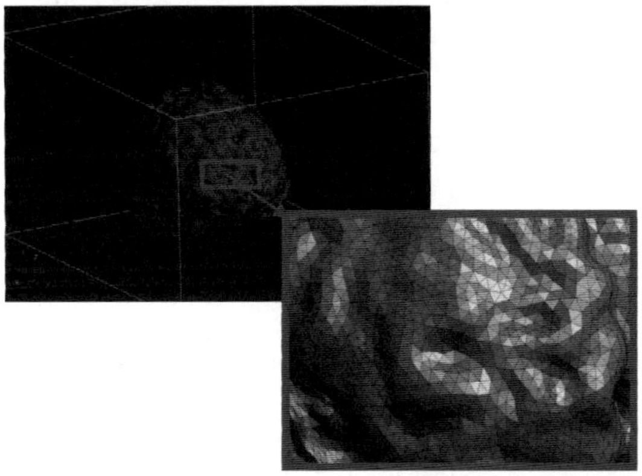

Figure 3. Example of FE surface discretisation.

3. Key parameters for finite element simulations

In FE simulations a number of key parameters are used which require careful analysis and optimisation for accurate and efficient solution. For FE solution of the forward problem presented in this paper the following parameters have been considered.

3.1. Overall mesh density (total number of elements, N and their distribution)

The accuracy of FE solution is inherently dependent upon the total number of FE nodes and hence, FE elements, N by which the problem domain is discretised for solution. For a given problem there always exists a maximum threshold number for N beyond which any further increase in elements does not result in corresponding increase in solution accuracy. Also the distribution of elements and the uniformity of mesh density in the problem domain play an important role in solution accuracy. A large number of FE elements beyond the above threshold number will lead to unnecessary computational overhead without essentially increasing solution accuracy. In this work the following geometric shapes are considered critical for surface meshing:

- Implemented current source surface
- White matter surface
- Grey matter surface
- Detection electrodes surface
- Surrounding air surface

The critical point of the problem is the variation of mesh size within the problem domain. As mentioned above, the uniformity of mesh density is considered important for FE analysis [11]. Thus the element sizes within the problem domain are kept almost the same except for the elements in and around the current sources where further refinement is made to account for their very small size. This leads to a clear dependence between the number of surface elements and the total number of elements in the model, N.

3.2. Mesh size near the current source and its discretisation

As was mentioned earlier, the current source in MFT is relatively small in size (in microns) compared to the total brain model. In some cases submodelling techniques can be used to improve solution accuracy [12]. However, to do this initial courser mesh must satisfy accuracy requirements and mesh refinement is needed within the region occupied by the current source. In this work, the number of elements within the current source region is given by the parameter M.

3.3. Number of anisotropic element properties in the model, P

In case of Section 4 below, conductivity of the brain material plays a major role and, therefore has to be carefully considered [13]. Due to high complexity and anisotropy level, each element has to have its own material properties which leads to a complicated structure of matrix for material properties. Reducing the number of material properties (P) in the model can result in significant reduction in computational time, especially at the first step of matrix assembly [9].

3.4. Time step size, T for transient cases

In case of transient solution, the temporal resolution is important because of integration procedure. Taking into account characteristic speed of signal propagation inside the neurons the optimal time step T for numerical integration can be evaluated.

4. Quasistatic magnetic field analysis I

For this analysis in linear FE solution domain, the solution accuracy to a large extent depends on the mesh quality, especially on the overall mesh density. The current source is given in terms of interface/boundary conditions, which does not require special considerations for meshing regions in and around the current source [2]. A typical FE mesh is shown in Figure 4 together with the position and configuration of the sensor detection surface. The current source was placed approximately in the centre of the brain model.

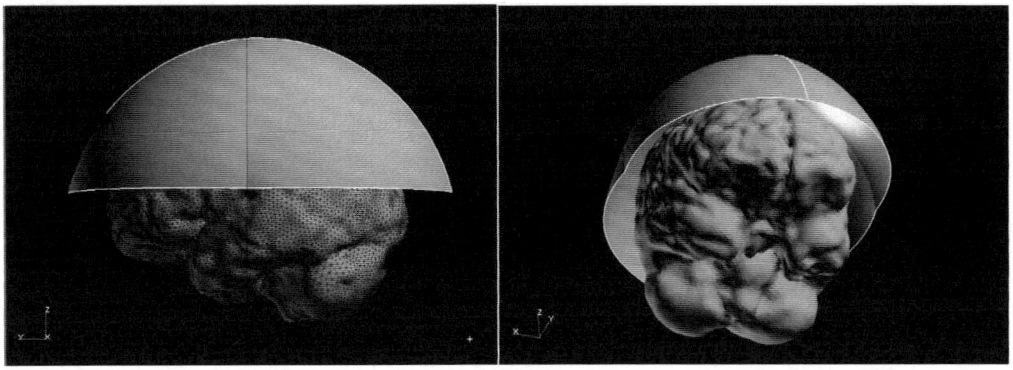

Figure 4. Surface mesh for the brain model and sensor detection surface corresponding to Section 4.

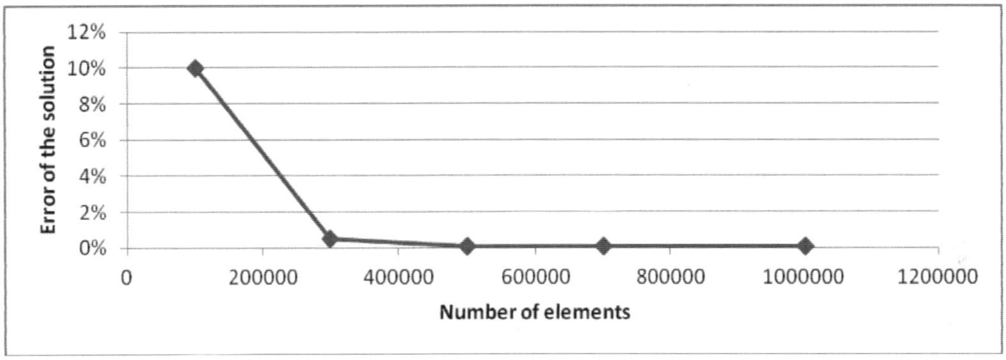

Figure 5. Determination of optimal number of FE elements, N for the test problem in Section 4.

In this case, the maximum value of the magnetic field flux density on the detection surface was taken as the solution parameter for testing modelling accuracy as a function of solution errors for successive solutions for various mesh densities. For comparison, precise analytical solutions were used to calculate FE solution error for a given mesh. The mesh convergence analysis shows a stable convergence as seen in Figure 5. The solution error was found to be minimum and stable at around 0.05% for the number of FE elements corresponding to about N=500k elements.

5. Quasistatic magnetic field analysis II

5.1. Description of the test problem

In this analysis the current source was approximated by a straight short conductor of quadratic cross-sectional area (Figure 6). The conductivity values of the current source were varied starting from the conductivity of the copper (6×10^7 S/m) in order to give high conductivity difference between the source and the surrounding brain matter. The conductivity value was then uniformly decreased to the realistic one (100 S/m, 10 S/m and finally 0.33 S/m). The current source was placed approximately at the centre of the brain model (the same as in the case of Section 4 above). In this case the brain model incorporates complex anisotropic conductivity tensor for brain material properties [6].

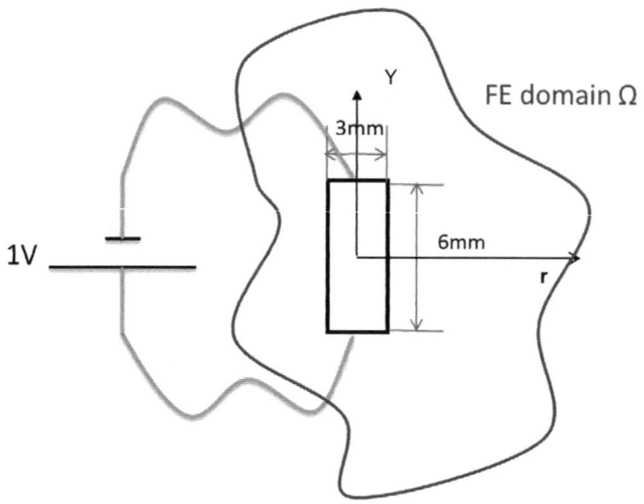

Figure 6. Current source approximation corresponding to FE analysis in Section 5.1.

5.2. Results of computer simulation for number of FE elements, N=500k

Initial simulations were performed which established the optimal number of elements for FE solution as *N*=500k. The characteristic number for source mesh density, *M* was taken as the number of elements along the longest edge of the current source; initially *M*=4. Results have been plotted as the variation of current density along the r-axis (Figure 6) of the current source (Figure 7). The corresponding analytical value for each solution was also calculated in form of total current flowing through the current source, and the solution error was calculated as the difference between analytical and numerical solutions (Figure 8).

The results show good agreement between analytical and numerical results for high conductivity ratios. However, in cases where conductivity of the current source was comparable to the average conductivity of surrounding media, FE results give unrealistic trend (Figure 7, bottom right graph). Here the current density peak is associated with inappropriate finite element mesh. Also this corresponds to a total current density difference of 500% between analytical and numerical solutions. These results show that both the total number of elements, *N* and the number of elements in the current source, *M* need to be increased together to improve solution accuracy.

5.3. Mesh convergence results

Full mesh convergence analyses were performed by varying the total number of elements *N*=500k-1600k and *M*=4-25. FE computations were performed for conductivity ratios from $1\text{-}10^8$. For each of the conductivity ratios the optimal number of elements, *N* together with characteristic current source mesh density, *M* were obtained which corresponded to a minimum mesh convergence ratio of 0.1%. These results are presented in Figure 9. It shows that for conductivity ratios near 1 the total number of elements required for accurate solution will exceed 1.5m. The current source in this case will be optimally divided into 20 elements along the longest edge. This corresponds to an approximate element size of 0.06 mm. This result essentially shows that the above FE approach for solving this particular problem is associated with high computational overhead in order to achieve the required accuracy of solution. However, as shown in [12] the same accuracy can be obtained by using submodelling approach which significantly decreases computational overhead.

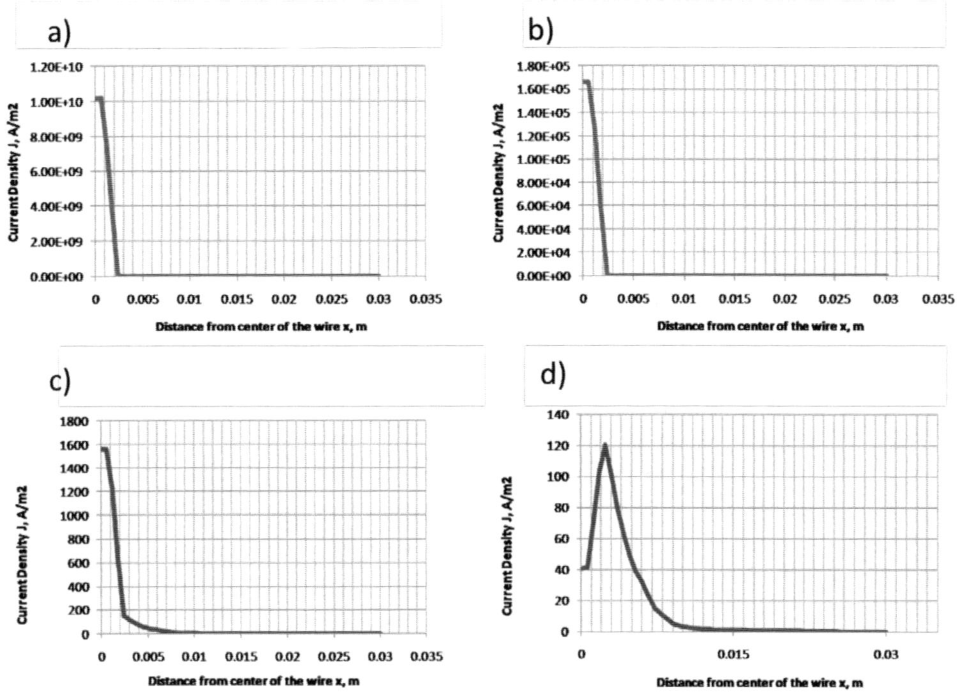

Figure 7. Variation of current density with distance for various conductivity values: a) Electric conductivity σ=6·10^7 S/m, resulting current I=8.15·10^4 A; b) σ=10^3 S/m, I=1.34 A; c) σ=10 S/m, I=2.45·10^{-2} A; d) σ=0.33 S/m, I=1.11·10^{-2} A.

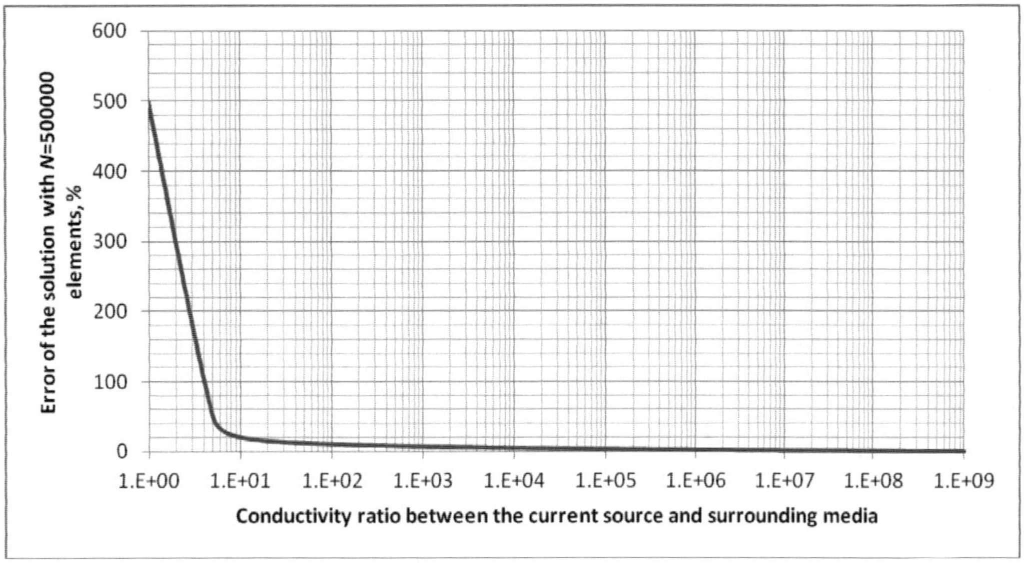

Figure 8. Variation of FE solution error with conductivity ratio for N=500k.

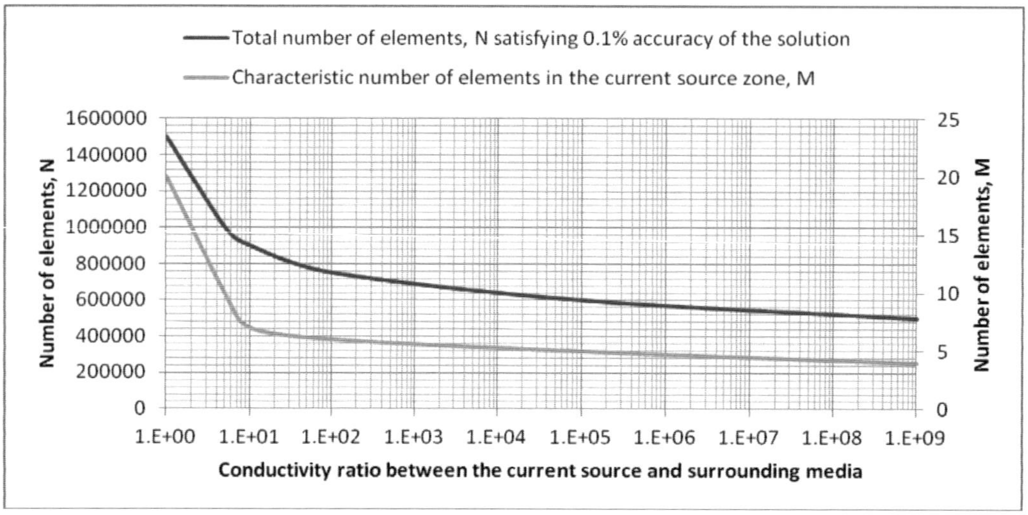

Figure 9. Optimal number of elements in the FE model N and characteristic number of elements in the current source M as a function of conductivity ratio.

5.4. *Optimal number of material properties*

The number of material properties P was also studied during mesh convergence analysis. The result is independent of the conductivity ratio. However, the number of common properties was found increased with increasing N. The variation of P as a function of N is shown in Figure 10.

6. Transient analysis

The transient analysis together with other parameters involves temporal resolution for numerical integration to be performed. The problem for transient cases was chosen exactly the same as in steady-state case, with the only difference in the application of voltage for source approximation as opposed to current elements. The voltage was applied to approximate the neuronal signal transmission with time along a given pathway. The exact parameters of this function can be found in [5]. The results obtained are presented in Figure 11.

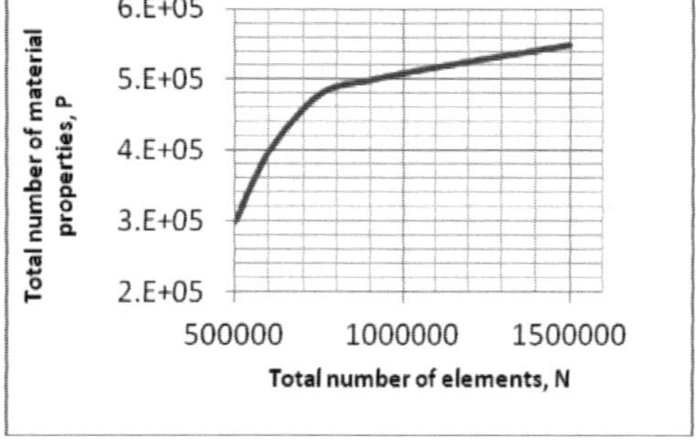

Figure 10. Optimal number of material properties in the model P as a function of total number of FE elements N.

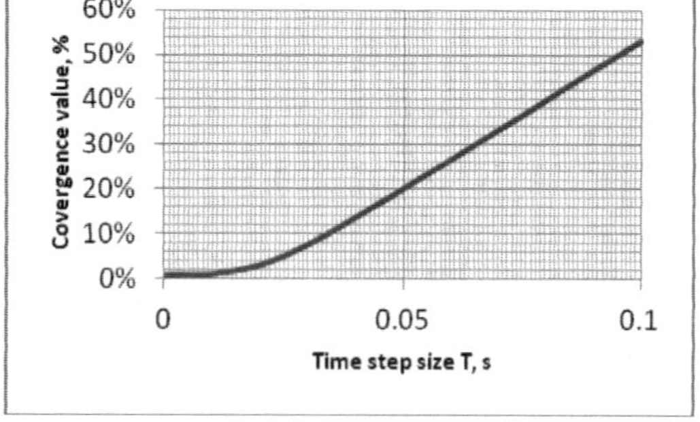

Figure 11. Convergence value as a function of the size of time step T in transient solutions.

7. Summary of optimal parameters and conclusions

The optimal parameters for FE solution of the forward problem in MFT based on MEG have been investigated in this paper. The summary of optimal parameters for each particular case is presented in Table 1. This can be used as a guide for forward solutions in MFT based on accurate FE modelling of the brain. In addition, parameters such as the number of surface elements and the average size of elements are also given to inform any follow-up simulations.

Table 1. Summary of optimal parameters.

Solution case	**Total number of FE elements, *N***	**Average size of elements, (mm)**	**Number of elements in the current source region, *M***	**Average size of elements in the current source region (mm)**	**Average number of surface elements in the grey matter**	**Average number of surface elements in the white**	**Time step, *T***
Quasistatic field analysis I	500000	3	n/a	n/a	10000	40000	n/a
Quasistatic field analysis II	1500000	0.5	20	0.06	50000	130000	n/a
Transient electromagnetic	1500000	0.5	20	0.06	50000	130000	0.01s

8. References

[1] Kemppainen P and Peters M 1991 On the forward solution of electroencephalography and magnetoencephalography *Clin. Phys. Physiol. Meas.* **12 Suppl A** 95-9
[2] Khan S H, Aristovich K Y and Borovkov A I 2009 Solution of the forward problem in magnetic field tomography (MFT) based on magnetoencephalography (MEG) *IEEE Trans. Magn.* **45** 1416-19
[3] Mosher J C, Leahy R M and Lewis P S 1999 EEG and MEG: Forward solutions for inverse methods *IEEE Trans. Biomed. Eng.* **46** 245-59
[4] Akalin Z, Acar C E and Gencer N G 2001 Development of realistic head models for electromagnetic source imaging of the human brain *Proc. of the 23rd Annual EMBS Int. Conf.*
[5] Bear M F, Connors B W and Paradiso M A 2007 *Neuroscience: Exploring the brain* (Baltimore, MD: Lippincott Williams & Wilkins)
[6] Aristovich K Y and Khan S H 2010 Automatic procedure for realistic 3d finite element modelling of human brain for bioelectromagnetic computations *J. Phys.: Conf. Ser.* **238**, 012065 (8p)
[7] Kwon Y W and Bang H 2000 *The finite element method using matlab* (Boca Raton ; London: CRC Press)
[8] Madenci E and Guven I 2006 *The finite element method and applications in engineering using ansys* (New York: Springer)
[9] Rao S S 2005 *The finite element method in engineering* (Oxford: Butterworth-Heinemann)
[10] Hamalainen M, Hari R, Ilmoniemi R J, Knuutila J and Lounasmaa O V 1993 Magnetoencephalography-theory, instrumentation, and applications to noninvasive studies of the working human brain *Rev. Modern Phys.* **65** 84
[11] Kraus J D 1991 *Electromagnetics*: McGraw-Hill)
[12] Aristovich K Y and Khan S H 2010 A new submodelling technique for multi-scale finite element computation of electromagnetic fields: Application in bioelectromagnetism *J. Phys.: Conf. Ser.* **238** 012050 (8p)
[13] Van Uitert R, Weinstein D and Johnson C 2003 Volume currents in forward and inverse magnetoencephalographic simulations using realistic head models *Ann Biomed Eng* **31** 21-31

Photoplethysmographic sensors for perfusion measurements in spinal cord tissue

J P Phillips and P A Kyriacou

School of Engineering and Mathematical Sciences, City University London, EC1V 0HB, UK

E-mail: Justin.Phillips.1@city.ac.uk

Abstract. Sensors for recording photoplethysmographic signals from the nervous tissue of the spinal cord are described. The purpose of these sensors is to establish whether perfusion is compromised in various states of injury which occur in certain animal models of spinal cord injury, for example compression injury. Various measures of perfusion are applicable such as the amplitude of the photoplethysmograph signal and the oxygen saturation, measured using a dual wavelength configuration. Signals are usually compared to baseline measurements made in uninjured subjects. This paper describes two types of probe, one based on optical fibres, and one in which optotes are placed in direct contact with the tissue surface. Results from a study based on a compression model utilising a fibreoptic sensor are presented.

1. Introduction

The mechanisms of tissue damage during and after spinal cord injury (SCI) are complex and poorly understood. Trauma causing SCI is classed as primary injury and can produce laceration, stretching or compression of the delicate nervous tissue of the spinal cord. Primary injury can give rise to hemorrhage, edema and ischemia which can cause further injury (known as secondary injury) after the initial impact (Castro-Moure *et al* 1997). This can cause serious permanent loss of function, however appropriate and rapid hospital treatment can potentially minimize secondary injury.

Several models have been developed to investigate various SCI mechanisms in adult rats. Such models mimic the effects of primary injury, mostly due to laceration and contusion (Crowe *et al* 1997). Other models, such as that used in the current study, provide quantitative information on the effect of compression on neuronal damage and subsequent recovery. Disruption of the blood supply, either by laceration, occlusion or other deformation of blood vessels can potentially lead to areas of tissue becoming seriously deprived of oxygen. The tissue of the spinal cord, like all central nervous system tissue, is extremely sensitive and irreversible injury can result from hypoxia within a very short time (Nemoto 1978). Laser Doppler flowmetry (LDF) measurements have been used to quantify perfusion in the spinal cord before and after injury so that the contribution of ischaemia to the total amount of secondary injury may be better understood (Westergren *et al* 2001).

Similar efforts to quantify the tissue perfusion have focused on photoplethysmography (PPG), which produces a signal whose amplitude gives an indication of the degree of pulsation of the arteries within the tissue (Allen 2007). Many commercial pulse oximeters report this variable as the 'perfusion index' of the tissue (Hager *et al* 2003). This paper describes two types of sensor, which were both designed specifically to make PPG measurements from the spinal cord tissue of rats studied using a compression injury model (Huang et al 2006, Nystrom 1988). Two types of sensor were designed for a series of PPG-based measurements made during compression model experiments. The first used optodes placed in direct contact with the tissue of the spinal cord, while the second utilised optical fibers to transmit light to and from the cord tissue. The performance of the two probes were compared and their suitability for their intended application is discussed.

2. Materials and methods

2.1. Contact probe

The non-fibre optic ('direct contact') probe is fabricated from surface-mount red and infrared emitters and a surface mount photodiode mounted onto a rectangular polypropylene strip. The red emitter is a light emitting diode (LED) of peak emission wavelength 660 nm, while the infrared emitter is an LED of peak emission 940 nm. The emitters are mounted on either side of the photodiode as shown in Figure 1 so that the distance between each emitter and the photodiode was 2 mm. The overall dimensions of the probe (*h* x *d* x *w*) were 10 x 4 x 3 mm Strips of black absorbing material placed between each optode reduce the effect of optical shunt (detection of light directly from the emitter). The sensor works in 'reflectance mode', i.e. the photodiode detects light re-emitted from the tissue surface (Mendelson *et al* 1988).

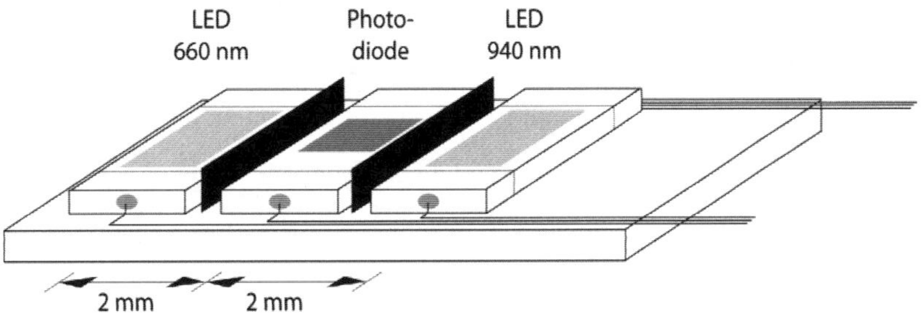

(a)

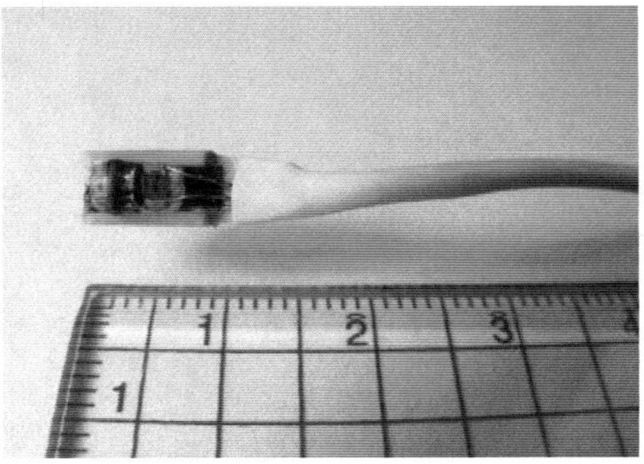

(b)

Figure 1. (a) Diagram of the contact reflectance mode probe. (b) photograph of the contact probe with scale in cm.

Although this sensor was found to produce reliable PPG recordings from the finger, the design was abandoned as it was found that the overall dimensions yielded this probe too large for practical use in rats, leading to development of a fibre optic probe (Phillips *et al* 2009).

2.2. Fibre optic probe

The probe consists of two parallel silica optical fibres (SpecTran Speciality Optics, Avon, CT, USA) with a core diameter of 400 µm, an outer cladding diameter of 730 µm and a numerical aperture (NA) of 0.39. Each fibre is terminated at one end with a subminiature version A (SMA) connector and the other end is cut and polished flat. The proximal end of one of the fibres is connected to the two light sources via a bifurcated optical fibre assembly (Ocean Optics Inc., Dunedin, FL, USA). The proximal end of the other fibre is coupled directly to the photodetector. For the compression experiments, the probe was adapted for use with the compression model apparatus. The distal ends were passed along a metal tube attached to a metal bar measuring 4 mm long by 1.5 mm wide. The fibre ends were inserted into the bar so they were flush with the lower surface of the bar as shown in Figure 2(a). The central axes of the two fibres were laterally separated by a distance of 1.0 mm The lower surface of the bar is curved to fit the contour of the rat spinal cord thus avoiding any localised compression of the tissue. The metal tube was supported within a stereotactic frame so that it could move in the vertical direction only.

2.3. Instrumentation.

This is housed in a metal box containing: a power supply (2 x 12 V lead-acid batteries); a circuit comprising two switchable regulated current sources connected to the LEDs and a differential transimpedance amplifier, a demultiplexing circuit and filters (to attenuate noise and to separate the ac and dc components of the PPG signal) connected to the photodiode. The ac signal was band-passed filtered with a passband ranging from 1 Hz to 19.4 Hz, while the dc signal was low-pass filtered with a cut-off frequency of 19.4 Hz. An additional amplifier for the ac signal was also used with a gain of 101. For use with the fibre-optic probe, light sources; SMA mounted red (660 nm) and infrared (850 nm) LEDs (The Optoelectronic Manufacturing Corporation Ltd, Redruth, UK); a photodetector; an SMA mounted PIN photodiode (The Optoelectronic Manufacturing Corporation Ltd) were added to the front panel of the instrumentation unit.

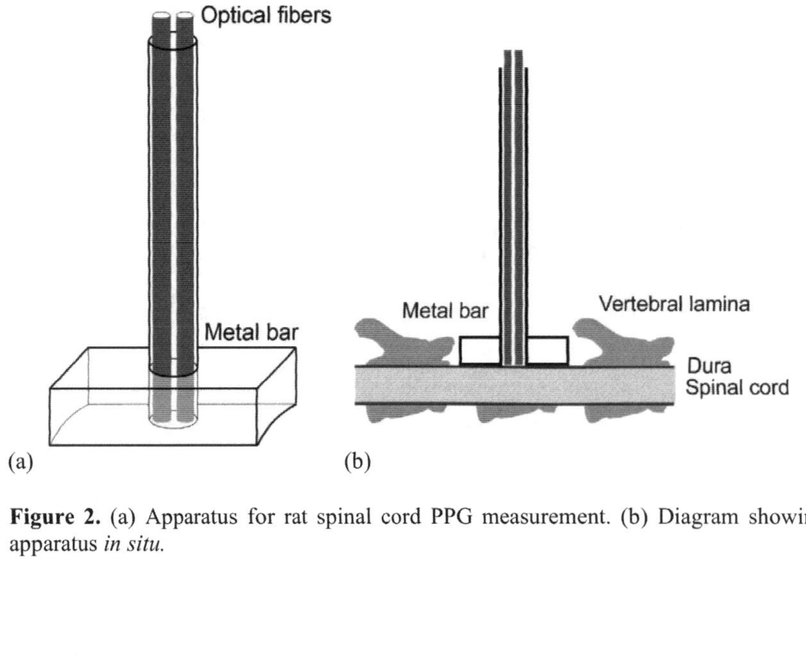

Figure 2. (a) Apparatus for rat spinal cord PPG measurement. (b) Diagram showing apparatus *in situ.*

2.4. *Measurement from rat spinal cord using fibreoptic probe*

All experimental protocols of this study were approved by the animal care committee of Queen Mary University of London, UK in accordance with the UK Animals (Scientific Procedures) Act 1986 and international guidelines on the ethical use of animals. The spinal cords of 6 female Sprague–Dawley rats weighing approximately 250 g were used for this study. The rats were deeply anesthetized in a fume box with a mixture of 5% Halothane (Meril, Essex, UK) in addition to a mixture of oxygen and nitrous oxide (1 : 1 ratio) at a flow rate of 750–1000 mL/min. Subsequent anesthesia throughout the procedure was maintained using 1.5–2% halothane with oxygen and nitrous oxide at unchanged ratio delivered through a nose-piece. The skin and muscle overlying the spinal column were incised and a laminectomy was then performed at T12, leaving the dura undisturbed. The compression bar was placed in light contact with the spinal cord by suspending the base of the compression platform onto the exposed T12 cord dura under microscopic control (see Figure 2(b)). Measurements were recorded from the PPG system for 5 minutes. A weight of 50 g was then applied statically to the platform for exactly 5 minutes, during which time PPG signal recording continued. The weight was then removed and PPG signals recorded for a further 5 minutes. The platform was then removed, the muscle layers were sutured and the skin layers closed with wound clips. Oxygen saturation values were not calculated for this study.

3. Results

Prior to applying any compressive loads, good quality PPG signals were achieved consistently from all six animals. Good quality PPG signals were achieved consistently from all six animals. Figure 3 shows a five-second sample of a PPG signal from the spinal cord dura of one subject.

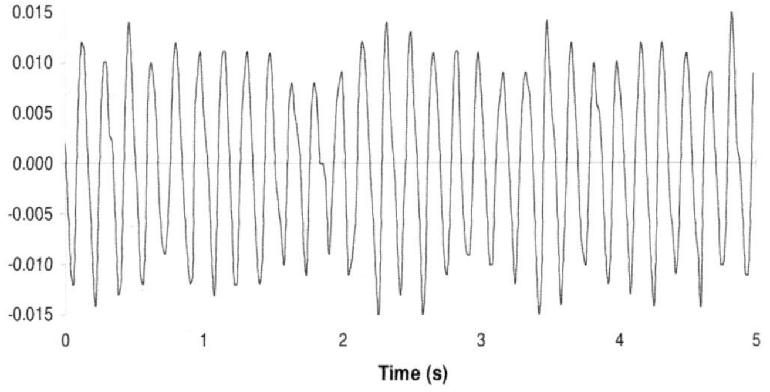

Figure 3. Five-second recording of infrared PPG from the spinal cord before compression.

An example of the waveform obtained for the entire 15-minute measurement period is shown in Figure 4. It can be seen that in this example, the amplitude of the PPG signal decreased dramatically on compression. The PPG signal was attenuated for the duration of the compression. When the compression was relieved, the PPG amplitude increased to roughly two-thirds of its baseline value and then gradually increased. After five minutes the PPG amplitude was approximately equal to the baseline value.

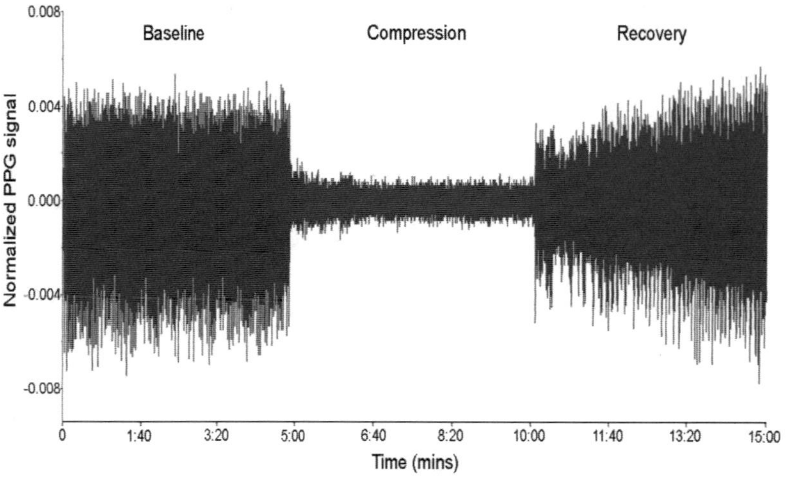

Figure 4. 15-minute recording of infrared PPG from the spinal cord before, during and after compression.

It was found that all six animals showed a reduction in average PPG amplitude from baseline following compression of the cord.

4. Discussion

Although both types of sensor potentially could be used for these measurements, the fibre-optic probe was more suitable as the are of contact with the tissue is smaller and it could be adapted more readily to the compression model apparatus. It was found that good quality infrared PPG signals could be obtained from the spinal cord dura using a fiber-optic probe during static compression model measurements in rats. Compression of the spinal cord resulted in reduced pulsation indicated by attenuation of the PPG signal amplitude. This suggests that ischemia may occur during compressive injury. The reduction in PPG amplitude persisted even after removal of the compressive load, suggesting a further potential mechanism for secondary injury.

The next stage in the development of this project will be a trial using a larger number of subjects exposed to different compression loads. A comparison of PPG amplitude attenuation and recovery of motor function may hopefully shed light on the likely extent of ischemia in compressive spinal cord injury.

References

Castro-Moure F, Kupsky W and Goshgarian H G 1997 Pathophysiological classification of human spinal cord ischaemia [Review] *J. of Spinal Cord Med.* **20** 74-87

Crowe M J, Bresnahan J C, Shuman, Masters J N and Beattie M S 1997 Apoptosis and delayed degeneration after spinal cord injury in rats and monkeys.[erratum appears in Nat Med 3 240] *Nature Medicine* **3** 73-6

Hager H, Reddy D, Kurz A 2003 Perfusion index-a valuable tool to assess changes in peripheral perfusion caused by sevoflurane? *Anesthesiology* **99** A593

Huang W L, Robson D, Liu M C, King V R, Averill S, Shortland P J and Priestley J V 2006 Spinal cord compression and dorsal root injury cause up-regulation of activating transcription factor-3 in large-diameter dorsal root ganglion neurons *European Journal of Neuroscience* **23** 273-8

Mendelson Y and Ochs B D 1988 Noninvasive pulse oximetry utilizing skin reflectance photoplethysmography *IEEE Trans. Biomed. Eng.* **35** 10 798-805

Nemoto E M 1978 Pathogenesis of cerebral ischemia-anoxia *Critical Care Medicine* **6** 203–14

Nystrom B, Berglund J E and Bergquist E 1988 Methodological analysis of an experimental spinal cord compression model in the rat *Acta Neurologica Scandinavica* **78** 460-6

Phillips J P, George K J, Kyriacou P A, Langford R M 2009 Investigation of Photoplethysmographic Changes using a Static Compression Model of Spinal Cord Injury *Proceedings of the 31st IEEE EMBC Annual International Conference, Minneapolis, MN, USA*

Westergren H, Farooque M, Olsson Y, Holtz A 2001 Spinal cord blood flow changes following systemic hypothermia and spinal cord compression injury: an experimental study in the rat using Laser-Doppler flowmetry *Spinal Cord* **39** 74-84

Non Invasive Microwave Sensor for the Detection of Lactic Acid in Cerebrospinal Fluid (CSF)

J. H. Goh, A. Mason, A. I. Al-Shamma'a
Liverpool John Moores University, Built Environment and Sustainable Technologies (BEST) Research Institute, RF and Microwave Research Group
Henry Cotton Building, 15-21 Webster Street, Liverpool, L3 2ET, UK

M. Field, M. Shackcloth, P. Browning
Liverpool Heart and Chest Hospital – NHS Foundation Trust
Thomas Drive, Liverpool, L14 3PE, UK

E-mail: j.h.goh@2006.ljmu.ac.uk

Abstract. This research involves the use of a low power microwave sensor for analysis of lactic acid in cerebrospinal fluid (CSF), an indicator of neurological impairment during aortic aneurysm surgery which could provide the basis for improved treatment regimes and better quality of care with more efficient use of resources. This paper presents initial work using standard lactate curves in water followed by lactate in "synthetic CSF". A multi-modal spectral signature has been defined for lactate, forming the basis for subsequent development of microwave sensor platform that is able to detect concentrations of lactic acid in CSF of volumes less than 1ml.

1. Introduction

The electromagnetic waves ranging from 0.3GHz up to 300GHz can be used to analyse the cerebrospinal fluid (CSF) from the patient based on the electrical properties for chemical specification contain in the fluid. Microwave analysis has a range of advantages for biomedical applications. It is a non-ionising technique utilising low power output at around 1mW (0dBm) but has good penetration depth and equipment can be portable for use at the bedside. The multi parameter nature of broadband microwave analysis can provide unique signal spectrum signatures which is a reflected signal $|S_{11}|$ and/or a transmitted signal $|S_{21}|$ based on parameters such as conductivity and permittivity [1]. Conductivity is a measurement of a material's ability to conduct an electric current. Permittivity is a measurement of how an electric field is affected by a dielectric medium, which is determined by the ability of a material to polarise in response to the field, and reduce the total electric field inside the material. Therefore, permittivity relates to a material's ability to transmit an electric field and is a complex value which varies with changing frequency, and accounts for both the energy stored by a material (ε') as well as any losses of energy (ε'') which might occur.

This project is work in partnership with Liverpool Heart and Chest Hospital (LHCH). LHCH has commenced with the analysis of CSF fluid. Patients who are undergoing surgical or endovascular aneurysm repair (EVAR) of acute and chronic thoraco-abdominal aortic disease [2] have an inherent risk of paraplegia. This is caused by restriction of the spinal cord blood flow and lack of oxygen during the procedures, referred to as spinal cord ischemia [3]. The current preventative measure is to insert a spinal drain to relieve pressure. At present this fluid is discarded as waste, but researchers at LHCH believe that analysis of the CSF can be used to indicate sub-clinical cord ischemia and compromise. If microwave analysis of CSF can be used to detect and monitor impending cord ischemia, intervention may be performed to reverse this process before irreversible ischemia and paraplegia are clinically apparent.

2. Sample Preparation

Serial dilutions of L(+)-lactic acid which were prepared to cover a range from low, physiological and supraphysiological levels (0-64 mmol/L) in distilled water and in a more complex solution prepared in-vitro containing some of the major constituents of CSF . An ultrafiltrate of plasma, CSF contains approximately 0.3% plasma proteins (15 to 40 mg/dL) of which low-molecular weight proteins predominate [4,5] "Synthetic CSF" was prepared from an ultrafiltrate of human citrated plasma [6] using Amicon Ultra-4 centrifugal filters to exclude proteins above a molecular weight of 100 kDa to mainly prealbumin, albumin and transferrin together with other low molecular weight constituents that may be found in CSF in-vivo. The plasma filtrate was diluted in Dulbecco's phosphate buffered saline (DPBS) (-CaCL2, -MgCL2, Ph 7.4) [7] to give approximate protein levels found in normal CSF samples. Sample concentrations were as follows: 0mM, 0.25mM 0.5mM, 1mM, 2mM, 4mM, 8mM, 16mM, 32mM, 64mM and 1M.

3. Electromagnetic Waves Fundamental and Waveguide

Electromagnetic (EM) waves [8,9] are waves of energy that travel through a vacuum at the speed of light, approximately c=3x10^8 m/s. EM waves consist of two primary components, an electric (E) field and a magnetic (H) field. The electric field and magnetic field oscillate in phase perpendicular to each other and perpendicular to the direction of energy propagation. EM waves can be visualised as a sinusoidal wave.

3.1. Maxwell's Equations

Maxwell's equations [10] describe the relation of the electric and magnetic field to each other and to the position and motion of charged particles. Maxwell's equations relate the electric and magnetic fields to their sources, charge density and current density. The equations can be combined to show that light is an electromagnetic wave. The general four fundamental electromagnetic equations are shown in equation (1) to (4),

$$\nabla \cdot D = \rho \tag{1}$$

$$\nabla \cdot \mathrm{B} = 0 \tag{2}$$

$$\nabla \cdot \mathrm{E} = -\frac{\partial \mathrm{B}}{\partial t} \tag{3}$$

$$\nabla \cdot \mathrm{H} = J + \frac{\partial D}{\partial t} \tag{4}$$

In equations (1) to (4), E represents the electric field in volts per metre (Vm^{-1}) and H represents the magnetic field in amperes per metre (Am^{-1}). D is electric displacement field in coulombs per square metre (Cm^{-2}) and B is the magnetic flux density in tesla (T). J is the electric current density in amperes per square metre (Am^{-2}) and ρ is the electric charge density in coulombs per cubic metre (Cm^{-3}).

3.2. Waveguide

The waveguide [11] is a particular form of transmission line; it is a hollow conducting tube to contain the electromagnetic energy. The different configuration of electric field and magnetic field could exist in the waveguide and are known as transverse electromagnetic (TEM) modes. The transverse electric (TE) and transverse magnetic (TM) are the two types of mode which usually exist in the waveguide.

Journal of Physics: Conference Series **307** (2011) 012017 doi:10.1088/1742-6596/307/1/012017

The electrical power is transported through a waveguide by means of electromagnetic waves, which can take several different forms (modes), depending on the frequency, the waveguide dimensions and the material properties inside. This determines how the electric and magnetic fields appear in the waveguide.

The two most important modes in the circular waveguide are known as TE_{111} and the TM_{010} modes which are shown in Figure 1. The red line in Figure 1 represents electric field and the green line represents magnetic field. These modes differ from electromagnetic waves in free space because they have to be higher than a certain minimum frequency in order to be able to propagate through the waveguide at all.

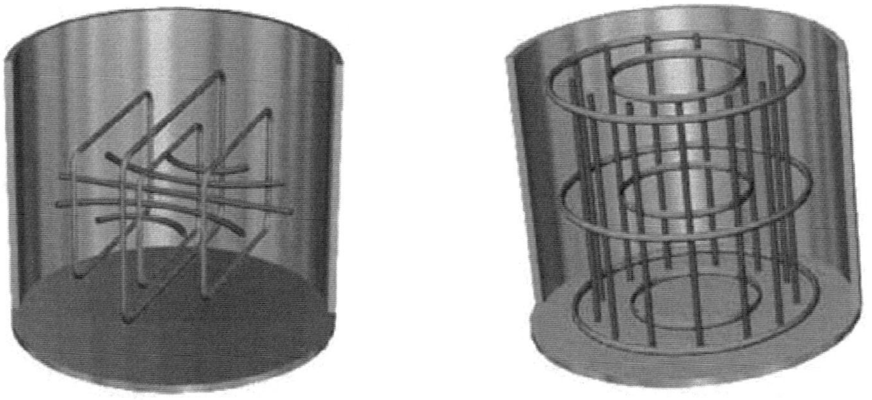

TE_{111} **TM_{010}**

Figure 1: Electromagnetic mode patterns in a circular waveguide

4. High Frequency Structure Simulator (HFSS)

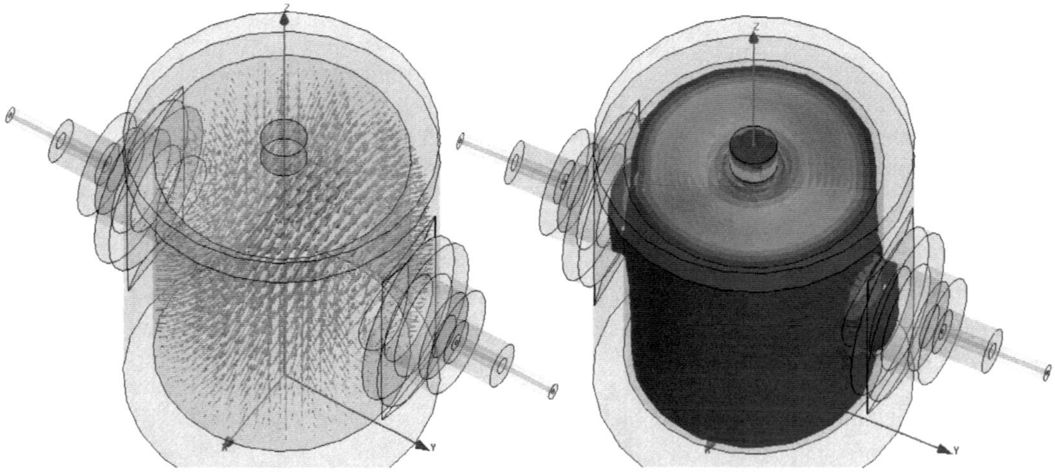

TE_{111} **TM_{010}**

Figure 2: TE and TM modes in a cylindrical cavity

High Frequency Structure Simulator (HFSS) is a 3D full wave electromagnetic field simulation software which can be used to design RF structures [12]. HFSS can be used to design IC packages,

PCB interconnects, antennas and RF and microwave components. The outputs of HFSS include S-parameter, Full-Wave SPICE extraction and 3D electromagnetic field simulation [13].

The TE and TM modes for cylindrical cavity which was used for this project can be simulated by using HFSS. Figure 2 shows the configurations for the first two modes in a cylindrical cavity which is used for this project. Figure 2 shows the first TE_{111} modes, and the second TM_{010} modes in the cavity. The red colour region in Figure 2 represents the maximum and the blue colour region represents the minimum of the electric and magnetic field.

5. Experiment and Results

The project begins with the evaluation of microwave technology using the simplest of a number of proposed biomedical applications, lactic acid as a marker of ischemia in CSF. This is partly due to lactate being inexpensive and also CSF provides a relatively simple background that can be synthesised in-vitro.

The LHCH laboratory prepares the samples, placing lactate in water and lactate in "synthetic CSF" with varying concentrations; 10ml volumes are placed in polypropylene centrifuge tubes. A ZVL-6 Rohde and Schwarz VNA was used, and temperature variations were minimised through the use of an environmental chamber. A cylindrical cavity sensor was used for the sample measurement which is shown in Figure 3. With the cavity attached to the VNA, it is possible to measure microwave spectrum parameters. The cavity was designed to accept 15ml polypropylene centrifuge tubes, which must contain identical fluid volumes since the cavity is volume sensitive. This system facilitates empirical analysis of materials presented to the cavity and is useful for determining specific areas of interest in the microwave spectrum.

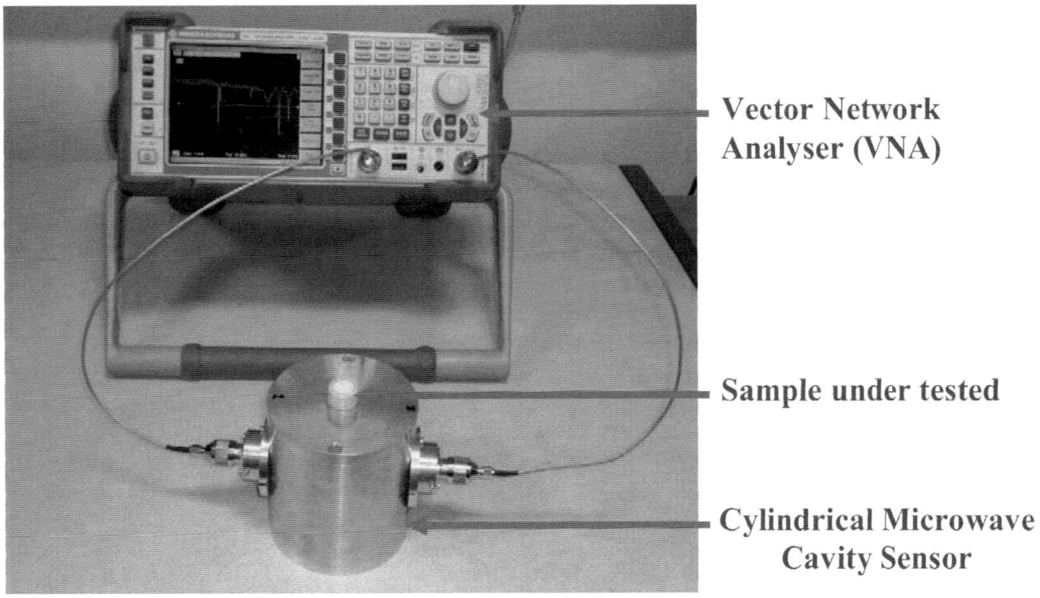

Figure 3: Experimental setup for lactate measurement

Figure 4 shows the $|S_{11}|$ for lactic acid in water. From Figure 4, one can see the level of amplitude decrease as the concentration of the lactic acid increases from 0mM (blank) to 64mM at around 1040MHz. These results also demonstrate a linear sensor response to varying concentrations of latic acid.

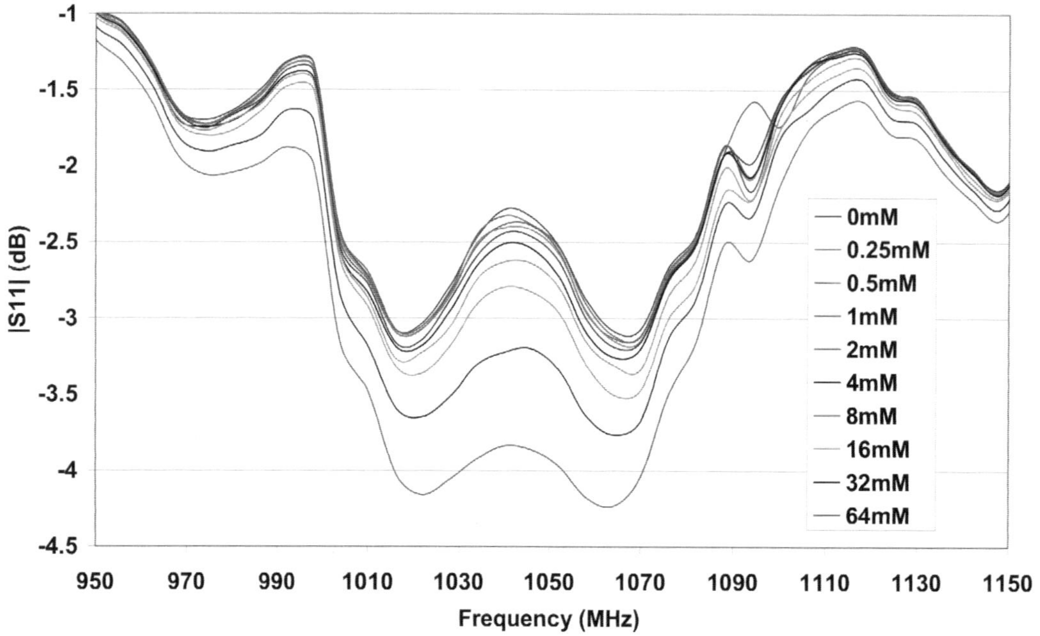

Figure 4: $|S_{11}|$ for lactic acid in water

Figure 5 shows the $|S_{11}|$ for lactic acid in "synthetic CSF". From Figure 5, one can see the signal amplitude decrease as the concentration of the lactic acid increases from 0mM to 64mM at around 1035MHz. These results also show the linearity for the lactic acid with different concentrations.

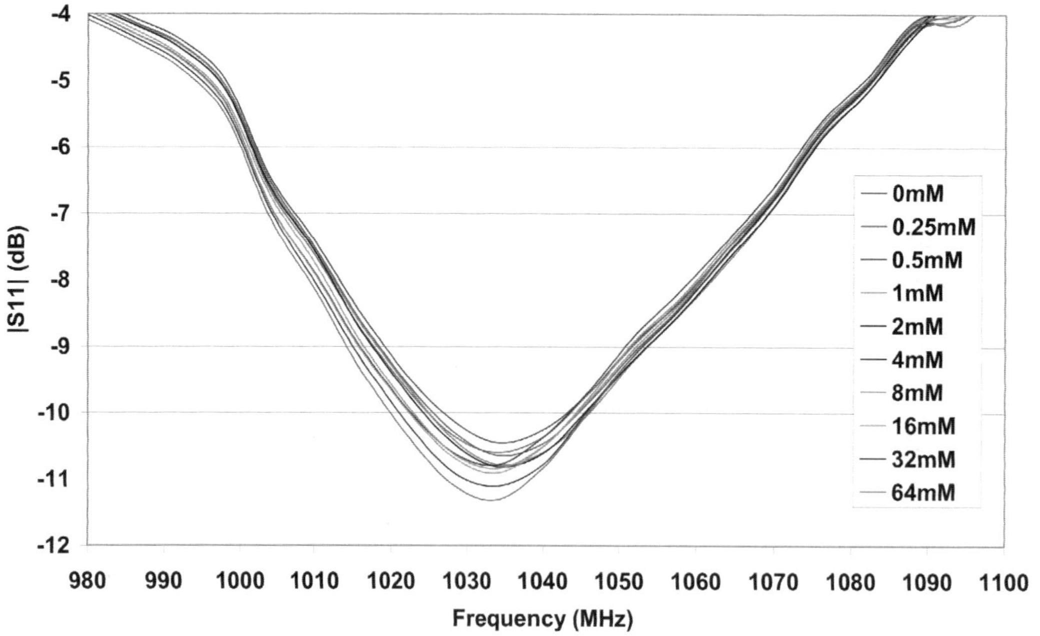

Figure 5: $|S_{11}|$ for lactic acid in "synthetic CSF"

6. Conclusion

Microwave detection in biology is an emerging new field with potential applications ranging from the non-invasive measurement of substances in blood such as glucose, alcohol and drugs, to the detection of infections and cancer. The sensor shown in this paper exhibits promising results for the lactic acid in water and "synthetic CSF". The work here serves as a means to identify the detection limits associated with the technique before progressing in defined steps towards validating the method for measuring lactic acid in patient CSF samples and then expanding the work to include other clinically useful indicators in CSF such as albumin, glucose, lactate dehydrogenase (LDH) and other biomarkers present at lower (μM) concentrations. This research aims to work towards multi-parameter biochemical measurements in a range of more complex biological fluids and to investigations with cellular material.

7. References

[1] White J F 2004 *High Frequency Techniques: An Introduction to RF and Microwave Engineering* A John Wiley & Sons, Inc pp.209-11

[2] Anderson R E, Winnerkvist A, Hansson L, Nilsson O, Rosengren L, Settergren G and Vaage J 2003 *Biochemical markers of cerebrospinal ischemia after repair of aneurysms of the descending and thoracoabdominal aorta* Journal of Cardiothoracic and Vascular Anesthesia 17(5): pp. 598-603

[3] Drenger B, Parker S D, Frank S M and Beattie C 1997 *Changes in cerebrospinal fluid pressure and lactate concentrations during thoracoabdominal aortic aneurysm surgery* Anesthesiology 86(1): pp. 41-47

[4] Felgenhauer K 1974 *Protein size and CSF composition* Klin. Wochenschr. 52 (24): 1158–64

[5] Tietz, NW, Ed: 1990 *Clinical guide to laboratory tests* 2nd ed. Philadelphia, WB Saunders

[6] TCS Biosciences Ltd *pooled normal plasma* Botolph Claydon Buckingham MK18 2LR UK

[7] Invitrogen Ltd 3 Fountain Drive Inchinnan Business Park Paisley PA4 9RF UK

[8] Staelin D H, Morgenthaler A W and Kong J A 1994 *Electromagnetic Waves* Prentice Hall International, Inc pp.1-11

[9] Kraus J D and Fleisch D A 1999 *Electromagnetics: with applications*, 5th ed, McGraw-Hill pp. 1-7

[10] Cheng D K 1989 *Field and Wave Electromagnetics*, 2nd ed, Addison-Wesley Publishing Company pp.460-2

[11] Kraus J D and Fleisch D A 1999 *Electromagnetics: with applications*, 5th ed, McGraw-Hill pp. 456-67

[12] Ansoft Corporation 1999 *Parametrics and Optimization using Ansoft HFSS* Product Feature, Horizon House Publications

[13] Park D M and Choi J J 2003 *Three-Dimensional Simulation of X-Band Coupled Cavity Traveling Wave Tube Amplifier* Journal of the Korean Physical Society 43(6):1150-1111

Instantaneous venous oxygenation estimation using the Photoplethysmograph (PPG) waveform

K Shafqat[1], P A Kyriacou[1], S K Pal[2], R M Langford[3]

[1]School of Engineering and Mathematical Sciences, City University London, London, UK

[2]St Andrew's Centre for Plastic Surgery & Burns, Broomfield Hospital, Chelmsford, UK.

[3]St. Bartholomew's Hospital, Bart's and The London NHS Trust, London, UK

Email: k.shafqat@city.ac.uk

Abstract. In this study oesophageal photoplethysmograph data from eight patients under positive pressure ventilation were analysed in order to test the hypothesis that the modulations created by the ventilation in the AC Photoplethysmograph (PPG) signal could be used to estimate venous oxygen saturation. In order to estimate the instantaneous arterial and venous oxygen saturation Smoothed-pseudo Wigner-Ville Distribution (SPWVD) was utilised. The result from this study showed that there was no significant different in the conventional (time domain) arterial saturation and the instantaneous arterial saturation. However, the instantaneous venous oxygen saturation estimated with the ventilator modulation were significantly lower then the conventional arterial saturation (P=0.008) and also from the instantaneous arterial saturation (P=0.008).

1. Introduction

In the last twenty years pulse oximeter has become one of the most commonly used medical devices in medicine [1]. In its most basic form the device is used to measure arterial saturation (SpO_2) and heart rate (HR). However, the advancement of digital signal processing has allowed further in-depth analysis of the Photoplethysmograph (PPG) signal resulting in extraction of information related to other physiological parameters such as respiration. In order to estimate SpO_2 pulse oximeters rely on two conditions, (1) pulsatile flow and (2) differential spectral absorbance [2]. Due to the pulsatile flow the spectral absorbance of tissue can be measured at maximum and minimum perfusion states. The change in the spectral absorbance profile of the tissue during this transition allows for the calculation of the spectral properties of the blood in motion, and hence its oxygen saturation.

In conventional pulse oximetry the pulsatile component (AC part of the PPG signal) is associated with arterial blood. However, in previous studies [3] it has been shown that motion of venous blood can also be detected using the PPG signal. In most cases, this phenomenon is seen as a source of artefact which interferes with the estimation of arterial oxygen saturation [3]. However, by using the two conditions mentioned above, for the estimation of SpO_2, with this moving venous component the saturation associated with venous blood can also be estimated. Reliable estimation of venous saturation simultaneously with arterial saturation will provide useful information relating to local

Published under licence by IOP Publishing Ltd

oxygen extraction. Such information may reflect adequacy of tissue perfusion in the area being measured. Real time non-invasive estimation of tissue perfusion would be beneficial in monitoring and detecting important clinical events such as early phases of shock. Also, by measuring venous oxygen saturation at different sites such as the earlobe, forehead, finger, might be possible to estimate the true mixed venous saturation which is known to provide valuable clinical information. This study presents the results obtained from the analysis carried out on oesophageal PPG signals in order to estimate the instantaneous arterial and venous saturation. Before presenting the analysis method and results, some of the previous studies dealing with the estimation of venous saturation through the use of PPG signal will be discussed in the section.

2. Previous studies on estimation of venous oxygen saturation using PPG signals

One of the main problems in the estimation of venous oxygen saturation is the continuous and reliable detection of pulsatile venous blood component. In the literature two different methods have been proposed to achieve this pulsatility, by mechanically inducing volume variation in the venous compartment. In one of these methods [4, 5] the volume in the venous compartment was increased by occluding the venous outflow using a pressure cuff. The saturation values were then estimated before and after the occlusion in order to detect the venous oxygen saturation. Significant correlation ($n = 19$, $r = 0.7$, $p < 0.0001$) was found when the venous saturation values obtained using this method were compared with the values obtained from co-oximetry (i.e., the gold standard *in vitro* method). With this method there are two major problems in estimating venous oxygen saturation. Firstly, the method does not allow continuous measurement of venous saturation as it relies on discrete interventions separated by enough time for the tissue to reach equilibrium. Secondly, the use of this method for an extended period could lead to potential complications such as venous stasis, venous thrombosis, and interference with intravenous access.

These deficiencies were addressed in another study [6] (second method) where a finger pressure cuff, driven by a pneumatic generator was used to create low pressure modulations in the finger at a frequency that does not overlap with the frequency of cardiac signal and its harmonics. A conventional PPG probe was placed on the same finger distal to the cuff. The pressure cuff frequency was set at approximately 7.5 Hz. Using conventional frequency domain algorithms the arterial oxygen saturation was estimated with the values at the cardiac frequency (~ 1 Hz) and venous saturation were estimated using the values at the pressure cuff frequency (~ 7.5 Hz).

The performance of this system was evaluated in patients undergoing cardiopulmonary bypass (CPB). The saturation values obtained with this method were compared with the saturation measured by the CPB machine on the blood that was travelling from the patient to the machine. Since the values obtained from the machine were indicative of central venous saturation it was not possible to perform a direct comparison with the values obtained from the pulse oximeter setup as they represented peripheral venous saturation. However, similar changes were observed in both venous saturation values with changes in temperature, VO2, and cardiac index.

In all the studies described so far an external source (pressure cuff) was used to create the pulsatile venous component however, in a more recent study [7] venous saturation was estimated by taking advantage of the fact that the PPG waveform is influenced by both positive pressure ventilation [8] and the peripheral venous pulsations [9]. In this study analysis was carried out on PPG signals that were collected using a special-purpose built PPG probe [10], in the oesophagus of ten patients undergoing coronary artery bypass surgery and postoperative care in the intensive care unit [11]. In order to estimate the arterial and venous saturation several time and frequency domain methods were employed. The time domain estimation was performed in the conventional manner by calculating R (ratio of ratios) as presented in Eq. (1). Using the R value, the saturation was estimated by employing the linear relationship presented in Eq. (2).

$$R = \frac{AC_{red}/DC_{red}}{AC_{ired}/DC_{ired}} \tag{1}$$

$$Saturation = 110 - 25R \tag{2}$$

In order to estimate the arterial saturation the numerator of the R (Eq. (1)) was calculated by dividing the peak-to-trough amplitude of the AC PPG waveform with the DC offset value for the red wavelength PPG signal while, for the denominator the corresponding values associated with the infrared PPG signal were used. In the case of the venous saturation the AC_{red} and AC_{ired} were calculated by using the peak-to-trough amplitude of the DC PPG signal. The calculations of venous saturation were performed in this way as the modulation in the PPG DC waveforms were dominated by the influence of the positive pressure ventilation. These modulations are thought to be due to the volume changes in the venous compartment. Thus, the saturation estimated using these values should be lower than the arterial saturation. Apart from these time domain measurements of arterial and venous saturation, an instantaneous saturation (InstSat) was also estimated. In this case at every time instance (t_k) a new saturation value was calculated by utilising an R value which was obtained as shown in Eq. (3).

$$R(t_k) = \frac{(AC_{red}\,(t_k) - AC_{red}\,\text{min})/DC_{red}(t_k)}{(AC_{ired}\,(t_k) - AC_{ired}\,\text{min})/DC_{ired}(t_k)} \tag{3}$$

Where $AC_{red}min$ and $AC_{ired}min$ in Eq. (3) represent the preceding trough values of the red and infrared AC PPG signal.

The InstSat was aimed to provide a moment-by-moment measurement of the average saturation of the blood in all the vascular compartments in the vicinity of the probe. One complication of InstSat was the fact that near the trough of the AC PPG signal the saturation values become unstable. This instability was dealt with in two steps. The first step was that the saturation values were not updated (kept constant) at time instances where the numerator or the denominator of Eq. (3).became less than 0.03 (3%). The first step produced an artefact in the saturation values at the point where the numerator or the denominator of Eq. (3) crosses the 3% criterion. Therefore, in the second step InstSat values within 0.5 s of this point were replaced by mean InstSat values. The resulting InstSat waveform was pulsatile with peaks approximately coinciding with the peaks in the AC PPG waveform. The envelope joining the peaks of the InstSat saturation was used as the arterial instantaneous saturation while the venous instantaneous saturation was obtained by creating the envelope by joining the troughs of the InstSat waveform. The results of the study showed that the venous saturation estimated using the time domain method and the instantaneous method both were significantly lower than the time domain arterial saturation.

These results provided evidence that it might be possible to utilize the modulations produced in the PPG signal due to positive pressure ventilation (breathing) to estimate venous oxygen saturation. However, the instantaneous saturation estimation in the time domain has stability issues. Therefore, to deal with these problems in this study the modulation due to the positive ventilation has been extracted from the AC PPG signal and then Smoothed-pseudo Wigner-Ville Distribution (SPWVD), which is a quadratic transform from Cohen's class [12], has been used to estimate the instantaneous arterial and venous saturations.

3. Methods

3.1. Subject and Protocol
In this study oesophageal PPG data from eight patients were analysed. This data were collected in previous studies [10, 13], after obtaining ethical approval, from ASA 1 and 2 adult healthy patients who underwent tracheal intubation as a routine part of general anaesthesia. Following induction of general anaesthesia the oesophageal probe was advanced under direct vision into the oesophagus until the end of the probe was 25 to 30 cm from the upper incisors. During these oesophageal PPG studies an identical reflectance finger probe was also placed on the index finger of the patient [10, 13].

3.2. Data Processing
This section presents the steps involved in separating the ventilator modulation from the AC PPG signal and the estimation of instantaneous saturation.

3.2.1. PPG peak detection
In order to detect the peaks of the AC PPG signals (at both wavelengths red (655 nm) and infrared (880 nm)), the first step was to detect the peak frequency related to the cardiac frequency and then find the frequencies on both sides of this peak where the amplitude dropped to 10% of the amplitude at the peak frequency. This frequency range was considered to be the cardiac frequency band. A digital bandpass FIR (Finite Impulse Response) filter was then used to separate the cardiac frequency component of the PPG signal from the raw signal. An adaptive threshold was made by convolving the absolute values of the filtered signal with a square window of 80 samples (which is slightly less than the width of the PPG pulse at a sampling rate of 100 Hz). This threshold was used to detect the peaks of the AC PPG signal. A possible peak was detected in the region where the filtered signal crosses the threshold. The peak was considered to be a true PPG peak if it does not occurs within 0.2s of the previously confirmed PPG peak and its amplitude lies within 0.2 and 3 times the average amplitude, calculated from the last five correctly detected peaks. A similar technique was also used to detect the troughs of the AC PPG signal.

3.2.2. Instantaneous arterial and venous oxygen saturation estimation
After detecting the peaks and troughs of the AC PPG signals an upper and lower envelope was constructed by joining the peaks and troughs using cubic spline interpolation. Using these upper and lower envelopes the mean envelope was constructed. This signal was used as the estimate of the ventilator modulation presented in the AC PPG signal and subtracted from the raw PPG signal to obtain the cardiac component which was used as the AC signal in the estimation of arterial saturation.
After separating the venous and cardiac component, SPWVD was used to estimate the instantaneous amplitude related to the AC venous, cardiac and the DC PPG signal. The derivation of the discrete SPWVD is presented in detail in the literature [14]. For a discrete sequence $s[n]$, the discrete SPWVD can be expressed as shown in Eq. (4).

$$SPWVD_x[n,k] = \sum_{l=-P+1}^{P-1} h[l] \sum_{lm=-Q+1}^{Q-1} g[m] \times r[n-m,l] e^{-j2lk\pi/M} \quad 4$$

Where $r[n,l] = s[n+l]s^*[n-l]$ is the instantaneous autocorrelation function. In this case the smoothing in the time direction is done using a window $g[m]$ of length $2Q+1$ and window $h[l]$ of length $2P-1$ is used for frequency smoothing. The signals were analysed in their analytical form. In this study both the time and frequency smoothing were done using Hamming windows 5.02 s and 10.02 s respectively.

3.3. Statistical test
A non-parametric test (Wilcoxon, signed rank test) was used to compare the time domain arterial oxygen saturation with the instantaneous arterial and venous oxygen saturation. Comparisons were also made between the instantaneous arterial and venous saturation. The statistical analysis was carried out using SigmaStat 2.03 (Systat Software Inc., USA). The significance level was set at $P<0.05$ for all tests.

4. Results
The peak and trough detection and the estimation of the mean envelop (ventilator modulation signal) results for the red and infrared AC PPG signals in one of the signals analysed in this study are shown in Fig. 1. The ventilator modulation signal was subtracted from the raw AC PPG signal as mentioned

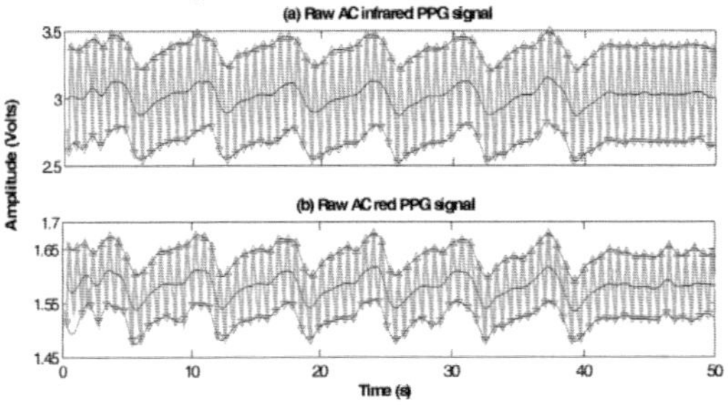

Fig. 1. Peak, trough detection and mean envelope detection results in one of the data set analysed in this study; (a) results from the infrared AC PPG signal; (b) results from the red AC PPG signal

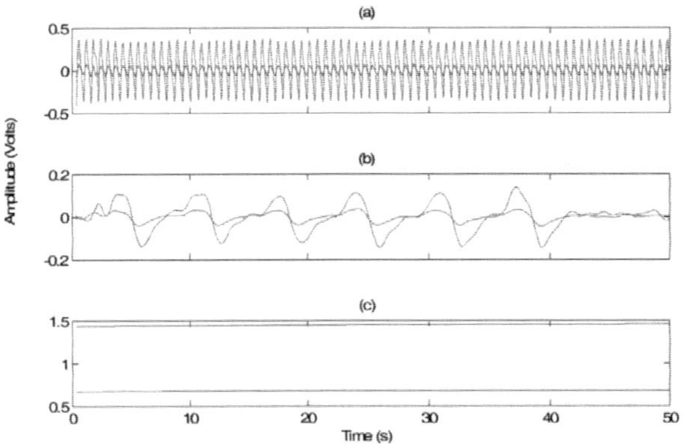

Fig. 2. (a) Signal obtained after subtracting the venous modulation from the raw AC PPG signals; (a) AC signals related to cardiac component; (b) Ac signal related to the venous component; (c) DC PPG signals. In Each part the infrared signal is presented with dashed lines while the red signal is presented with solid lines.

before. The resulting AC PPG signals associated with the cardiac and ventilator modulation and the DC PPG signals which were used for the estimation of the instantaneous arterial and venous oxygen saturation are shown in Fig 2.

The instantaneous arterial and venous saturation estimated from the signals shown in Fig. 2 are presented in Fig. 3. Figure 3(a) also shows the conventional time domain estimation of arterial oxygen

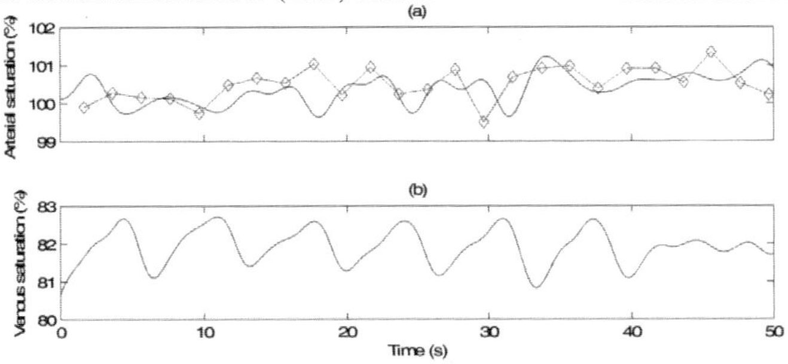

Fig. 3. Saturation results obtained from the signals shows in Fig. 2; (a) Conventional (time domain) arterial saturation (diamond marker with dashed line) and instantaneous arterial saturation; (b) instantaneous venous saturation

saturation. From the results presented in this figure it can be seen that there is a close match between the time domain (average) and the instantaneous arterial oxygen saturation values. Also the instantaneous venous oxygen saturation values shown in Fig. 3(b) are lower than the arterial oxygen saturation values. It is also interesting to see that the venous oxygen saturation is showing slight changes synchronous with the ventilator frequency, which also shows that the venous blood is moving due to the positive pressure ventilation.

4.1. Statistical test results

A statistical test was carried out using Wilcoxon Signed Rank Test. For the time domain arterial oxygen saturations the median and percentile (25% -75%) values were 98.1 (96.4 - 99.6) while the values for the instantaneous arterial and venous saturation were 98.9 (95.3 - 100.4) and 82.5 (81.6 - 84.1) respectively. The results of the test showed that there was no significant difference between the time domain and instantaneous arterial oxygen saturations while the venous oxygen saturation was significantly lower compared to the time domain arterial oxygen saturation (P=0.008). The venous saturation was also significantly lower compared to the instantaneous arterial saturation (P=0.008).

5. Conclusion

In this study oesophageal PPG data were analysed in order to test the hypothesis that modulations produced on the AC PPG signals caused by positive pressure ventilation could be used to estimate venous oxygen saturation. The results of the study showed that the ventilator modulation produced venous oxygen saturations which were constantly lower than the arterial oxygen saturation. Unlike a previous study [7], where the instantaneous venous oxygen saturation estimated in the time domain required thresholding in order to deal with the instabilities near the troughs of AC PPG signal, in this analysis the instantaneous venous oxygen saturation estimated using SPWVD had no such issues. By avoiding this instability, the slight changes caused in the venous oxygen saturation due to ventilation could be seen easily and such information could be of great importance to clinical experts. Also, this study provides strong evidence of the hypothesis that positive pressure ventilation causes movement in the venous compartment.

References

[1] Welch J 2005 *Biomed Instrum Technol* **39** 125–130

[2] Shelley K H and Shelley S 2001 Pulse oximeter waveform: photoelectric plethysmography. *Lake CL, Hines RL, Blitt CD, eds, Clinical monitoring: practical applications for anesthesia and critical care. W.B. Saunders Company, Philadelphia PA,* pp 420–428

[3] Shelley K H, Tamai D, Jablonka D, Gesquiere M, Stout R G and Silverman D G 2005 *Anesth Analg* **100** 743–7, table of contents

[4] Yoxall C W and Weindling A M 1997 *Med Biol Eng Comput* **35** 331–336

[5] Nitzan M, Babchenko A, Khanokh B and Taitelbaum H 2000 *J Biomed Opt* **5** 155–162

[6] Echiadis A S, Crabtree V P, Bence J, Hadjinikolaou L, Alexiou C, Spyt T J and Hu S 2007 *Physiol Meas* **28** 897–911

[7] Walton Z D, Kyriacou P A, Silverman D G and Shelley K H 2010 *J Clin Monit Comput* **24** 295–303
[8] Natalini G, Rosano A, Franceschetti M E, Facchetti P and Bernardini A 2006 *Anesth Analg* **103** 1182–1188 http
[9] Wardhan R and Shelley K 2009 *Curr Opin Anaesthesiol* **22** 814–821
[10] Kyriacou P A, Moye A R, Gregg A, Choi D M, Langford R M and Jones D P 1999 *Med Biol Eng Comput* **37** 639–643
[11] Kyriacou P A, Powell S L, Jones D P and Langford R M 2003 *Anaesthesia* **58** 422–427
[12] Cohen L 1989 **77** 941–981 ISSN 0018-9219
[13] Kyriacou P A, Moye A R, Choi D M, Langford R M and Jones D P 2001 *Physiol Meas* **22** 223–232
[14] Richman M, Parks T and Shenoy R 1998 **46** 1517–1527 ISSN 1053-587X

Highly Sensitive and Selective NiO Film / K^+-Exchanged Glass Composite Optical Waveguide Sensor for Chlorobenzene Vapor Detection

Hairisha Abulaiti, Abulizi Yimiti*, Dilinuer Talifu, Adalaiti Abudureheman
College of Chemistry and Chemical Engineering, Xinjiang University, Urumqi 830046, P.R.China

E-mail: ablizyimit@xju.edu.cn

Abstract. In this paper, we propose a new and simple NiO-based Chlorobenzene vapor optical waveguide (OWG) sensor. The highly sensitive element of this sensor was made by coating the nickel oxide film over a single-mode potassium ion (K^+) exchanged glass OWG. Sensitivity of the NiO film composite OWG sensor was estimated by the change of absorption properties of the film. The sensor has a linear response to Chlorobenzene concentration, and sensitivity up to 1.8ppm. Moreover, it shows a satisfactory repeatability and a low sensitivity to common interfering gases such as benzene, toluene, xylene and ethanol vapors.

1. Introduction

Environmental pollution has increased considerably during the last few decades [1]; Volatile organic compound (VOC) emissions from industrial and commercial processes are very dangerous for human health and environment [2]. Thus, it is very important to develop chemical sensors that allow the measurement of volatile organic compounds in a fast, in situ and continuous way. Optical waveguide (OWG) sensor have received particular interest for biomedicine [3, 4], environmental monitoring [5], and process control and safety, particularly for the detection of gaseous analytes in trace concentrations [6-11]. In this paper, we propose a new and simple NiO-based Chlorobenzene OWG sensor, and then report several characteristics of the device; finally, provide results for the detection of parts-per-million levels of Chlorobenzene vapor as an example of VOC gases detection.

2. Experimental Section

Nickel oxide powder was synthesized by the coprecipitation method [12]. The NiO film was

Published under licence by IOP Publishing Ltd

prepared by vacuum evaporation of NiO powder onto K^+ exchanged glass substrates. The thickness of the NiO film was about 160nm. To taper the ends (slopes) of the thin film, a mask with a rectangular window was mounted at a distance of several mm from the substrate during vacuum evaporation. The mask can be made of a metal plate, an alumina ceramic plate, or slide glass. For a typical experimental configuration, see Figure 1. The K^+-exchanged glass OWG device is made on a soda-lime glass substrate (76 mm×26 mm×1 mm), which were immersed in molten KNO_3 salt at 400 °C for 40 min. Since the polarizability of K^+ is considerably greater than that of Na^+ on glass, K^+ can be easily incorporated in glass [13]. After ion exchange process, the substrate was taken out of the molten KNO_3, cooled at room temperature, and then washed with distilled water in order to remove the solid KNO_3 on its surface.

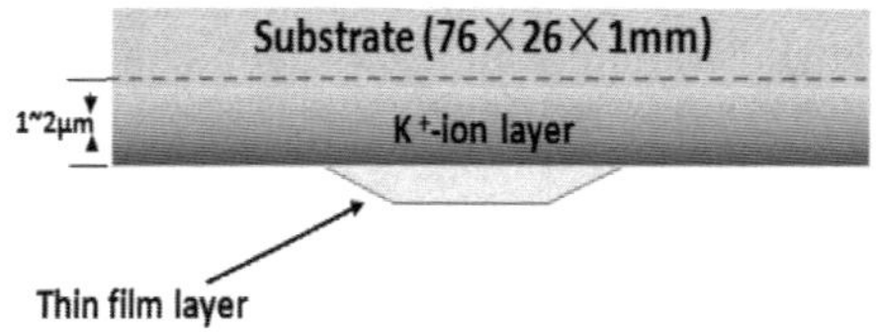

Figure 1. NiO Film / K^+-Exchanged Glass Composite Optical Waveguide

3. Results and discussion

The VOC gases-testing apparatus (Figure 2a) was constructed from compressed air sources, a flow meter, a diffusion tube, and gas that contained VOC gases. A gas mixing manifold was used to mix the air stream that contained VOC gases with a stream of pure air and to introduce the mixture into the flow cell, which enclosed the waveguide sensor. The NiO film/K^+-exchanged glass composite OWG gas sensor device in a flow cell (2 cm × 1 cm × 1 cm) was mounted on a rotational stage equipped with X–Y–Z translation. A He–Ne laser (λ= 632.8 nm) beam was introduced into the OWG using a prism coupler (glass prism: n = 1.78 and a matching liquid, diiodomethane: n = 1.74), and it emerged from another prism coupler. The intensity of the output light was monitored by using a radiometer, and the signal was recorded by a computer program. In every measurement, a new syringe was used to inject 10 ml of the Chlorobenzene vapor sample into the flow chamber, and then out from the vent (Figure 2b). Pure air that functioned as a carrier and dilution gas flowed through the cell at a constant rate of 30 ml/min in order to transfer the Chlorobenzene vapor to the sensor. All measurements were made at room temperature.

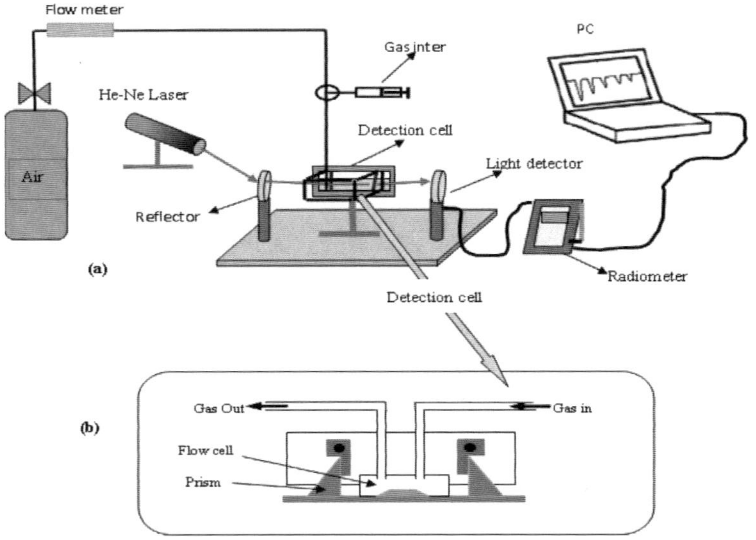

Figure 2. The schematic view of VOC gases sensor system (a) and Amplificatory diagram of Detection cell (b).

The typical reversible response of the NiO film/ K^{+}-exchanged glass composite OWG sensor to Chlorobenzene vapor is given in Figure 3. As the NiO film was exposed to Chlorobenzene vapor the sensor baseline steadily decreased. Upon exposure to Chlorobenzene vapor, the sensor response returning to dry air results in partial recovery of sensor signal. The extent of reversible signal decrease seems to increase as the Chlorobenzene vapor concentration increases and constitutes a dosimetric response. The response and recovery times of the NiO film/K^{+}-exchanged glass composite OWG Chlorobenzene sensor were fast.

As can be seen from Figure 3 the intensity of output light decreases with the increase in the concentration of Chlorobenzene vapor in the range of 1.8-18500 ppm and decreased when the concentration of Chlorobenzene vapor was increased. The recovery time was found to increase with an increase in Chlorobenzene vapor concentration, by means of dry air. The recovery time of the sensor was longer than its response time because the oxidant (oxygen) concentration was lower and a longer time was required for a sufficient amount of oxidant to reach the indicator in the flow chamber. The response and recovery times of the Metal oxide planar OWG Chlorobenzene sensor were fast and less than 6 s and 17 s, and highest output light intensity of this vapor is 1.8ppm.

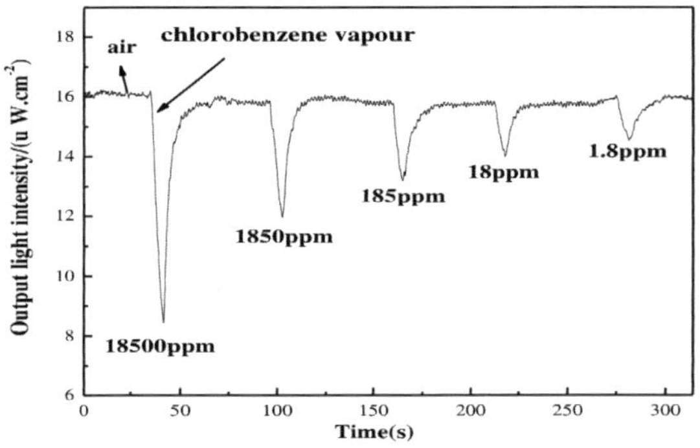

Figure 3. Typical response of NiO film / K^+-exchanged glass OWG sensor when exposed to Chlorobenzene vapor in air

Several OWG sensors were fabricated and tested for response to gaseous Chlorobenzene. The repetitive responses of typical sensor exposed alternately to pure air and 1.8 ppm and 18500ppm Chlorobenzene vapor is shown in Figure 4. At this concentration the Relative Standard Deviation of output light intensity is ± 0.75%. The response time was rapid (6s) and fully reversible (17 s) after two times consecutive injected 1.8 ppm and 18500ppm of Chlorobenzene vapor, respectively. These results demonstrated that the OWG sensor was fully reversible and reproducible for Chlorobenzene vapor sensing.

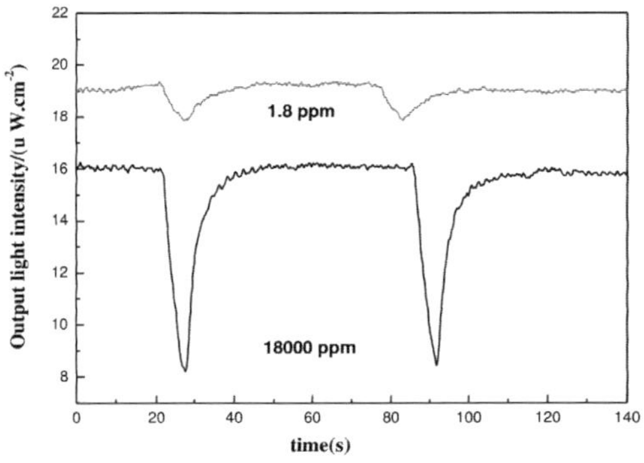

Figure 4. Repetitive response curve of OWG sensor to two different concentration of Chlorobenzene vapor

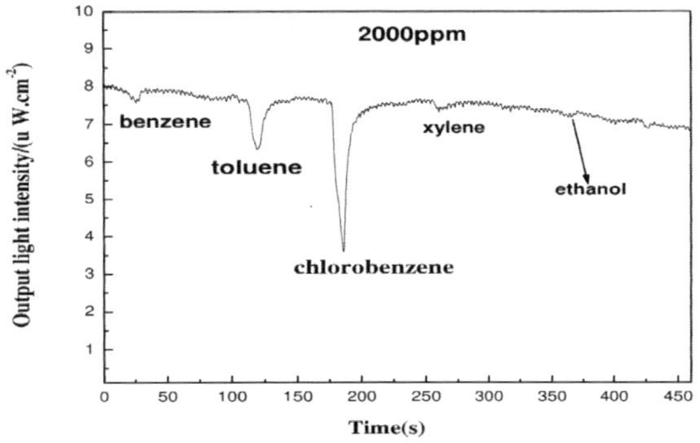

Figure 5. Response curve of NiO film/ K^+-exchanged glass composite OWG sensor to various gases of 2000ppm

For applications in environmental monitoring, a selective fast response to NiO film/ K^+-exchanged glass composite OWG in the presence of some interference like benzene, toluene, Xylene, Ethanol and Chlorobenzene is describable. The OWG sensor was tested for the experimental result shows that the OWG sensor has not any considerable response to benzene, Xylene, Ethanol, even at same concentrations. Figure 5 shows the response of OWG sensor to 2000 ppm benzene, toluene, Xylene, Ethanol and Chlorobenzene Among same consistence, the sensor has higher response to Chlorobenzene vapor.

4. Conclusion

A novel simple-structure optical waveguide sensor for Chlorobenzene vapor detection has been proposed and characterized. We measured the detection limit of the device under operating conditions that are determined according to the investigations described above. We succeeded in detecting parts-per-million Chlorobenzene vapor, which is a typical example of VOC vapor; it exhibits general sensitive and reversible response to the Chlorobenzene vapor. The detection limit for Chlorobenzene vapor is <1.8 ppm, it shows very good characteristics in terms of response repeatability, response time and a low sensitivity to common interfering gases. Moreover, it has a simple structure, and requires a simple front end optical waveguide.

This OWG system proved to be fast in response, reversible, and highly sensitive for Chlorobenzene vapor, and this NiO film/K^+-exchanged glass composite OWG sensor is inexpensive and reusable. It will be possible to achieve a parts-per-billion detection limit by controlling the adsorption properties of the sensing layer and its absorption properties. We believe the proposed technique should be very attractive for the eventual realization of a compact and inexpensive distributed alarm VOC optical waveguide sensor.

Acknowledgment
The authors acknowledge the Natural Science Foundation of China for the support of this project under the 20965008 grants.

References

[1] Spaeth K, Gauglitz G 1998 *Mater. Sci. Eng. C* **5** 187

[2] Boeglin M L,Wessels D and Henshel D 2006 Environ. Res. **100** 242

[3] Yimit A, Axel G R, Amemiya T and Itoh K *Talanta* **65** 1102

[4] Lukosz W 1995 *Sens. Actuators, B* **29** 37

[5] Umemura T et al 2006 *Anal. Chem.* **78** 7511

[6] Yimit A, Itoh K and Murabayashi M 2003 *Sens. Actuators, B* **88** 239

[7] Yang L, Saavedra S S and Armstrong N R *Anal. Chem.* **68** 1834

[8] Székács A et al 2003 *Anal. Chim. Acta.* **487** 31

[9] Ablat H, Yimit A, Mahmut M and Itoh K 2008 *Anal. Chem.* **80** 7678

[10] Abdurahman R et al 2010 *Anal Chim Acta* **658** 63

[11] Kader, R et al 2009 *Environ. Sci. Technol* **43** 5113

[12] Zhang C X and Xiao Y. 2005 *J Chinese Function Material* **4** 36

[13] Ramaswamy R V and Srivastava R 1988 *J. Lightwave Technol.* **6** 984

Detection of Alkylating Agents using Electrical and Mechanical Means

Yulia Gerchikov[1], Elena Borzin[1], Yair Gannot[1], Ariel Shemesh[1], Shai Meltzman[1], Carmit Hertzog-Ronen[1], Shay Tal[2], Sara Stolyarova[3], Yael Nemirovsky[3], Nir Tessler[3] and Yoav Eichen[1]

[1]Schulich Department of Chemistry, Technion-Israel Institute of Technology, Technion City, 32000, Haifa (Israel).
[2]Present address: Systems Biology Department, Harvard Medical School, Boston, MA 02115 USA.
[3]Department of Electrical Engineering, Technion-Israel Institute of Technology, Technion City, 32000, Haifa (Israel).

E-mail: chryoav@tx.technion.ac.il

Abstract. Alkylating agents are reactive molecules having at least one polar bond between a carbon atom and a good leaving group. These often simple molecules are frequently used in organic synthesis, as sterilizing agents in agriculture and even as anticancer agents in medicine. Unfortunately, for over a century, some of the highly reactive alkylating agents are also being used as blister chemical warfare agents. Being relatively simple to make, the risk is that these will be applied by terrorists as poor people warfare agents. The detection and identification of such alkylating agents is not a simple task because of their high reactivity and simple structure of the reactive site. Here we report on new approaches to the detection and identification of such alkylating agents using electrical (organic field effect transistors) and mechanical (microcantilevers) means.

1. Introduction

Alkylating agents, such as methyl iodide, dichloromethane and epichlorohydrin are commonly used in organic synthesis as reagents and solvents. Similar alkylating agents, such as methyl bromide, are still used as soil sterilizers,[1] while nitrogen mustards are being used as anticancer drugs.[2] Alkylating agents are also dangerous byproducts of the water purification process.[3] Unfortunately, due to their ease of production and storage, similar systems are also used as chemical warfare agents.[4] Owing to their ability to react with nucleophiles in our body, most of these materials are toxic, carcinogenic and/or mutagenic.[5] The combination of simple synthesis with aggressive reaction with biological tissues makes many alkylating agents perfect potential "poor people" chemical warfare agents and thus a perfect terror inflicting weapon. Therefore, there is a clear need for simple, sensitive and informative tools for the detection and identification of such agents, especially in the gas phase.

Tailoring selective receptors for small and reactive molecules is a non trivial and challenging goal. Additionally, it is much easier to synthesize a new toxic alkylating agent than to tailor a selective receptor for its detection, rendering this approach inefficient. Previous attempts to address this challenge resulted with two basic approaches. The first approach gave up selectivity, detecting all alkylating agents in a non specific way, relying on a color change of a nucleophile upon reacting with an alkylating agent.[6] The second approach focused on a limited family of alkylators by detecting a specific mass or the presence of a specific atom it contains (for example the sulfur atom in sulfur-based mustard gas).[7]

One possible solution to the detection of alkylating agents could be the use of a nucleophile as a trap for alkylating agents as an efficient nucleophile, such as an amine group, is expected to react with all the reactive alkylating agents as they are essentially electrophiles, Scheme 1. This reaction guarantees very high sensitivity, as the binding is no longer reversible.

Scheme1: General equation for the reaction of a nucleophile (Nu:) with an alkylating agent.

The reaction of the nucleophile with the alkylating agent needs to be reported and this may be achieved through one on a wealth of methods developed along the years. For example, we have recently reported on using luminescence turn-on as well as luminescence shifts for the detection and identification of different alkylating agents.[8]

Here we report on the coupling of the reaction between a nucleophile and alkylating agents to electrical (organic field effect transistors, OFETs) and mechanical (microcantilevers) signal transducers for the detection, risk assessment and identification of alkylating agents. The three complimentary methods demonstrate the potential of using such reactions for the detection of reactive substances.

2. Results and Discussion

2.1 Detecting alkylating agents using organic field effect transistors

Coupling of a nucleophile to an organic semiconductor was achieved by co polymerizing 4-[2,5-bis (2-thienyl)-3-thienyl] pyridine, **1**, with 3-hexylthiophene, **2**, to obtain co-polymer **3**, Scheme 2.

Scheme 2: Preparation of co-polymer **3**. i: $Ba(OH)_2 \cdot 8H_2O$ + $Pd[Ph_3P]_4$; ii: NBS in acetic acid; iii: $Pd[Ph_3P]_4$; iv: $FeCl_3$ in chloroform; v: $H_2N\text{-}NH_2$.

3-(4-pyridyl)-thiophene, **6**, was prepared in 75% yield using palladium catalyzed cross coupling of thiophene-3-boronic acid, **4**, with 4-bromo pyridine, **5**. **6** was di-brominated at the 2 and 5 positions using NBS in acetic acid, affording 2,5-dibromo-3-(4-pyridyl)-thiophene, **7**, in 60% yield. Palladium catalyzed cross coupling of **7** with two equivalents of 2-(tri-n-butyl stannyl) thiphene, **8**, afforded 4-[2,5-bis (2-thienyl)-3-thienyl] pyridine, **1**, in 80% yield. Co-polymer **3** was obtained as a bright red solid, Mn=37000, Mw=97000 (GPC), in 17% yield by co-polymerization of a 1:10 solution of **1** and **2** in chloroform in the

presence of $FeCl_3$. NMR characterization reveals that co-polymer 3 is composed of $\mathbf{2}_n\mathbf{3}_m$ where n=0.95±0.005 and m=0.05±0.005.[9]
Co-polymer **3** reacts with alkylating agents according to Scheme 3 both in solutions and as a film when treated with their vapors.

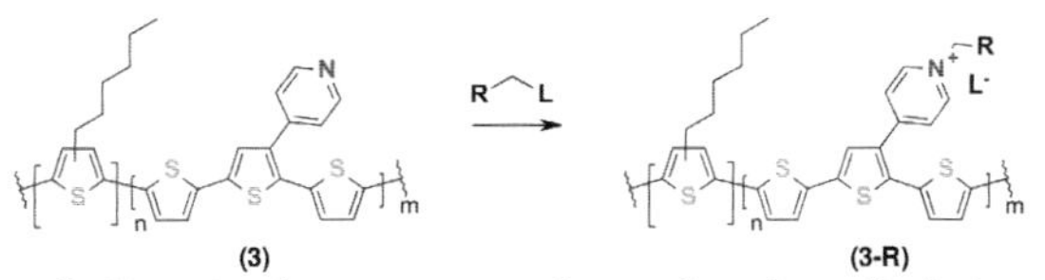

Scheme 3: Reaction between co-polymer **3** and an alkylating agent.

Co-polymer **3** is an excellent film forming polythiophene organic semiconductor. Bottom contact organic field effect transistors (OFETs) were fabricated by spin-coating a 50nm film of the co-polymer atop the FET structure using a previously described procedure.[10]
Figures 1a and b depict the electrical characteristics of the OFET having co-polymer **3** as the channel. Figure 1a shows the drain-source currents of the OFET as a function of drain-source voltage for several gate-source voltages (V_{GS}=0, 10 and 20V) before (black lines and inset) and after (gray lines) exposure to the alkylator methyl iodide (CH_3I, **9**). Figure 1b shows the drain-source current as a function of gate-source voltage for drain-source voltage of V_{DS}= 10V measured the same OFET having co-polymer **3** as the channel and under the same conditions.
As can be clearly seen in the figure, prior to exposure to the alkylating agent the device exhibits standard FET output characteristics with a slight indication of doping, probably due to the exposure to atmospheric oxygen. The post exposure curves show considerably higher currents and a resistor like response while the effect of the gate bias is lost. This is a clear indication for high level p-type doping associated with the exposure to methyl iodide.

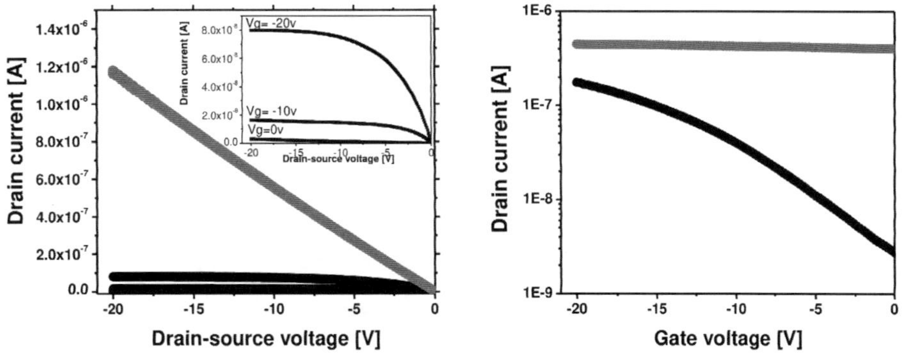

Figure 1: Output (a) and transfer (b) characteristics measured on a FET having co-polymer **3** as the channel. The black lines and the inset show the device response prior to exposure while the gray (overlapping in a.) lines show the response after exposure to methyl iodide, **9**.

Consistent with the results presented in Figure 2A, the transfer characteristics of the OFET prior to exposure show current modulation due to the gate-source voltage. The post exposure gray horizontal line shows almost no dependence on the gate voltage. At zero gate bias, V_{GS}=0, the exposure of the OFET to methyl iodide enhances the current by almost two orders of magnitude while at V_{GS}=-20V the factor is only 2-3. Assuming that the mobility is charge density independent and not affected by the doping one can use the above results to estimate the doping density according to Equation 1, where Cox is the gate capacitance [F/cm^2], VGS is the gate source bias [V], q is the elementary charge and ρ_D is the dopant surface density.

Assuming the doping is uniformly distributed across the ~50 nm film the doping density is ρ_D=1.75x10^{18} cm^{-3}.

$$\rho_{2D} = 2 \cdot \frac{1}{q} C_{ox} V_{GS} = 2 \cdot \frac{1}{1.6 \times 10^{-19}} 35 \times 10^{-9} \cdot 20 = 8.75 \times 10^{12} cm^{-2} \quad (1)$$

The detection scheme of co-polymer **3** based OFETs involves the formation of a new covalent bond between the analyte (the alkylating agent) and the polymer in the channel. As this process is not reversible, all available reaction sites are bound to react. The number of reaction sites (pyridine rings, see Scheme 3) represents also the upper detection limit of the device. The total weight of the polymer embraced by the source and drain electrodes, M, is 10^{-9} g. (assuming a film density of 1g/cm^3), Equation 2, where W is the width of the device, L is the length and h is the thickness of the polymer film.

$$M = W \times L \times h \times \rho = 1 \times 2 \cdot 10^{-4} \times 5 \cdot 10^{-6} \cdot 1 = 10^{-9} g \quad (2)$$

Co-polymer **3** consists of 5% (5x10^{-11}g) monomer **1** which are equivalent to ~1.5x10^{-13} moles. Consequently, reacting the OFET with ~9x10^{10} analyte molecules will bring the system to saturation of all the active sites of the polymer.

The OFET of co-polymer **3** produced similar results when exposed to a variety of alkylating agents, such as chloromethoxy ethane, **10**, 1-chloro-2-ethylsulfanyl ethane (a close analog of sulfur mustard gas), **11**, and benzyl bromide, **12**. In contrast, exposure of the OFET to vapors of non reactive materials that are incapable of forming covalent bonds with co-polymer **3**, such as methanol, **13**, acetone, **14**, water, **15**, and isopropanol, **16**, resulted with the deterioration of the output and transfer characteristics, as can be seen in Figure 2. A similar behavior of deterioration of the field effect characteristics could also be observed in OFETs lacking the ability to chemically react with alkylating agents such as OFETs having region regular poly-3-hexyl thiophene, **rr-poly-2**, as the active channel.

The effect of alkylation on FETs having co-polymer **3** as the channel can be rationalized by following the effect of alkylation on the charge distribution in the polymer. Co-polymer **3** is composed of electron rich polythiophene skeleton with pendant pyridine groups. Upon reacting with an alkylating agent, the pyridine group is transformed into an electron deficient pyridinium group. These electron poor groups attract charge density from the polythiophene skeleton, thus partially doping it.

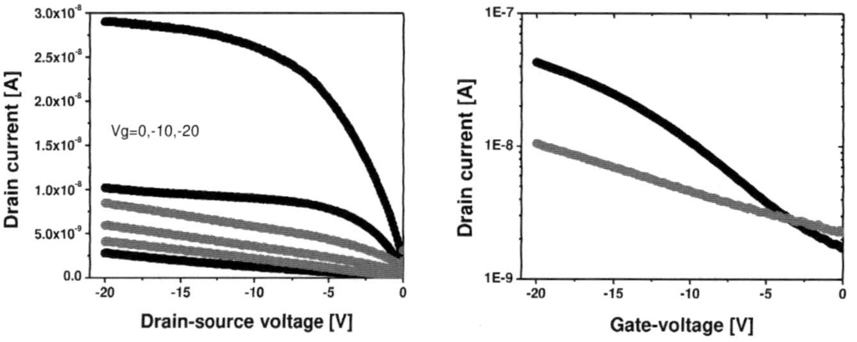

Figure 2: Output (a) and transfer (b) characteristics of OFET having co-polymer **3** as the channel. The black lines show the device response at ambient conditions, while the gray lines show the response after exposure to methanol, **13**.

As indicated above, the OFET readily reacts with any alkylating agent, producing the same type of reaction and thus cannot provide any straightforward information regarding the identity of the reacting alkylator.

Nevertheless, the OFET can provide valuable information regarding the reactivity of the alkylator. Figure 3 depicts the drain-source current as a function of the exposure time to vapors of methyl iodide, **9**, at a constant gate-source voltage of -10V. When methyl iodide vapors are introduced to the atmosphere surrounding the OFET (marked with an arrow) the drain-source current drops, indicating the process of absorption of the vapors by the polymer and associated changes in its morphology and packing.
The drop in the drain-source current is expected in view of the similarity between this process and what is found for non-reactive substances with OFETs containing co-polymer **3** and OFETs consisting of poly-3-hexyl thiophene, **rr-poly-2**, with any of the substances. Subsequently, the source-drain current starts increasing with time indicating that methyl iodide, **9**, starts reacting with the polymer.

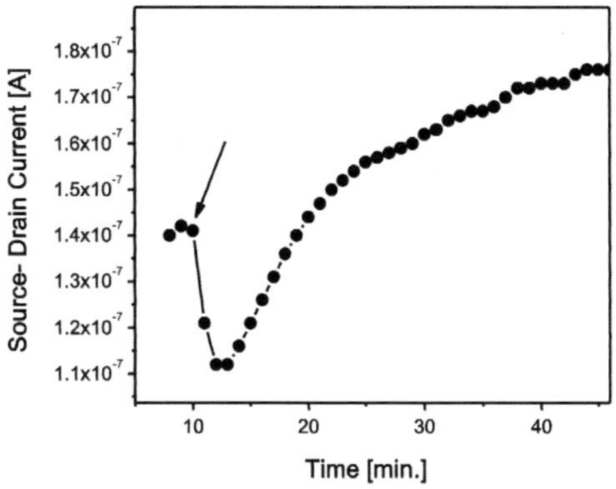

Figure 3: Time response of an OFET having co-polymer **3** as the channel. Methyl iodide, **9**, is introduced at T=10min (arrow).

2.2 Detecting alkylating agents using chemomechanical asymmetrized microcantilever systems
Alkylation of nucleophilic nitrogen derivatives is able to induce irreversible effects other than changes in conductivity. In the next stage we have explored to ability to harness chemo mechanical effects to report an alkylation reaction.

Scheme 4: Asymmetrized Si microcantilevers: a: picture of the roughened surface of the beams. b: SEM image of the intact face. c: SEM image of the roughened face. The length of the white bar is 500 nm.

For this purpose we have manufactured asymmetric silicon microcantilevers that are capable of interacting with alkylating agents on one of their faces while their other surfaces are indifferent to the presence of alkylators, Scheme 4. The microcantilever of Scheme 4 is selectively etched on one of its faces according to a previously published procedure.[11] The asymmetrized microcantilever is then immersed in a solution of poly-4-vinylpyridine, **17**, in DMF for 20 minutes then removed from solution and dried. The asymmetrized microcantilever was exposed to vapors of the analyte and its guest induced bending was studied in a previously described system.[12]

Figure 4 depicts the time dependent bending curves of the same microcantilever first exposed to vapors of methanol, **11**, in a nitrogen atmosphere then to a pure nitrogen atmosphere, then to a nitrogen atmosphere containing vapors of methyl iodide, 9, then to a pure nitrogen atmosphere again.

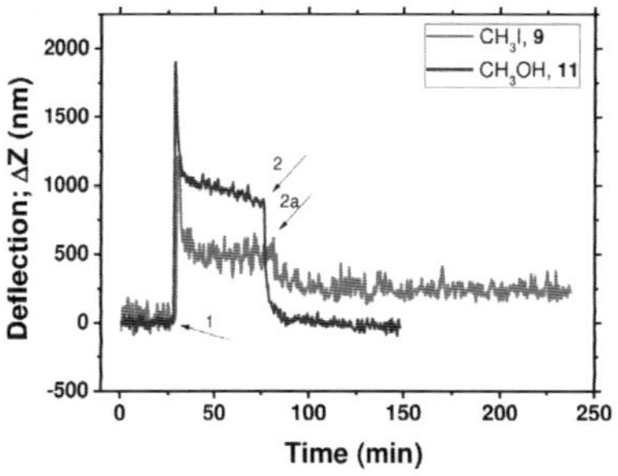

Figure 4: Time dependent guest induced bending of a poly-4-vinylpyridine coated Octensis microcantilever in the presence of methanol, **11**, then methyl iodide, **9**. Arrow 1 denotes start of guest flow, arrows 2 and 2a denote end of guest flow.

As can be clearly seen in Figure 4 the absorption of methanol, **11**, to the polymer coated microcantilever is fully reversible. In the presence of the guest the microcantilever bends while in the absence of the guest the polymer bound methanol molecules are desorbed and the beam returns to its original bending. In contrast, upon exposure of the same microcantilever to methyl iodide, **9**, a reaction occurs between the polymer and the guest. This reaction links the guest to the polymer through a new and stable bond. This cannot be reverted at room temperature and the deflection is partially irreversible, probably due to the desorption of some fraction of unreacted guest molecules.
The magnitude of bending depends on the nature of the alkylating agent, on its concentration and on the exposure time. Efforts are currently directed towards the identification of the alkylator through different properties of the bendogrammes.

3. Conclusions:
Organic field effect transistors and microcantilevers were harnessed to the detection of alkylating agents. Reaction of gas phase alkylators with a nucleophilic organic semiconductor in an OFET structure yields dramatic and irreversible changes in the transistor properties while other innocent molecules inflict reversible changes to the OFET parameters. In a similar manner, gas phase alkylators that react with a nucleophulic polymer at the surface of a microcantilever induces its irreversible bending while for innocent materials, such as alcohols, the bending is reversible and depends on the momentary concentration of the guest in the atmosphere.

[1] Yagi K, Williams J, Wang N Y and Cicerone R J 1993 *Proc. Natl. Acad. Sci. USA* 90 8420.

[2] Tomasz M and Palom Y 1997 *Pharmacol. Ther.* 76 73.

[3] Daniel F B, Schenck K M, Mattox J K, Lin E L C, Haas D L and Pereira M A 1986 *Fund. Appl. Tox.* 6 447.

[4] Burke R 2007 Counter-Terrorism for Emergency Responders (2nd edition Taylor & Francis Group) p. 686.

[5] Bolt H M, Laib R J, Peter H and Ottenwälder H 1986 *J. Cancer Res. Clin. Oncol.* 112 92

[6] Hertzog-Ronen C, Borzin E, Gerchikov Y, Tessler N, Eichen Y 2009 *Chem. Eur. J.* 15, 10380 – 10386.

[7] Smith P A, Lepage C R J, Koch D, Wyatt H D M, Hook G L, Betsinger G, Erickson R P and Eckenrode B A 2004 *Tren. Anal. Chem.* 23 296.

[8] a) Borzin E, Shemesh A, Hertzog-Ronen C, Gerchikov Y, Tessler N and Eichen Y 2010 *J. Phys. Org. Chem.* 23 1108. b) Tal S, Salman H, Abraham Y, Botoshansky M, Eichen Y, 2006 Chem. Eur. J., 12, 4858-4864.

[9] Gannot Y, Hertzog-Ronen C, Tessler N, Eichen Y, 2010 *Adv. Funct. Mat.* 20, 105-110.

[10] Shaked S, Tal S, Roichman Y, Razin A, Xiao S, Eichen Y, Tessler N, 2003 *Advanced Materials* 15, 913.

[11] Stolyarova S, Cherian S, Raiteri R, Zeravik J, Skladal P and Nemirovsky Y 2008 *Sens. Actuators B: Chemical* 131 509.

[12] Shemesh A, Stolyarova S, Nemirovsky Y and Eichen Y 2011 *J. Mater. Chem.* 21 2070.

Lipid droplet detection by the cavity perturbation method

R T Blakey[1], A Mason[1], C E Rolph[2], G Bond[3] and A I Al-Shamma'a[1]

[1] School of Built Environment, Liverpool John Moores University, Liverpool L3 3AF, UK
[2] School of Pharmacy and Biomedical Sciences, University of Central Lancashire, Preston PR1 2HE, UK
[3] School of Forensic and Investigative Sciences, University of Central Lancashire, Preston PR1 2HE, UK

Email: r.t.blakey@2010.ljmu.ac.uk

Abstract. There are currently no point-of-care diagnosis strategies available to indicate the presence of neoplasmic growth. This research aims to develop a novel diagnostic strategy based on detecting TAG accumulation in cells. This element of the research is a preliminary experiment to prove the concept of detecting TAG lipid droplets in YEPD media. It was found that a change in mono-unsaturated concentration can be detected by the frequency shift in a resonant cavity. The dielectric constant of TAG vegetable oils was calculated at 2.34-2.39. It was also found that concentrations of lipid droplet can be differentiated up to 5% (v/v).

1. Introduction

Despite decades of research, cancers or malignant neoplasms continue to significantly contribute to patient morbidity and mortality. Neoplasms occur when cell proliferation exceeds cell apoptosis (naturally initiated cell death) which results in the formation of a tumour or increased blood cell volume depending on the origin of the cancerous cell. Tumours can be problematic to detect in the early stages due to symptoms not arising until later stages of development. When tumours are detected in earlier stages treatments become less invasive, simpler and cheaper resulting in increased patient comfort and prognosis. Screening programs have been extremely successful in lowering the incidence of late stage cancers (e.g. breast [1], cervical [2]) but are limited to a specific type and site of cancer also requiring specialised facilities, staff and transport (i.e. laboratories). Currently there are no quantitative or qualitative point-of-care diagnostic assays available to indicate neoplasmic tissue. This communication proposes a novel diagnostic strategy that can detect a tumour in the body from any origin.

This is proposed through the measurement of triacylglycerol (TAG) stores within cancerous cells and specialised cells of the immune system called dendritic cells. Under normal circumstances, most cells store small amounts of TAGs in organelles called lipid droplets (lipid bodies, adiposomes) and are used to synthesise lipid membranes, other molecules of the lipid family and as a source of

[1] r.t.blakey@2010.ljmu.ac.uk

Published under licence by IOP Publishing Ltd

chemical energy. However, due to cellular mutations and/or abnormal cell to cell signalling, the metabolism of the cell is altered resulting in TAGs being accumulated in such lipid droplets. Depending on the mutations that lead to the neoplasm, a lipid droplet can constitute a significant sum of cell volume [3]. Hence, the cell concentration of TAG is an excellent indicator or biomarker of cell metabolism and proliferative state. Furthermore, excess TAGs released into circulation are "mopped up" by cells of the immune system. Dendritic cells are specialised cells of the immune system which scavenge for non-self or otherwise foreign molecules and present digested material to other cells of the immune system (T-cells). Recent research has found that dendritic cells from tumour bearers accumulate TAG due to exacerbated macrophage scavenger receptor 1(MSR1) mediated uptake [4]. This relationship is complex and it is unclear whether TAG accumulation inhibits the immunogenic actions of dendritic cells facilitating the growth of tumours or the result of a tumour being present. Dendritic cells have the advantage of being present in the blood stream while migrating to specific sites around the body allowing cells to be easily isolated from the body (e.g. blood sample) to be characterised, removing the need for invasive diagnostic strategies such as surgery or biopsy. Lipid droplet accumulation also occurs in a range of cells specifically related to conditions of infection and inflammation [3].

Many different impedance or dielectric spectroscopy techniques have been developed in an array of cancer diagnostic applications and show promising results. Approaches vary depending on the site and type of cancer from characterising isolated individual cells to non-invasive *in-vivo* imaging techniques [5-7]. Dielectric spectroscopy analyses of cellular material are based on dielectric properties at different frequencies. Biological cells exhibit a characteristic decrease in permittivity as a function of increasing frequency known as α, β, δ and γ dispersions [8]. The aforementioned review is particularly useful to the biologist and engineer alike who may not be familiar with electromagnetic biological cell interactions.

The following experiment provides a simple preliminary proof of concept trial to determine if different conformations of TAGs and different concentrations of TAG droplets can be differentiated based on their dielectric properties at microwave frequencies. This is calculated using the small perturbation method whereby a sample introduced to a resonating cavity shifts the frequency of resonant modes and decreases the quality factor dependant on the dielectric characteristics of the sample under test [9-11]. Resonant based sensors have proved to be extremely efficient in detecting the small changes in the biochemical make up and biological characteristics of a cell which can be calculated based on the measured electromagnetic properties of the cell sample [12].

2. Methods

2.1. Chemicals and reagents

Yeast extract, bacto-peptone, dextrose (YEPD) were purchased from Sigma-Aldrich UK. Grapeseed, Groundnut, Olive and Sunflower Oils were purchased as non-branded products from a well known U.K. supermarket.

2.2. Preparation of TAG and TAG lipid droplet samples

Samples of 100% grapeseed, groundnut, olive and sunflower oil were prepared in sample tubes ready for dielectric measurement. Secondly, olive oil was suspended in YEPD media to yield concentrations of 50, 10, 5, 1, and 0.5 (% v/v) oil. YEPD was prepared to the following protocol: 1 % (w/v) of yeast extract, 2 % (w/v) of bacto-peptone and 2 % (w/v) of dextrose diluted in distilled water. Before measurement, the tube to be sampled would be vortex until a homologous suspension was achieved.

2.3. Dielectric measurement and calculation

Measurements were performed using an Agilent Technologies (Hewlett Packard) 8720 ET Vector Network Analyser (VNA). The instrument was set to generate a signal between 1.5-1.9 GHz over 1601 data points and calculate the S_{21} (transmission) parameters. Measurements were made through a custom cylindrical resonating cavity, designed and constructed in house. The cavity was designed to resonate at 1.7 GHz at mode TM_{010}.

The dielectric constant, dielectric loss and loss tangent of the sample were then calculated using the following equations [11].

$$\varepsilon' = \frac{V_c\,(f_c - f_s)}{2V_s\,f_s} + 1 \tag{1}$$

$$\varepsilon'' = \frac{V_c}{4V_s}\left(\frac{1}{Q_s} - \frac{1}{Q_c}\right) \tag{2}$$

$$\tan\delta = \frac{\varepsilon''}{\varepsilon'} \tag{3}$$

where ε' and ε" is the dielectric constant and dielectric loss respectively, f_c and f_s are the resonant frequencies of the empty cavity and loaded cavity respectively, V_s and V_c are the volumes of the sample and cavity respectively and Q_c and Q_s are the quality factors of the empty cavity and the loaded cavity respectively.

3. Results

Figure 1 shows the S_{21} parameters for the vegetable oil standards and empty cavity as a function of frequency illustrating frequency perturbation for groundnut oil of 17.1 MHz, olive oil of 17.4 MHz, sunflower oil of 17.5 MHz and grapeseed oil by 17.7 MHz to 1.69 GHz.

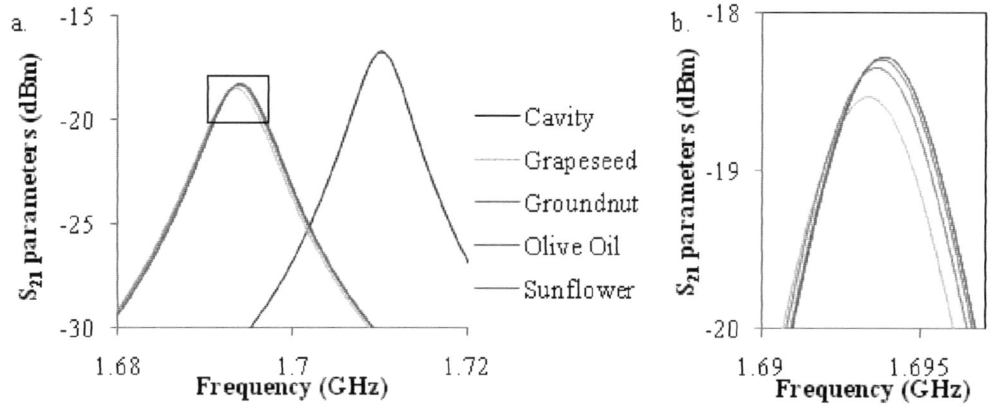

Figure 1a. S_{21} parameters of TAG samples (magnified in figure 1b.) and the unloaded cavity. S_{21} parameter frequency spans from 1.68 to 1.72 GHz.

Figure 2 shows the S_{21} parameters of differing concentrations of Olive Oil suspended in YEPD. It can be seen that as the concentration of olive oil decreases, the resonant frequency of the sample decreases or shifts to the left away from the resonant frequency of the empty cavity. Results can be seen in table 1. The total frequency shift between the empty cavity and 0.5% oil is 135 MHz to 1.576 GHz (Peak E). Peak D and E consist of two curves on the below graph, 10% -5% and 1%-0.5% respectively. The quality factor of the detected parameters decreases as YEPD media is introduced into the sample.

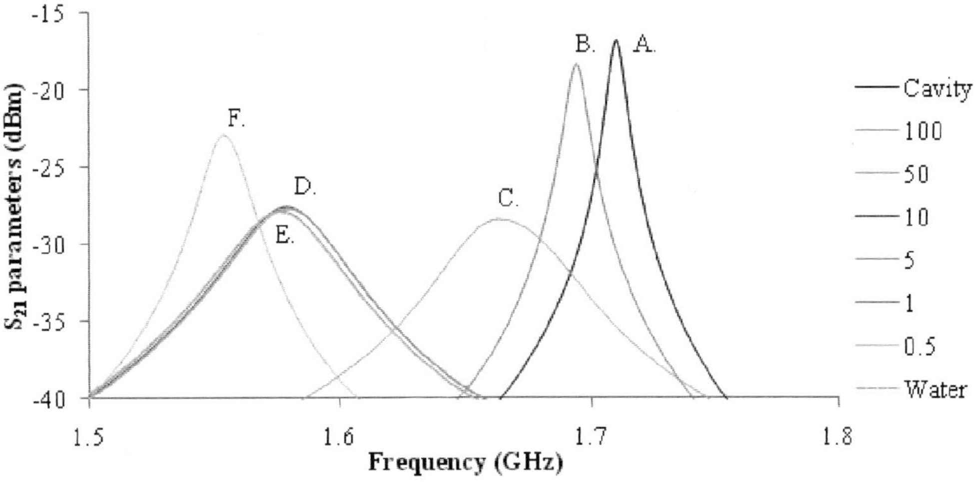

Figure 2. S_{21} parameters of lipid droplets present in differing concentrations (olive oil % v/v). Peaks from right to left are A: Cavity. B: 100%. C: 50%. D: 10% and 5%. E: 1% and 0.5%. F: Water.

3.1 Dielectric calculations

The dielectric constant, dielectric loss and loss tangent of the samples were calculated using (1-3). The same parameters were also calculated for the TAG-YEPD suspensions. The results can be seen below in table 1. It can be seen that grapeseed and sunflower oil have the greatest shift in frequency resulting in a higher dielectric constant. This is due to the levels of unsaturated acyl chains present on TAG molecules that make up the oil. Groundnut and Olive Oil are composed of approximately 53% (w/w) and 72% (w/w) monounsaturated fatty acids respectively as compared to 21% (w/w) and 16% (w/w) of sunflower and grapeseed oil. This changes the dielectric characteristics of the sample resulting in a frequency shift. This is in agreement with previous research that states the dielectric constant of TAGs increases as the level of unsaturated fatty acids (FA) increases [13]. The Q-factor of the oil samples decreases indicating an increase in dielectric loss.

Table 1. Calculated dielectric properties of oil samples and suspensions.

	Frequency shift (MHz)	Q-factor	Dielectric constant	Dielectric loss	Loss tangent
TAG					
Grapeseed	17.7	199.21	2.387	0.100	0.003
Groundnut	17.1	199.28	2.340	0.100	0.004
Olive	17.4	199.25	2.364	0.100	0.004
Sunflower	17.5	201.61	2.372	0.006	0.004
Olive Oil (% v/v)					
100	17.0	199.25	2.364	0.100	0.003
50	47.0	35.03	4.749	1.572	0.331
10	132.5	38.04	12.142	1.422	0.117
5	132.0	38.99	12.096	1.380	0.114
1	134.5	38.93	12.324	1.382	0.112
0.5	135.0	39.40	12.370	1.362	0.110

The dielectric constant of the oil suspensions increases as the concentration of YEPD increases. This is due to the water content of the media shifting the frequency closer to that of the water standard. Dielectric losses also increase as YEPD is introduced. This is due to the ion content of the media resulting in energy being lost as dielectric heating.

4. Conclusions and Discussion

This experiment has proved that differing conformations of lipids and different concentrations of lipid droplets can be distinguished based on the electromagnetic properties of the suspensions which are apparent in the frequency shift of the samples at this frequency range. However, limitations include the sample size being too large for this method. Accuracy of the technique increases as the sample size decreases. The technique also has the drawback of being inaccurate on low loss samples [11]. The current resonator and sample volumes are sensitive enough to detect lipid droplet concentration up to 5% (v/v). Further modification of the resonator cavity used to detect the lipid droplets will positively advance the accuracy of the calculations involved with this method. The homogeneity of the sample may also have introduced errors. As the sample is introduced to the cavity the TAG droplets begin to converge into larger droplets until the TAG begins to separate into a different phase resting on the YEPD media. This results in the lipid droplet concentration being measured actually being lower than the newly vortex suspension. Stabilisation of the lipid droplets using phospholipids would improve the homogeneity of the suspensions improving measurement accuracy.

The next stage of the research will look to detect lipid droplets within living yeast cell cultures. A number of complex theoretical models exist that can calculate the size and shape of the cell and composition of the cytoplasm including organelles from the dielectric characteristics of a sample. Such models will prove to be crucial in the correct determination of TAG accumulation within living cells.

References

[1] Wyld L and Ingham C E 2007 Screening of the population for breast cancer *Surgery* **25** 254–6
[2] Peto J, Gilham C, Fletcher O and Matthews F E 2004 The cervical cancer epidemic that screening has prevented in the UK *Lancet.* **364** 249–56
[3] Bozza P T and Viola J P B 2010 Lipid Droplets in Inflammation and Cancer *Prostag. Leukotr. Ess.* **82** 243–50
[4] Herber D L *et al.* 2010 Lipid accumulation and dendritic cell dysfunction in cancer *Nat Med* **16** 880–6
[5] Sun T P *et al.* 2010 The use of bioimpedance in the detection/screening of tongue cancer *Cancer Epidemiol.* **34** 207–11
[6] Kerhet A, Raffetto M, Boni A and Massa A 2006 A SVM approach to microwave breast cancer detection *Eng. Appl. Artif. Intel.* **19** 807–18
[7] Bellorofonte C, Vedruccio C, Tombolini P, Ruoppolo M and Tubaro A 2005 Non-invasive detection of Prostate Cancer by Electromagnetic Interaction. *Eur. Urol.* **47** 29–37
[8] Markx G H and Davey C L 1999 The dielectric properties of cell at radiofrequencies: Applications in biotechnology *Enzyme Microb. Tech.* **25** 161–71
[9] Montgomery G A 1947 *Techniques of microwave measurements* (New York, NY: McGraw-Hill)
[10] Murthy V R K and Raman R 1989 A method for the evaluation of microwave dielectric and magnetic parameters using rectangular cavity perturbation technique *Solid State Commun.* **70** 847–50
[11] Sheen J 2005 Study of microwave dielectric properties measurements by various resonance techniques *Measurement* **37** 123–30
[12] Fang Y, Ferrie A M, Fontaine N H, Mauro J and Balakrishnan J 2006 Resonant Waveguide Grating Biosensor for Living Cell Sensing *Biophys. J.* **91** 1925–40
[13] Rudan-Tasič D and Klofutar C 1999 Characteristics of vegetable oils of some Slovene manufactures *Acta Chim.* **46** 511–21

An integrated hybrid system for genetic analysis combining EWOD sample preparation and magnetic detection.

Des Brennan[1], Dorothee Jary[2], Christine Peponnet [2], Filipe Cardosa [3], Paolo Freitas[3], Mihai Dinca[1], Margaret Aherne[1], Paul Galvin[1].

[1] Tyndall National Institute, University College, Cork, Lee Maltings, Prospect Row, Cork, Ireland.

[2] CEA-LETI, Department of Technology for Biology and Health, 17 Avenue Des Martyrs, 38054 Grenoble, France.

[3]INESC-MN, Rua Alves Redol, 9, 1000-029 Lisbon, Portugal.

des.brennan@tyndall.ie

Abstract. Over the last decade microelectronic technologies have delivered significant advances in devices for point of care diagnostics. Complex microfluidic systems integrate components such as valves, pumps etc. to manipulate liquids. In recent years, the drive is to combine biochemical protocols in a single system, delivering "sample in answer out". An Electrowetting on Dielectric (EWOD) device offers the possibility to move and manipulate 64nl volumes implementing biochemical processes, while the magnetic sensor facilitates hybridisation detection. We outline an injection molding approach where EWOD and magnetic devices are integrated into a hybrid microfluidic system with the potential to implement "sample in answer out" biological protocols.

1. Introduction

In a recent review Haeberle et al. [1] described how microfluidic platforms should implement defined unit operations (e.g. valving, pumping, mixing etc.). The system should exploit low cost well defined fabrication techniques and be capable of implementing bioassays with ease and flexibility. They reviewed a diverse range of fluidic handling approaches including wicking, electrowetting, centrifugal technologies and microdroplet systems. Simple lateral flow systems are limited in the assays they can be applied to, however emerging platforms integrating a number of technologies such as the Cepheid GeneXpert [2] implement a series of complex assays such as cell lysis and PCR amplification. Many publications have demonstrated individual biochemical protocols in lab on chip systems, but relatively few have combined sample preparation and detection in a single system, Whooley et al. [3] implemented a monolithic system with capability to amplify and detect specific DNA using capillary electrophoresis. Anderson et al. [4] described an integrated monolithic device capable of nucleic acid extraction and PCR amplification. Quake et al.[5] demonstrated a fully integrated microfluidic system implementing DNA sequencing by synthesis, discriminating up to four consecutive base pairs, suggesting this approach can be integrated with upstream protocols e.g. DNA extraction and PCR amplification. Capillary electrophoresis (CE) detection has been used extensively with integrated all silicon or glass systems, however CE systems don't lend themselves to multiplex detection a requirement for high throughput biological screening [6].

Three strategies exist for integration and packaging of hybrid microfluidic systems: (i) vertical stacking where chips are bonded to substrates containing fluidic and electrical interconnects [7], (ii) monolithic horizontal integration where all components are fabricated on a single substrate with compatible fabrication materials and processes. The third approach involves using hybrid integration as illustrated in the commercially available i-STAT blood analyser, where a polymer microfluidic cartridge is combined with an electrode substrate for electrochemical detection.

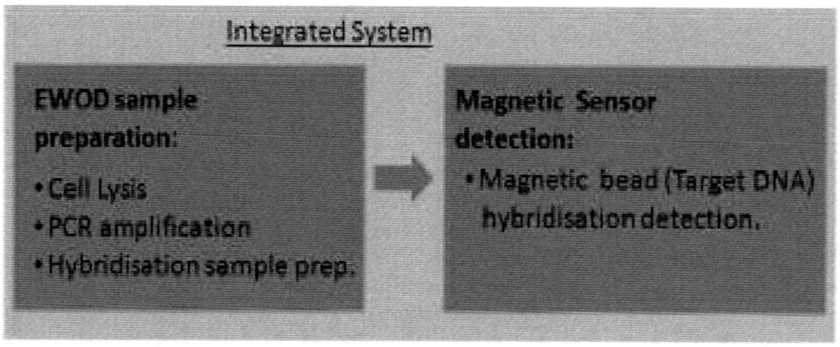

Figure 1. The functionality of the EWOD component in sample preparation and the magnetic sensor in hybridisation detection is outlined within the integrated platform.

The implementation of the biochemical protocols within the system are outlined in figure1, with the EWOD device potentially implementing cell lysis, PCR amplification and sample preparation for hybridisation detection. While the magnetic sensor array detect's mutations associated with specific diseases. The EWOD device has been used to successfully implement PCR amplification on a mutation associated with Cystic Fibrosis (CFTR ΔF508) with 200nM yield for amplicons 79bp in length, while the magnetic sensor has demonstrated hybridisation detection sensitivities in the fM target DNA range.

2. Device Integration

This work outlines a hybrid integration approach to devices fabricated in silicon and glass using injection molded polymer microfluidics. The system is designed to implement a sequence of biochemical protocols on a biological sample i.e. cell lysis, PCR amplification and hybridisation. Our approach integrates a number of devices, (i) electrowetting on dielectric (EWOD) silicon device [8] to move and mix liquid droplets, (ii) copper micro-coils on silicon [9] to manipulate magnetic beads for biomolecule capture, (iii) a magnetic sensor device [10] for mutation detection. Each device type is fabricated at wafer level, diced, packaged, interconnected and mounted on a modified printed circuit board (PCB) specific to each component. The devices are then "sandwiched" (figure 2) in a polymer (Zeonor) structure providing leak proof packaging, with horizontal microfluidic interconnects between modules. The microfluidic design contains reservoirs for reagents and reactions, with microchannels for sample input and waste. Each device is positioned and fixed within a plastic carrier substrate containing pockets, the microfluidic lid and device carrier substrate are then aligned and the system sealed.

Our integration strategy addresses key challenges including: (i) maintaining the functionality and performance of each device during the integration process and (ii) maintaining the functionality of biomolecules immobilised on the magnetic and optical devices during integration.

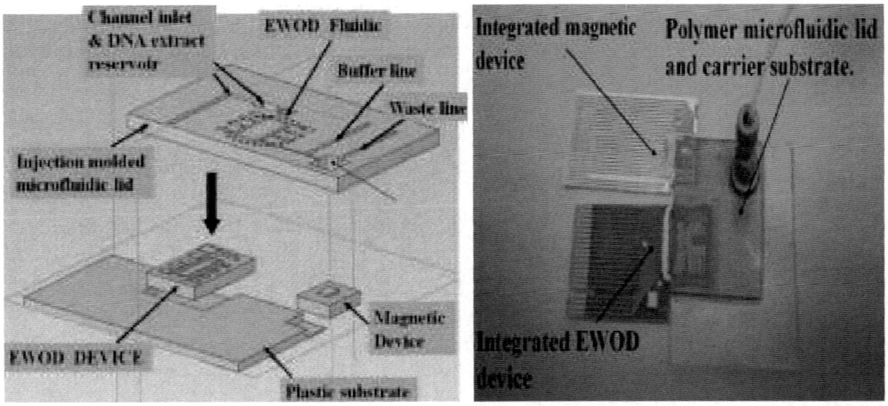

Figure 2. The concept of hybrid device integration with specific functions within a POC system is illustrated (left). The "sandwich" structure is outlined with carrier substrate and microfluidic lid contains reservoirs and interconnects between devices (right).

2.1. Micro-fluidic Lid design and fabrication

The microfluidic lid is injection molded using a Babyplast 6/10P, a low cost highthroughput fabrication technique [11], using Zeonor® (Zeon Europe GmbH), a popular thermoplastic with low auto-fluorescence used extensively to fabricate optical devices. Negative relief SU8 masters (figure 3) are defined on silicon substrates using photolithography, the master is placed in the injection mold cavity for fabrication (figure 3). This offers a low cost approach where multiple microfluidic design iterations are evaluated. The microfluidic channel depth was typically 100µm with feature width of 1mm, suitable for acetate mask photolithography. Design layout was carried out on Mentor graphics CAD package and the masks printed by JD Phototools (2000 dpi). A 100µm thick SU8 (Microchem Ltd) layer was spin coated onto a silicon wafer (Single side, 4 inch, 750µm thick). To achieve the fluidic channel a three spin process of 300rpm (60 seconds) 600rpm (30 seconds) and 1500rpm (4 seconds) is used to deposit the SU8 on the wafer surface. This is followed by a 30 minute bake at 90°C and 105°C respectively and a hard bake at 95°C. The substrate is exposed using a mask aligner (Zeiss) with the acetate mask defining the microfluidic design, following exposure the substrate is hard baked and developed using EC solvent. The substrate is finally diced into individual masters. For injection molding, Zeonor® pellets are dried in an oven for 40minutes at 50°C removing any moisture before being loaded onto the injection molder for fabrication. The system temperatures were set to, (i) Injection nozzle 195°C, (ii) injection chamber 190°C, (iii) plastic melt chamber 220°C, with injection parameters of 60bar pressure for a 3 second injection. Thermal shock and sonication in DI water was used to remove the SU8 master from the molded plastic part. After fabrication the lid was coated firstly with 180nm of Indium Tin Oxide (ITO) followed by 1000nm of SiOC. The ITO layer forms a planar ground electrode while the SiOC forms a hydrophobic layer for EWOD operation. To prevent plastic melting we repeated three ITO depositions of 60nm allowing the substrate to cool between evaporation cycles.

To fabricate the polymer carrier substrate, a brass mold was CNC machined. The injection cavity was 40mm x 45mm containing negative relief structures forming pockets in the polymer substrate within which the devices were inserted. The dimensions of the EWOD device pocket was 11mm x 12mm x 750µm, the dimension of the magnetic sensor pocket was 5mm x 6mm x 750µm, for the magnetic micro-coil pocket dimensions varied but the thickness was also 750µm. To compensate for polymer shrinkage with component cooling during fabrication, the microfluidic relief structures were 5% oversized.

Figure 3. SU8 microfluidic master structures were fabricated on a 4" silicon wafer (left) which was diced and used for component fabrication within the Babyplast injection molding system (right).

2.2. Device Preparation (EWOD, Microcoil, Magnetic Sensor)

In most biosensors a probe biomolecule is immobilised on a substrate surface to capture a specific target biomolecule of interest from a biological sample. For genetic screening assays, probe DNA is immobilised on the sensor substrate upon which a hybridisation reaction takes place. Where specific binding occurs between probe and target DNA a genetic mutation of interest is identified. Thus surface chemistry and DNA immobilisation must be carried out on the magnetic/waveguide substrate before integration with microfluidic components. Following wafer level fabrication the EWOD and magnetic sensor devices are diced and mounted on PCB's, wire bonded and encapsulated, each packaged device is then tested verifying electrical performance. In the case of the magnetic sensor chip DNA probe molecules are immobilised on the sensor surface. For the magnetic detector an array of 32 spin valve sensors screen for specific mutations of interest, a thin gold film (20nm) patterned upon each sensor acts as a base for thiol surface chemistry to which a specific combination of probe DNA is immobilised [9], DNA probes are spotted by hand using a 2µl pipette. The sensor chip is then inserted into the carrier substrate before the microfluidic cover lid is aligned and sealed, hybridisation detection takes place in a single reservoir. In a genetic assay the magnetic beads are functionalised with a surface chemistry to capture target DNA released from lysed blood cells, these then hybridise to probe DNA immobilised on magnetic sensor surfaces.

A number of bonding approaches were evaluated, UV cured glue (Norland 68) gave a robust seal but the curing process had an adverse effect on the immobilised probe DNA. Low temperature cured epoxies also gave good bonding results, but blocking of fluidic channels was an issue. We found a thermal bond process using partially cured PDMS polymer to give the best result in terms of bond performance and maintaining the functionality of the immobilised DNA probe. A 10:1 mixture of PDMS to curing agent is formulated and dispensed on the microfluidic device, it's partially cured in an oven at 75°C for 10 minutes forming a thin gasket layer. The microfluidic lid is aligned and sealed to the carrier substrate containing the EWOD and magnetic sensor device, the integrated system is placed in an oven for 30 minutes at 75°C to complete the bond process. Integration of the magnetic coil into the lid was implemented by injection molding over the device, following evaporation of ITO and SiOC the performance of the copper microcoil is maintained as illustrated by its ability to manipulate magnetic beads.

3. Integrated Component test

3.1. Integrated hardware

Following EWOD integration, tests were undertaken to verify device performance. The EWOD microfluidic reservoirs (Figure 4) were preloaded with silicone oil using a 2µl pipette, the integrated system was connected to the EWOD drive electronics. This instrumentation delivers a sinusoidal signal (120V RMS 3Khz) selectively to EWOD electrodes via a Labview™ (National Instruments) interface. As EWOD electrodes are energised the surface property at the solid/liquid interface changes from hydrophobic to hydrophilic moving droplets in a controlled fashion. Figure 4 illustrates droplets dispensed from a reservoir onto the EWOD electrode bus with test liquid TRIS buffer containing 5% TWEEN 20. The smallest size droplet dispensed and moved using the EWOD system was 64nl. Dispensing, moving and mixing reagent droplets allows biochemical protocols to be implemented within the system without the need for externally actuated pumps or valves.

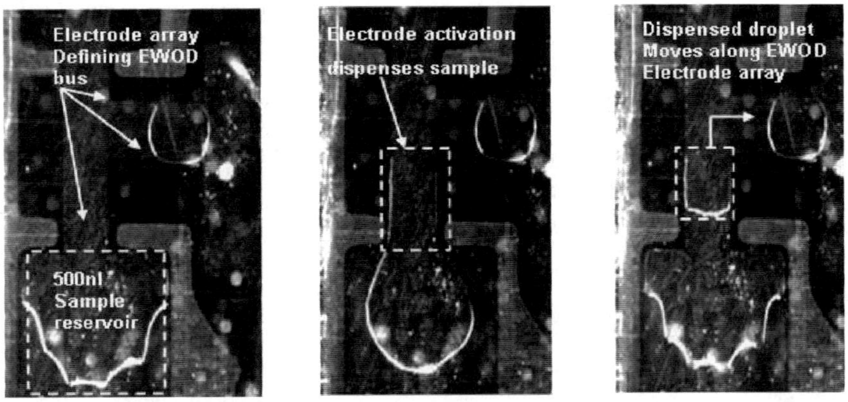

Figure 4. The integrated EWOD device was tested by dispensing and moving liquid droplets (TRIS Buffer containing 5% TWEEN 20) on the EWOD device.

The integrated magnetic sensor is positioned within a hybridisation reservoir with fluidic interconnect to the EWOD device, an external magnetic field biases the sensors generating a stable baseline signal. The magnetic sensors were biased in with a 1mA DC and the magnetic polarising field was set to 30 Oe DC and 13.5 Oe rms at 211Hz. When paramagnetic nano-particles (250nm diameter) come in close proximity to the sensor surface a change in magneto-resistance is observed. To test the integrated sensor device, a reference liquid of TRIS buffer was introduced to the sample reservoir as a blank measurement giving a constant signal over time. A solution containing magnetic nano-particles was introduced causing a sharp signal "dip", with signal change observed for all sensors, when the beads were washed away the signal returned to its original reference value in TRIS buffer (figure 5).

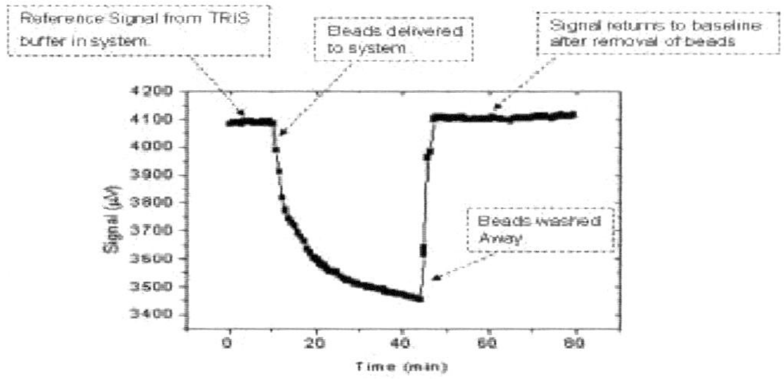

Figure 5. Signal from an integrated magnetic sensor, the plot outlines magneto-resistance over time, with the signal "dip" occurring where magnetic beads were introduced to the chamber. The signal returns to its baseline after the beads are washed away.

4. Conclusion

The goal of "Lab on Chip" systems is to implement biochemical protocols on whole biological samples, removing the need for time consuming sample preparation by skilled laboratory staff. The challenge is to combine liquid handling techniques on a platform to implement the required biochemistry, combining this with sample detection. We have integrated silicon and glass devices using a low cost, manufacturable hybrid approach. The EWOD device moves and manipulates liquid droplets with the capability to implement biochemical protocols and the magnetic sensor detects specific mutations of interest in biological samples. The performance of each device after integration has been demonstrated, maintaining functionality and biochemical viability. In future work, a full biochemical sample preparation and hybridisation protocol will be implemented on integrated platforms using biological samples (i.e. blood, saliva or urine specimens) as a demonstrator of this platform with point of care applications.

Acknowledgements

This work was financed by the EU Framework 7 programme in the Snip2Chip project (Contract:NMP4-CT-2005-01683) also supported in part through the NANOTEIRE and NBIP initiatives funded by the Irish Higher Education Authority PRTLI4 program.

References

[1] Haeberle S and Zengerle R 2007 Microfluidic platforms for lab on a chip applications *Lab chip* **7** 1094-1110
[2] www.cepheid.com.
[3] Woolley A T, Hadley D, Landre P, DeMello A J, Mathies R A and Northrup M A 1996 Functional integration of PCR amplification and capillary electrophoresis in a microfabricated DNA analysis device *Anal. Chem.* **68** 4081 – 4086
[4] Anderson R C, Su X, Bogdan G J and Fenton J 2000 A miniature integrated device for automated multistep genetic assays *Nucleic Acids Research* **15** 28-40
[5] Kartalov E P and Quake S R 2004 Microfluidic device reads up to four consecutive base pairs in DNA sequencing by synthesis *Nucleic Acids Research* **32** 2873-2879

[6] Brennan D, Justice J, Corbett B, McCarthy T and Galvin P 2009 Emerging optofluidic technologies for point-of-care genetic analysis systems: a review *Anal. Bioanal. Chem.* **395** 621–636
[7] Zbeblick M J, Anderson R, Jankowski J, Kline-Schoder B, Christel L, Miles R and Weber W

1994 Solid State Sensor and actuator workshop 251

[8] Berthier J , Clementz P, Raccurt O, Jary D, Claustre P, Peponnet C and Fouillet Y 2006 Computer aided design of an EWOD microdevice *Sensors and Actuators A: Physical* **127** 283-294

[9] H. Rostaing, H. Chetouani, M. Gheorghe and P. Galvin 2007 A micromagnetic actuator for biomolecule manipulation *Sensors and Actuators A* **135** 776–781

[10] Germano J, Martins V C, Cardoso F A, Almeida T M , Sousa L, Freitas P P and Piedade M S 2009 Portable and autonomous magnetic detection platform for biosensing *Sensors* **9** 4119-4137

[11] Becker H and Gartner C 2008 Polymer microfabrication technologies for microfluidic systems *Anal. Bioanal. Chem.* **390** 89-111

Amperometric Immunosensors for screening of Polycyclic Aromatic Hydrocarbons in water

A Ahmad[1,2], A Paschero[1] and E Moore[1,3]

[1]Life Science Interface, Tyndall National Institute, Lee Maltings, Dyke Parade, Cork, Ireland

[2]Department of Chemical Sciences, Faculty of Science and Technology, Universiti Malaysia Terengganu, Mengabang Telipot, 21030 Kuala Terengganu, Malaysia

E-mail: azrilawani.ahmad@tyndall.ie

Abstract. An amperometric immunosensor with low limit detection was developed for the screening of polycyclic aromatic hydrocarbons (PAHs) in water. The system was based on detecting the specific substance using an immunological reaction by measuring the chemical responses to specific antibodies. An integrated biochip with a three electrode system was fabricated. Gold was used as the working electrode with platinum was used as the counter electrode. A modified Ag/AgCl reference electrode was employed to enhance the stability of the immunosensors. Indirect competition enzyme-linked immunosorbent assay (ELISA) was carried out within the electrode using alkaline phosphatase (AP) as the labelled-enzyme. The system shows acceptable reproducibility and good stability. The immunosensor exhibited a wide linear response to PAHs. A limit of detection for this sensor was in the range of 1 to 10 ng ml^{-1} in aqueous sample.

1. Introduction

The technique of immunoassay has been widely used in the field of clinical science. Recent developments however have seen its application extended to areas such as food science, health screening, determination of pollutants in environment and for illicit drug detection.

In comparison with other immunoassay detection methods, electrochemical enzyme immunoassay offers the advantages of sensitivity and low detection limit [1]. Moreover, this technique introduced the miniaturized device that allow for simple and low cost instrumentation with only small sample volume required [2]. There are a number of transducers that have been used in electrochemical detection including amperometric, potentiometric and conductometric. These are all widely used in the measurement of pollution in water.

Amperometric detection is based on the measurement of current when a potential is applied to the working and reference electrodes of the system [3]. The species of interest is either oxidised or reduced at the working electrode causing a transfer of electrons which generates a measureable current. Amperometric probes are highly suitable when oxidase or dehydrogenase enzymes are employed to generate electrooxidizable peroxide or NADH species [4].

[3] Corresponding author: eric.moore@tyndall.ie

Published under licence by IOP Publishing Ltd

immunosorbent assay (ELISA). ELISA is based on the ability of antibodies to bind specifically to target antigens for monitoring of PAHs. In ELISA, immobilization and incubation steps are applied to bind the reagents to the surface of the electrodes [5]. The binding of target antigens to antibody within the substrate will lead to enzymatic interaction with the production of a current signal.

2. Experimental

2.1 Fabrication of biochips

The fabrication of biochips was carried out at the Central Fabrication Facility in the Tyndall National Institute. The electrodes were fabricated on a 100 mm pyrex wafer using e-beam lithographic techniques for electrode patterning. The wafer underwent lift-off processes where the gold (Au) working electrode was deposited first, followed by the platinum (Pt) counter electrode and the final stage was the silver (Ag) reference electrode. To coat the Ag/Cl layer onto the reference electrode, the wafer was dipped in 50 mM ferric chloride solution for 50 seconds [6]. The metal lift-off process was carried out by immersing the electrodes in hot acetone, isopropanol and finally rinsing under running deionized water. The electrodes were packaged using flip-chip technology where epoxy resin was used to attach the electrodes to the printed-circuit board (PCB) [7]. Figure 1 shows the interface for connecting the biochip with the electrochemical instrument of PalmSens potentiostat (Palm Instruments BV, Netherlands).

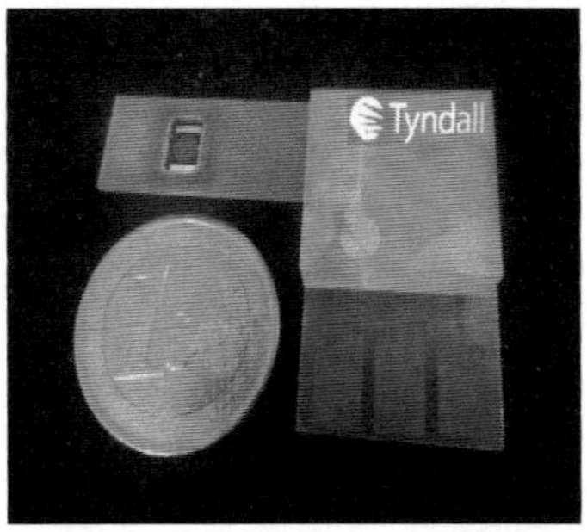

Figure 1. The image of an integrated biochip with the interface to be used with PalmSens potentiostat.

2.2 Competition Assay

The assay for amperometric detection for biochip followed the ELISA format. BSA-Pyrene coating conjugate was synthesized by coupling the adipic acid dihydrazide with bovine serum albumin (BSA), 1-ethyl-3-(3-dimethylaminopropyl) carbodiimide) (EDC) and 1-pyrene butyric acid [8]. Immobilization of antibodies was carried out directly on the electrode surface to enhance signal for maximum sensitivity. 5 µl of sample was required to completely cover the working area of the biochip. Three replicates were carried out for each concentration of the antigen (n=3). The immobilized biochip for each antibody was allowed to incubate for one hour at 37°C. The inhibition ELISA was carried out with the use of alkaline phosphatase as labelled-enzyme. Primary monoclonal 4D5 (α-pyrene) antibody was allowed to compete with a known concentration of PAHs. The enzyme used specifically converts the substrate (*para*-aminophenyl phosphate) into a detectable signal. The *p*APP is dephosphorylised to form *p*-aminophenol (*p*AP) by reaction with enzyme, which is then oxidised at 0.3 V to form the imino-quinone, thus two electrons are generated (figure 2). This substrate was chosen rather than other common substrates as the product of *para*-aminophenol can be oxidized at much less positive potential than the nitrophenol, an enzymatic product from nitrophenyl phosphate which would usually be used as an AP substrate for spectrophotometric detection [9].

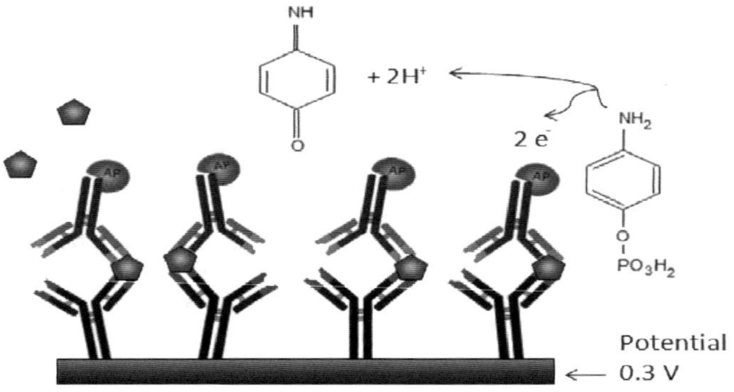

Figure 2. Competition assay format with electrochemical reaction occurs at the surface of gold electrode when 0.3 V potential is applied (primary antibody; analyte; and AP-labelled antibody).

3. Results and Discussions

3.1 Scanning Electron Microscopy (SEM)

The morphology of the electrodes surface was examined by using SEM to study the stability of the electrodes. The surface of the reference electrode was covered with an AgCl layer. Different sizes of AgCl grain could be seen on the surface but very few pores are observed. The AgCl layer with the fewest pores functioned as a stable and durable Ag/AgCl reference electrode [10]. Smooth and clear surfaces were obtained for the platinum electrode and gold electrode.

(a) (b) (c)

Figure 3. SEM photographs of the surfaces of (a) Ag/AgCl reference electrode, (b) gold working electrode and (c) platinum counter electrode.

3.2 Electrochemical Detection

Investigation of the stability of the reference electrode was carried out by placing the reference electrode in 1M KCl for 1 hour. The potential of the integrated Ag/AgCl reference electrode was measured against the commercial reference electrode (figure 4). The potential shifted about 20 mV after 15 minutes of operation and continued stable for 1 hour. This is acceptable when compared to the 13 mV shifted potential for glucose sensors reported by Se-Ik et al [11]. Kinetic study was tested to determine the concentration of pAPP substrate that would yield a signal of 95 % of I_{max}. Figure 5 represents the Michaelis-Menten plot for different concentrations of pAPP in reaction with AP labelled. The K_m for the pAPP substrates was 1.7 mg/ml and the corresponding concentration of substrate at 95 % was 2.4 mg/ml (figure 5).

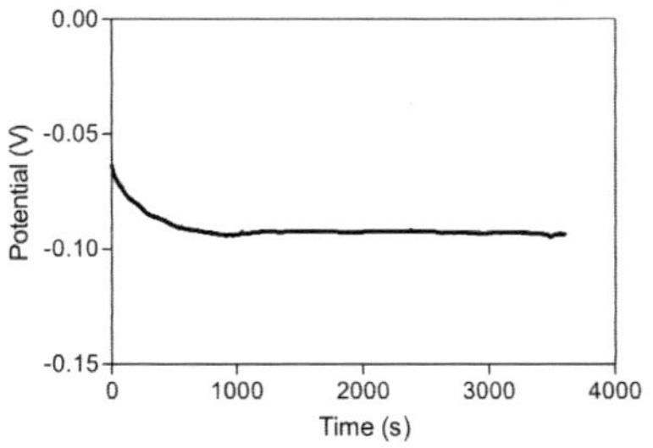

Figure 4. Potentiometric detection of on-chip Ag/AgCl versus commercial reference electrode in 1M KCl.

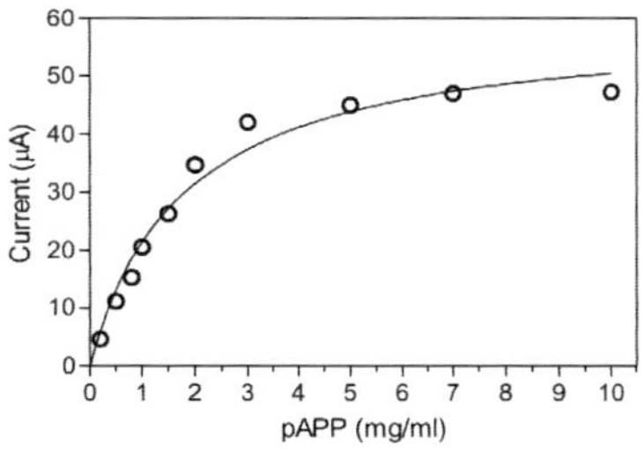

Figure 5. Michaelis-Menten plot for different concentrations of the substrate pAPP.

The amperometric detection of the Ag-Ab immobilized biochip was measured by detecting the amount of the antibodies that react with the benzo[a]pyrene. The detection time was set to 60 seconds where *p*APP was injected at 40 seconds when the solution reached equilibrium. The scan rate applied was 20 mV s^{-1}. The observed trends are illustrated in figure 6.

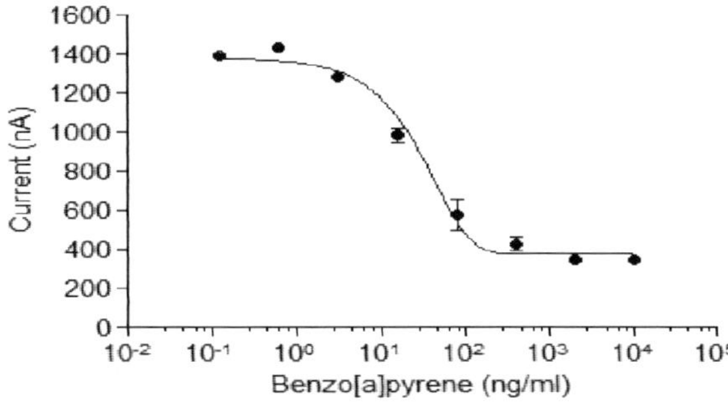

Figure 6. Benzo[a]pyrene indirect competition for 4D5 monoclonal antibodies.

The limit of detection (LOD) for this sensor was found to be 5.6 ng ml^{-1} with an IC_{50} value of 9.82 ng ml^{-1}. This sensor allowed detection in a linear range from 6 to120 ng ml^{-1}. The R^2 value of 0.9589 was lower than that reported by other investigators but is still within acceptable limits [12]. The future studies will be performed in order to enhance the sensitivity of the sensor by modifying the surface of the gold working electrode.

4. Conclusions

A portable amperometric immunosensor based on an integrated three electrode system was developed. This system provides fast response times, on-field measurement and requires only small volumes of sample. Further investigation will focus on modification of surface electrodes to enhance the binding of antigen to antibodies thus improving the sensitivity of the sensor towards PAHs. Analysis on real sample water will be performed using this immunosensor.

References

[1] Diaz-Gonzalez M, Gonzalez-Garcia M B and Costa-Garcia A 2005 *Electroanalysis* **17** 1901
[2] Wang J 2002 *TrAC Trends in Analytical Chemistry* **21** 226
[3] Monroe D 1990 *Critical Reviews in Clinical Laboratory Sciences* **28** 1
[4] Wang J 2000 *Analytical Electrochemistry* (New York, John Wiley & Sons)
[5] Moore E J, Kreuzer M P, Pravda M and Guilbault G G 2004 *Electroanalysis* **16** 1653
[6] Polk B J, Stelzenmuller A, Mijares G, MacCrehan W and Gaitan M 2006 *Sensors and Actuators B: Chemical* **114** 239
[7] Moore E, Rawley O, Wood T and Galvin P 2009 *Sensors and Actuators B: Chemical* **139** 187
[8] Li K, Chen R, Zhao B, Liu M, Karu A E, Roberts V A and Li Q X *Analytical Chemistry* **71** 302
[9] Duan C and Meyerhoff M E 1994 *Analytical Chemistry* **66** 1369
[10] Kim H R, Kim Y D, Kim K I, Shim J H, Nam H and Kang B K 2004 *Sensors and Actuators B: Chemical* **97** 348
[11] Se-Ik P, Sang Beom J, Sejin P, Hee Chan K and Kim S J *Sensors Journal IEEE* **3** 267
[12] Fahnrich K A, Pravda M and Guilbault G G 2003 *Biosensors and Bioelectronics* **18** 73

Sensors & their Applications XVI IOP Publishing
Journal of Physics: Conference Series **307** (2011) 012026 doi:10.1088/1742-6596/307/1/012026

Multi-parameter system for marine environmental applications

K.Twomey[1], W.Moujahid, V.I. Ogurtsov, G. Herzog
Tyndall National Institute, Lee Maltings, University College Cork, Ireland

E-mail: karen.twomey@tyndall.ie

Abstract. The contamination of sea waters by pollutants is a subject of growing public concern. The existing monitoring process is costing an estimated 350 million Euros/year in the EU. Most of the detection methods are laboratory-based, which require extensive sample preparation prior to analysis and, hence are not suitable for on-site analysis. Here a multi-sensor system is described for eventual deployment in a port. The system can provide an accurate assessment of the water quality through monitoring a variety of different parameters. Copper detection, conductivity measurements and redox potential measurements are described, and the stability of each sensor in artificial seawater is estimated.

1.Introduction

The established methods for the detection of pollutants in waters are based on sampling and analysis of discrete water samples. The analysis is performed in laboratories located remotely away from the sampling sites and frequently the chemical analysis is carried out with expensive apparatus, such as bench-top spectrometers and chromatographs. The issue with these methods is that, although they function well and provide reliable chemical data, they are laboratory-based, personnel-dependent, time-consuming and expensive. However an emerging research trend in the last 1-2 decades has been the growing interest in research and development of chemical sensors and biosensors which can meet the monitoring needs in polluted water at the point of need. Electrochemical techniques are low-cost, sensitive and both instrumentation and sensors can easily be miniaturised [1-3] which facilitates the development of portable analytical systems. Moreover, by using microelectrode arrays in such systems, it is possible to detect low levels of pollutants without special procedures for sample preparation [4].

The described system includes three types of electrochemical sensors. The first type of sensor relates to detection of chemical pollutants. They are based on amperometric / voltammetric measurements and used for the detection of heavy metals, phenols and dissolved oxygen. The remaining sensors are concerned with monitoring of physical parameters of the seawater including conductivity and the oxido-reduction potential. An impedimetric sensor is used for theseawater conductivity measurements and potentiometric sensor is used for potentiometric measurements of the oxido-reduction potential. All of these sensors have a fast response time which is suitable for a rapidly changing water environment. Integration of all these sensors in the complete system provides not only manifold environmental control of seawater quality but allows also application of data fusion approaches for improving performance of the system, The sensors have initially been fabricated on separate chips; however, the final device will incorporate all devices on one silicon chip as shown in fig. 1. It consists of: working electrode arrays (noted as (1) on Fig. 1 (amperometric/voltammetric measurements,), interdigitated electrode ((2), conductivity) and Pt electrode ((3) oxido-reduction potential) and includes counter electrodes – CE to provide current to the working array and a reference electrode RE to define working array potential. Due to conditions of working in seawater all electrodes (excluding RE) made of platinum, the reference electrode made of silver | silver chloride paste.

This paper presents the detection of copper using voltammetric sensor, the conductivity of artificial seawater using impedimetric sensor and the redox potential in seawater using potentiometric sensor.

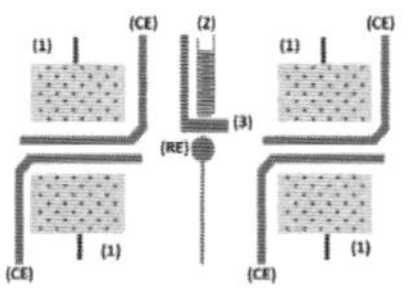

Figure 1. Targeted multi-sensor chip

2.Materials and Methods

1.1. Fabrication

Arrays of platinum disks were fabricated by standard deposition, etching and lithographic techniques used in silicon microfabrication technology [2]. The first step involved growth of a 1μm thermal silicon oxide layer on a silicon wafer. Titanium (Ti) adhesion layer and platinum (Pt) microelectrodes were then patterned on the surface by lift-off process. This was followed by plasma enhanced chemical vapour deposition of a 500 nm silicon nitride layer. A photoresist was then spun on the silicon nitride passivation layer and the recessed microelectrode arrays on the microelectrochemical cell (500nm recess depth) were then obtained using a photolithographic etch process (Pt-EKC solvent). The silicon nitride over the electrodes was then removed via an octafluorocyclobutane Deep Reactive Ion etch. Following fabrication, the wafers were diced and the electrodes packaged on printed circuit boards (PCB) by attaching the individual chips to the PCB with gold wire bonds and finally protecting the wire bonds and chips edges by covering in a polymeric selective encapsulant (Amicon 50300 HT, Emerson and Cuming) cured at 150°C for 2 hours [2]. To allow a robust connection during testing of the sensor die, a 10 way straight header from AVX interconnect was soldered to the PCB via holes. The corresponding reusable socket had leads attached which connected to the potentiostat channels.

3.RESULTS AND DISCUSSIONS

3.1 Copper detection

Copper is both an essential nutrient and a water contaminant. Its occurrence in seawater is due to natural processes as well as man-related activities such as mining and petroleum refining. Furthermore, copper has replaced tributyl tin compounds in the composition of antifouling paint. Underpotential deposition – stripping voltammetry (UPD-SV) is suitable for the detection of heavy metals in environmental samples [5]. The sensor prototype for Cu detection is shown in Figure 3. The suggested analytical procedure for the detection Cu^{2+} consists of four steps. The first step is the preconditioning of the electrode; its role is to prepare the surface before the cleaning step (second step), which ensures that all the metal ions deposited from previous experiments are dissolved. The third step is the preconcentration stage during which the deposition of metal ions from the seawater onto the surface of the electrode is occurring. Finally, the last step is the detection of the metal ions by redissolution of the metal ions into the seawater. The signal recorded during this step allows the quantification of the metal ions.

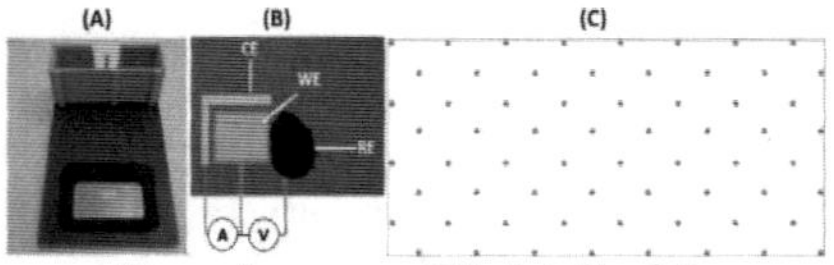

Figure 2. (A) Pictures of type 1 sensor. (B) detail of the sensor area. CE stands for counter electrode, WE: for working electrode array and RE for reference electrode. (C) detail of the working electrode array made of 64 microdiscs of 10 μm in radius arranged in a honeycomb layout and with a centre to centre separation of 200 μm.

Figure 3 shows the signal recorded during the detection step for increasing concentration of copper. The peak current was extracted from the different voltammograms and calibration curves for copper were plotted. The peak current values increase linearly with the copper concentration.

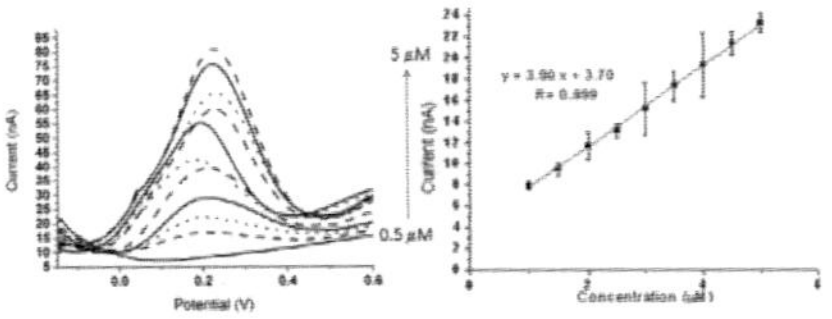

Figure 3. Underpotential deposition – stripping voltammetry and calibration curve for copper in artificial seawater

The limit of detection which was calculated from this calibration curve was estimated at 157 nM which is within the range of measured levels for copper in polluted port areas.

3.2. Stability of Voltammetric sensor

The stability of the sensors over time is essential for operation of autonomous and remote sensing systems. Stable and reliable electrodes indeed allow in-situ detection and monitoring of pollutants levels. Repetitive cyclic voltammetry of 1 mM ferricyanide in artificial seawater was investigated by scanning in the potential window from +0.6 to 0.0 V at a scan rate of 10 mVs^{-1} every hour for 72 hours. The investigated microelectrochemical cell was immersed in the artificial seawater during the whole duration of the experiment. The stability of the cell was evaluated by observing the limiting current measured, the slope of the sigmoidal wave and the variation of the current at given potentials. The limiting current density versus time is shown in Figure 4. It drops from 302.2 nA.cm^{-2} at t= 0 hour down to 270.6 nA.cm^{-2} at t= 6 hours. The limiting current density response then drifts slowly to reach 278.8 nA at t= 72 hours. The drift can be due to two factors: the impact of the artificial seawater medium or/and to electrode degradation as a result of repetitive use.

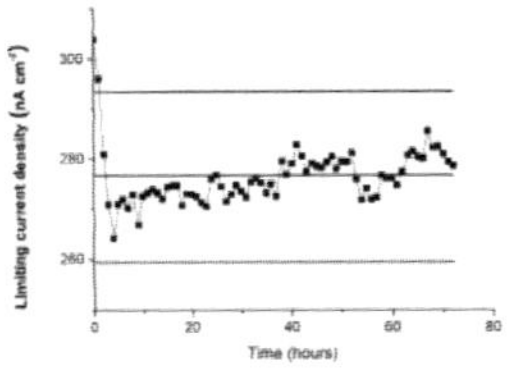

Figure 4. 73 repetitive cyclic voltammograms of 1 mM ferricyanide in artificial seawater Rest time between experiments was 1 hour. Inset: Variation of the limiting current versus time.

3.3. Conductivity Detection

The impedimetric sensor was characterized using conductivity standard solutions and artificial seawater and impedance measurements where real and imaginary impedance values were recorded over a frequency range 1MHz to 0.1Hz. Figure 5 shows a Nyquist plot of the sensor for three standard solutions with KCL concentrations 0.01M, 0.05M and 1M.

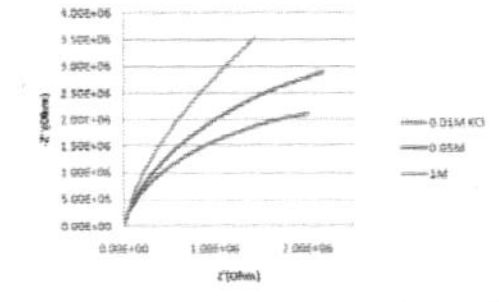

Figure 5. Nyquist plot for impedimetric sensor in different KCl solutions

3.4. Stability of the Impedimetric sensor

Short term stability in artificial seawater was determined through repeated impedance testing. The variation at increasing frequency was determined using the relation

$$Var = \frac{Z_i - Z_{avg}}{Z_{avg}} \% \qquad (1),$$

where Z_{avg} is the average impedance value. A maximum variation of 5.4% was observed at 0.1Hz.

Table 1 Variation over frequency sweep for impedimetric sensor

Variation in total impedance at 1MHz (%)	Variation in total impedance at 10KHz (%)	Variation in total impedance at 100Hz (%)	Variation in total impedance at 10Hz (%)	Variation in total impedance at 0.1Hz (%)
1.3	1.58	3.56	4.33	5.41

3.5. The Redox Potential sensor

The oxidation-reduction potential is another indicator of water quality and can provide additional information regarding the seawater pollution and be used to compensate readings from the main chemical sensors against interference.

3.6. Stability of the redox sensor

The redox sensor was used to measure the oxidation-reduction potential over 72 hours. A drift of 2.9 mV / hour was observed over the time-frame, indicating that efforts should be concentrated on the stability of the reference electrode.

4. CONCLUSIONS

This paper has described the fabrication of microelectrode arrays of microdisk and interdigitated sensors using UV photolithographic and e-beam lithographic techniques. The microdisk arrays have been characterised for copper detection with a LOD of 157nM. The stability of the arrays has been investigated through a 72 hour repetitive cyclic voltammograms of 1 mM ferricyanide in artificial seawater. A 23nA.cm^{-2} drift was recorded which could have been due to either the impact from the seawater or electrode degradation with repeated use. The stability of the conductivity sensors was determined through repeated impedance testing in artificial seawater. Dependent on the frequency conditions, a variation of 5.41% at 0.1Hz and 0.5% at 1MHz was determined. The stability of the redox sensors was measured over 72 hours and a drift of 2.9mV/hr was observed. Future efforts will focus on improving the stability of the reference electrode.

Acknowledgements

The authors gratefully acknowledge the European Commission (SHOAL) ICT- 231646 for funding of this work.

References

(1) Beni V, Ogurtsov VI, Bakunin NV, Arrigan DWM, Hill M 2005 *Anal. Chim. Acta*, **552**, 190-200.

(2) Berduque A, Lanyon YH, Beni V, Herzog G, Watson YE, Rodgers K, Stam F, Alderman J, Arrigan DWM 2007 *Talanta* **7** 1022-1030.

(3) Moujahid W, Eichelmann-Daly P, Strutwolf J, Ogurtsov VI, Herzog G, Arrigan DWM 2011 *Electroanalysis* **23** 147-155.

(4) Beni V, Arrigan DWM 2008 *Curr. Anal. Chem.* **4** 229-241.

(5) Herzog G, Arrigan DWM 2005 *Trac-Trends in Analytical Chemistry* **24** 208-217.

Journal of Physics: Conference Series **307** (2011) 012027 doi:10.1088/1742-6596/307/1/012027

Real-time Optimisation of a Microwave Plasma Gasification System

B. Kabalan, S. Wylie, A. Mason, R. Al-khaddar and A. Al-Shamma'a
RF Microwave Group, Built Environment and Sustainable Technologies Research Institute, School of Built Environment, Liverpool John Moores University, Liverpool, UK

C. Lupa, B Herbert and E Maddocks
Stopford Energy & Environment Ltd, The Gordon Manley Building, Lancaster Environment Centre, Lancaster University, Lancaster, UK.

B.kabalan@2009.ljmu.ac.uk

Abstract. A microwave plasma gasifier has been designed to produce syngas from waste. Gasification using microwave plasma has various controllable parameters to achieve optimal syngas production. These parameters include the microwave power applied, the reflected power from the microwave plasma jet, the EH tuner arm position, the gas flow and pressure, in addition to the temperature inside the gasifier. A variety of sensors are required to provide feedback and control for each of these parameters. This paper discusses the benefits of gasification, particularly via microwave plasma techniques, the first steps toward the optimisation of such a system and some preliminary results of this optimisation.

1. Introduction

The general definition of gasification is the conversion of any carbonaceous feedstock "fuel" to gaseous products using partial oxidation. The main product of this partial oxidation is synthesis gas or syngas; it consists of hydrogen (H_2) and carbon monoxide (CO). The Syngas is later cleaned-up and further processed to produce chemicals, fertilisers, liquid fuel and electricity [1]. The feedstock that is converted to syngas can be derived from hydrocarbon materials. Coal was the first feedstock to be used in the gasification process, but due to demand of sustaining the earth resources, other types of feedstock are being used; biomass is in use at the moment which includes corn stover, sawdust, wood [2], and most importantly food waste. Feedstocks undergo several chemical reactions to produce syngas, some of which are exothermic and some are endothermic [3]. Gasification technology gives the advantage of getting energy from a low cost and otherwise useless waste, and recycles it at the same time. However, the novelty of this procedure is getting maximum energy levels from the system when using minimum input (i.e. power), thus optimising this procedure would be vital for efficiency and this is the aim of our study.

Plasma is defined as the fourth state of matter. If the temperature of a solid substance increased to or above a certain degree, it eventually transforms to plasma (passing through liquid and gas); the bonds between the electrons and ions are broken in that state of matter and the gas becomes electrically conducting plasma [4]. Plasma discharge has three main categories, and can be listed as follows depending on the source used [5-6];

- DC plasma
- AC plasma
- Microwave plasma

The differences between DC – AC and microwave plasma are that both AC and DC plasma need electrodes and high power to ensure the electron jump from the cathode surface to the anode, thus ionising a gas atom while in microwave plasma the technique is different, the electric field ionises the gas atoms without the need of the electrode.

Microwaves are electromagnetic waves that have frequency range from (0.3-300) GHz [7]. In order to transmit a high frequency wave, a waveguide is needed. Waveguides are the transmission lines for microwaves and have different shapes; each one has its own characteristics and mode of propagation [8]. However, the shape used to create plasma in this paper is rectangular waveguide.

The key component of the system is the microwave plasma jet which has advantages over typical DC plasma jet. These advantages can be listed as [9]:

- The plasma jet properties and quality is stable comparing to the DC plasma, as it has no electrode and therefore no electrode vapour contamination.
- The nozzle electrode erosion is ignored as the nozzle stays relatively cold since the plasma is operated in the open air at the tip of the nozzle.
- Cheaper requirements to create the plasma; the magnetron power supply is robust and cheap (any commercial microwave oven would do).

2. Experimental System

The gasification system is shown in figure 1(left). It consists of a magnetron power supply, a circulator, a water-cooled matched load, a tuning system (E-arm and H-arm), a plasma cavity (which contains a nozzle where the plasma is created using argon) and the gas chamber. The creation of the plasma is achieved by using a microwave power source (1.1KW, 2.45 GHz) and argon gas passing through the nozzle inside the chamber. When the electric field is strong enough at the nozzle, argon molecules are ionised and Argon is turned to plasma [10]. The adjustment of the tuning system affects the electric field at the nozzle and the optimal position of the arms is when the electric field is at its maximum at the nozzle. Figure 1 (right) shows the E and the H arm of the tuner system; they are automatically controlled by a stepper motor mounted on the top of each arm.

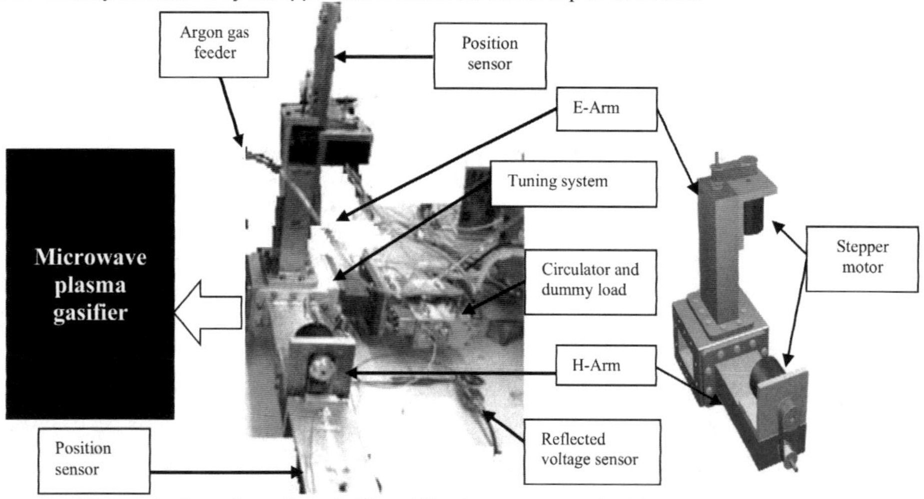

Figure 1. Overview of the (left) gasification system and (right) the tuning section.

In order to optimise the gasification procedure and increase the efficiency of the syngas production, a control and sensing system is implemented. This system's block diagram is shown in figure 2 and includes;

- Position sensor and control,
- Voltage sensor (Microwave detector diode),
- Gasmet gas analyser,
- LabVIEW interfaces and control circuit.

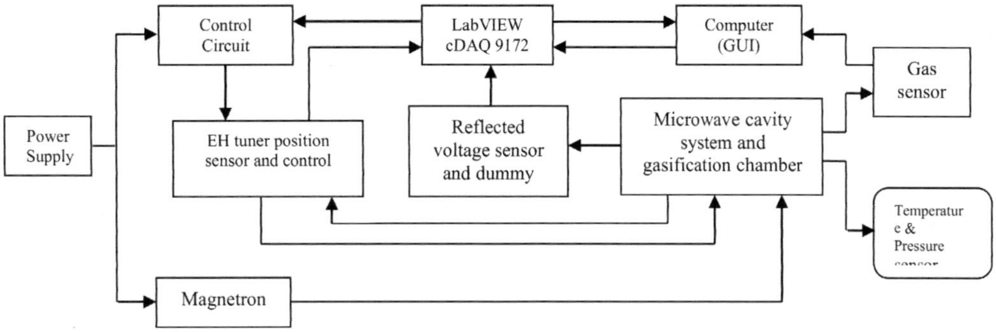

Figure 2. The control and the sensing system's block diagram.

The gasification procedure is a thermo-chemical one, thus a relatively hot microwave plasma jet is used to break the feedstock into its basic components. The plasma is created by an argon flow through a high electromagnetic field generated by the magnetron using a frequency of 2.45GHz. However, this plasma needs to be optimised in order to use its full power and generate syngas. This is achieved by the implemented sensing and control system, where the reflected power is sensed by a microwave detector (S-team diode detector) and accordingly the tuning system is adjusted to decrease this reflected power to its minimal levels. The tuning system has two arms (E and H arms), these arms control the electric and the magnetic field respectively and can be sit in a range between (0-100)mm. The tuning arms are used to adapt the impedance change when the plasma first ignites. Experiments show that once the plasma has struck, it affects the system load matching and thus the reflected power is increased. In order to keep the reflected power to a minimum level, the tuning system via the two arms plays a key role by changing the inner dimensions of the waveguide.

The generated gases from the gasified feedstock are sensed using Gasmet gas analyser (figure 3) and monitored by a LabVIEW program (figure 5) which is programmed to monitor and control the gasification procedure. The LabVIEW GUI is connected to the system via DAQ's and control circuits (figure 4). The data acquisition boards are compatible with LabVIEW programming and they feedback to the GUI the arms position, the reflected power and sends to the system control signals. The pressure inside the gas chamber is monitored by a gas sensor gauge and has a relief security valve to release the pressure when it rises above 300 mbar. Temperature effects on optimising the production of syngas, by introducing temperature sensors and pressure control, are ongoing and will be reported upon in a subsequent publication

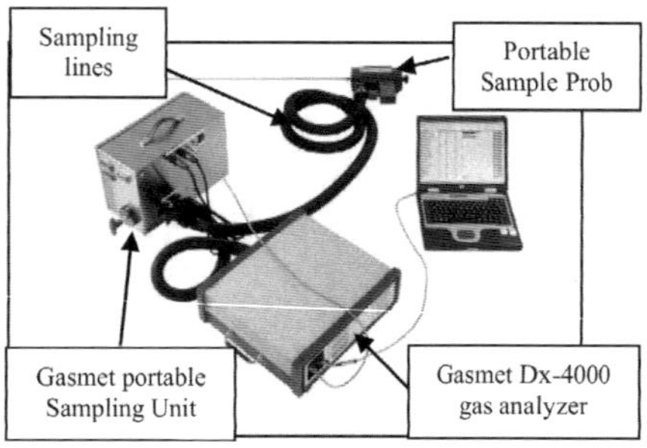

Figure 3. Gasmet gas analyser

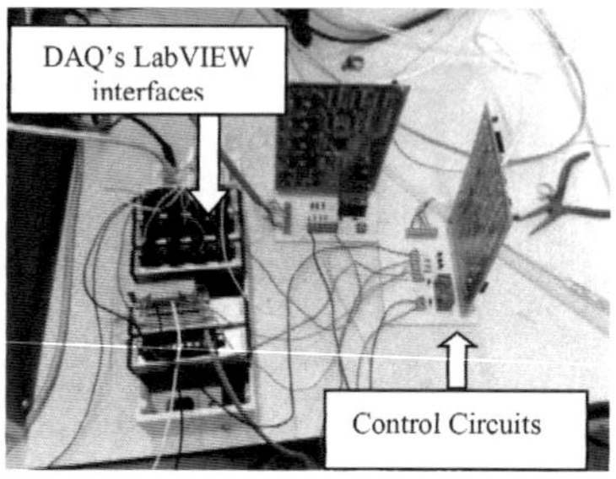

Figure 4. Control circuits

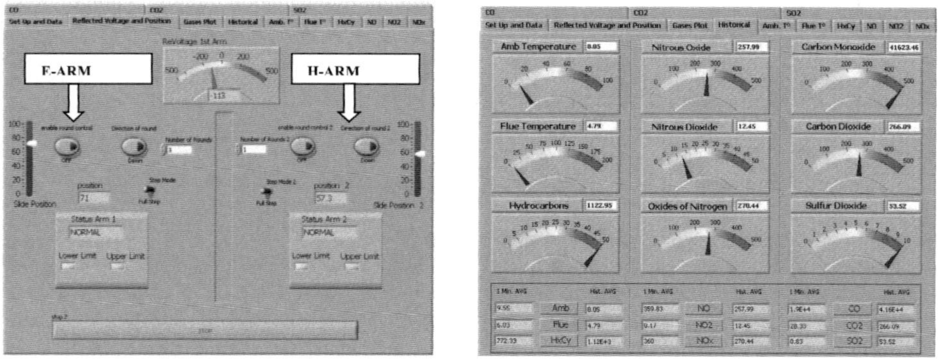

Figure 5. LabVIEW GUI for controlling and monitoring.

3. Experimental Results

Experiments were conducted using 30g pieces of dried wood and the results analysed with and without optimisation which is achieved by the control system to compare the efficiency. The pressure inside the chamber was fixed on 100 mbar and argon flow fixed on 1 L/min.

Figure 6 shows a sweep of E-arm along all possible positions to study the behaviour of the plasma. For each position of the arm, a reflected voltage is recorded and converted to reflected power. It can be noted that the plasma first strikes when the arm position is 82 mm where the reflected voltage is recorded at -360 mV giving reflected power of 4.2 dBm. Directly after the plasma is struck, the reflected power is increased and the arms start to change their position until the reflected voltage is minimised and this optimal position occurs at 54 mm with reflected voltage around -100 mV.

Figure 7 shows a study of two samples of wood, the first sample was gasified without using the optimisation system while sample 2 was gasified using the control system to track the optimal position. We studied the efficiency of the gasification procedure depending on carbon monoxide concentration and CO/CO_2 ratio in addition to the time needed for the gasification procedure. The sample gasified without automatic control took more time (460 seconds) and gave less CO percentage (1.04 %) compared to the optimised automatically controlled gasification method (table I).
The average gas concentrations using the same piece of wood are shown in figure 6. A small amount of greenhouse gases and acid rain gases (SO_2, NO_2, N_2O) are emitted using this procedure which satisfy the environmental limits and give a clean and green way to produce energy.

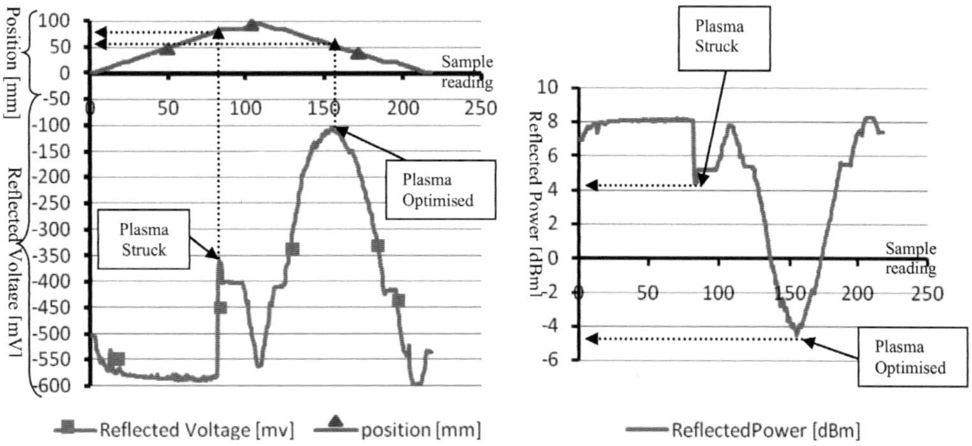

Figure 6. E-arm sweep and plasma behaviour.

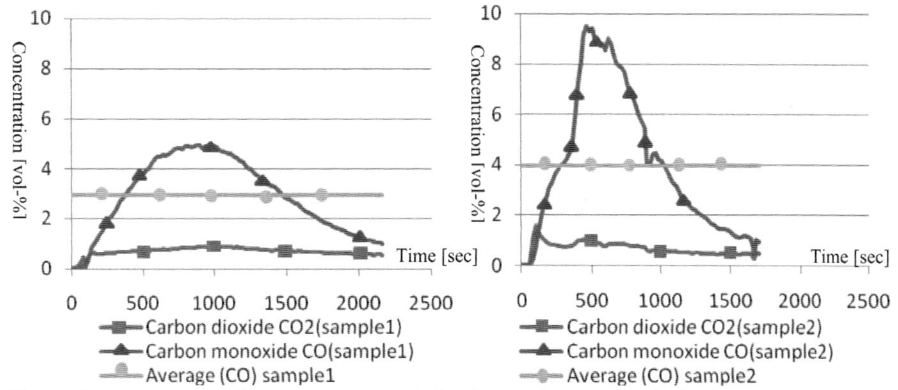

Figure 7. A comparison between a non-optimised (sample 1) and optimised (sample 2) gasification.

These figures are a clear evidence of the need and the benefit of the control and optimisation system in the gasification process. It helps in obtaining larger amounts of syngas components in shorter time, thus increases the efficiency of the process.

Sample	Condition	Weight before gasification [g]	Weight after gasification [g]	Time elapsed [s]	CO concentration [vol-%]	Reflected voltage [mV]	Reflected power [dBm]	CO/CO_2 ratio
Sample 1	No automatic control	30	11	2160	2.91	140-145	-2.5,-2.2	4.2

Sample 2	Automatic control	30	9	1700	3.95	90-95	-5.4,-5.1	5.81

Table 1. Sample comparison

4. Conclusion and Future Work

Gasification is the future green energy source as it can provide different kinds of energy products using waste feedstock, which in return would help solving the landfill problem and save the earth's resources. However, this procedure needs various controls and monitoring to achieve best results. This paper presented the sensing and the control system and the outcome gases analysis. It showed the benefit of using the control system to optimise the gasification procedure and compared the results with and without the control system. However, this system still needs some other sensing and control; a temperature sensor is still to be introduced to the system and a pressure control system would make the monitoring easier. It also needs a gas flow control as controlling argon flow affects the plasma density. Moreover, a hydrogen sensor should be attached to the system to give a full image about the syngas components. These sensors would be fitted at the next stage of the system development and a full gas analysis then can be provided.

5. References

[1] Gasification Technologies Council, *'Gasification: Redefining Clean Energy'*, available online: http://www.gasification.org/ [accessed Oct 2010]
[2] R. Mark Bricka, Dave C. Swalm, *'Energy Crop Gasification and Gasification Issues'*, School of Chemical Engineering Mississippi State University
[3] Christopher Higman, Maarten van der Burgt, *Gasification*, Gulf Professional Puplishing, second edition 2008, pp. 12-15, 33-36
[4] John Howard, *Introduction to Plasma Physics*, 2002, Australian National University
[5] A. Ganguli & R.D. Tarey, *Understanding Plasma Sources*, 2002, Current Science, Vol 83, issue 3, pp 279-290.
[6] Eric Seymour, *Preformance Considerations of High-Power AC Plasma Deposition Power Supplies*, 2004, Advanced Energy Industries.
[7] David M.Pozar, *Microwave Engineering*, 1993, Addison -Wesley publishing Company
[8] K.F. Sander & G.A.L. Reed, *Transmission and Propagation of Electromagnetic Waves*, 1986, Cambridge University Press
[9] A. Al-Shamma'a, S.R. Wylie,J. Lucas, & C.F. Pau, *Design and Contruction of a 2.45 GHz Waveguide-based Microwave Plasma Jet At Atmospherice pressure for Material Processing*, 2001, Journal of Physics : Applied Physics, vol 34, pp 2730-2735.
[10] S. Eliezer & Y. Eliezer, *The Fourth State of Matter, An Introduction of Plasma Science*, 2001, IOP Publishing Ltd.

Acknowledgments

The authors would like to thank the Technology Strategy Board, Stopford Energy & Environment Ltd and BEST Research Institute from Liverpool John Moores University for their continuous help and support.

Upload of Dead Reckoning Measurements for Improved Navigational Efficiency on Embedded Robotics

Andrew J. Tickle[1] and Paul K. Harvey

School of Electrical Engineering, Electronics and Computer Science, University of Liverpool, Liverpool L69 3GJ, United Kingdom

Email: prouction_leader@hotmail.com

Abstract. The process behind Dead Reckoning (DR) is simple in that a robot can know its current location via a record of its starting position, direction and speed without the need to look for landmarks or follow lines. This process allows a robot to drive around a known environment such as indoors and heavy urban areas where traditional GPS navigation would not be an option. Discussed in this paper is an improvement of a previously designed DR mechanism in DSP Builder where now the user enters the DR measurements and commands as a sequence via a keypad. This replaces the need for user to programme the details into the system by altering numerous value tags within the design one-by-one, thus making it more user-independent and easier to alter for different environments. The paper shows updated simulations for repeatability, how the keypad links to the system and where this work will lead.

1. Introduction

The material presented in this paper continues the work presented in [1] and uses all of the same assumptions (such as no wheel slip), the same simulated test environment (first figure in [1]) and the same robot parameters (such as the wheel diameter and step size of the motors). The way that this is achieved is by modifying the previous design so that each of the counters and comparators can be easily adapted so that their inputs can be obtained from an external source. In this case a keypad will be used to simulate the inputs to the system in order to enter the Dead Reckoning (DR) measurements and commands into the system. The design will follow the same general design methodologies and simulation procedures as presented in [2].

The main reason why DR is used is as it is a well proven method for robotic navigation when GPS or landmark based navigation is not effective, such as in indoor environments or when the robot is required to be small and compact. The mathematics behind DR differ from robot to robot due to physical parameters and the type of steering mechanism used. The two main steering types used in conventional mobile robotics are either a car-like steering system, this allows the robot to turn in a curvature of constant and known radius, whilst a differential drive robot (the case considered in this paper) can make a very narrow turn and spin on the spot. The process of knowing its current location deals with two variables: The distance travelled by the robot (x) and the angle (θ) through the directional axis of the robot, this means that the robot will travel a set distance x_1, then orientate itself by θ_1 before moving along its next distance, x_2 [3]. This process is repeated many times until the robot

[1] Andrew Tickle is now a Lecturer in Electrical Engineering at Coventry University

Published under licence by IOP Publishing Ltd

reaches its final position where it will spin on the spot and then return to the starting position by retracing its steps.

2. Modifications to the DSP Builder Dead Reckoning System

Several of the blocks from the original system will remain unchanged by these updates as they do not require any further modification. These are the independent calculator that allows the speed of the motors to be calculated by using the onboard clk from the FPGA and the Pulse Width Modulation (PWM) blocks that create the signals to be sent to the stepper motors. Therefore, the only existing block that needs modification will be the DR navigational protocols and a new keypad interface block will be added to allow the keypad data to be processed and sent to the DR block. The new Register Transfer Level (RTL) [1] diagram is shown in Fig. 1, however the data validation has been omitted on this figure as it outside the context of this work. The new keypad interface block connects the multiple signals that will send data to the counters that make up the DR navigational protocols in the form of a bus, the inputs to this block are the keypad's numerical data, stating the distance to travelled every odd value followed by a command operation for every even operation, the total number of commands and distances entered and an "OK" signal to allow the DSP Builder system know when the value has been entered. The operations are as follows; the old Movement E operation to stop the robot has now been hardwired into the code and so is not included in the list of usable operations:

- **Command 1 (Movement A):** The robot performs a point turn left.
- **Command 2 (Movement B):** The robot performs a point turn right.
- **Command 3 (Movement C):** The robot performs a spin on the spot.
- **Command 4 (Movement D):** The robot will increase its forward speed and deactivate object avoidance sensors.
- **Command 5 (Movement F):** The robot will wait a preset period of time.

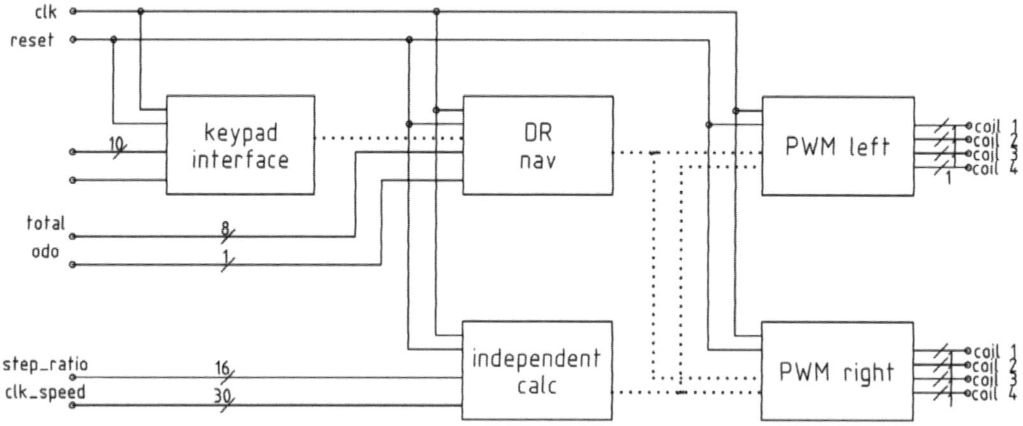

Figure 1: RTL schematic of DR system

Therefore when Command 3 is entered into the system, the robots DR sequence is complete as it will spin on the spot and travel back along the route it has just travelled. When this command is entered by the keypad it will send in the total number commands and distances into the system, this value is used to determine the return route for the turning commands as if the robot turned left at a corner before the end of the sequence, then on the return route it would have to turn right at that corner. As a result, a useful equation has been developed to perform this function:

$$\textit{Return Operation} = (\textit{Total No of Operations} - \textit{Operation No}) + \textit{Total No of Operations} \quad (1)$$

Journal of Physics: Conference Series **307** (2011) 012028 doi:10.1088/1742-6596/307/1/012028

The above equation allows the location of the inverse movement to be calculated and correctly inserted into the system. Now looking at the keypad interface block shown in Fig. 2, this is a screenshot from the DSP Builder environment but shows only a fraction of the circuitry as the whole system to too complex to understand from a single screenshot. The data from the keypad enters the system where the distance parameters shown in the top of the figure, have value tags of '1', '3', '5' etc linked to comparator blocks in order to detect these odd values from a counter which uses the "OK" signal to keep track of the entered values. If an odd match is found then this is multiplied by a co-efficient of 90.9 obtained from [1] that will convert the distance into a step pulse for these given robot parameters. This value is then held via an infinite OR-loop (an OR gate with its output feeding an input) before being output to its corresponding sequence block, this is directly fed into the comparators on the output of the counter to replace the value tag that was manually entered on the previous version. The distance inputs have a one-to-one relation with counters in the DR block as the counters will need new distance measurements and so cannot be easily reused. The operation commands on the other hand can easily be reused, this will reduce the amount of counters in the system and in the end, save resources on the FPGA for when the system is ported into VHDL for full system implementation. There are five commands blocks, each one representing an operation from the bullet points shown previously. Moving back to Fig. 2, these commands (two of which are shown) can be seen in the bottom part of the figure, these look to see if the counter produces an even result and if so, the product lets the data pass through to the next set of comparators where it searches for a match to the entered value and when one is found that output line goes high. The counter from the keypad "OK" signal is also output and sent to the DR block.

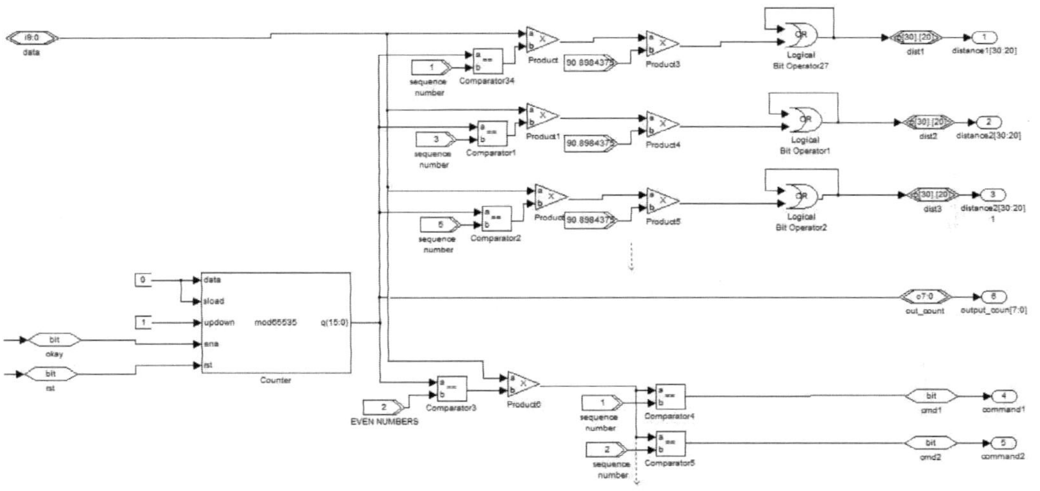

Figure 2: Gate level diagram of aspects of the keypad interface

The process by which the command operations are assigned is slightly different to that of the distances as all are used by the same operation block over and over again and values are stored in infinite OR-loops that identify the position in the sequence to be used by the master clock to activate that block when it is required in the sequence of movements. Each of these operations has the number of steps pre-coded into it via a value tag as in [1] and just uses a multi-input OR gate to send the information into the actual count mechanism of the operation. The overall process is displayed in Fig. 3(a) to avoid the confusion of a graphical approach, this can be done numerous times for different sequence positions via a multi-input OR gate and a number of infinite OR-loops retaining sequence values that feed the multi-input gate. The master clock remains largely the same as it did previously apart from now, like all the other counter blocks, is set to a very high value unlike previously and has a comparator set to $\leq$ so that it does not exceed the total number of operations. The comparator limits on the second counter have an upper limit of the total value so that after this value, the "updown" signal on the first counter will fall to '0' and it will start to count down and thus allow the robot to retrace its steps. This can be easily visualized by looking at figure 7(b) in [1]. The overall structure of the DR block can now be seen in Fig. 3(b), the blocks at the top represent the numerous number of blocks that can have distances sent to them for travelled forward, the blocks below that are the command operators, the blocks in front of them represent the selection process described in Fig. 3. The latter is also linked to some mathematics blocks that perform Equ. 1 using values from Commands 1, 2 and the total number of operations so that the reverse sequence can be completed. Finally, like in the previous system, the master counters send and obtain data from all counters in the sequence to ensure the sequence of events is kept, the reverse turn operations have two signals from the master counters as one is from the second counter for when the values are over the total in order to override the original command and make it turn the opposite way on the return journey.

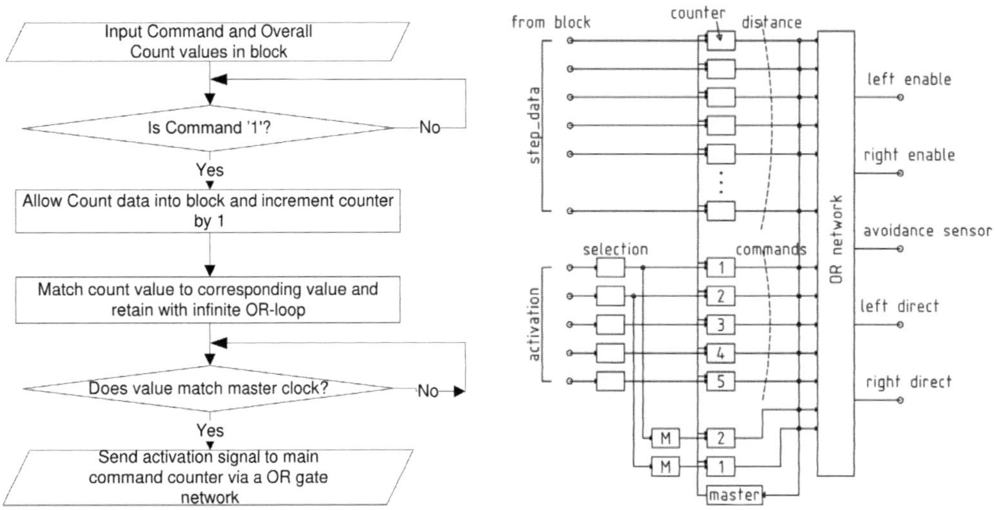

Figure 3: (a) Flowchart showing the selection process for the command operations, (b) Gate level block diagram of the update DR navigational protocol

3. Results and Keypad Interface

The keypad was always going to the intended way of entering the data into the system, the difficulty was deciding on how this data would be used. The initial method investigated was to just input a long string of data and using a series of comparators or bit extractors in order to extract sets of numbers from this in order to make the numbers for the counters. This however failed to live up to expectations and so another method of using dividers to reduce the data down was devised, this was partially successful but the results still suffered from data contamination. Finally the chosen method was to enter a value into the keypad and press the '#' key before entering the next value was devised, this also allowed for easy interchange between distances and command operators.

The keypad code has been designed to interface directly with the DSP builder circuitry, and therefore can be used as a complete user interface solution for implementation onto a physical device. The top-level block diagram of the keypad can be seen in Fig. 4. The keypad completes the design by providing input error checking and entry cancellation via a dedicated cancel key (star '*' key). The design as shown in the aforementioned figure occupies 250 logic elements, of which takes only 3% of available resources on an APEX 20K200EFC484-2X FPGA, and may certainly be further cut down by removing the test and debugging outputs. This portion of the robot design works by first scanning for input keys and stores the keys into a synthesized RAM buffer. The circuit checks for overflow on the input and also clears the buffer on the keying of the 'clear' key. The keypad design can be seen working in Fig. 5(a), where the system correctly registers the distance key presses and converts the binary-coded decimal key buffers into a single 10-bit word that can be directly passed on to the DSP-builder section of the design, giving the robot a range of 1023 m per movement command, with a resolution of 1 m. Note that for the purposes of the simulation, the keypad scanner was removed and data entered directly into the main state machine. When interacting with the real-life implementation, the keypad scanner was used.

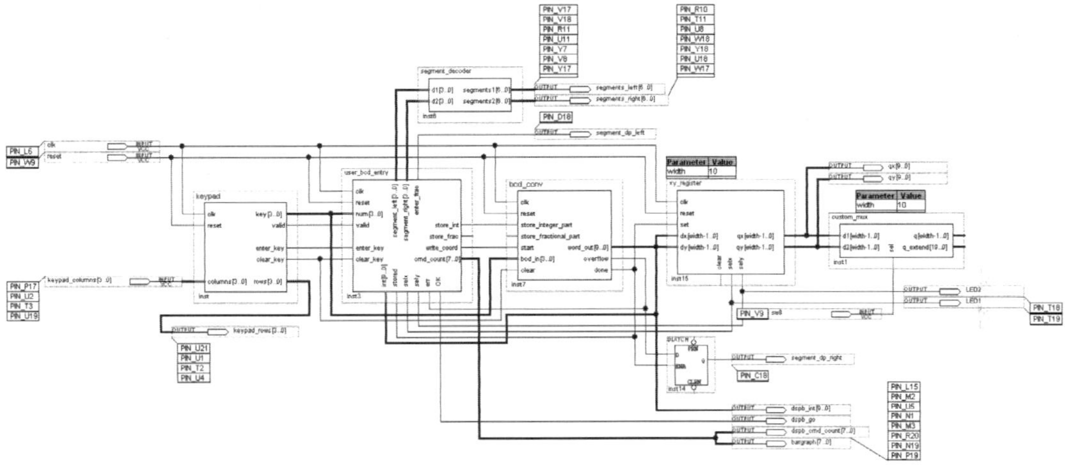

Figure 4: Top-level block diagram of the keypad interface.

4. Conclusions and Future Work

As seen from the experimental results, the keypad and the robot simulations work effectively together in order to help programming the DR aspect of a mobile robot easier for the non-expert by means of an easier keypad interface. One aspect that will be changed in future versions of this work will be to use the distance measurements in millimeters rather than meters, this would increase the bit usage from 10-bits for the current range of 1023 m to 20-bits and then this will provide more accurate robot movement in the real world whilst still lying within the 50-bit operation range within the DSP Builder software. Then by a simple process of expanding the registers to handle larger numbers to therefore increase the operational range of the robot. Future work will consist of adding stereo vision based navigational techniques into the design for object avoidance as well as PID controls for compensation of errors such as wheel slip which have so far not been included in any designs. By including these, the mobile robot envisaged will be more robust and adaptable to changes within its environment and be a better tool for applications such as security, condition monitoring and human-robot interactions.

Acknowledgements
Dr Tickle and Mr Harvey would like to thank the University of Liverpool for providing the resources to undertake this work and to Prof. J S Smith and Em Prof. J Lucas for their support and guidance during the course of this work.

References

[1] Tickle A J, Meng Y and Smith J S 2010 *Simulation of a Dead Reckoning Embedded System Security Patrol Robot for Deployment inside Structures and Buildings* Proc. SPIE 7833: Unmanned/Unattended Sensors and Sensor Networks, Toulouse, France, September 2010 (12pp)

[2] Tickle A J, Smith J S and Wu Q H 2007 *Development of Morphological Operators for Field Programmable Gate Arrays* Proc. IOP Conference: Sensors and their Applications XIV, Liverpool John Moores University, Liverpool, UK

[3] Kamal I 2008 *WFR, a Dead Reckoning Robot – A Practical Application to Understand the Thoery* IKA Logic: Electronics Solutions Online Documents pp 1 – 13

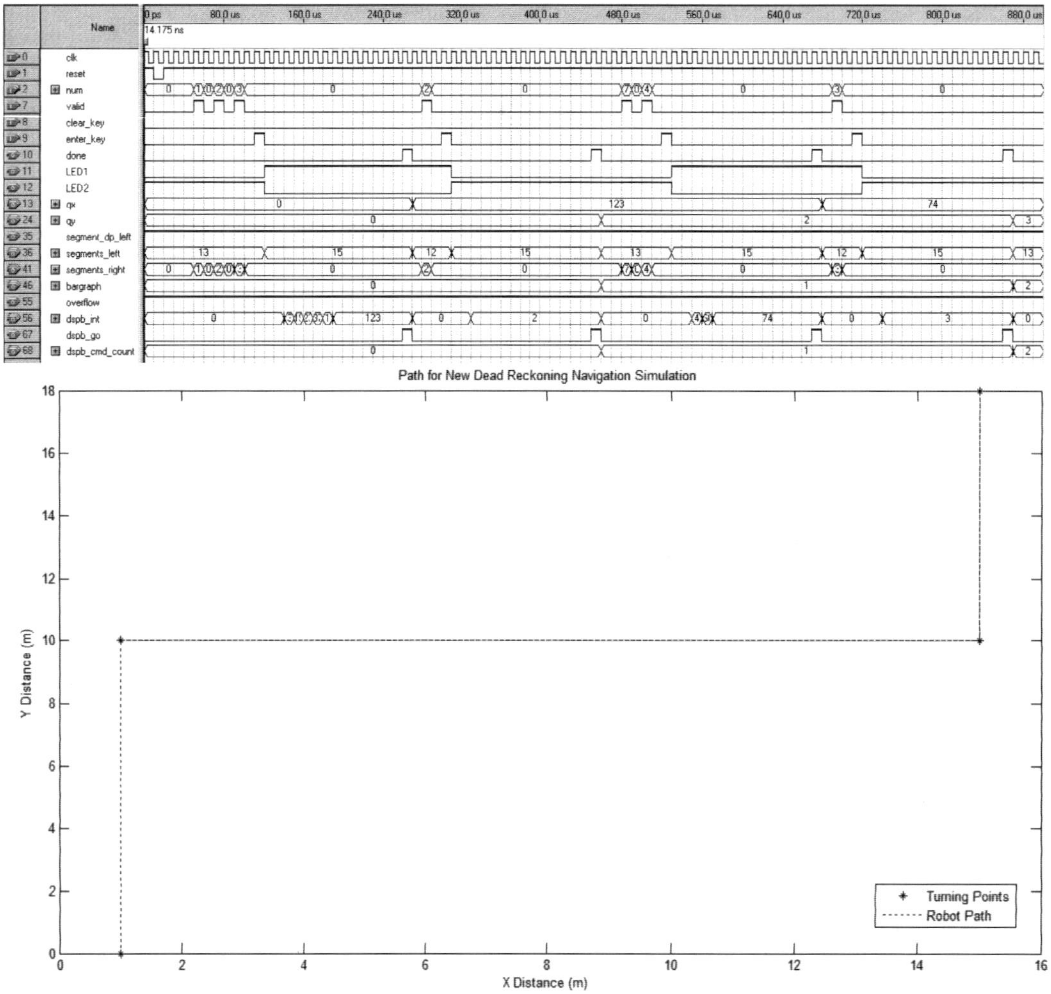

Figure 5: (a) Simulated keypad signals, showing the system reacting to the entry of two distance values and two commands, (b) MATLAB plotted results from DSP Builder for the robots route in the virtual environment.

Journal of Physics: Conference Series **307** (2011) 012029 doi:10.1088/1742-6596/307/1/012029

Arch-bridge Lift Process Monitoring by Using Packaged Optical Fibre Strain Sensors with Temperature Compensation

M. R. Mokhtar [1,7], T Sun[1], K T V Grattan[1], K Owens[2], J Kwasny[2], S E Taylor[2], P A M Basheer[2], D Cleland[2], Y Bai[2], M Sonebi[2], G Davis[2], A Gupta[3], I Hogg[3], B Bell[4], W Doherty[5], S McKeague[5], D Moore[6] and K Greeves[6]

[1] School of Engineering and Mathematical Sciences, City University, London, UK
[2] School of Planning, Architecture and Civil Engineering, Queen's University of Belfast, UK
[3] Macrete Ireland Ltd, Belfast, UK
[4] Network Rail, UK
[5] Creagh Concrete, Belfast, UK
[6] Arups, UK
[7] Faculty of Engineering, Multimedia University, 63100 Cyberjaya, Malaysia

E-mail: t.sun@city.ac.uk

Abstract. This paper presents a novel sensor design and packaging, specifically developed to allow fibre grating-based sensors to be used in harsh, in-the-field measurement conditions for accurate strain measurement, with full temperature compensation. After these sensors are carefully packaged and calibrated in the laboratory, they are installed onto the paragrid of a set of flat-packed concrete units, created specifically for forming a small-scale, lightweight and inexpensive flexi-arch bridge. During the arch-bridge lifting process, the sensors are used for real-time strain measurements to ensure the quality of the construction. During the work done, the sensors have demonstrated enhanced resilience when embedded in concrete structures, providing accurate and consistent strain measurements during the whole installation process and beyond into monitoring the integrity and use of the structure.

1. Introduction

Fibre Bragg gratings (FBGs) are well established as important sensing elements for the measurement of strain, temperature, and a wide range of other parameters. In the civil engineering sector, this is primarily attributed to the fact that the sensing signals from such fibre Bragg gratings are wavelength encoded, thus making them well suited for multiplexing: this allows multiple sensing points to be included within a single length of fibre, thus showing advantages over using a large number of conventional strain gauges for large-scale structural monitoring. However, FBG sensors can suffer a significant limitation in real engineering applications due to their cross-sensitivity to several parameters, especially strain and temperature. To overcome this limitation, a number of techniques have been reported, such as dual-wavelength superimposed gratings [1], hybrid Bragg grating/long-period gratings [2], dual-diameter FBGs [3], FBGs super-imposed with polarisation-rocking filters [4], FBGs combined with EDFAs [5], FBG/Fabry-Perot cavities [6], the use of long-period gratings [7], FBGs inscribed on the splice joint between two different fibres [8], and a chirped Bragg grating fabricated in a fused fibre taper [9]. Most of these designs are based on modifying the fibre gratings structurally or physically, which imposes additional complexity in the fabrication process for the resultant sensor head and thus increases the overall system cost and complexity. Additionally, bare FBGs can easily be damaged when being handled improperly or subjected to harsh working conditions. Therefore, it is necessary to protect the FBGs in a way that allows for use under normal working conditions. In this paper, a special sensor packaging is reported, developed for a specific and novel application in structural health monitoring, to enable discrimination between strain and temperature while still providing enhanced mechanical resilience for an important sensing application.

2. Sensing principle

Fibre Bragg Gratings (FBGs) [10] reflect light centred at a particular wavelength, termed as the Bragg wavelength, which is equal to twice of the product of the effective refractive index of the fibre core and of the grating period. Consequently, a change in temperature and/or in strain can cause a shift in the reflected wavelength through effects such as mechanical and thermal expansion, the photoelastic and the thermo-optic effects. By having two independent fibre gratings exposed to the same surroundings, the strain and temperature can be distinguished by solving the following matrix [11],

$$\begin{bmatrix} \Delta T \\ \Delta \varepsilon \end{bmatrix} = \frac{1}{D} \begin{bmatrix} K_{\varepsilon 2} & -K_{\varepsilon 1} \\ -K_{T2} & K_{T1} \end{bmatrix} \begin{bmatrix} \Delta \lambda_{B1} \\ \Delta \lambda_{B2} \end{bmatrix} \tag{1}$$

where $K_{\varepsilon i}$ is the effective strain coefficient, K_{Ti} is the effective temperature coefficient for the fibre gratings, $i = 1,2$ referring to gratings 1 and 2, $D = K_{T1}K_{\varepsilon 2} - K_{\varepsilon 1}K_{T2}$ and $\Delta\lambda_{Bi}$ is the Bragg wavelength shift. These coefficients can be determined experimentally as they are related both to the optical fibre glass materials and to the influence of the packaging medium which is used to protect the gratings.

3. Experimental setup and results

3.1. Sensor Fabrication and Packaging

The authors and other have demonstrated the use of optical fibre Bragg gratings in a number of applications to civil engineering structures [12] but in this work the aim was to provide a suitably packaged grating that could readily be installed on a structure exposed to considerable strain during the arch lifting process. Thus the aim of the packaging process for the gratings employed was to create a sensor system that would allow independent measurement of strain and temperature and be ruggedized for use in-the-field, employable during the construction and evaluation of a novel arch bridge design. To achieve this, as shown in figure 1 (a), two adjacent fibre Bragg gratings (set 30mm apart) were inscribed in a piece of a B/Ge co-doped photosensitive fibre before being sandwiched between two plastic slabs (polypropylene) by means of cyanoacrylate adhesive. To enable one grating to measure temperature only (and thus allow for correction for temperature effects on the grating measuring temperature *and* strain), a special packaging scheme was developed in which one fibre grating is slightly bent in order to prevent strain transfer from the package. The grating was enclosed by a thin metal envelope to protect and isolate the grating from any transverse mechanical stresses and lateral strain surge. By contrast, the other grating was glued to the package material to sense *both* strain and temperature so that the combined signal can be used to separate strain from temperature by using equation (1). The actual construction of the sensor is shown in figure 1(b). The length of the package was 80mm and the width of the sensing length was 6mm. The ends of the package are intentionally made slightly wider to function as anchors on the structure.

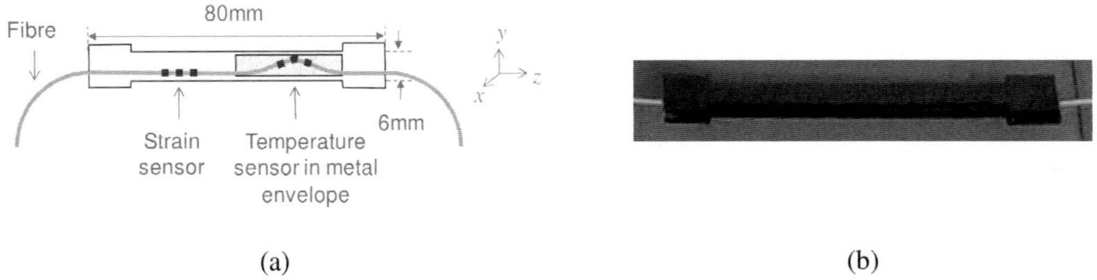

Figure 1. (a) Schematic diagram, and (b) photograph, showing the actual construction of the special sensor packaging design.

3.2. Sensor calibration

The effective strain and temperature coefficients for both sensors were obtained by measuring separately the wavelength shifts when they are subjected to controlled strain or temperature variations. To achieve this, the strain calibration was performed by using a purpose-built calibration rig with a 1μm-resolution and the temperature calibration was performed by using a climatic chamber (Binder, KBF 115 (E2)). The individual strain and temperature coefficients of both sensors, used together in the packaging discussed, were obtained through calibration and are shown in figure 2. The differences in the coefficients between the paired gratings were clearly evident.

It was verified through a series of experiments carried out that the fibre grating used for temperature measurement in the metal envelope exhibits no response at all to the applied strain. This enables it to act as a temperature-only sensor, one which is immune to the influence of strain when the sensor package is used. The calibrated data from the temperature sensor is subsequently utilized to remove the temperature effect from the dedicated strain sensor (i.e. the other fibre grating) by using equation (1).

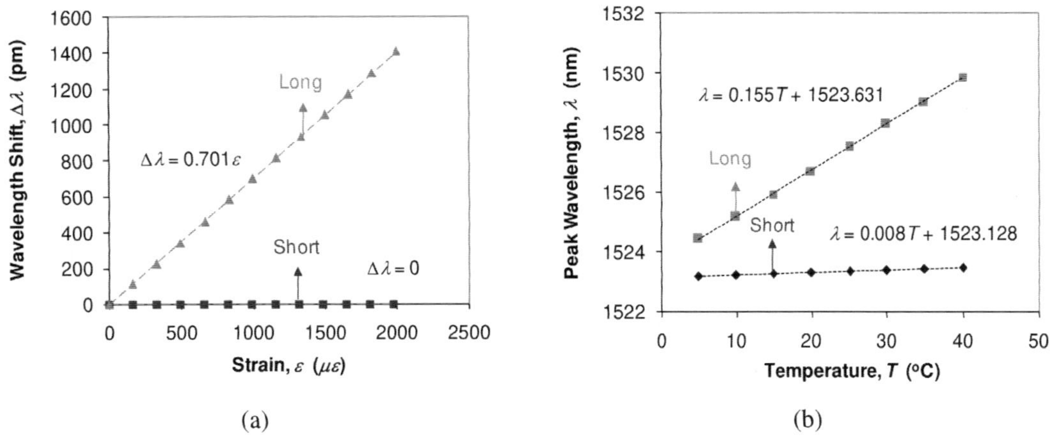

Figure 2. The response of packaged sensor for (a) strain calibration, and (b) temperature calibration.

3.3. Sensor installation in a model arch bridge

The above sensor package had been designed specifically for use on an arch bridge to allow strain to be determined when the bridge was being constructed and subsequently when loaded. The arch bridge is designed for convenience of movement to a site for construction and figure 3 shows a 'flat pack' assembly, comprising a number of concrete units, connected by a paragrid layer on the top. In this work, a total of 10 fibre optic sensors was specially packaged and then installed on the paragrid, following which the flexi-arch was formed from the 'flat pack' by lifting the concrete units. The sensors are constructed both to survive vigorous handling in the process of arch formation and to allow the measurement of strains in real time during the lifting process.

Figure 3. Flat pack assembly for arch bridge construction, which is ready for lifting to form an arch bridge

This is a unique application of fibre Bragg grating based sensors and, as may be expected, a number of the sensor package connections were broken during the initial phases of construction and lifting of the arch: in the end 3 fibre-optic sensors, as shown in figure 4, survived until the end of the experiment. One of the reasons for the failure of the sensor link was due to an unexpected forceful breakage of the screed layer during the arch formation and the avoidance of this problem should be planned into future installations. However in spite of the level of breakage, there proved to be a sufficient number of sensors to gain valuable data on the bridge and to establish a protocol for future tests to limit the level of breakage of the sensors and their network connections.

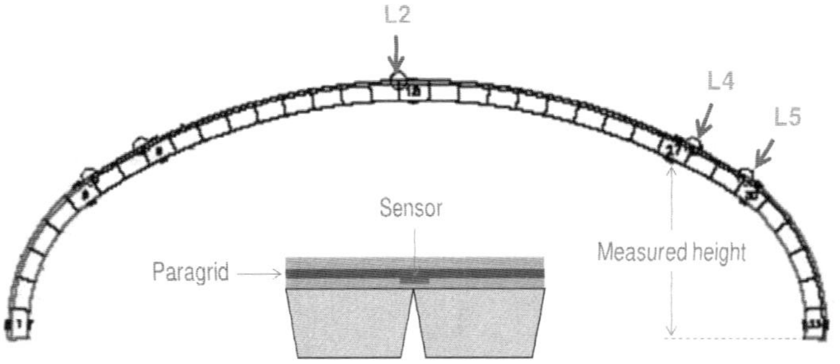

Figure 4. The arrows indicate the positions of fibre-optic sensors along the paragrid of the flexi-arch bridge, and the inset shows the placement of the fibre-optic sensors underneath the paragrid.

As shown in figure 4, the sensors were fixed to the paragrid by using cyanoacrylate adhesive, which has proved in previous work to be highly successful in achieving a good level of strain transfer and stability. The top surfaces of the sensors were then covered loosely by the same paragrid material in order to decouple the sensors from the screed layer, while their anchor points were intentionally exposed to be attached to the screed material.

4. Experimental results obtained

As indicated in figure 4, in this initial test only three sensors, i.e. L2, L4 and L5 survived the entire construction and lifting process; therefore as the flexi-arch was lifted, the strain was measured at 3 different locations and the measurement data obtained are presented as a function of the vertical height of the one-third point. Figure 5 shows the results obtained from sensor L2, located at the middle of the flexi-arch bridge, where the necessity for enhanced temperature compensation is illustrated.

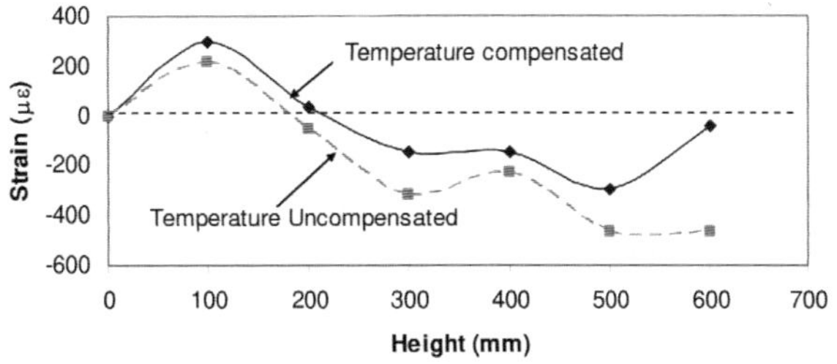

Figure 5. Strain measurement on the paragrid at the centre of the flexi-arch bridge during lifting by the middle sensor with and without temperature compensation.

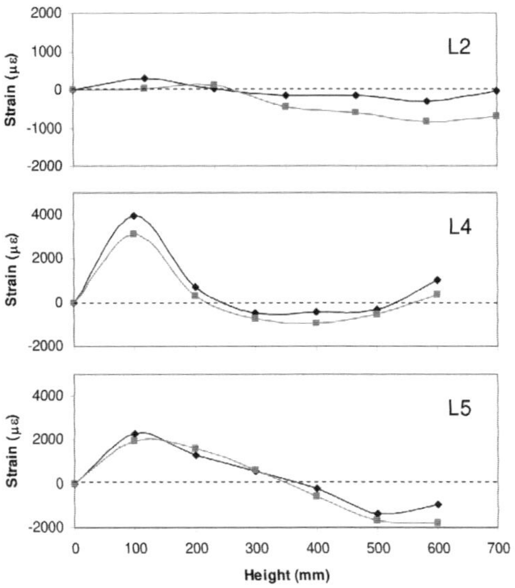

Figure 6. Overlay of two different sets of strain measurement on paragrid during lifting by three fibre-optic sensors at various joint locations.

Two tests were undertaken sequentially, allowing two sets of measurement data for all three fibre-optic sensors on the paragrid at various joint locations to be obtained. Figure 6 shows the two sets of strain measurement data obtained from the three fibre-optic sensors at various joint locations, as shown in figure 4, during process of lifting the arch into position. The results obtained have confirmed that the sensors have not just been able to survive the harsh conditions of construction and lifting but also to provide consistent readings of the strains experienced during the experiments, with the temperature effect being fully compensated.

5. Discussions and conclusion

The work has uniquely tackled an application of measurement of strain on the arch as it is both lifted into place and has shown the value of the specially designed package of a combined strain/temperature sensor. This has been demonstrated in a challenging application to a new type of light weight arch bridge which has significant potential in construction processes. To do so, a compact design for the temperature compensated fibre grating-based strain sensors has been demonstrated and shown to achieve enhanced mechanical resilience and to allow the accurate strain/temperature measurement data required, both through laboratory calibration and field tests. The work, undertaken by civil and electrical engineers working together on developing, placing and installing the packaged fibre grating sensors has shown that they could be successfully installed on the paragrid of a model flexi-arch bridge and tested extensively for strain/temperature measurements, even though a number were broken on this initial trail. Based on the positive experimental results obtained, it can be concluded that the novel sensor system has been successfully integrated into the concrete material and the performance of the sensors has been validated through a series of successful tests of the sensors installed in the paragrid of the arch model bridge when it is lifted. The novel design of the packaging helps the sensors to survive harsh handling both during the sensor installation and during the bridge tests, and even when embedded in concrete.

On-going work will tackle the issues of enhanced survivability of the sensor package and allow the testing and evaluation of the packaged sensors 'in-the-field' on a larger version of the arch bridge on a commercial site.

Acknowledgements

The authors would like to thank the support from the Technology Strategy Board in the UK and from the Royal Society of Engineering and the Leverhulme Trust for the Senior Research Fellowship to one of the authors.

References

1. Xu MG, Archambault JL, Reekie L and Dakin JP, Electron. Lett., 30(13), 1085-1087 (1994)
2. Patrick HJ, Williams GM, Kersey AD, Pedrazzani JR, and Vengsarkar AM, IEEE Photonics Technol. Lett., 8(9), 1223-1225 (1996)
3. James SW, Dockney ML and Tatam FP, Electron. Lett., 32(12), 1133-1134 (1996)
4. Kanellopoulos SE, Handerek VA and Rogers AJ, Opt. Lett., 20(3), 333–335 (1995)
5. J. Jung, H. Nam, JH Lee, N. Park, and B. Lee, "Simultaneous measurement of strain and temperature by use of a single-fiber Bragg grating and an erbium-doped fiber amplifier," Appl. Opt. 38(13), 2749-2751 (1999)
6. Du, WC, Tao, XM, and Tam, HY, "Fiber Bragg Grating Cavity Sensor for Simultaneous Measurement of Strain and Temperature," IEEE Photonics Technology Letters, 11(1), pp. 105–107 (1999)
7. Bhatia V, Campbell D, Claus RO and Vengsarkar AM, "Simultaneous strain and temperature measurement with long-period gratings," Opt. Lett., 22(9), 648–50 (1997)
8. B. O Guan, HY Tam, SL Ho, WH Chung, and XY Dong, "Simultaneous strain and temperature measurement using a single fiber Bragg grating," Electron. Lett. 38, 1018-1019 (2000)
9. Frazao O, Melo M, Marques PVS and Santos JL, "Chirped Bragg grating fabricated in fused

fibre taper for strain–temperature discrimination," Meas. Sci. Technol. 16, 984-988 (2005)

10. Grattan K T V and Meggitt B T (Eds). "Optical Fibre Sensor Technology: Advanced Applications - Bragg Gratings and Distributed Sensors", Kluwer Academic Publishers, 2000
11. Frazao O. and Santos J L, "Simultaneous measurement of strain and temperature using a Bragg grating structure written in germanosilicate fibres," J. Opt. A: Pure Appl. Opt., 6, 553–556 (2004)
12. Kerrouche A, Boyle W J O, Sun T and Grattan K T V, "Design and in-the-field performance evaluation of compact FBG sensor system for structural health monitoring applications", Sensors and Actuators A: Physical, 151, 107-112 (2009)
13. Jin W, Craig Michie W, Thursby G, Konstantaki M and Culshaw B, "Simultaneous measurement of strain and temperature: error analysis," Opt. Eng., 36, 598–609 (1997)

Bulk Material Based Thermoelectric Energy Harvesting for Wireless Sensor Applications

W S Wang, W Magnin, N Wang, M Hayes, B O'Flynn and C O'Mathuna

Tyndall National Institute, Dyke Parade, Cork, Ireland

E-mail: wensi.wang@tyndall.ie

Abstract. The trend towards smart building and modern manufacturing demands ubiquitous sensing in the foreseeable future. Self-powered Wireless sensor networks (WSNs) are essential for such applications. This paper describes bulk material based thermoelectric generator (TEG) design and implementation for WSN. A 20cm^2 $Bi_{0.5}Sb_{1.5}Te_3$ based TEG was created with optimized configuration and generates 2.7mW in typical condition. A novel load matching method is used to maximize the power output. The implemented power management module delivers 651μW to WSN in 50 °C. With average power consumption of Tyndall WSN measured at 72μW, feasibility of utilizing bulk material TEG to power WSN is demonstrated.

1. Introduction

While wireless sensor networks (WSN) have found and created various new applications in recent years, the fundamental problem of short lifetime remains unsolved. New breakthroughs in ultra-low power circuits have reduced the power consumption of wireless sensor modules from mW to the μW level in the last 10 years. Novel network protocols with focus on low power operation further enhance the trend. However, despite of the efforts, coin-sized high energy density lithium batteries can only supply an average load of 100 μW for less than a year [1]. One simple way to obtain longer lifetime is to use larger or multiple batteries. Although this may extend the lifetime in some cases, it is not commonly acceptable for many applications, especially for wearable and medical wireless sensor networks. Even for applications with little dimension requirement the inevitable battery replacement always brings unfavourable costs and maintenance efforts to the end users.

Harvesting environmental energy is often an effective approach to prolong the battery lifetime or eliminate the battery from the system. Common energy sources can be harvested in daily environment includes heat, light and vibration energy. The thermal energy can be found and used in various locations and conditions. Using thermoelectric energy harvesting to power WSN with aerospace sensors is demonstrated by EADS in [2]. A wrist watch formed thermoelectric generator (TEG) shows ultra-low power wireless communication module powered from human body heat [3]. In the design of TEGs, it has been discovered that the energy conversion efficiency in thermoelectric module and conditioning circuits both play essential roles. The voltage generated from TEG is inherently low due to the limited Seebeck coefficient of materials. With 1 Kelvin temperature difference applied on two sides of one commercial off-the-shelf Bismuth Telluride thermopile, less than 1mV is measured. A MIT team presents a design [4] with 35mV start-up voltage to allow the ultra-low voltage energy harvested with comparatively high efficiency. A similar work is presented by Carlson et al. in [5]. Both of the power management devices require comprehensive and high cost ASIC designs.

In this paper, a different approach is presented. With a simple but essential investigation on bulk material based TEG module, the design is optimized to allow higher voltage output compares to most Peltier cooler modules. A charge pump and switching regulator based power management circuit obtains a start up voltage at 300mV, delivers 650 μW power. Electric double layer capacitor energy storage is implemented to replace the battery element. A first prototype is demonstrated in section 4. The prototype evaluation was conducted on Tyndall WSN module with integrated temperature and vibration sensing capabilities. The proposed thermal energy powered WSN module can be used to monitor temperature and other parameters in industry and office environment when a 50 degree Celsius temperature is available.

2. Tyndall Wireless Sensor Networks

2.1. Tyndall 25mm Wireless Sensor Module

The Tyndall wireless sensor module is a miniaturized, low-cost multi-function solution for wireless sensing applications. The core of the node is an ATMEL ATmega1281 microcontroller with an 8MHz external crystal oscillator. The ATmega1281 provides eight Analogue-to-Digital Converters (ADC), an Inter-Integrated Circuit bus (I^2C) and eight interrupt pins. This WSN module integrates EM2420 radio chip from EMBER which is 2.4 GHz IEEE 802.15.4 compliant RF transceiver. It interfaced to the microcontroller with a 10 pins bus including a Serial Programmable Interface (SPI).

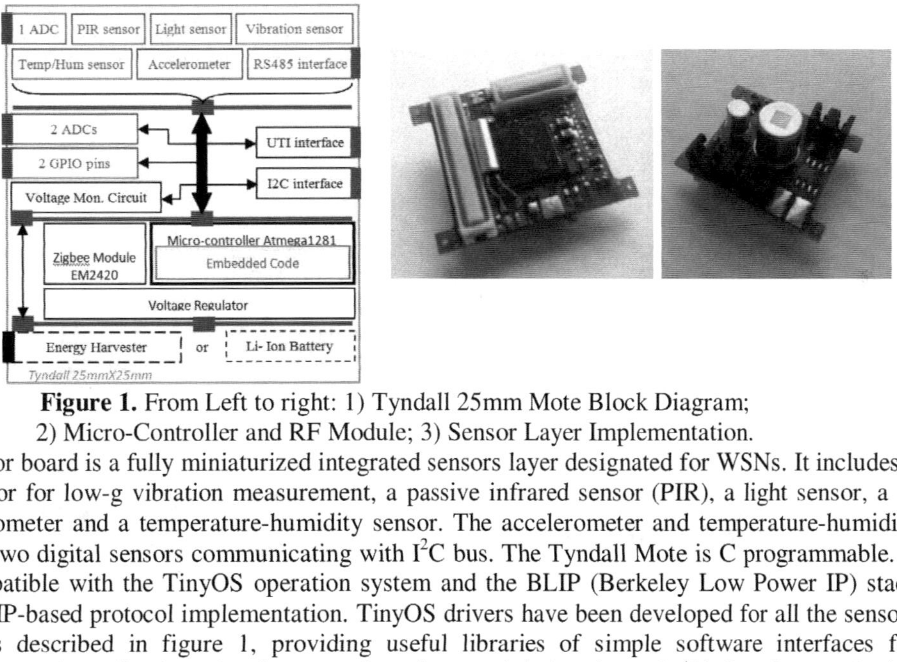

Figure 1. From Left to right: 1) Tyndall 25mm Mote Block Diagram;
2) Micro-Controller and RF Module; 3) Sensor Layer Implementation.

The sensor board is a fully miniaturized integrated sensors layer designated for WSNs. It includes a motion sensor for low-g vibration measurement, a passive infrared sensor (PIR), a light sensor, a 3-axis accelerometer and a temperature-humidity sensor. The accelerometer and temperature-humidity sensors are two digital sensors communicating with I^2C bus. The Tyndall Mote is C programmable. It is also compatible with the TinyOS operation system and the BLIP (Berkeley Low Power IP) stack which is an IP-based protocol implementation. TinyOS drivers have been developed for all the sensors and systems described in figure 1, providing useful libraries of simple software interfaces for embedded network applications development. In order to minimize the embedded code complexity, sensor drivers provide unprocessed data only.

2.2. Power Consumption

The wireless sensor modules operate in duty cycle manner to reach monitoring frequency requirement while conserve the energy by entering low power mode. In every service cycle T_{SP}, the sensing and transmission mode only take place once. The energy consumption in sensing and transmission mode and sleep mode is expressed in Equation 1.

$$E_{Total} = E_{Sleep} + E_{ST} = E_{DTx} + E_{ARx} + E_{Init} + E_{Sen} \quad (1)$$

Where E_{Sleep} and E_{ST} are the energy consumptions in sleep mode and sensing-transmission mode. E_{DTx} = 4.6mJ, E_{ARx} = 0.06mJ, E_{Init} = 0.4mJ and E_{Sen} = 0.4mJ are the data transmission, acknowledgment receiving, board initialization and sensing energy consumptions, respectively. The measurement is taken on Tyndall mote with temperature/humidity sensor and Analog Devices 3-axis accelerometer activated. Both of the sensors require short warm up time, enables an ultra-low duty cycle operation. The detailed energy consumption measurement is presented in Table 1.

Table 1. Energy Consumption Measurement

	Energy(mJ)	Power(mW)	Time(ms)
Sensing and Transmission (ST)	5.46	85.98	63.5
Data Transmission	4.6	131.4	35
Data Receiving	6.1	174.3	35
Mote Initialization	0.4	20	20
Sensing	0.4	53.3	7.5
Acknowledge Transmission	0.05	111.1	0.45
Acknowledge Receiving	0.06	133.3	0.45
Sleep	3.23	0.054	59,847

The time interval between each measurement is one minute in this case. The duty cycle of the operation is measured at 0.1%. The average power consumption is 72μW in this typical test condition.

3. Thermoelectric Energy Harvesting

3.1. Thermoelectric Materials

To design a practical thermoelectric energy harvester, several types of thermoelectric materials are investigated for optimized energy conversion rate. The efficiency of a TEG is determined by the dimensionless figure of merit, ZT, The correlation between thermoelectric energy conversion efficiency and ZT figure of merit is shown in Equation 2:

$$\eta = \frac{\Delta T}{T_H} \frac{\sqrt{1+ZT}-1}{\sqrt{1+ZT}+\frac{T_C}{T_H}} \tag{2}$$

A typical cost efficient with relatively high energy conversion efficiency bulk material features a ZT around 0.50-0.75 when located in 300K temperature condition. This was achieved by BixSb2-xTe3 p-type material in 1960s. A typical TEG implementation based on bulk material is shown in Figure 2.

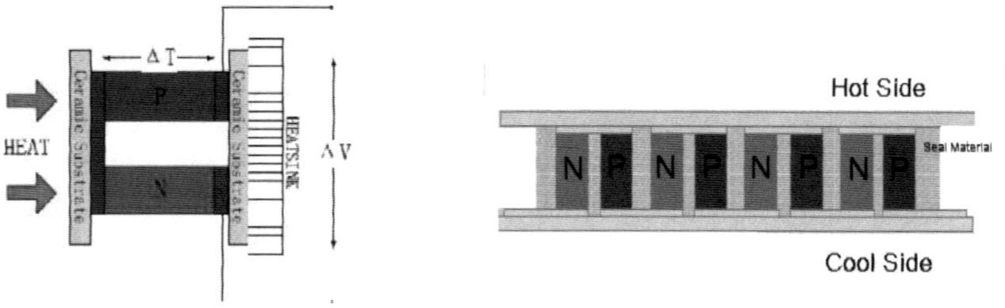

Figure 2. Thermoelectric Module based on Bulk Material

Recent breakthroughs in nano-technology based thermoelectric materials have demonstrated great advantages for ZT improvement. Phonon-blocking superlattices structure material shows the highest ZT figure of merit at 2.4 [6]. These advances certainly show the possibility of high efficiency energy conversion in future TEGs. However, ZT is not the only parameter should be considered when design a TEG. The actual output voltage is a key parameter in TEG design, shown in equation 3. Where α is Seebeck coefficient, N is the number of thermopile pairs, ΔT is the temperature difference between the outer surfaces of two substrates, k and k_C are the thermal conductivities of thermopile and substrate. L and L_C are the thickness of thermopile and substrate.

$$V_o = \frac{\alpha N \Delta T}{1+2\frac{L_C}{L}\cdot\frac{k}{k_C}} \tag{3}$$

With the nano material used in TEG, the thickness of thermopiles L will decrease significantly, while the substrate thickness L_C remains the same as in bulk material based design. This effect may substantially reduce the output voltage on single thermopile. Although high number of thermopiles N will be available in nano technology based TEG design. The inevitable high internal resistance makes it impossible to directly utilize the energy without load matching and other power management. The complexity and associated high cost lead to the decision of using bulk material based thermoelectric materials. In this work, $Bi_{0.5}Sb_{1.5}Te_3$ based thermoelectric material with a ZT at 0.7 is used to design thermoelectric modules.

3.2. Thermoelectric Module Design

Thermoelectric module shows behaviors similar to a typical current controlled voltage source when various temperatures applied across it. Similar to other practical voltage source, internal resistance also exists in TEG. To deliver the maximum power output to the load, impedance matching is important. However, different from conventional load matching with fixed internal resistance, in this case the TEG can be modified with different configurations of thermopiles. This leads to a configurable internal resistance. 2 mm^3 sized P type and N type $Bi_{0.5}Sb_{1.5}Te_3$ thermoelectric cubes are used as the basic units of the TEG. With a series connecting configuration, the output power of the TEG can be derived from equation 3. The output power P_{out} is presented in the following equation.

$$P_{out} = \frac{(\alpha N \Delta T)^2 \cdot R_L}{[(1+2\frac{k}{L}\cdot\frac{L_C}{k_C})\cdot(N\cdot R_{TEG}+R_L)]^2} \tag{4}$$

Where R_L is the load resistance and R_{TEG} is the internal resistance of single thermopile, N is the number of series connected thermopiles. The internal resistance of a pair of 2 mm^3 thermopiles is measured at 0.396 Ω, while the load resistance in the supercapacitor is 30 Ω when operate in low frequency. Without forced convection cooling, the temperature difference between thermopiles only measured at less than 5 Kelvin when 50 degree Celsius is applied on high temperature side. Due to the small form factor nature of the WSN, the size of the TEG should no larger than 60mm by 60mm. This factor limits the number of thermopiles pairs to less than 200. A simulation is conducted with above predetermined conditions. The output power to the supercapacitor is presented in the following figure.

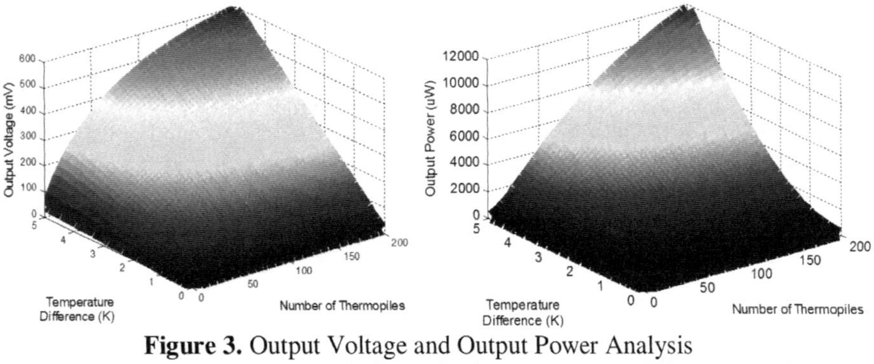

Figure 3. Output Voltage and Output Power Analysis

The temperature difference shown in this figure is the temperature difference on the two ceramic substrates, instead on the thermopiles directly. The simulation results illustrated in figure 3 show the required 72μW can be obtained from the TEG when 36 thermopiles connected in series at one Kelvin temperature difference. However, under the same condition, the output voltage on the supercapacitor is only 52.4mV. To operate without ASIC based DC/DC conversion circuit, the minimum voltage output should be higher than 300mV. In the worst case scenario, when the temperature on the hot side is 50 degree Celsius, the temperature difference on the substrates is 3.5 Kelvin when minimum airflow is measured in laboratory condition. In this case, to obtain a 300mV output voltage, the minimum number of thermopile is 120. With this number of thermopiles integrated in the TEG, the output power is 2.96mW. Based on this analysis, a 16 by 16 thermopile legs (P/N type cubes) array is created, given a thermopile number at 128. The device is measured at 45mm by 45mm.

3.3. Power Management

In this implementation, low cost off-the-shelf components are utilized to step up the output voltage of TEG to usable voltage for WSN. A Texas Instruments TPS61200 boost converter and a Seiko Instruments S-882Z18 charge pump are selected due to their high efficiency in ultra-low voltage conditions. To step up the output voltage from 300mV, directly utilizing a boost converter is not feasible since the starting voltage of boost converter is substantially higher. The power management circuit is shown in Figure 4.

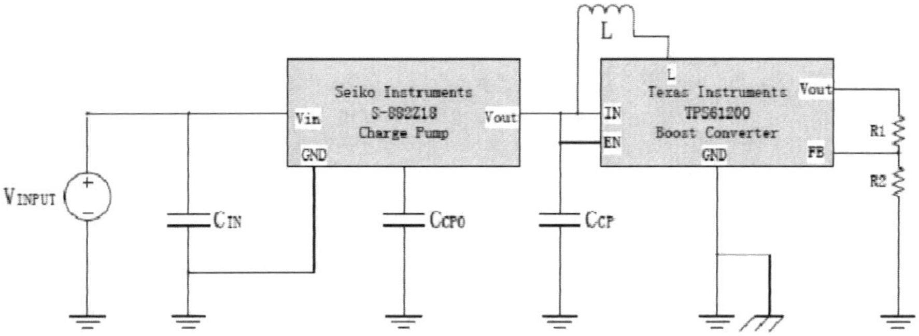

Figure 4. Simplified Step-up DC/DC Conversion

The ultra-low input voltage of Seiko Instruments S-882Z enables a start up voltage at 300mV. Once the voltage of TEG boosts to above the threshold, the voltage on the output capacitor of the charge pump increases until it reaches the start up voltage of the boost converter at 700mV. The Enable pin is controlled manually in this case to assist the start up process. It also can be connected to a comparator based self start controller similar to a design presented in previous work [1]. It employs a MAX931 ultra-low power comparator to detect the V_{CCP} until it settled at the 700mV and enabled the DC/DC converter to step up to 3.3V.

4. Thermoelectric Generator Powered WSN Implementation

Based on the proposed TEG design, the assembling and packaging was conducted by [6]. Due to the fact WSN operates in duty cycling manner, energy buffer is essential to avoid periodic power failure. Electric double-layer capacitor (supercapacitor) is the selected energy storage in this work. Low equivalent series resistance (ESR) energy storage unit is used to avoid sudden voltage drop in the WSN active mode. While high density type supercapacitor is connected in parallel to increase the capacity. Additionally, energy failure detection circuit is integrated using a voltage comparator to detect low voltage condition in supercapacitor. The failure detection is based on Seiko Instruments S-89530A ultra-low power (5 pA) comparator. The prototype is shown in Figure 5.

Figure 5. Thermoelectric Powered WSN Prototype

The viability of the TEG design and the application on WSN were tested through a set of experiments. The prototype was placed on hotplate in various temperatures to test the start up performance, continuous operation and finally energy storage lifetime when no thermal energy is available.

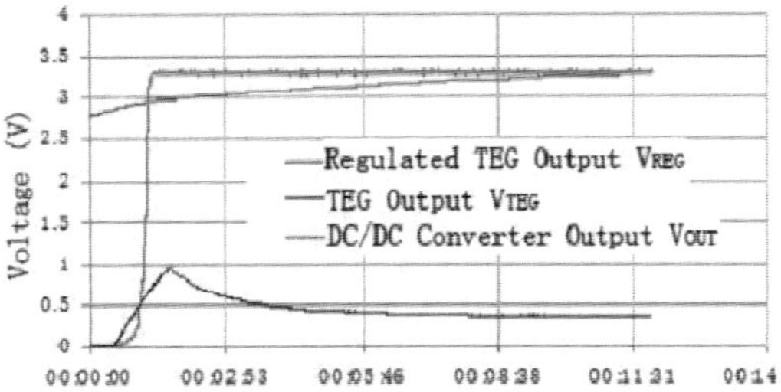

Figure 6. TEG Start-up Performance

The above figure shows the start up performance when prototype deployed on 50 degree Celsius temperature. The initial elevated temperature difference was clearly observed when output voltage peaked at 1 minute 25 second from the beginning of the test at 0.97V. This effect automatically starts up the boost converter. The switching regulator operates in continuous mode providing a stable output at 3.3V to the WSN when output voltage of TEG drop back to 350mV in the later stage of start up. The supercapacitor was pre-charged to 2.7V in this experiment. The charging lasts 11 minutes and 30 seconds to reach 97% of the full charge. The TEG obtains an output power at 2.7mW, which consistent with the simulation results with an 8.7% error. The power management circuit features average conversion efficiency at 55%. The leakage current on the supercapacitor and the output end power conversion further decreases the conversion efficiency. The available power delivered from TEG to WSN is measured at 651μW.

5. Conclusions

A thermoelectric powered wireless sensor module which eliminates the need for a battery is developed. The relative simplicity of the design and utilizing bulk material thermopiles plays a major role in reducing the system cost. The verified implementation, with $Bi_{0.5}Sb_{1.5}Te_3$ based thermopiles arranged in a 16 by 16 configuration deliver 2.7mW when 50 degree Celsius temperature is applied upon. The power management features a charge pump and switching regulator design. It steps up the voltage from 300mV to 3.3V. The power delivered to WSN is measured at 651 μW in this typical condition. The generated power is substantially higher than the average power consumption of WSN at 72 μW. The results demonstrated the feasibility of using bulk material based TEG to achieve power autonomous in many WSN applications.

Acknowledgments

This work has been funded by Scientific Foundation Ireland through ITOBO (398-CRP) project and EU FP-7 through Nanofunction (257375) project.

References

[1] Wang W S, O'Donnell T, Wang N, Hayes M, O'Flynn B and O'Mathuna C 2010 *J. Emerg. Technol. Comput. Syst.* **6** 1-26
[2] Becker T, Kluge M, Schalk J, Tiplady K, Paget C, Hilleringmann U and Otterpohl T 2009 *Sensors. J.* **9** 1589-95
[3] Leonov V, Torfs T, Fiorini P and Van Hoof C 2007 *Sensors. J* **7** 650-57
[4] Ramadass Y K and Chandrakasan A P 2011 *Solid-State Circuits* **46** 333-41
[5] Carlson E, Strunz K and Otis B P 2010 *Solid-State Circuits* **45** 741-50
[6] Venkatasubramanian R, Siivola E, Colpitts T and O'Quinn B 2001 *Nature* **413** 597-602.
[7] Thermonamic: http://www.thermonamic.com/

Improvement of dielectrophoretic impedance measurement method by bacterial concentration utilizing negative dielectrophoresis

R Hamada[1,2], H Takayama[2], Y Shonishi[2], T Hisajima[2], L Mao[2], M Nakano[2], J Suehiro[2]

1 R & D Center, Panasonic Healthcare Co., Ltd., 2131-1, Minanikata, Toon, Ehime, Japan

2 Department of Electrical and Electronic Engineering, Graduate School of Information Science and Electrical Engineering, Kyushu University, 744 Motooka, Nishi-ku, Fukuoka, Japan

E-mail: hamada.ryo@jp.panasonic.com

Abstract. In this study, the concept design for the improvement of the bacterial detection sensitivity of the DEPIM (Dielectrophoretic Impedance Measurement) method has been proposed. The cells in the micro-chamber are repelled and concentrated by n-DEP (negative dielectrophosesis). The concentrated cells are captured by p-DEP (positive DEP) and detected by measuring the change in the electrical impedance. The numerical simulations and the preliminary test were performed to investigate the effectiveness of the n-DEP concentration. When n-DEP concentration was employed, the increase in the rate of the conductance became approximately two times higher than that obtained without n-DEP.

1. Introduction

To inspect the level of bacterial contamination in the production process and products is a crucial matter for the food and beverage industry because an outbreak of food poisoning results in health hazard to consumers, and the company may lose its image. Conventionally, bacterial count using the cultivation and colony counting method is adopted as a standard method for quantification of bacterial contamination. The culturing method is well established but the cultivation process requires typically one to several days to complete the inspection, so the development of a rapid bacterial quantification method is required in the industrial fields. In recent years, several rapid bacterial counting methods alternative to conventional cultivation methods, including Adenosine Tri-Phosphate (ATP) bioluminescence [1] and the direct-count technique using epifluorescence microscopy (EFM) [2] have been proposed. In most cases, alternative methods require operation with a reagent, so the process of bacterial detection is rather complicated. The authors have previously proposed DEPIM (Dielectrophoretic Impedance Measurement) method utilizing DEP (dielectrophoresis) as a simple and rapid technique for detection of bacteria without any reagents [3]. DEP is a motion of dielectrically polarized particles in non-uniform electric fields [4]. As most bacteria behave as polarizable particles in external AC electric fields, DEP allows the manipulation of the bacteria in a liquid suspension. The direction of the DEP force depends on dielectric properties of particles and the surrounding medium, as well as frequency of applied electric field. In case a particle moves toward the high electric field region it is referred to as positive DEP (p-DEP), while it is referred to as negative DEP (n-DEP) for a particle repelled from the high field region. The DEPIM utilizes p-DEP in order to capture bacteria onto an electrode chip. Since bacteria have intrinsic electrical impedance, so the trapping of bacteria using p-DEP at the electrode results in a change in the electrical impedance of the electrode. Higher cell population results in faster bacteria trap and the electrode impedance change. By monitoring the temporal variation of the electrode impedance, the cell population can be quantitatively evaluated. DEPIM can realize simple and rapid bacterial inspection because it utilizes only the electrical phenomenon, also it can realize high sensitive detection combined with electropermeabilization [5] and selective detection of bacteria based on viability [6] or species [7].

In the DEPIM inspection, the lower the bacterial concentration, the longer it takes for the bacteria trapping process to sense the measureable change in the electrical impedance. Therefore, improving bacteria trapping efficiency by p-DEP is essential to achieve more rapid bacterial count by DEPIM. In the conventional DEPIM method, bacteria suspended in an aqueous medium are trapped onto a thin microelectrode, which is fabricated on a glass substrate to generate high electric field. The p-DEP force acting on a particle decreases rapidly in accordance with the increasing distance between the electrode and the particle. In these situations, one strategy to increase the trapping efficiency is to decrease the distance between the electrode surface and the lid aligned at the opposite side of the electrode. However, the narrower the flow path, the smaller the amount of bacteria which can pass through the chamber. This results in an increase in the time to trap sufficient amount of bacteria, which causes a measureable electrical impedance change. Another strategy is using n-DEP to concentrate bacteria to enhance trapping efficiency at p-DEP electrode [8].

In this study, an improved technique of the DEPIM for higher bacterial detection sensitivity was proposed and its effectiveness was experimentally confirmed.

2. Principle and theoretical prediction

Figure 1 shows the concept design of the bacterial concentration by n-DEP combined with the conventional DEPIM method. The concentrator consists of the microelectrode, which induces the n-DEP force on the cells passing through the chamber. The cells experience the n-DEP force and move towards the lower substrate. The detector is aligned at the downstream of the flow, and the structure is the same as the conventional DEPIM electrode system. As the cells reach the detector area to be concentrated in the vicinity of the lower substrate by the preceding n-DEP electrode, the cells are efficiently captured onto the microelectrode fabricated on the lower substrate and an impedance change arising from the captured cells will become larger in comparison with the conventional DEPIM method.

The numerical simulations were carried out to investigate the effectiveness of the bacterial concentration using n-DEP combined with the conventional DEPIM method. The DEP force acts on a spherical particle of radius r suspended in a medium of permittivity ε_s is given by [9]

$$\boldsymbol{F}_{DEP} = 2\pi r^3 \varepsilon_s \mathrm{Re}[\boldsymbol{K}(\omega)]\nabla E^2 \tag{1}$$

where E is the magnitude (RMS) of the applied field and $\mathrm{Re}[\boldsymbol{K}(\omega)]$ is the real component of the Clausius–Mossotti factor given by

$$\boldsymbol{K}(\omega) = \frac{\varepsilon_p^* - \varepsilon_s^*}{\varepsilon_p^* + 2\varepsilon_s^*} \tag{2}$$

where ε_p^* and ε_s^* are the complex permittivity of the particle and the surrounding medium, respectively. For a real dielectric, the complex permittivity is defined as

$$\varepsilon^* = \varepsilon - j\frac{\sigma}{\omega} \tag{3}$$

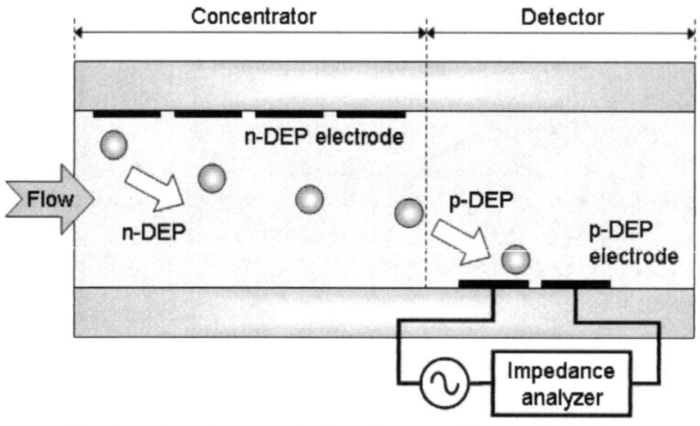

Figure 1. The concept design of the n-DEP concentrator combined with DEPIM system.

where ε is the permittivity, σ is the conductivity of the dielectric, and ω is the angular frequency of the applied field.

A particle moves in an aqueous medium by the DEP force, it experiences the drag force, $\boldsymbol{F}_{\text{drag}}$, which can be written as

$$\boldsymbol{F}_{drag} = -6\pi\eta a v \tag{4}$$

where η is the dynamic viscosity of the surrounding medium, $\boldsymbol{v}$ is the particle velocity relative to the surrounding medium. Assuming Brownian motion and buoyancy force can be negligible, the equation of motion for a particle with mass m in a medium can be written as

$$m\frac{dv}{dt} = \boldsymbol{F}_{DEP} + \boldsymbol{F}_{drag} \tag{5}$$

The electric field distribution, DEP force and bacteria motion trajectory were numerically calculated by a finite element method using COMSOL Multiphysics. The following values of the geometry of the interdigit electrode inducing n-DEP were employed on the numerical simulations: the electrode width was 50 μm, gaps between adjacent electrodes were 5 μm, and thickness of the electrode was 100 nm. We assumed the geometry of bacteria as a double-shelled structure, i.e. the conductive cytoplasm is covered by the thin insulation membrane. The values of the dielectric properties and geometry of bacteria were obtained from literature [10].

In order to study influences of the height of the chamber h, and the applied voltage V, the simulations were carried out at two different heights of the chamber; h = 10, 20 μm, and two different applied voltage; V = 20, 50 $V_{\text{peak-peak}}$. Figure 2 shows calculation results of the electric field distribution and bacteria trajectories. In the case of the thinner chamber, h = 10 μm and lower voltage, V = 20 $V_{\text{peak-peak}}$, bacteria trajectories were focused within about 5 μm above the lower substrate of the chamber after bacteria pass through the four electrode gaps (Figure 2(a)). Increasing voltage to 50 $V_{\text{peak-peak}}$ results in the enhancement of bacterial concentration so that bacteria were focused within 1 μm above the lower substrate after they passed the three electrode gaps (Figure 2(b)). On the other hand, increasing height of the chamber decreased concentration efficiency. In the case of the higher chamber, h = 20 μm, the applied voltage was lower, V = 20 $V_{\text{peak-peak}}$, bacteria could not arrive the region within 10 μm above the lower substrate even after passing the four electrode gaps (Figure 2(c)). At higher voltage, V = 50 $V_{\text{peak-peak}}$, the bacteria was barely concentrated within 10 μm above the lower substrate after passing the four electrode gaps (Figure 2(d)).

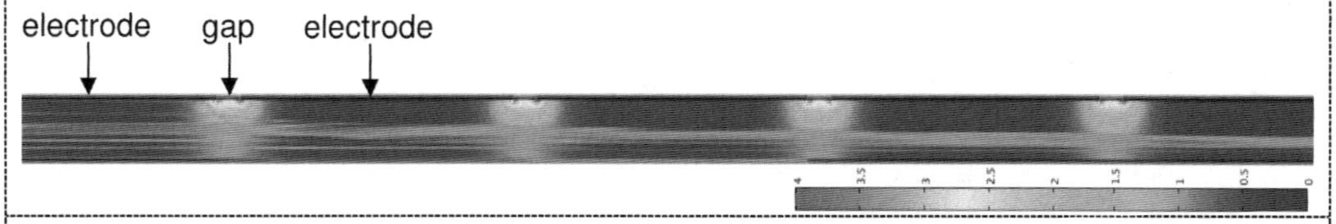

(a) h = 10 μm, V = 20 $V_{peak\text{-}peak}$

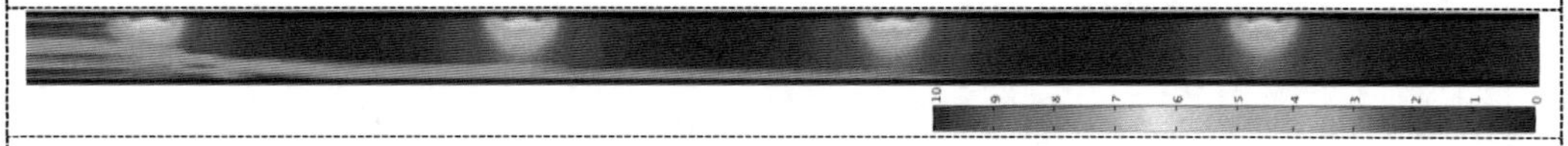

(b) h = 10 μm, V = 50 $V_{peak\text{-}peak}$

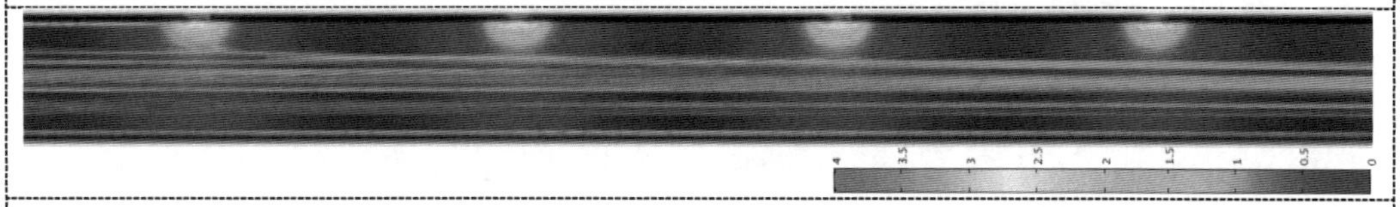

(c) h = 20 μm, V = 20 $V_{peak\text{-}peak}$

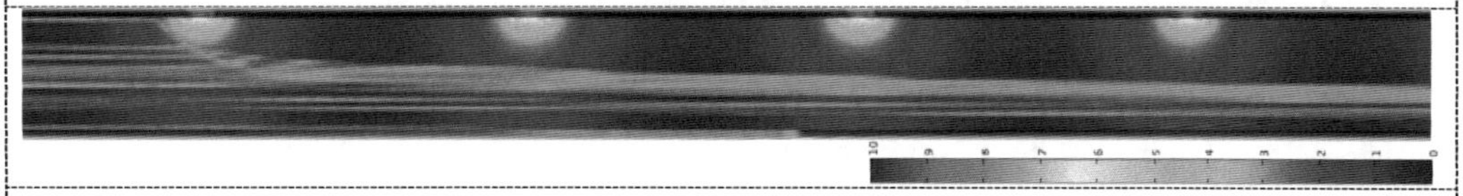

(d) h = 20 μm, V = 50 $V_{peak\text{-}peak}$

Figure 2. Numerical calculation results of the electric fields distribution and bacteria trajectory near the n-DEP electrode.

3. Materials and Methods

Figure 3 depicts an outline of the DEPIM system combined with the n-DEP concentrator. We employed the 66-gap n-DEP electrode followed by the 4-gap p-DEP electrode arranged at opposite sides of each other. Due of the limitation of the manufacturing of the chamber, the height of the chamber was 200 μm, which was larger than that assumed in the theoretical model. The frequencies of the applied electric field to n-DEP and p-DEP electrodes are 1 kHz and 100 kHz, respectively, which were determined by the investigation of Re[$\boldsymbol{K}(\omega)$] calculated in our previous work [11]. These values were suitable for n-DEP and p-DEP in a low conductivity medium. The cell suspension liquid was stored in a reservoir tank and circularly fed to the test chamber using a peristaltic pump. Sinusoidal AC voltage was generated by a function generator (WF 1945, NF Corporation, Japan) and applied to the n-DEP electrode for repelling the bacteria towards the p-DEP electrode. AC voltage was applied to the p-DEP electrode by a lock-in amplifier, which also monitors the current passing through the p-DEP electrode via shunt resistance (1 kΩ), and the data was transferred to the PC to calculate the impedance change. *E. coli* strain K-12 (NBRC3301), which has a high growth rate and has been successfully employed in previous work [3, 5-7], was employed in order to improve the efficiency of the experiments. *E. coli* was incubated at 30°C on agar plates for 24 hours. Before each measurement, cells were harvested from the agar and suspended in a 0.1 M mannitol solution. After several washings by centrifugation, they were finally resuspended in a 0.1 M mannitol solution (conductivity 1 μS/cm) at various diluted concentrations as determined by a colony counting method.

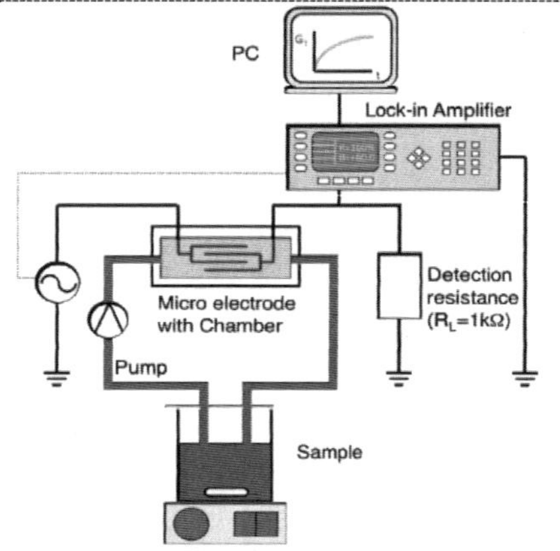

Figure 3. Outline of the DEPIM system combined with the n-DEP concentrator.

4. Results and discussion

Figure 4 depicts the DEPIM results with and without the n-DEP concentration. When the n-DEP concentration was employed, the increase rate of conductance was approximately two times faster compared to that obtained without the n-DEP concentration. This result proves that the number of bacteria trapped onto the p-DEP electrode increases as a result of n-DEP concentration as theoretically predicted in Figure 2.

In the present study, we aimed to improve DEPIM sensitivity by utilizing n-DEP concentrator, in which bacteria are repelled downward from the n-DEP electrode surface towards the p-DEP electrode surface. It is expected that energizing the n-DEP electrode with higher electrical potential can enhance the n-DEP force and the resultant concentration of cells. According to numerical calculation results shown in Figure 2, it was revealed that an increase in the applied voltage enlarged high field regions near the n-DEP electrode and could drive more bacteria toward the p-DEP electrode. In a real situation, however, there is an upper limit of applied voltage due to electrochemical stability of the thin metal electrode. More n-DEP concentration may be alternatively achieved by increasing the number of electrode gap, where high electric field and strong n-DEP force arises. In order to confirm if this idea works, computer simulation of cell trajectory was conducted again for various number of the electrode gap. As shown in Figure 5, a cell can be driven closer to the bottom surface of the chamber as the number of the electrode gap. This result shows that an increase in the number of the gaps of the n-DEP electrode can enhance the efficiency of the bacteria concentration by n-DEP.

According to Figures 2 and 5, it would seem that making the chamber height smaller has an advantage because bacteria can be driven closer to the chamber bottom with less number of electrode and lower electrode energizing voltage. However, lower chamber height or smaller channel cross section decreases the number of bacteria flowing into the chamber per unit time for a constant flow rate of bacteria suspension and bacteria concentration. If the suspension flow rate is increased to compensate the negative effect of the smaller height, the drag force will be comparable with n-DEP force so that n-DEP effect might be suppressed. In the future work, these parameters should be optimized to make DEPIM sensitivity as high as possible.

5. Concluding remarks

In this study, we proposed a new concept of the n-DEP bacterial concentration combined with the conventional DEPIM method to improve DEPIM sensitivity for on-site inspection of bacterial contamination. Results of numerical simulations and preliminary DEPIM tests show that the concept of the n-DEP concentration works.

This work was partly supported by a Grand-in-Aid for Scientific Research from the Japan Society for the Promotion of Science (No. 20360184, No. 21651063 and No. 22110510).

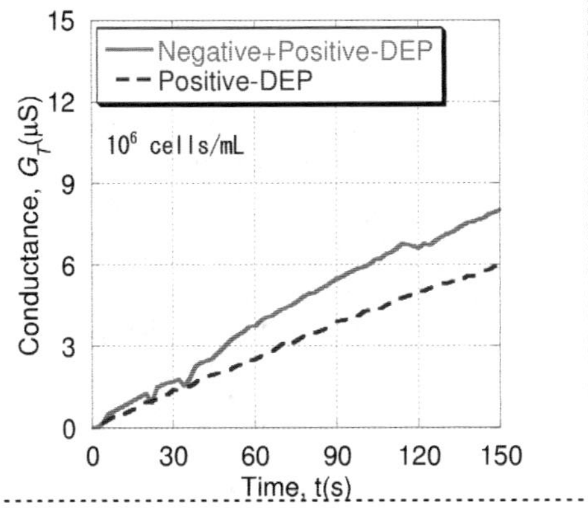

Figure 4. DEPIM results with and without n-DEP concentration.

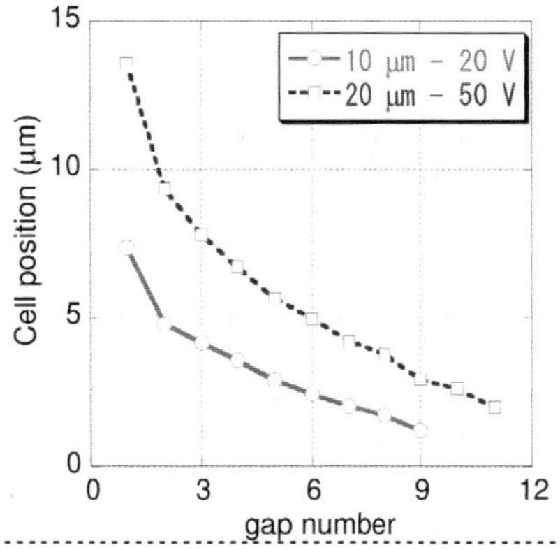

Figure 5. Effects of the number of n-DEP electrode gap on cell position.

References

[1] Jinping Luo, Xiaohong Liu, Qing Tian, Weiwei Yue, Jing Zeng, Guangquan Chen and Xinxia Cai 2009 Disposable bioluminescence-based biosensor for detection of bacterial count in food *Analytical Biochemistry* **394** 1–6

[2] L. KEPNER JR. and JAMES R. PRATT 1994 Use of Fluorochromes for Direct Enumeration of Total Bacteria in Environmental Samples: Past and Present *MICROBIOLOGICAL REVIEWS* Dec. 603-615

[3] J. Suehiro, R. Yatsunami, R. Hamada and M. Hara 1999 Quantitative estimation of biological cell concentration suspended in aqueous medium by using dielectrophoretic impedance measurement method *J. Phys. D: Appl. Phys.* **32** 2814-2820.

[4] R. Pethig 1996 Dielectrophoresis: using inhomogeneous ac electrical fields to separate and manipulate cells *Crit. Rev. Biotechnol.* **16** 331-348

[5] J. Suehiro, T. Hatano, M. Shutou and M. Hara 2005 Improvement of electric pulse shape for electropermeabilizationassisted dielectrophoretic impedance measurement for high sensitive bacteria detection *Sensors and Actuators B* **109** 209-215

[6] J. Suehiro, R. Hamada, D. Noutomi, M. Shutouand and M. Hara 2003 Selective detection of viable bacteria using dielectrophoretic impedance measurement method *J. Electrostat.***57** 157-168

[7] J. Suehiro, D. Noutomi, M. Shutou and M. Hara 2003 Selective detection of specific bacteria using dielectrophoretic impedance measurement method combined with an antigen–antibody reaction *J. Electrostat.* **58** 229-246

[8] F. Aldaeus, Y. Lin, J. Roeraade and G. Amberg 2005 Superpositioned dielectrophoresis for enhanced trapping efficiency *Electrophoresis* **26** 4252-4259

[9] T. B. Jones 1995 Electromechanics of Particles *Cambridge University Press*

[10] M. Llamas, V. Giner and M. Sancho 1998 The dynamic evolution of cell chaining in a biological suspension induced by an electrical field *J. Phys. D: Appl. Phys.* **31(21)** 3160-3167

[11] R. Hamada, J. Suehiro, M. Nakano, T. Kikutani and K. Konishi (inpress) Development of rapid oral bacteria detection apparatus based on dielectrophoretic impedance measurement method *IET Nanobiotechnol.*

An Imaging Method for Electrical Capacitance Tomography Based on Projections Multiplication

A. Martínez Olmos*, E. Castillo, F. Martínez-Martí, D.P. Morales, A. García, and J. Banqueri

ECsens group, Department of Electronics and Technology of Computers, University of Granada, 18071 Spain

* e-mail: amartinez@ugr.es

Abstract. In this work, a novel method for the formation of Electrical Capacitance Tomography (ECT) images is presented. This method is based on the modification of the classical algebraic superposition of sensitivity maps, weighted with the inter-electrode capacitance measurements, which is the basis of most of the existing image reconstruction algorithms. The proposed approach replaces this superposition with the scalar multiplication of partial images generated considering only the data corresponding to fixed source electrodes. Results obtained with this proposed method show that the quality of the images can be improved without any image post-processing, which is very interesting especially for real-time applications.

1. Introduction

Electrical Capacitance Tomography (ECT) is an imaging technique for the visualization of the instantaneous distribution in a multiphase flow composed of elements of different permittivity [1]. ECT is based on the measurement of the electrical capacitance between all the different electrode pairs surrounding a pipe when an electric field is applied. From these data, a reconstructed image of the permittivity distribution inside the vessel can be obtained [2].

The reconstruction algorithm selected for a particular application implies a trade-off between imaging velocity and image quality, since improving the quality of a reconstructed image usually requires the image to be processed by filtering techniques, error minimization protocols, etc., which increases the computation time required for obtaining that image. Therefore, in real-time applications, where the elapsed time for the image reconstruction is a limiting factor, simple algorithms are usually selected, for example the Linear Back Projection (LBP) or the Tikhonov regularization [3, 4], which generate images very rapidly, but also with poor quality.

In this work, an image composition method is proposed, which is suitable for being applied with any reconstruction algorithm while providing a better image quality and saving computation time without any additional

2. Image Composition Method

As described above, the aim of any ECT system is to obtain images of the instantaneous phase distribution inside a vessel through a set of capacitance measurements. The forward problem in ECT, i.e., the calculus of the capacitance values for a given permittivity distribution, can be written as a

Published under licence by IOP Publishing Ltd

linearized and discrete form [5] $\Lambda = S\times G$, where Λ is the normalized capacitance vector, G is the normalized permittivity vector, and S is the sensitivity matrix, formed by all the sensitivity maps corresponding to the different electrode pairs in the sensor [6]. The inverse problem in ECT consists of the generation of the instantaneous permittivity distribution from the measured capacitance values, i.e., the resolution of the forward problem stated above, being G the unknown parameters that describe the distribution. The direct solution could be expressed as $G=S^{-1}\times\Lambda$ However, in general, the inverse of the sensitivity matrix does not exist, and the solution must be obtained as

$$g = \hat{S}\times\lambda \tag{1}$$

where g is the approximation to the real solution and $\hat{S}$is a modified sensitivity matrix that represents the inverse of S. Equation (1) is a common expression for many reconstruction algorithms, being the way $\hat{S}$is constructed what establishes the difference between them. In the LBP algorithm, $\hat{S}$is taken as the transposed sensitivity matrix, i.e., $\hat{S}$=ST [5]. In the Tiknhonov regularization, this matrix has a more complex expression [7] $\hat{S}$=$(STS+\beta I)ST$, where I is the identity matrix and $\beta$is a regularization parameter. For the Singular Value Decomposition method, this matrix is written as $\hat{S}$=$V\Sigma^{-1}UT$, where the matrices V, U and $\Sigma$are obtained by decomposing S in singular values [8]. If Equation (1) is further developed, it can be written as:

$$g = \hat{S}^{1}\times\lambda^{1} + \hat{S}^{2}\times\lambda^{2} + \ldots + \hat{S}^{n}\times\lambda^{n} = g^{1} + g^{2} + \ldots + g^{n} = \sum_{k=1}^{n}\hat{S}^{k}\times\lambda^{k} = \sum_{k=1}^{n} g^{k} \tag{2}$$

where n is the number of electrodes in the sensor. The term $\hat{S}^k$ represents the modified normalized Jacobian matrix, i.e., the modified sensitivity matrix, corresponding to the k-th source electrode with respect to the rest of the detecting electrodes, and λ^k is the normalized capacitance vector between the same source electrode k and the rest of electrodes. Each factor in the form $\hat{S}^k\times\lambda^k$ can be seen as a partial image reconstruction for a fixed source electrode, as expressed in Equation (2), which is added to the other partial images in order to generate a final reconstruction. It is obvious that a partial image, formed considering only the data corresponding to a fixed source electrode, will contain an important amount of incorrect pixels, compared to the original distribution. Thus, most of the partial image will be wrong, and only some pixels will represent the real distribution, as it can be seen in Figure 1, where some partial images obtained with the LBP algorithm for a simulated distribution are depicted. If these images are added to form the final reconstruction the original permittivity distribution will be formed by the sum of the pixels of the partial images with a high value (marked in black in Figure 1A), whereas the addition of the pixels with low values will generate a background noise, i.e., a background color different from white, as illustrated in Figure 1A (d).

This noise is inherent to the so-called non-iterative reconstruction methods and should be removed to generate clearer images, and this can be done using some iterative method, which is usually based on the minimization of the image error [4], or applying some image filtering. In any case, this image processing implies an increment of the computation time required for obtaining the image, which can be unacceptable, especially in real-time applications. Moreover, iterative algorithms or filtering techniques are usually experimentally fitted, and the fitting parameters can be useful for certain distributions but not for others, which introduces a serious problem in real applications where the phase distribution is unknown, and can also vary with time.

To solve this problem, an image composition method is proposed in this paper, which implies a modification of Equation (2). Instead of forming the final image by superposing the partial images, that is, by adding them, the permittivity reconstruction can be generated by the scalar multiplication of these partial images, as expressed in the following:

$$g = g^{1}\cdot g^{2}\cdot\ldots\cdot g^{n} = \left(\hat{S}^{1}\times\lambda^{1}\right)\cdot\left(\hat{S}^{2}\times\lambda^{2}\right)\cdot\ldots\cdot\left(\hat{S}^{n}\times\lambda^{n}\right) = \prod_{k=1}^{n}\left(\hat{S}^{k}\times\lambda^{k}\right) = \prod_{k=1}^{n} g^{k} \tag{3}$$

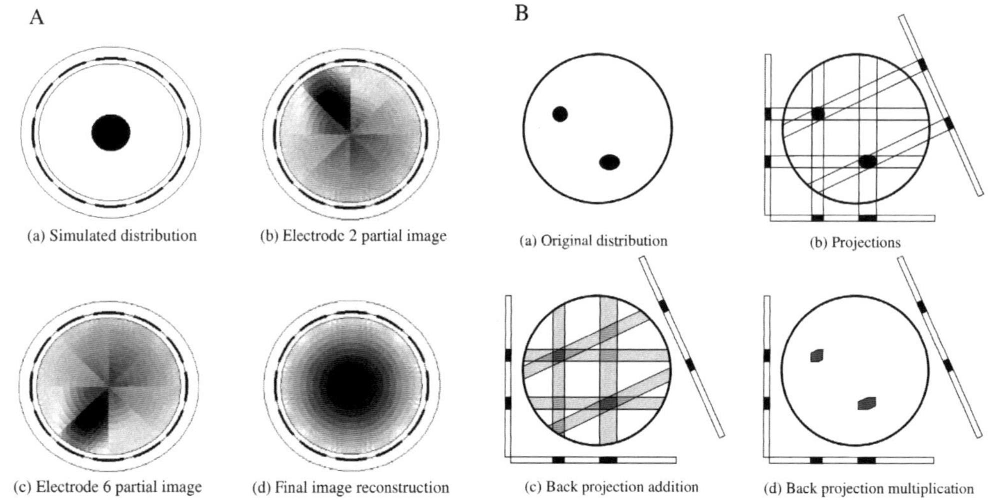

Fig. 1(A) Reconstruction of a simulated distribution obtained with the LBP algorithm. (B) Back projection scheme

where the scalar product, denoted as a dot, means the multiplication of the elements in different matrices occupying the same position. Thus, every pixel in the final image is obtained as the product of the pixels with the same index in the partial images, instead of being obtained as the sum of these, as it was done in Equation (2).

This approach implements a process of mask-based image filtering, in which every partial image obtained for fixed electrode positions acts as a mask for the rest of partial reconstructions. This can be understood if the concept of back-projection reconstruction is revised. This technique typically consists of the addition of all the capacitance measurements, or back-projections, normalized through a sensitivity matrix, as stated in Equation (2) [9, 10]. By summing up all these partial images over the cross section of the pipe, the grey level in the areas where objects with higher permittivity are found will be enhanced, while in other areas the grey level will be lower. This can be observed in Figure 1B, where the scheme of a back-projection reconstruction of two objects is shown. In this simplified example, only three measurements, or projections, are considered, as illustrated in Figure 1B(b). The addition of these back-projections results in the reconstruction depicted in Figure 1B (c), where the areas with higher grey levels correspond to the positions of the original objects in Figure 1B (a), while the rest of the low-grey zones are artifacts or reconstruction errors that do not correspond to any real object. If the normalized back-projections are pixel-to-pixel multiplied, as expressed in (3), instead of added, the reconstruction of Figure 1B (d) is obtained. In this way, only the common grey areas for the three projections remain in the final image, showing the positions of the real objects, while the rest of the pipe cross section presents a white tone, i.e., it is free of reconstruction errors.

This simple idea has been translated to back-projection based reconstructions in capacitance tomography, and it is carried out following two steps: firs, the classical back-projection reconstruction is applied to every set of capacitance measurements corresponding to a fixed position of the source electrode, obtaining the partial images g^k, defined in (2). Secondly, these partial images are pixel-to-pixel multiplied in order to eliminate the areas with low grey level, as defined in (3) and illustrated in Figure 1B. Thus, these areas with low grey level act as masks for the reconstruction, resulting in an enhanced final image free of background noise. This characteristic leads to a better quality of the image reconstruction, since the background image error is removed by the reconstruction algorithm itself, while little or no further processing is required. Therefore, the image processing required to obtain a better resolution and removing the background noise is included in the reconstruction algorithm itself, as it is discussed in the following section.

3. Results and Discussion

The image composition method described in the previous section has been applied for the reconstruction of different permittivity distributions, which have been processed through simulations based on the Finite Element method [11]. Results presented in Figures 2 and 3 compare the images obtained with both methods (addition and multiplication composition), corresponding to equations (2) and (3), respectively, and for three non-iterative and one iterative image reconstruction algorithms, namely the Linear Back Projection, the Singular Value Decomposition algorithm, the Tikhonov regularization and the Landweber iteration. The parameters α and β correspond to the regularization parameters of the SVD and Tikhonov algorithms, respectively. It must be remarked that the images shown in these figures correspond to the direct result of the reconstruction algorithms, with no additional processing. These simulations have been carried out using a 12-electrode sensor.

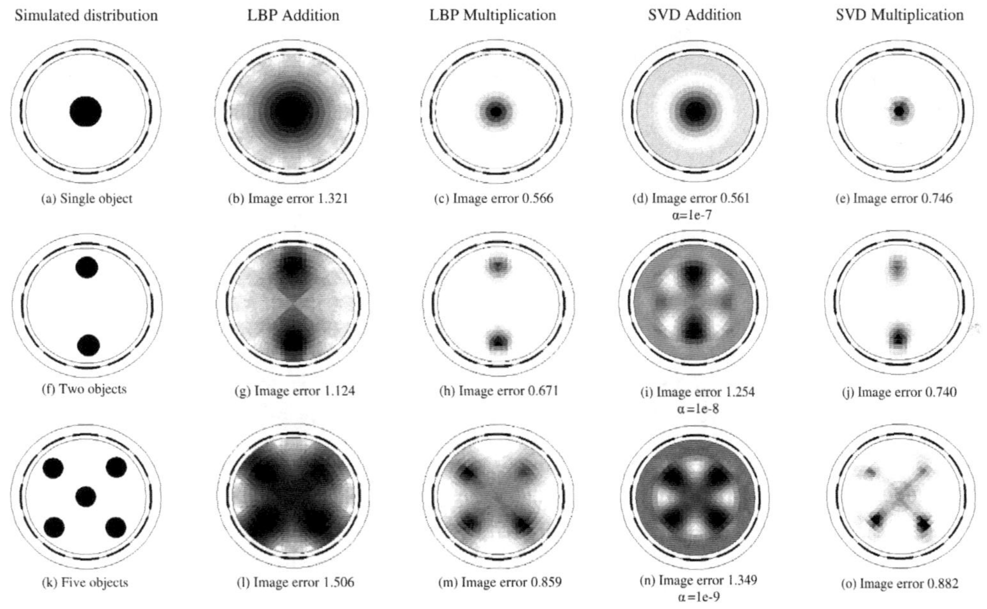

Fig. 2. Image reconstructions using the LBP and SVD algorithms

The reconstruction error is indicated below every image, and it has been calculated as the relationship between the norms of the vectors corresponding to the original simulated distribution G and the obtained with the reconstruction algorithm g [8]:

$$\text{Image error} = \frac{\|g - G\|}{\|G\|} \tag{4}$$

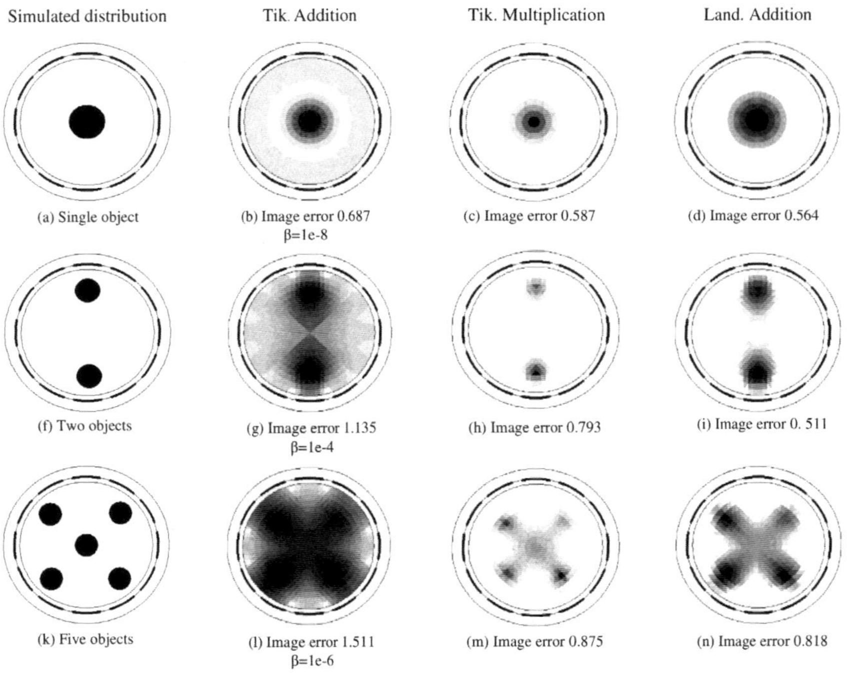

Fig. 3. Image reconstructions using the Tikhonov and Landweber algorithms

From the results shown in figures 2, and 3, it can be seen that the reconstructions obtained with the proposed image composition method are free of the background noise described above, while this noise is observed in every reconstruction carried out with the traditional composition methods. This fact leads to a better quality of the images, as it is confirmed by the lower reconstruction error generated in the multiplication-based reconstructions. Moreover, as indicated in the previous section, the images obtained with the SVD and Tikhonov algorithms have a strong dependence on the value of the regularization parameter selected, and it should be adjusted for every permittivity distribution, as it can be seen in Figures 2 and 3, where the best value generating the lowest image error has been selected in every case. Nevertheless, for the new image composition method, results are independent of these parameters and the same image is obtained when the corresponding value is changed. This characteristic represents a great advantage for real applications, where there is no previous information on the real distribution available for adjusting the regularization parameter. The iterative Landweber algorithm has been applied with 300 iterations and, as it can be derived from results of Figure 3, the image errors have decreased down to similar values to those provided with the one-step LBP reconstruction based on the proposed multiplication-based image composition. Nevertheless, the computation time required to achieve this image quality is higher when using the Landweber iteration; in this case, the reconstructions of Figure 3 have been carried out using custom-developed Matlab® code based on Finite Elements, which run on a PC laptop platform with a Centrino® Duo (2.2 GHz) processor and 2 Gbytes of RAM memory. The elapsed time for the generation of images based on the Landweber iteration is 3.2s, while the images reconstructed using the proposed multiplication-based composition can be completed in 45 ms. Therefore, although the same image quality, or even better, can be achieved using an iterative reconstruction algorithm, it could not be applied if real-time images are required, while the proposed image composition method is able to provide images with low error in a short time, which makes it appropriate for on-line applications.

4. Conclusions

The proposed image composition method, which represents a modification of the classic algebraic combination of the partial images that form the final reconstruction, is a technique suitable for application in every reconstruction algorithm, and generates better images with no need of additional data processing, such as error minimization or filtering methods. This quality improving is the result of the image filtering inherent to the proposed algorithm. Moreover, this composition method is completely independent of the value of the fitting parameters corresponding to any reconstruction algorithm based on regularization, such as the Singular Value Decomposition or the Tikhonov algorithm. All these benefits make the proposed method suitable for real applications, especially in situations where real-time images are required, since higher quality images are produced with no need of periodically adjusting or fitting the reconstruction algorithm.

References

[1] Z. Huang, B. Wang, H. Li, *Application of Electrical Capacitance Tomography to the Void Fraction Measurement of Two-Phase Flow*, IEEE Transactions on Instrumentation and Measurement. vol. 52, pp. 7–12, 2003.

[2] H. Wegleiter, A. Fuchs, G. Holler, B. Kortschak, *Analysis of Hardware Concepts for Electrical Capacitance Tomography Applications*, IEEE Sensors vol. 3, pp. 688-691, 2005.

[3] H.G. Wang, P.R. Senior, R. Mann, W.Q. Yang *Online measurement and control of solids moisture in fluidised bed dryers*, Chemical Engineering Science vol. 64, pp. 2893-2902, 2009.

[4] A.G. Polak, *An Error-Minimizing Approach to Regularization in Indirect Measurements*, IEEE Transactions on Instrumentation and Measurement, vol. 59, pp. 379-386, 2010.

[5] C.G. Xie, S.M. Huang, B.S. Hoyle, R. Thorn, C. Lenn, D. Snowden, M.S. Beck *Electrical capacitance tomography for flow imaging - system model for development of image-reconstruction algorithms and design of primary sensors*, IEE Proceedings-G Circuits Devices and Systems. vol. 139, pp. 89-98, 1992.

[6] H. Yan, F.Q. Shao, S. Wang, *Fast calculation of sensitivity distributions in capacitance tomography sensors*, Electronic Letters vol. 34, pp. 1936-1937, 1998.

[7] J. Lei, S. Liu, Z.H. Li, M. Sun, *An image reconstruction algorithm based on the extended Tikhonov regularization method for electrical capacitance tomography*, Measurement, vol. 42, pp. 368-376, 2009.

[8] W.Q. Yang, L. Peng *Image reconstruction algorithms for electrical capacitance tomography*, Meas. Sci. Technol. vol. 14, pp. R1-R13, 2003.

[9] C.G. Xie, A. Plaskowski, M.S. Beck, *8-electrode capacitance system for two-component flow identification. Part 1: Tomographic flow imaging*, IEE Procedings, vol. 136, no. 4, pp. 173-189, 1989.

[10] S.M. Huang, A.B. Plaskowski, C.G. Xie, M.S. Beck, *Tomographic imaging of two-component flow using capacitance sensors*, J. Phys. E: Sci. Instrum. vol. 22 pp. 173-177, 1989.

[11] S.H. Khan, F. Abdullah, *Finite element modelling of electrostatic fields in process tomography capacitive electrode systems for flow response evaluation*, IEEE Transactions on magnetics vol. 29 pp. 2437-2439, 19933.

Journal of Physics: Conference Series **307** (2011) 012033 doi:10.1088/1742-6596/307/1/012033

Inspection of Samples using a fast Millimetre Wave Scanner

A Hommes, D Nüssler, P Warok, C Krebs, S Heinen, H Essen, [1]
Fraunhofer Institute for High Frequency Physics and Radar Techniques, *Dep. MHS*, Neuenahrer Str. 20, D 53343 Wachtberg, Germany

E-mail: Alexander.Hommes@FHR.Fraunhofer.de

Abstract. Millimeterwaves and terahertz sensors can cover a broad field of applications ranging from production control to security scanners. The outstanding features are the transparency of many materials like textiles , paper and plastics in this frequency region, the good contrast of any humid or dense dielectric material and the capability to employ miniaturized RF systems and small antenna apertures or dielectric probes. A stand-alone-millimetre-wave-imager, SAMMY, was developed and built, to demonstrate the outstanding features of this part of the electromagnetic spectrum for material inspection.

1. Introduction

Millimetre waves and the lower Terahertz region exhibit very good penetration capabilities for any type of dielectric material. Especially security applications demand such features for scanner technology against persons carrying hidden threat objects like weapons and explosives [1] or for the inspection of parcels and letters. Millimetre wave technology can be much cheaper and is completely free of radiation hazard in comparison with available x-ray technology. Moreover a considerably better contrast can be achieved for glass-like or dielectric material as well as for explosives.
The same technology can also be used for quality control of food within its package as long as it is non metallic or for bulk material characterization, e.g. on a conveyor belt with recycling material.
As a first step in the development of a millimetre wave based high speed inspection system a miniaturized stand-alone scanner operating at 77 GHz was developed and tested for several inspection applications.

2. Millimetre Wave Scanner Concept

The scanner should allow the non contacting detection of inhomogenities in dielectric material. Quasi transparent samples are inspected in transmission geometry at a frequency of 77 GHz. The scanning is achieved mechanically by rotating probes. The transmit-receive circuits are based upon commercial millimetre wave integrated circuits from automotive applications. Steering, data acquisition and quick-look evaluation is done by FPGA based electronics built into the compact apparatus. Further evaluation can be done on a PC, which is linked to the scanner.
Quick-look images are processed and depicted in two ways: The first gives the amplitude image for the relative attenuation in the sample, the second shows the phase. Very often phase changes are more significant if the dielectric constant of the sample changes locally due to a foreign object.

[1] Alexander.hommes@fhr.fraunhofer.de

2.1. Mechanical Scanner Concept

The scanning devices are dielectric millimetre wave probes [2] positioned opposing each other on two rotating discs covering the scanning volume in between. The sample is positioned on a sample holder from plexiglass, which moves linearly through the scanning plane. During the circular scanning process each position of the sample holder and thus of the samples upon the holder plate is illuminated by the collimated millimetre wave beam. The total scanning area is 30 cm x 30 cm. Scanning time for this area is < 30 s, which is only limited by the mechanical constraints. From an electric standpoint of view, much faster scanning is possible. For a future development with endless scanning area a speed of 3 m / s is envisaged. As a further perspective the scanning discs can be equipped with more sensor couples to increase the scanning speed or to use different frequencies simultaneously, which may be of importance for material characterization. Figure 1 demonstrates the total mechanical set-up, figure 2 shows a part of the scanning disk.

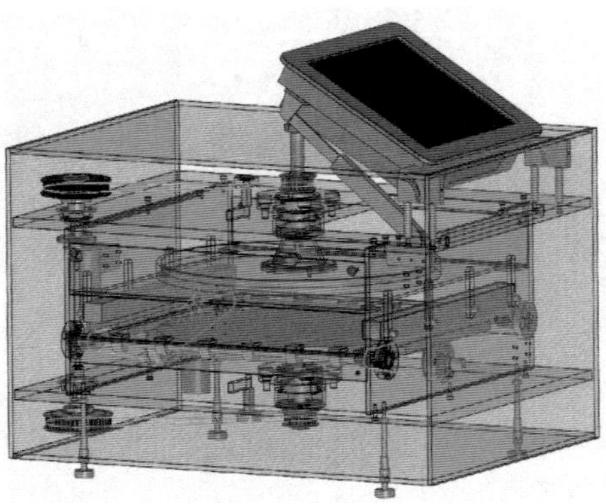

Figure 1. Illustration of the Mechanical Concept of the Scanner

Figure 2. Scanning Disk with part of mmW Hardware

It has to be noted, that all RF hardware is mounted on the two disks. IF signals and DC lines are coupled to the disks using rotary joints.

2.2. Millimetre Wave Circuits

Figure 3 shows a block diagram of the RF section. The transmission system is operating fully coherent at 78 GHz, which allows to evaluate the phase of the transmitted signal in addition to the attenuation. All signals are derived from a highly stable quartz oscillator to which phase lock oscillators are coupled. The millimetre wave frequency is generated by consecutive doubling and tripling with adequate filtering of the sidebands and amplification. The phase comparison is done at the up/down-converted 13 GHz level .

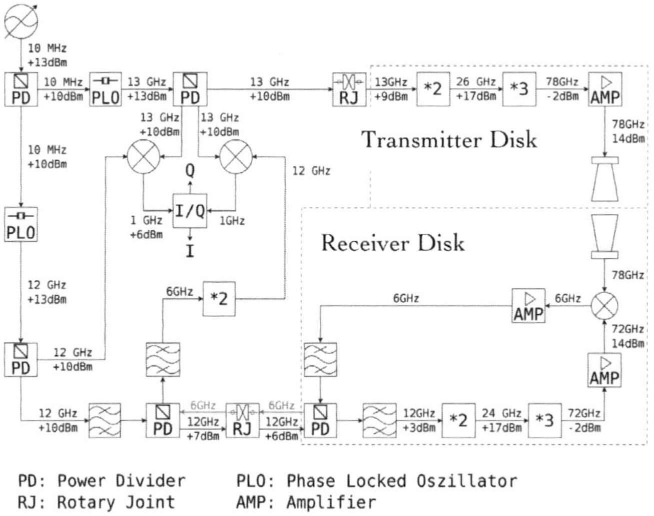

Figure 3. Block Diagram of the Millimetre Wave Circuit

The Transmitter is located at the upper scanning disk, the receiver at the lower disk. The IF signal is transferred via a rotary joint. Further processing is done consecutively.

2.3. Production Prototype

A production prototype was assembled and can be used for demonstration purposes. Figure 4 shows a photo of the prototype.

Figure 4. Production Prototype of SAMMY Scanner

For the prototype a laptop is used instead of the embedded PC to give more flexibility for evaluation tests. The prototype has been tested for different applications like food chain control, letter inspection and quality control. It is fully certified concerning to electrical and mechanical requirements and electromagnetic interference.

3. Results

Typical Samples were scanned with SAMMY. Figure 4 shows an amplitude image of an envelope with gun powder and some metal screws. As the scanner has a fully coherent transmission system, it is also possible to extract the phase changes induced by inhomogenities in the envelope. Figure 5 also shows the respective phase, which is unwrapped and coded into pseudo colours.

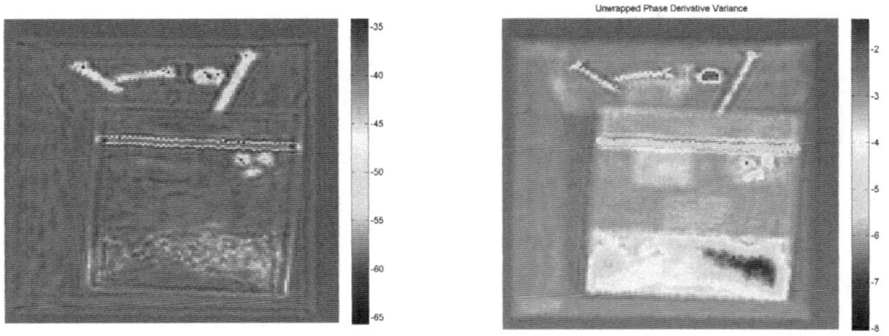

Figure 5. Pseudo Colour Images for Attenuation and the respective unwrapped Phase for an Envelope

The content of the envelope is clearly visible, different materials can be discriminated. The phase image allows a considerable enforcement of border lines compared o the amplitude attenuation image. Figure 6 shows the scanning result for a chocolate bar containing some glass particles. Also for this object the phase image gives a much more detailed representation. The glass particles are very easy to identify.

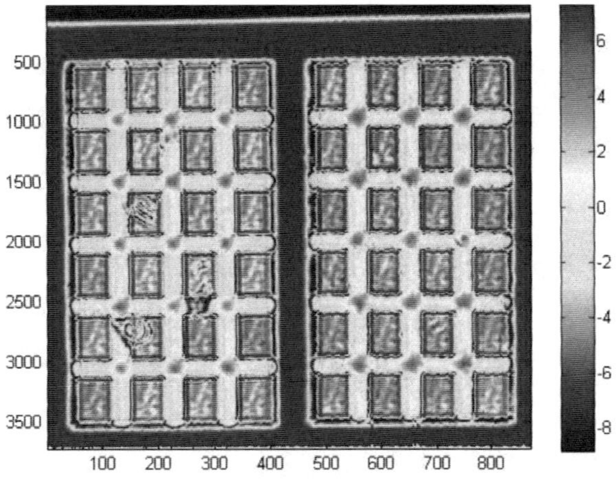

Figure 6. Pseudo Colour Images for Attenuation and the respective unwrapped Phase for Chocolate with Glass Particles

The SAMMY scanner could demonstrate, thatit is possible to give a quite detailed analysis of envelopes for letter inspection purposes and also easily detect foreign particles within food packages. The instrument can be used as it is now for demonstration purposes and inspection in small quantities. The same principle can be employed also for bigger scanners, e.g. for parcel inspection. The frequency of operation allows the use of commercial mmW chips and microstrip technology. In principle also higher operation frequencies can be used, if necessary to gain a better contrast. Going to hiher frequencies, however, would require classical waveguide technology, which is not a principle problem, because alle RF parts are located on the scanning disks and only low frequency signals have to be transferred via the rotary joint.

References

[1] Essen H, Hägelen M, Johannes W, Sommer R, Wahlen A, Schlechtweg M, Tessmann A 2008 *High Resolution Millimetre Wave Measurement Radars for Ground Based SAR and ISAR Imaging*. IEEE Radar Conference, Rome

[2] Hofmann A, Weinzierl J, Schmidt L-P, Brand H 2003 *Flexible Low-Loss Dielectric Waveguides for THz Frequencies with Transitions to metal Waveguides,* GeMic Conference Munich

Non-contact measurement of local conductivity variations in carbon fibre based composite materials

S. Mukherjee, P. Watson, R. J. Prance

Centre for Physical Electronics and Quantum Technology, School of Engineering and Design, University of Sussex, Brighton, BN1 9QT, United Kingdom

r.j.prance@sussex.ac.uk

Abstract. An ultra high input impedance sensor is used to monitor the electric field above a sample of carbon composite. The working distance, spatial resolution and diameter of the sense electrode are all of the order of 50 μm. The sample is excited with an a.c. current over a range of frequencies up to 11 MHz. The resultant potential gradient gives rise to the electric field near the surface of the sample. Results are presented for different excitation frequencies showing the internal structure of the composite material. A gradient subtraction technique is used to enhance the variations in local conductivity.

1. Introduction

Electric Potential Sensor (EPS) technology has been developed at Sussex over a number of years for a wide variety of applications. These span the fields of body electrophysiology [1], security [2], nuclear magnetic resonance [3], forensics and materials testing [4]. EPS technology may be most easily understood by assuming that it behaves as a perfect voltmeter, with infinite input impedance, requiring no external bias current for stable operation. Previous publications on the microscopic imaging applications of EPS have focused on either using an a.c. potential drop method on stainless steel samples [5] or on low frequency low spatial resolution measurements on carbon fibre [6].

Other non destructive testing methods such as acoustic [7], eddy current [8] X-ray and contact surface resistance [9] are of importance, but all suffer from problems when applied to carbon based composites. The most commonly used methods in fault detection are acoustic and eddy current probes. For acoustic measurements the sample generally has to be immersed in a liquid to enhance the coupling between the sensor and the sample. As for the eddy current method, this is best suited to highly conducting materials and so far from ideal for carbon composite. In general x-ray techniques are extremely inconvenient and in many cases will require the section to be tested to be removed from the structure. In addition, for carbon based materials the image contrast is generally poor.

2. Method

The principles of operation of the EPS have been described previously by the authors in detail in the literature [10]. A combination of positive feedback techniques such as guarding and bootstrap are applied together with a mechanism for providing a stable d.c. bias current. The significant development of the technology reported here is a dramatic increase in the operational bandwidth. Previous versions of the EPS, used for microscopic imaging applications, have been restricted to audio

Published under licence by IOP Publishing Ltd

frequency (<100 kHz). The sensor used for the results reported in this paper has an upper frequency -3 dB point of >30 MHz. This has been achieved, without compromising the input impedance, by the use of phase compensation networks in the feedback circuitry to ensure that the positive feedback (guard) signal remains precisely in phase at all frequencies.

The sensor electronics is attached to a coaxial electrode structure with an inner sense electrode diameter of 50 μm. The complete sensor assembly is then placed on the movable carriage of a stepper motor driven XYZ stage. This system allows the position of the sensor to be controlled in three dimensions and raster scans to be performed at a fixed height above a sample with a minimum step size of 6 μm. For the initial laboratory tests reported in this paper the carbon composite sample consisted of a 19 mm wide strip 3 mm thick and was directly connected to an a.c. signal source via metal end electrodes. It was driven by a current (I) of 146 mA, as shown in figure 1, corresponding to a current density of 2.6 mA/mm^2. For a practical implementation of this system it is envisaged that an induction excitation coil, similar to that used for eddy current testing, would be employed. The electric field above the surface of the sample corresponding to the potential drop created by the current flow is detected, as a function of position, by the sensor during a raster scan. The gradient of the potential should be linear for a sample of homogeneous conductivity. In order to visualize the small variations in conductivity within the material this constant linear background is subtracted from the data.

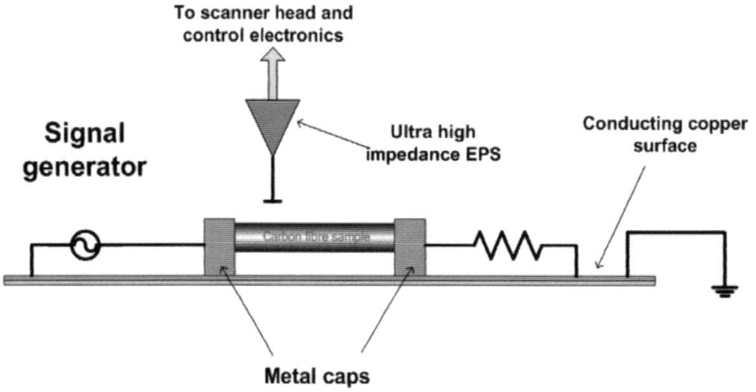

Figure 1. Schematic of the experimental setup.

3. Results

The broadband nature of the EPS used for these measurements is demonstrated by the frequency response shown in figure 2. This was acquired using a calibrated and guarded coupling capacitor of 500 fF, somewhat higher in value than the coupling capacitance usually encountered during a scan. A deliberate decision was made to restrict the low frequency response of the sensor to a few kHz, at the design stage, in order to reduce the settling time required when conducting a raster scan. The data of figure 3a shows the raw output from the sensor when used to raster scan a 12.6 mm x 12.6 mm section of the sample. An excitation frequency of 100 kHz was used in this case. The potential gradient from left to right, the direction of current flow, may be clearly seen in the raw data of figure 3a. In figure 3b a linear background has been subtracted from the data enabling the local conductivity variations to be visualised easily over the whole of the scan area. The rhomboid shaped features are coincident with the weave of the carbon fibre mat within the sample. It is possible that the speckling seen at the top left is indicative of imperfections in the epoxy bonding to the fibre, but we are unable to confirm this at this stage. It should be noted at this point that the composite was fabricated by hand at Sussex in order to produce a sample containing significant imperfections. It is to be expected from simple skin

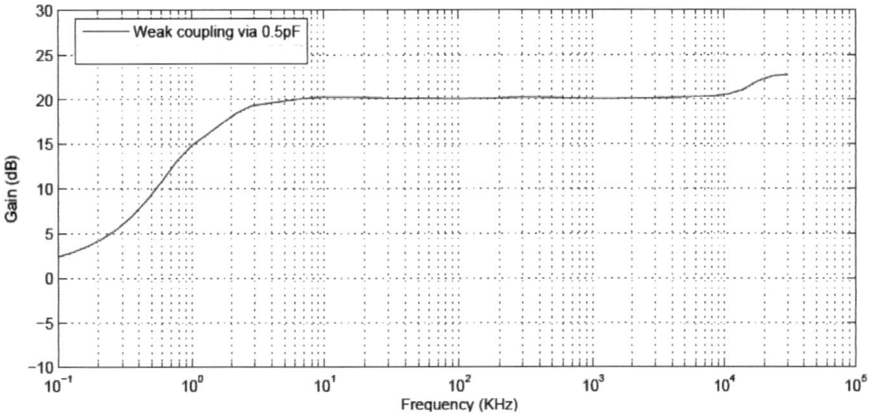

Figure 2. Frequency response of sensor measured using a 500 fF coupling capacitor.

depth arguments that as the signal frequency increases the image will be dominated by surface effects. However, simple skin depth calculations can be misleading for multilayered composites of finite sample size [11]. Initial results obtained at a range of signal frequencies show similar results for this sample, with apparent defects in the top left corner section of the scan area. This may be seen in the results shown in figure 4a at a frequency of 10 kHz and figure 4b at a frequency of 1 MHz.

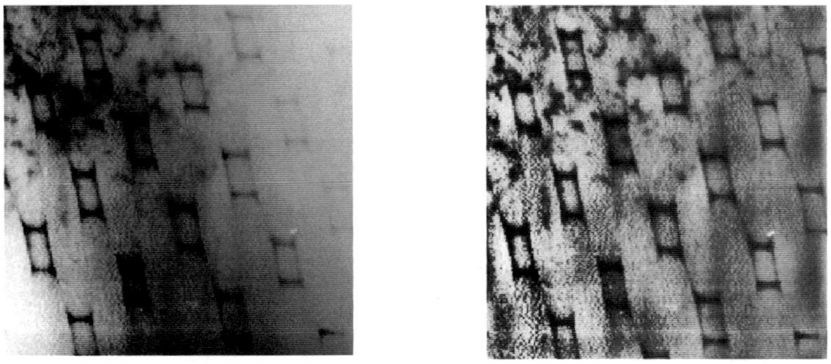

(a) (b)

Figure 3. EPS raster scan of composite sample at 100 kHz, (a) raw data and (b) processed data with potential gradient subtracted.

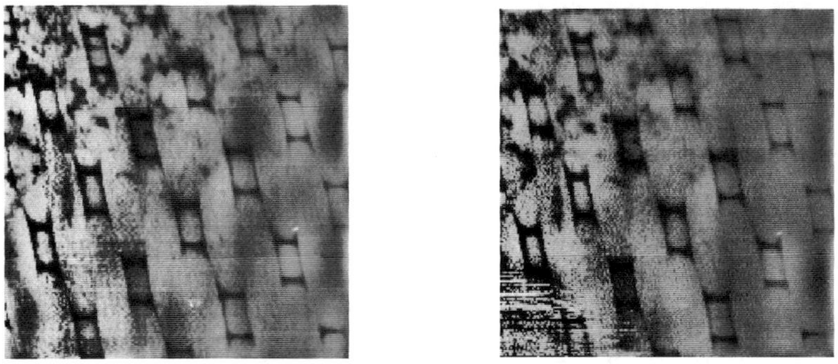

(a) (b)

Figure 4. EPS raster scan of composite sample at, (a) 10 kHz and (b) 1 MHz.

Comparing these initial results taken over a large range of frequencies seems to indicate that the apparent defects are not a surface feature. It would be expected that surface or near surface features would dominate the response with higher signal frequencies. This is not the case with the data presented here. Indeed, if we scan the sample at the highest frequency at which we can currently operate (11 MHz), due to instrumental limitations, we see not dissimilar results occurring. A smaller scan area corresponding to the left hand side of the other scans is shown in figure 5 for comparison purposes. Potential gradient subtraction has been applied to all of the data shown in figures 4 and 5.

Figure 5. EPS raster scan of composite sample at a frequency of 11 MHz.

4. Conclusions

A broadband version of the electric potential sensor has been applied to the measurement of local conductivity variations in a carbon based composite material over a range of frequencies from 10 kHz to 11 MHz. The sensor was used as a non-contact detection system with high spatial resolution, of the order of 50 μm. The well known a.c. potential drop method was employed, but the potential gradient induced by the current was monitored via the electric field it generated close to the surface of the sample, instead of the usual contact probe. Initial results were presented which indicate that local variations in conductivity due to the structure of the composite may be imaged by this technique. In addition, there is some indication that defects within the material may also be imaged. These are seen to be present at the same physical location in each case and at all signal frequencies used. This opens up the possibility of a new technique, similar to eddy current testing, but suitable for lossy poor conductors. In practice such a system would require a complementary non-contact inductive excitation coil. As such, it would also be capable of monitoring local conductivity variations through insulating surface layers.

Acknowledgements

The authors would like to thank the UK Basic Technology Programme and in particular, the EPSRC, for its generous support of this work.

References

[1] Harland C J, Clark T D and Prance R J 2002 Electric potential probes—new directions in the remote sensing of the human body. *Meas. Sci. Technol.* **13** 163–169.

[2] Beardsmore-Rust S T, Watson P, Stiffell P B, Prance R J, Harland C J, Prance H 2009 Detecting electric field disturbances for passive through-wall movement and proximity sensing. Proceedings of SPIE '*Smart biomedical and physiological sensor technology VI*' 16-17 April 2009, Orlando USA' 7313, 7313OP, ISBN 9780819475794

[3] Prance R J, Aydin A 2007 Acquisition of a nuclear magnetic resonance signal using an electric field detection technique. *Appl Phys Letters* **91** DOI: 10.1063/1.2762276

[4] Beardsmore-Rust S T, Watson P, Prance R J, Harland C J, Prance H 2009 Imaging of charge spatial density on insulating materials *Meas. Sci. Technol.* **20** 095711 doi:10.1088/0957-0233/20/9/095711

[5] Prance R J, Gebrial W, Antrobus C 2008 Depth profiling of defects in stainless steel using electric potential sensors and a non contact a.c. potential drop method. *Insight* **50**(2) 95-98.

[6] Gebrial G, Prance R J, Harland C J, Stiffell P B, Prance H 2006 Non-contact imaging of carbon composite structures using electric potential (displacement current) sensors *Meas Sci & Tech*nol **17**(6) 1470-1476.

[7] Chen A S, Almond D P and Harris B 2001 Acoustography as a means of monitoring damage in composites during static or fatigue loading. *Meas. Sci. Technol.* **12** 151–156.

[8] Tavrin Y, Krause H J, Wolf W, Glyantsev V, Schubert J, Zander W and Bousack H 1996 Eddy current technique with high temperature SQUID for non-destructive evaluation of nonmagnetic metallic structure. *Cryogenics* **36** 83–86.

[9] Irving P E and Thiagarajan C 1998 Fatigue damage characterization in carbon fibre composite materials using an electrical potential technique. *Smart Mater. Struct.* **7** 456–66.

[10] Prance R J, Debray A, Clark T D, Prance H, Nock M and Harland C J 2000 An ultra low noise electric potential probe for human body scanning. *Meas. Sci. Technol.* **11** 291–297.

[11] Stiffell P, Prance R J, Gebrial W, Harland C J, Prance H 2008 Defect detection using a non-contact electric potential drop method for multilayer carbon composite materials *Emerging Technologies in NDT* eds. G Busse et al, Taylor and Francis, London, 251-255.

Performance Studies on the Transmitted Light Drop Analyser

M. O'Neill[1], N.D. McMillan[1], S.R.P. Smith[2], S. Riedel[1], K. Arthure[1] and M. Morris[3]

[1] Drop Technology, Tallaght Business Park, Whitestown, Dublin 24, Ireland
[2] Physics Centre, Essex University, Wivenhoe Park, Colchester CO4 3SQ, UK
[3] Spectrecology, 460 Boulder Falls Lane, Jasper, GA 30143, USA

E-mail: martina.oneill@droptechnology.com

Abstract. The testing of a new microvolume spectrometer instrument, the Transmitted Light Drop Analyser (TLDA) is based on drop physics. The advantages, benefits and features of this spectroscopic accessory are elucidated for the first time. The performance of the instrument with regard to drop evaporation, carry-over, calibration standards, photometric accuracy and photometric reproducibility are presented. The paper ends with a brief discussion on the practical significance of the test programme taken as a whole.

KEYWORDS: TLDA; sessile-drop; advantages; features; benefits; evaporation; carry-over.

1. Introduction

The Transmitted Light Drop Analyser (TLDA) provides an elegantly simple 'drop physics' approach to UV-visible spectrophotometry as can be seen by Figure 1. The parameters varied in the modelling were vertical position z of the input fibre above the top of the quartz base; (ii) horizontal position x of the input fibre away from the (vertical) centre line of the system; (iii) drop volume V. This model and preliminary testing of the TLDA is described by McMillan et al [1]. The ray-tracing computer model is based on the Laplace model of the sessile drop and properly describes in a quantitative way the physics of the TLDA optical system. This model provides the tool for engineering the optimized system and ensuring the Spectroscopic signal has the best signal-to-noise.

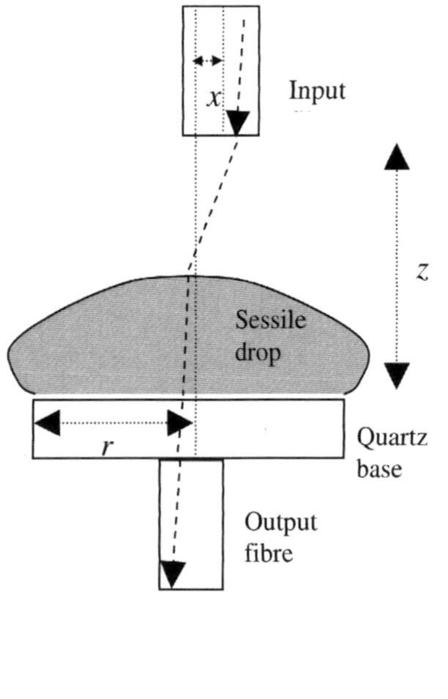

Figure 1. Optical arrangement for sessile drop.

The design of the TLDA seeks to provide ultra-accurate, repeatable measurements on microvolume samples. The TLDA can work as an instrument accessory as shown with a Agilent Cary 50 spectrophotometer in Plate 1 or as part of a collection of components as shown in Plate 2. When working with the Cary50 unit it operates with the Cary external source for this market leading spectrometer or the unit can be adapted to work with any existing UV-visible spectrophotometer with SMA fibre ports. When part of the latter compact system measures it can operate for example with an Ocean USB 4000 spectrometer inside the housing, or alternatively with two fibres to be used with any standard UV-visible spectrometer with SMA fibre plug-in source and detector access.

Plate 1: The TLDA accessory for the Agilent Cary 50.

Plate 2: The TLDA accessory on the workstation showing the pipette holder, pipette tip box, Ocean Optics source and Ocean spectrometer.

The TLDA, micro-volume spectrophotometer, is a patented 'real drop physics' technology that allows for sample volumes as small as 1 μL. Because of this mastery of drop physics it can offer unequalled quantitative performance[1]. The TLDA offers good quantitation for a range of analytical science applications. It has many advantages over other spectrophotometric hardware. Direct, easy measurements can be made in a few seconds – just pipette & wipe it has improved sample handling as the TLDA employs only the drophead surfaces, two optical surfaces are used in some other commercial instruments. The sample placed on this single drophead surface is presented to the source as a discrete and pristine sessile drop. Full spectral output from the spectrometer can be obtained with the Drop Technology TLDA software or working simply and directly with the standard software of the relevant fiber spectrophotometer. Measures of DNA, RNA and Protein concentrations and sample purity can be obtained. Pre-configured methods for common applications such as Nucleic Acid, Protein A280, Bradford, BCA, Biuret, Lowry and Pierce 660nm re incorporated into the TLDA software. The TLDA has an impressive concentration dynamic range without dilutions as drop volumes vary the average pathlength and thereby optimising the concentration range that can be measured. Regular and very simple calibration checks using CRM (certified reference material) standards are provided for those who are concerned about accuracy and minimising measurement error. A full set of Drop Technology accessories including special 1mm optically polished reference cuvettes for absolute calibration of pathlength and absorbance measurement on microvolume samples are available. No instrument adjustments are required and accurate average pathlengths are determined by the pipette setting of the drop sample volume and of course there is low-cost operation – no plates or other consumables.

The TLDA has important features that are worth briefly outlining. The rotating sample loading operation uniquely and importantly can be used with equal facility by either right or left handed users; the top of the instrument rotates in opposite directions. One important practical issue is that the TLDA operates with low cost pipettes with proven unparalleled dispense volume accuracy and surprisingly deliver outstanding measurement quantitation. The TLDA has minimal maintenance and set-up requirements given its simple optical configurations and the drop samples can be viewed directly by the operator after they are placed on the drophead. It provides considerable time saving in sample preparation and the customised work station for positioning of components and ensuring efficient work practice.

There are a number of benefits associated with the TLDA not least the fact that the accessory can be employed with many user's own existing hardware and software. The multi-functioning of this drop instrument/accessory provides access to a variety of analytical techniques. The TLDA can be used in a wide-range of applications areas e.g. bio, pharma, medical, chemical, environmental, food and engineering laboratories. Drop Technology Windows based software can be designed for your specialist applications and includes Custom Methods and data export capabilities to Excel. In addition, the TLDA has some important features.

2. Experimental

2.1 Evaporation

The potential evaporation of a drop put on a substrate is a recurrent issue met by all the experimenters who measure sessile drops. Therefore, a study was conducted to investigate the evaporation effect of small volume drops within the TLDA system. The TLDA evaporation ring creates a high humidity closed system where evaporation from the liquid surface in the reservoir occurs and saturates the air space, in preference to loss from the drop.

A reference water drop was placed on the measurement plinth. The absorbance over a twenty minute time scale was recorded. A constant zero absorbance would be expected if no evaporation occurred. Three repeat measurements were carried out in both cases of (a) not filling the evaporation ring with liquid and (b) filling the reservoir. The new drop in each case was saved as the reference drop. The absorbance was measured at 250nm as this would be considered the worse case scenario as the signal slightly drifts with time due to the solarisation of the fibres and also the grating performs least effectively at this wavelength.

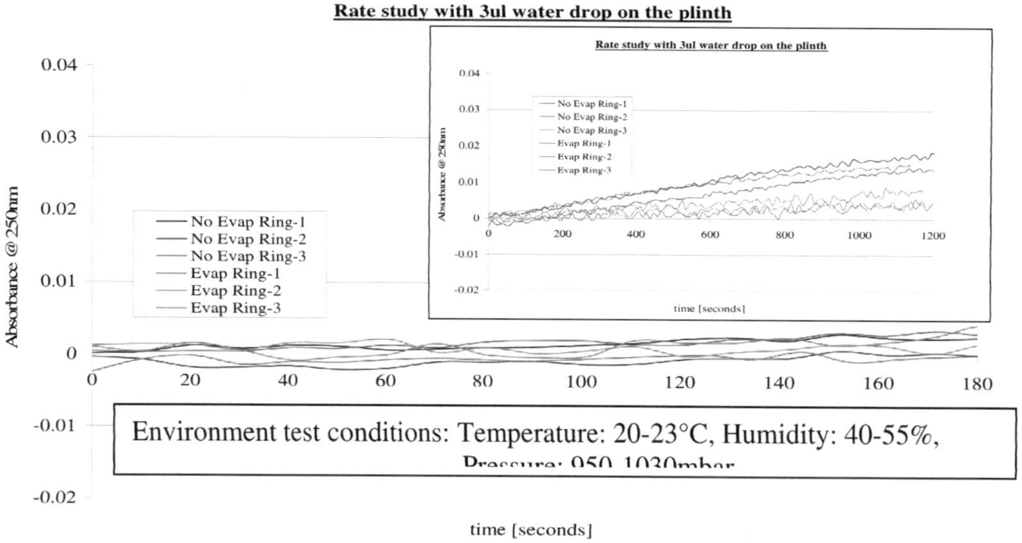

Figure 2. Evaporation rate study for a 3μl water drop.

The graph shows that provided the measurement is completed within the first three minutes, no significant evaporation occurs. However, even over a twenty minute period, the results show that by filling the reservoir with liquid reduces the evaporation effect to acceptable levels.

2.2 Photometric Testing

Drop Technology supplies two certified Reference Materials (DTRM-2SG & DTRM-4SG) for validation of the absorbance scale of your TLDA in a screw-top vial format. These CRMs are certified for both peak wavelength and Absorbance values in a calibrated 1mm pathlength cuvette, together with appropriate expanded uncertainty budgets (*k=2*). Certification is in compliance with ISO Guide 34 & ISO/IEC 17025 protocols. The approach to the calibration of photometric accuracy and reproducibility is illustrated by the statistics of the experimental studies. The wavelength reproducibility and accuracy can also be obtained with these standards.

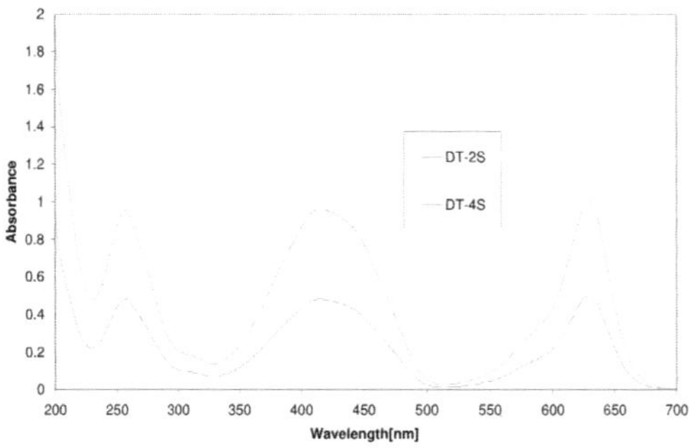

Figure 3. Spectra of the calibration standards

Table 1. Typical Absorbance Values

Wavelength	DTRM-2SG	DTRM-4SG
257.5nm	0.482	0.960
415.8nm	0.481	0.964
629.5nm	0.502	1.006

The absorbance of the standards should be measured at the three peak wavelengths specified on the certification. In the example shown below, these wavelengths are 257.5nm, 415.8nm and 629.5nm. The absorbance scale is verified by using the certified pathlength (supplied with each TLDA plinth) and applying the Beer-Lambert law to the certified reference values. The spectra of the two reference standards are shown in Figure 3 along with the data in Table 1. These pathlength 'corrected' values are then compared to the actual measured values.

The TLDA software carries out this test under the calibration tab. It requests 10 measurements to be made and determines whether the calibration has passed or failed. Regular repeated use of these CRMs will not only provide 'Evidence of Control' but also the raw data for additional Quality Assurance (Q.A.) procedures such as Control Charts, etc.

2.3 Carry-over Study

Two tests were run to demonstrate the lack of sample carryover. A high concentration sample was measured followed by measurements of water samples or a higher concentration sample. The tests indicated no significant presence of sample carryover by simply cleaning with a lab tissue in between sample loading. The DNA absorbance was read at 260nm and was converted to its equivalent concentration value. Table 2 shows the concentration values obtained for the five repeat readings and Figure 4 shows the repeat DNA spectra.

Table 2. Concentration values to investigate carryover with DNA samples

Sample ID	Concentration [ng/µl]
Water	0.0
Water	0.8
Water	0.4
Water	0.4
Water	0.8
Sample A	539.7
Sample A	540.5
Sample A	539.3
Sample A	542.2
Sample A	540.5
Water	0.8
Water	0.4
Water	0.4
Water	0.0
Water	0.0

Figure 4. 5 replicate DNA readings showing reproducibility of spectra

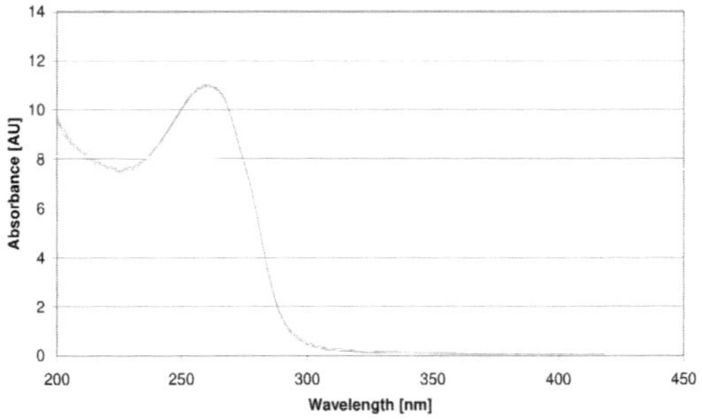

The absorbance of the potassium dichromate solutions was recorded at 350nm. Table 3 shows the concentration values obtained for the ten repeat readings and Figure 5 shows the spectra.

Table 3. Absorbance values of the UV standards.

	$K_2Cr_2O_7$ Concentrations		
	Sample2	Sample1	Sample2
	1.275	0.127	1.225
	1.233	0.152	1.275
	1.309	0.177	1.182
	1.284	0.186	1.309
	1.292	0.152	1.292
	1.242	0.144	1.309
	1.368	0.118	1.275
	1.242	0.127	1.258
	1.225	0.118	1.267
	1.292	0.152	1.258
Average	1.276	0.145	1.265
Std. Dev.	0.044	0.023	0.039

Figure 5. Spectra of the high and low concentration UV standards.

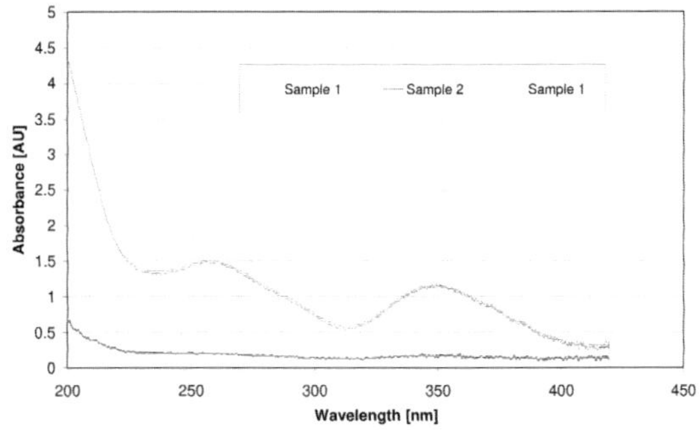

The spectra of the UV standards show insignificant carry over as sample2 is the same after running Sample1 in between.

2.4 The Development of Microclimate Control of Drop Evaporation

One of the authors in a test programme [2] found problems with the evaporation of drops causing noticeable variation in the signal over times frames of minutes. To obviate this problem a circular reservoir was fitted to the drophead which can be seen in Plate 3. The drophead is in fact removable and the reservoir with drophead can be popped out of the TLDA using the Drop Technology tool and a new drophead put in its place. The photograph shows the procedure of drophead removal. The reservoir generates a microclimate around the drophead and reduces evaporation to a negligible amount.

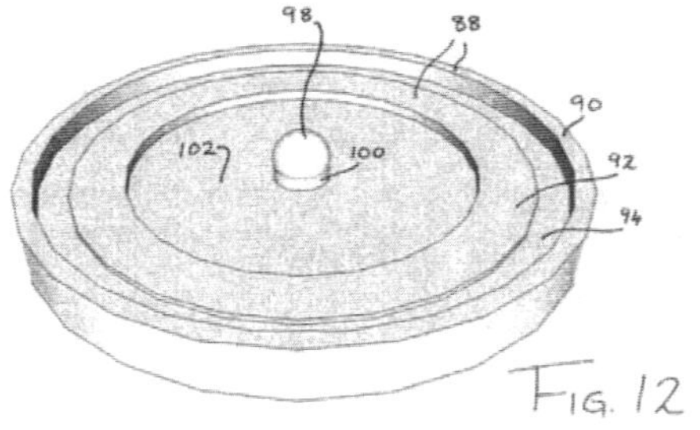

Figure 7. Evaporation ring to scale showing sessile drop sample (98), high vapour retaining wall (90), low inner channel-wall (88), black quartz (102) and clear quartz drophead (100)
Plate 3. The removal of the drophead with the tool.

3. Conclusions

This study demonstrates very clearly the value of product beta-testing. The effects of evaporation of the microdroplets were not in any way significant in Ireland. Consequently, the instrument tested well in Dublin. However, the important testing conducted in Jasper, USA showed that drop evaporation was a practical limitation to the performance of the TLDA when operated in dry atmospheric conditions such as found in desert regions.

The evaporation ring was a simple practical answer and then shown by testing to get over this problem. This plastic ring is one that has the added virtue of allowing dropheads to be simply removed as it easily clicks in or out. A special purpose tool has been designed to make the removal a trivial operation. New TLDA drophead are delivered in packaging that in fact acts as a tool for the insertion of the new drophead and avoids any contact by fingers of the measurement surface.

The testing on carry-over and the photometric testing show very satisfactory results when compared to published results of other microvolume instruments.

The study here reports some of the beta-testing trails undertaken to prepare for the launch of this product as it transfers from research into development and onto a full commercial entity. A golden rule in engineering is to over test to guard against product recall. The presentation of this material will no doubt be useful to those with a very specific interest in microvolume spectroscopy but also perhaps to those who can see the general lessons to be drawn from the experience of the authors.

References

[1] McMillan N.D, O'Neill M, Smith S.R.P, Hammond J, Riedel S, Arthure K and Smith S 2009 *Reliable TLDA-microvolume UV spectroscopy with applications in chemistry and biosciences for microlitre analysis and rapid pipette calibration* J. Opt. A: Pure Appl. Opt. pp 11-22 doi:10.1088/1464-4258/11/5/054016.

[2] Morris M 2010 Private Testing Report of TLDA

Journal of Physics: Conference Series **307** (2011) 012036 doi:10.1088/1742-6596/307/1/012036

Researches towards establishing procedures and protocols for testing instrumental accuracy and reproducibility of UV-visible microvolume drop spectroscopy instrumentation

N.D McMillan[1], M. O'Neill[1], J. Hammond[2], S. Riedel[1], K. Arthure[1] and S.R.P. Smith[3]

1 Drop Technology Ltd., Tallaght Business Park, Whitestown, Dublin 24

2 Starna Scientific Limited, 52/54 Fowler Road, Hainault, Essex IG6 3UT, UK

3 Physics Centre, Essex University, Wivenhoe Park, Colchester CO4 3SQ, UK

Email: norman.mcmillan@droptechnology.com

Abstract. The Reference Materials (DTRM-2SG & DTRM-4SG) for validation of the absorbance scale for a Transmitted Light Drop Analyser (TLDA) have been developed specifically for microvolume spectroscopy. The approach to the calibration of photometric accuracy and reproducibility is briefly illustrated. These materials can be used also for wavelength checks. The study addresses however a wider range of issues for microvolume spectroscopy quality assurance. This study concludes with a general discussion of this quality programme on the important practical and philosophical issues needed for drop science quality systems.

1. Introduction

Reference Materials (RMs) for microvolume spectroscopy need to be liquids, or conceivably solids that could be dissolved up to make a reference solution. The first and most important issue in making proposals for RMs is the requirements concerning health and safety standards for the handling of these materials. The RMs have to be chemically safe in all regards including: environmental and health effect including many aspects. The list includes physicochemical characterisation; properties of the materials; identification of any explosion risks; hazard information like acute and chronic toxicity, toxicokinetics, ecotoxicity, bioaccumulation, etc.; safety with regards to cell and tissue penetration, potential circulatory effects, mutagenicity and genotoxicity and many other such issues. Given the absence of established external quality assessment schemes for such a new technique as drop spectroscopy, the aim is here to advance the discussion on RMs for microvolume spectroscopy. Obviously, the drop spectroscopy RM must have very accurate and reproducible drop spectra that can be obtained in a simple and well-established measurement procedure, and must deliver property values suitable for its intended use, up to its expiration date. The property value can therefore be used as a reference value for inter-comparison purposes.

In summary the following criteria should be considered in assessing the fitness of a RM are:

(i) they should be chemically safe.

(ii) their property values should be stable for an acceptable time period, under realistic conditions of storage, transport and use.

(iii) they should be sufficiently homogeneous that the property value(s) measured on one portion of the batch shall apply to any other portion within acceptable limits of uncertainty.

(iv) be manufactured in a well-defined and regulated manner.

(v) provide good inter-laboratory correlation.

(vi) generate reliable test data from automatic, semi-automatic and manual equipment.

(vi) the tensiographic property values of the RM should have been established with a precision and accuracy sufficient to meet the needs of the end users and should include spectroscopic, photometric and volume standards.

The solution is happily at hand for the UV-visible region of drop spectroscopy as there exists a full gamut of reference materials in this field of spectroscopy which is the most established of conventional spectroscopy disciplines. These materials are available commercially from a number of suppliers but the authors have worked here with Starna Scientific in the development of Reference Materials (DTRM-2SG & DTRM-4SG) [1]. Well-established techniques can be readily adapted for drop spectroscopy and thereby be validated for these calibrations. There are of course numerous existing patents, procedures, methods and assays in existence for UV-visible which actually need to be updated as the theory of 'true' drop science and physics requires a new theoretical basis for this field of spectroscopy. These photometric reference materials have been employed in this present study to demonstrate that the technique is accurate and precise, and provide the essential check on instrument performance on a regular basis to ensure that it is within satisfactory parameters i.e. "under control", and to allow for corrective action to be taken when found to be outside these limits. Their whole range of certified reference materials for absorbance/transmission, wavelength, resolution/bandwidth and stray light could be generally adapted to drop science measurement systems. It is possible that given the use of such materials applications of drop spectroscopy could be approved for use in GLP or ISO 9000 controlled environments, and in due course be accredited to ISO/IEC 17025. Therefore, using the above statement, it is conceivable that drop spectroscopic CRMs with a defined traceable path to national or internationally recognised primary materials or procedures could be made available. By routinely verifying the performance of an instrument it is possible to ensure consistent quality that meets expected performance demands, and drop spectroscopy will have to face up to the need to provide not only RMs but in due course CRMs, as more and more market sector laboratories come under enforced regulatory control. For example, it is a technique that will shortly have to face the most demanding field with respect to CRMs, namely forensic science [2] and this gives an excellent if ambitious guide to considerations for the adoption of RMs for drop science.

2. Experimental

2.1 TLDA Testing with RMs

Drop Technology supplies two certified Reference Materials (DTRM-2SG & DTRM-4SG) for validation of the absorbance scale of your TLDA in a screw-top vial format. These CRMs are certified for both peak wavelength and absorbance values in a calibrated 1mm pathlength cuvette, together with appropriate expanded uncertainty budgets ($k=2$). Certification is in compliance with ISO Guide 34 & ISO/IEC 17025 protocols. The absorbance of the standards should be measured at the three peak wavelengths specified on the certification. In the example shown below, these wavelengths are 257.5nm, 415.8nm and 629.5nm. The absorbance scale is verified by using the certified pathlength (supplied with each TLDA plinth) and applying the Beer-Lambert law to the certified reference values. These pathlength 'corrected' values are then compared to the actual measured values. The TLDA software organises the execution of this test procedure through the data recording under the calibration tab. It requests 10 measurements to be made and determines whether the calibration has passed or failed.

2.2 Thoughts on Quality Programme for Drop Science

Regular repeated use of RMs, and or CRMs, will not only provide 'Evidence of Control' but also the raw data for additional Quality Assurance (Q.A.) procedures such as Control Charts, Calibration Test Records, etc. The work undertaken by the authors is a start towards investigating some of these issues with respect to microvolume spectroscopy. It is important here to keep in mind issues related to Good Laboratory Practice (GLP) and in particular to understand the requirements for regulatory overview dealt with by Burgess and Knowles[3].The first volume of the Burgess and Knowles publication[4] dealt with the crucial issue of cell cleanliness in detail, which as a consequence was only dealt with in a cursory fashion in the second volume [4]. This issue is of course of absolute importance to microvolume spectroscopy and here the conditioning of the quartz drop optics with absorbing lanolin to assist in drops adhering to the plinths has been recommended for some assays and raises serious issues regarding drophead cleanliness. Indeed, the recent development of microvolume spectroscopy raises many new issues of 'Quality Systems' in this area of analytical measurements[5]. The performance characteristics and quality are of course central now to good laboratory procedures [5]. This present work here seeks to begin discussion on some new facets of microvolume quality standards, accreditation, methodologies and protocols that will advance quality systems for these new areas of spectroscopic work for drop physics techniques.

Figure 1. Spectra of the calibration standards

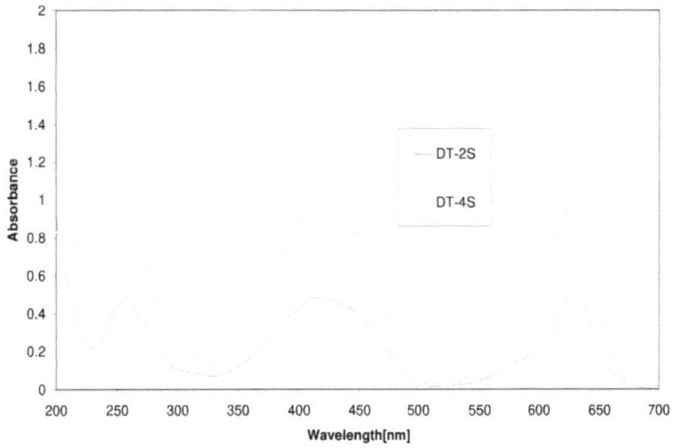

Wavelength	DTRM-2SG	DTRM-4SG
257.5nm 415.8nm	0.482 0.481	0.960 0.964
629.5nm	0.502	1.006

Typical Absorbance Values

The first check beyond the obvious standard RM or CRM procedures for photometric accuracy and reproducibility is to provide a validation to determine the pathlength for microvolume drop spectroscopy. This determination required a reference instrument with proper known performance characteristics. When working with 'drop physics' methods a development of standard Beer's law methods is necessary in considering these RM issues as shown by recent work [6]. The basis for these 'true drop physics' techniques is a modified Beer's law developed on the basis of a ray-tracing Laplace-Young model of the drop. The absorbance/pathlength data from the Specord spectrophotometer used here was selected for the present work because of clear metrological considerations. Importantly, the **SPEKOL®** systems operating in the range of 190-1100 nm has been referenced against diverse quality standards such as Ph.Eur., USP, TGA and ASTM. The demand of analysis reliability of course requires a regular validation and given microvolume spectroscopy is a new and developing area, it is recognised immediately that it will be necessary to begin defining appropriate quality procedures for microvolume spectroscopy. Starna Scientific have developed recently manufactured a 1 mm pathlength cell. This special cell has been designed to provide a reference quartz cuvette with a pathlength appropriate to microvolume measurements. As with all such Starna cells, these 1mm cuvettes are polished to interferometric precise dimensions. Reference data from the 1cm and 1mm Starna cuvettes was then used to estimate the pathlength for a 3μl drop in the TLDA system by comparative measurements with an RM.

The testing involves applying the modified Beer's law. The absorbance measure against the pathlength (A/ℓ) should have a constant and indeed predictable value for a RM. For RM the molar absorptivity has a well established value and the pathlength should be constant in the reference cuvette. From this modified Beer's law it can be seen that an individual sample, should deliver an equal εc (Beers law) no matter what pathlength is used. For microvolume spectroscopy of course the pathlength is an average of various ray paths through the drop and will in the first instance depend on the microdrop sample volume. Comparative measurement between the reference and microvolume measurements should deliver experimentally the TLDA pathlength that agrees with that predicted theoretically from the Laplace-Young model. The value obtained for the path length we have defined as the 'effective pathlength' (ℓ_{eff}). The results are shown below to illustrate the thinking involved in developing our quality assurance procedures.

Martina: Average specord absorbance/pathlength value @257.5nm divided by TLDA abs at 257.5nm

Martina: Average specord absorbance/pathlength value @λp and 416nm divided by TLDA abs at 415.8nm

Martina: Average specord absorbance/pathlength value @257.5nm divided by TLDA abs at 257.5nm

Martina: Average specord absorbance/pathlength value @λp and 416nm divided by TLDA abs at 415.8nm

			l=1cm		l=1mm	
Sample	TLDA-257.5nm	TLDA-415.8nm	257nm	415.8nm	257nm	415.8nm
	0.061	0.061	0.126516	0.126795	0.122179	0.123989
	0.061	0.063	0.126516	0.130953	0.122179	0.128054
2S	0.062	0.063	0.12859	0.130953	0.124182	0.128054
	0.06	0.061	0.124442	0.126795	0.120176	0.123989
	0.061	0.061	0.126516	0.126795	0.122179	0.123989
Average	0.061	0.0618	0.126516	0.128458	0.122179	0.125615
Std Dev	0.000707107	0.001095445	0.001467	0.002277	0.001416	0.002227
	0.115	0.118	0.119743	0.122369	0.113505	0.119379
	0.119	0.121	0.123908	0.12548	0.117453	0.122414
4S	0.117	0.12	0.121825	0.124443	0.115479	0.121403
	0.117	0.12	0.121825	0.124443	0.115479	0.121403
	0.122	0.127	0.127032	0.131702	0.120414	0.128485
Average	0.118	0.1212	0.122867	0.125687	0.116466	0.122617
Std Dev	0.002645751	0.003420526	0.002755	0.003547	0.002611	0.003461

Figure 2. Spreadsheet comparison showing approach to the comparative approach required to determine the effective pathlength of a drop sample measured by the TLDA.

The same measurements were repeated on two other TLDA instruments (TLDA2 & TLDA3). This comparative approach is very useful and indeed important in properly determining the experimental 'effective pathlength' of the TLDA. Such a series of comparative intra-instrumental measurements are indeed vital in clarifying these quality issues. This TLDA instrument was developed by the authors and detailed in Drop Technology [7]. It may be useful therefore to provide a brief discussion of the work involved in validating this new instrument as part of a more thorough going series of measurements and tests aimed at establishing top-down and bottom-up methods. The Excel spreadsheet from Dr Martina O'Neill who conducted these tests is shown to visually present the procedures to used here to check the predicted theoretical Laplace-Young model values. This is of course a top-down approach from quantitative theoretical values to experiment. The familiar top down RM and CRM traceable measurement issues are well known, but we are here clearly working also with a completely new range of bottom–up issues. Explicit use research knowledge of the instrument's manufacture should allow extrapolation so as to confirm that all outcomes, that have not specifically individually checked, will still be met. In short, the new 'drop physics' demands the TLDA developers to re-evaluate every issue once again and not to simply assume that because these matters have been resolved for traditional Beer's law instruments that these issues have been laid to rest. Such an assumption would be very dangerous given a new theoretical foundation has been defined for 'drop physics'. The results presented here are meant to highlight new practical issues and begin this debate. This discussion is meant to underline the importance of developing these new quality approaches which deserve some greater attention than space allows here.

2.3 Theoretical Issues on Drop Science

Collimation of light is in non-laser systems really an approximation that in reality cannot seriously be stood over as an absolute achievable reality. When working with cuvettes, there is the vexed matter of multiple reflections from the four cuvette surfaces that further complicate the 'collimation' situation.

What is seen either in the spectroscopy associated with either pendant or sessile drop physics is that serious consideration must be given to what is meant by ***pathlength***. This issue was theoretically addressed by McMillan, Smith et al [8] for pendant drops and then later specifically for the sessile drops of the TLDA. There is theoretical integral pathlength defined by $\ell_1 = \int \ell \mathrm{P}(\ell)\, \mathrm{d}\ell$ computed from the integral of a series of rays passing through the drop of length ℓ, each path of which has a probability of $\mathrm{P}(\ell)$, and it is to be remembered the sum of these probabilities for all ray-traced paths is unity. Thus, the probability that the pathlength lies in the range ℓ to $\ell+d\ell$ be $P(\ell)\ d\ \ell$, and normalized such that $\int P(\ell)d\ell = 1$. Above we have explained we can experimentally determine the effective pathlength ℓ_{eff}. It should be noted that $\ell_1 \neq \ell_{eff}$. The idea of an rms pathlength ℓ_2 which is defined from $\ell_2^2 = \int \ell^2 \mathrm{P}(\ell)\, \mathrm{d}\ell$ is a useful quantity where the rms pathlength is determined as the integral of the square of these paths. The variance in pathlength is defined mathematically as usual as $\Delta\ell^2 = \ell_2^2 - \ell_1^2$. If there is a truly collimated light source the $\ell_1 = \ell_2$, but this is obviously not the case in a microvolume drop spectrophotometer. It is found that Beer's law applies in pendant and sessile drops only as an approximation and thus $A_{Drop} = \varepsilon c \ell_1$. For what would be the theoretically and most exacting situations $A_{Drop} = \varepsilon c \ell_1 - k\varepsilon^2 c^2 \Delta\ell^2$ for the pendant drop where ε is the molar absorptivity. A similar first order approximation is required in the TLDA. A small correction applies to situations were we have deviating from the linear Beer's law relationship. In summary, the effective pathlength is determined experimentally and can be compared with theoretical values derived from the ray tracing model of drops whose complex dropshapes are derived from a Laplace-Young predicted shape. Summarizing, drop physics will require a great deal more rigorous thought than has been required in traditional UV-visible spectroscopy with regards to the thinking connecting theory and practice in 'drop physics' based spectroscopy.

3. Conclusions

This are many recent developments pushing forward the analysis small liquid volumes, e.g. <3 μL, due to the increasing demand to preserve in non-destructive testing method of what are precious samples and especially in the biotechnology/pharmaceutical markets. The challenge of making reproducible measurements has of course been recognised by all the commercial leaders in the field. The theoretical and practical task of delivering such systems perhaps however falls to those developing 'real drop physics' such as commercialized in the TLDA. In theory, such 'drop physics instruments' should provide improved accuracy and reproducibility coupled with the ability to measure micro volumes in a user friendly piece of equipment. The corollary here is these instruments raise a whole raft of new quality issues. This paper has raised some initial thoughts and discussion on procedures for 'drop physics' quality requirements and it is clear that this demands attention to both proper definitions of pathlength and measurement protocols.

The user wants simplicity and increased productivity and if you consider our next generation of users as the young people now entering laboratories it is clear that they all live and breathe technology. They will like all new generations welcome the new demands of the technology. Advances in sample handling, associated with micro volume measurements are a good example of an important area of advance in the oldest laboratory technology, namely UV-visible. As applications and demand expand for lower volume sample measurements, new advances approaches are appearing in the field of micro measurements. Of course rapid advances in microelectronics resulting in the development of new light sources, such as high intensity LEDs covering the UV to the NIR region of the spectrum, will contribute to not only the advance of this area of spectroscopy, but also to an opening up of new combinations of spectroscopic instrumentation. Lab-on-a-chip will see new developments in conventional spectrophotometers and help open the door to lower cost, handheld UV–visible measurement systems. Handheld and disposable instruments and measurement systems will be common. What is obvious that in these new developments there must be greater attention to standards and procedures of calibration and importantly traceability. The demand for UV–visible instruments will continue to increase as it is an inexpensive and non-destructive technique for measuring liquid and solid materials. The work here should open out the discussion of fundamental quality issues for drop physics and a new application field of analytical science.

References

[1] Starna review see http://www.optiglass.com/ukhome/d_ref/starna_colourimeter_accessor09.pdf

[2] Epstein D.M, Lothridge K.T and Tilson W.J 2001 *The use of CRM in Forensic QA13*, INTERPOL Forensic Science Symposium, Lyon, France, Oct. 16-19

[3] Hammond J 1999 *Regulatory Overview* in 'Standards and Best Practice in Absorption Spectrometry' eds Burgess C and Frost T, Oxford pp. 120-130

[4] Burgess C, Knowles A (eds) 1981 *Techniques in Visible and Ultraviolet Spectrometry Vol. 1, Standards in Visible and Ultraviolet Spectrometry*. Chapman and Hall, K, Smith London.

[5] Currell G 2000 *Analytical Instrumentation: Performance Characteristics and Quality* Wiley, New York pp.45-66.

[6] McMillan N.D, O'Neill M, Smith S.R.P, Hammond J, Riedel S, Arthure S, 2009 *Reliable TLDA-microvolume UV spectroscopy with applications in chemistry and biosciences for microlitre analysis and rapid pipette calibration*, J. Opt. A: Pure Appl. Opt. **11** 054016 (11pp) doi:10.1088/1464-4258/11/5/054016

[7] Drop Technology website 'www.droptechnology.com'

[8] McMillan N.D, O'Neill M, Pringuet P, O'Mongain E, O'Rourke B, Riedel S, Neill M, Augousti A, Wüustneck N, Wüstneck R, McMillan D. D. G, Colin F, Hennerbert P, Pottecher G and Kennedy D, 2008 *Quantitative drop spectroscopy using the drop analyser: theoretical and experimental approach for microvolume applications of non-turbid solutions,* Meas. Sci. Technol. 19 055601 pp.18pp

Handheld colorimeter for determination of heavy metal concentrations

N. López Ruiz*[1], M. Ariza[2], A. Martínez Olmos[1], J. Vukovic[3], A.J. Palma[1], L.F. Capitan-Vallvey[2]

ECSens [1]ETSIIT, Departamento de Electrónica, Universidad de Granada. C\ Periodista Daniel Saucedo s/n 18071 Granada, Spain. [2]Department of Analytical Chemistry. Faculty of Sciences. University of Granada, E-18071 Granada, Spain. [3]Department of Analytics and Control of Medicines, Faculty of Pharmacy and Biochemistry, University of Zagreb, A. Kovačića 1,HR-10000 Zagreb, Croatia.

* e-mail: nurilr@ugr.es

Abstract. A portable instrument that measures heavy metal concentration from a colorimetric sensor array is presented. The use of eight sensing membranes, placed on a plastic support, allows to obtain the hue component of the HSV colour space of each one in order to determinate the concentration of metals present in a solution. The developed microcontroller-based system captures, in an ambient light environment, an image of the sensor array using an integrated micro-camera and shows the picture in a touch micro-LCD screen which acts as user interface. After image-processing of the regions of interest selected by the user, colour and concentration information are displayed on the screen.

1. Introduction

The detection and determination of heavy metals in the area of analytics and drug control are important and demanding tasks, and, therefore, drug analysis and quality control require increasingly better analytical methods and procedures to solve specific problems in pharmaceuticals and in the industrial production of medicines [1]. A recent trend to obviate the difficulties of obtaining single analyte sensors based on specific receptors – problems of selectivity – is to combine non-specific sensors in an array that produce a set of analytical signals (electronic tongue for gaseous species/electronic nose for species in solution). This information with higher dimensionality, encoded in the overlapping measurements, can be processed to extract qualitative and quantitative information through advanced mathematical procedures of pattern recognition and/or multivariate analysis [2,3], improving in this way the performance of the sensors [4].

Different principles have been used for the preparation of sensor arrays for electronic tongues, with the most widely used being: tin oxide catalytic sensors, conducting polymer sensors, acoustic wave sensors, quartz crystal microbalance sensors, sensors based on MOSFET technology, systems based on ion mobility spectrometry as well as mass spectrometry techniques like API and PTR, and lastly, optical techniques, principally fibre optics and fluorescence [5].

The basic requirements for the optical tongue systems, in reference to the sensing array, which is the key of the system, is that sensors should present low selectivity or high cross-sensitivity, and that they have reproducible analytical characteristics. Moreover, it is necessary to look for sensors to be more sensitive and more robust, which is contradictory to a certain degree, since usually the more sensitive a sensor is, the less robust it becomes [6]. One solution is to use disposable sensors which are, therefore, not integrated into the device. Additional requirements refer to cost and portability.

Most different optical approaches used for tongues are based on the measurement of colour using different colour spaces obtained from the image captured principally by imaging devices, mainly CCD cameras or scanners. By far, the most commonly used space is RGB, whose coordinates are used for processing with multivariate techniques. However, in this work, we used the HSV colour space, whose main characteristic is that it represents the useful information of the colour in one single parameter, the coordinate H (hue). Previous studies from our group have shown that the use of the H coordinate in bitonal sensors that produce a change in colour by reaction yields a substantial improvement in resolution and repeatability [7].

We approached the analysis of heavy metals using the concept of an optical tongue using an array of membranes placed on a transparent support that contains immobilized conventional chromogenic reagents [8]. Their usual drawback, i.e. their lack of selectivity, is used here as the source of non-selective information about the metals present, considering their usual good sensitivity.

2. Preparation of the sensing array

Different chromogenic reagents for heavy metals were tested for immobilization using different strategies such as ion-exchange, adsorption, and entrapment using as criteria for selection the conditions of no leaching, change in tonal colour coordinate by reaction and non-selective behavior.

As chromogenic reagents for this proof of concept study, 1-(2-pyridylazol)-2-naphtol (PAN) and zincon (ZIN) were used and the needed amounts of polymers, reagents, plasticizers and lipophilic salts were optimized. For characterization of sensing films, individual membranes were prepared by using a spin-coating technique from the cocktails containing all the reagents in THF or in ethanol:water mixture on the sheets of Mylar-type polyester. In addition sensing arrays were prepared on a Mylar support covered by an adhesive and transparent PVC layer with 12 holes, 5 mm in diameter each. The array was produced by placing 8 μL of the corresponding cocktail in each hole.

3. Description of the instrument

The portable instrument here developed is a microcontroller-based system that allows to calculate the H component of the HSV colour space from an image of a colorimetric sensor array obtained with a micro-camera. The colour information is used to determinate the concentration of metals in a solution. The block diagram of the prototype described is in Figure 1.

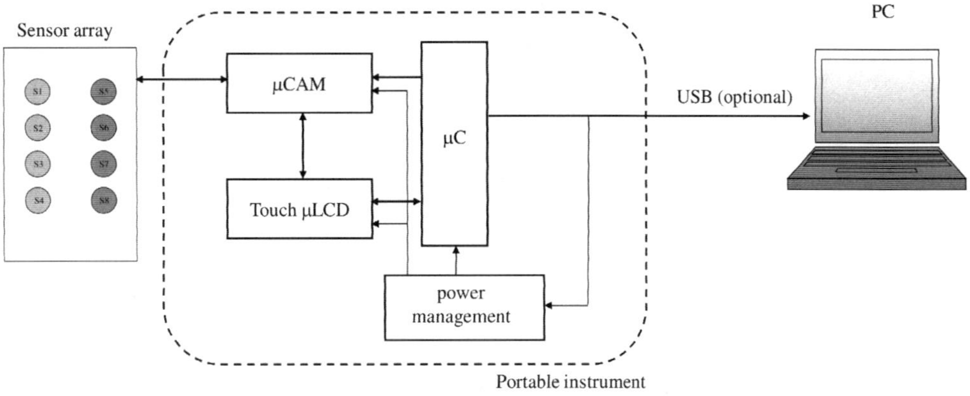

Figure 1: Block diagram of the instrument.

The model of microcontroller used in this instrument is the PIC18F2550 (Microchip Technology Inc., USA) which is a low cost device with an integrated USB module that has been used for developing and calibration purposes. The user can interact with the instrument through a touch micro-

LCD screen model μLCD-32PT-SGC (4DSystems, Australia) included in the prototype, which also shows the colour information obtained from the sensors. Therefore, there is no need of external elements to handle the instrument. The touch micro-LCD screen integrates 4-wire resistive touch panel which has 240x320 QVGA of resolution and 65K colours. The dimensions of the screen are 56 x 77.6 x 3.7 mm and 3.2'diagonal, which are high enough to present an easy and clear interface to the user. Moreover, the screen integrates an on-board micro-SD memory card adaptor for multimedia storage and data logging purposes. The instrument makes use of a micro-camera model TTL-uCAM (4DSystems, Australia) to capture an image from the sensor array which is shown in the micro-LCD screen. The micro-camera is a module with serial output that allows to acquire images in JPEG or RAW format in different sizes and colour or gray-scale resolutions. Both the micro-camera and the touch-screen provide a set of embedded commands for an easy communication and programming. The main reasons for choosing this model of LCD-screen are the touch function to avoid external elements for user interaction and the high colour resolution which is enough for the explained purpose, achieving a handheld colorimeter.

For this work, a 16-bit colour RGB and RAW format was chosen because of the simplicity in the structure of this format which does not use any compression algorithm. Thanks to the compatibility between the micro-camera and the micro-LCD screen the RAW format permits direct communication between them with no action of the microcontroller or any external memory. However, since RAW is an uncompressed format, the received data from the micro-camera are heavier than JPEG format which increases the time required to obtain the image, but it makes possible a direct treatment of the pixels information with no need of a previous processing.

In Figure 2, a photography of the prototype described is presented. The micro-LCD screen shows an image of the sensor array in which are presented different membranes. The user can select, by touching the screen, the individual sensors to calculate the H coordinate and metals concentration, so that a defective sensor can be ignored. The sensor array is placed at the bottom of the instrument, in a hole built facing the micro-camera, covered by a transparent Mylar polyester sheet, to allow to capture the picture of the membranes which is processed in hindsight.

Figure 2: Photography of the instrument.

The prototype has multiple ways to obtain the power supply: from a PP3 battery of 9V through a voltage regulator to generate 5 V, from a AC/DC external adaptor or via the USB port.

3.1 Measurement procedure

The concentration of metals through colour determination of the sensor array is carried out after image processing from the picture of the membranes captured with the micro-camera. The image is taken in an ambient light environment; therefore the prototype has not need of a specific light source to obtain a fine image. To avoid overexposure in the image, a thin white film of PVC is positioned between the sensor array and the light source to act as an intensity attenuator of the ambient light.

The sensor array is placed by the user in a holder situated facing the micro-camera. The aperture has a piece of transparent Mylar to place the sensor array for taking the image. After the sensor array is positioned in the instrument, the user starts the capture of the picture by touching the micro-LCD screen which was in a sleep mode.

Once the image is presented on the micro-LCD screen, the user can select a particular sensor to obtain its H component and the correspondent metal concentration, as it can be seen in Figure 3. The touch option offers the user the possibility of discarding defective membranes of the image and only takes into account good quality membranes to calculate the concentration of metals. On the other hand it is possible to determine the H component of every membrane at the same time. As the user can select a specific sensor by touching the micro-LCD screen, the position of the sensor array respect to the micro-camera is not critical since the H parameter is determined using the area selected by the user.

The algorithm used to obtain the H value consist of a pixel-by-pixel treatment of the image in which the RGB coordinates are loaded and processed to generate the hue component for the regions of the image corresponding to the metal sensitive membranes. A statistic of these pixels is carried out to determine the mode of the H data which is taken as the value for the whole membrane. After image processing to estimate the hue component, the concentration of copper and zinc can be calculated and shown in the micro-LCD screen.

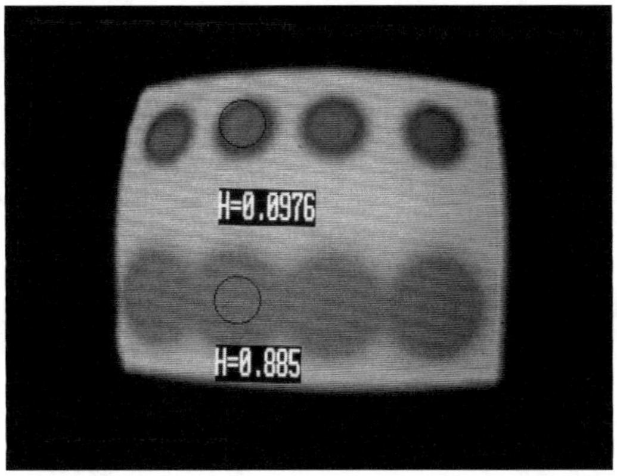

Figure 3: Micro-LCD screen showing obtained H components for two membranes.

4. Results

The characterization of the sensor array was performed using four replicates for each kind of membrane, PAN or ZINCON, using solutions containing the metals ions in a range from 0.06 to 65 mg/l for each metal. The reaction conditions (equilibration time, pH) were studied and optimized using single membranes against each metal ion. A comparison between the response of the prototype and a commercial scanner ScanMaker i900 (Microteck, Taiwan) is shown in Figure 4. The used scanner has a 6400 x 3200 dpi resolution and a maximum optical density of 4.2 and 24 to 48 bits of colour. The H parameter is calculated from the scanned image using the software Matlab r2007b (The MathWorks, Inc, Natick, MA, USA) [7].

As depicted in Figure 4, the response curves obtained with the scanner and the prototype for copper and zinc concentrations present the same behaviour and in both cases are obtained through image-processing of the captured image. Nevertheless, there are some discrepancies on the responses, caused by the different light sources and the lower resolution of colour in the micro-camera. The experimental curves for copper and zinc were modelled by splitting them into linear sections in order to obtain the metal concentration from the H parameter previously determined [9].

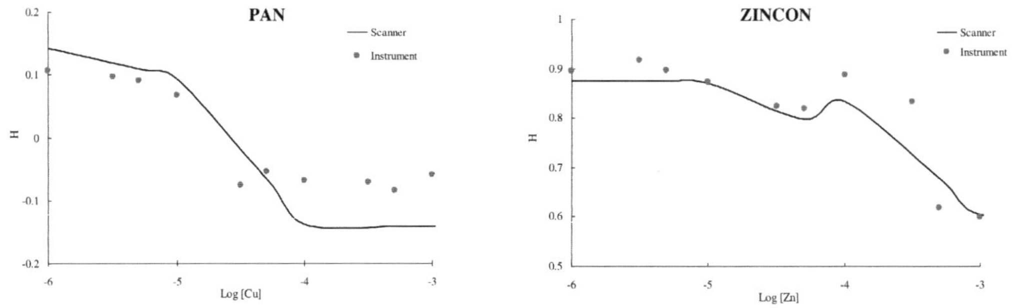

Figure 4: Response of the sensor array for copper and zinc.

5. Conclusions

A handheld colorimeter which uses the hue component of HSV colour space for a colorimetric sensor array to calculate the concentration of metals in solutions such as zinc and copper is presented. The analytical procedure consists of determining the colour changes in two different kind of sensing membranes, PAN or ZINCON, and relating them with variations of the metal concentration.

The H component of the colorimetric sensor is obtained from an image captured with an integrated micro-camera and displayed in a in a touch-micro LCD screen. The main advantage of the direct communication between the micro-camera and the LCD screen is to avoid the necessity of using an external memory to save the image, since the colour information of each pixel can be directly read from the screen. The prototype works in an ambient light environment; therefore there is no need of a specific light source integrated on the prototype. Because all of this, the result is a simple electronic design with a reduced number of electronic components.

A study of the membranes response in solutions with different metals concentration was carried out to obtain a complete characterization of the instrument. The obtained curves have a good agreement with the acquired models using a reference instrument, in this case, a commercial scanner. Although in this work the membranes response have been modelled only for a unique specific metal in a solution, future work will be focused in determine metals concentration in mixtures.

References

[1] J. Vuković, S. Matsuoka, K. Yoshimura, V. Grdinić, R. Jurišić Grubešić, O. Županić, 2007 *Simultaneous determination of traces of heavy metals by solid-phase spectrophotometry*, Talanta, vol 71 pp 2085-2091

[2] Yu. Vlasov, A. Legin, A. Rudnitskaya, C. Di Natale, A. D'Amico, 2005 *Non-specific sensor arrays («electronic tongue») for chemical analysis of liquids*, Pure and Applied Chemistry, Vol.77 pp 1965-1983

[3] P. Ciosek, W. Wróblewski, 2007 *Sensor arrays for liquid sensing - electronic tongue systems*, Analyst, vol 132 pp 963 - 978

[4] J. Gallardo, S. Alegret, R. Muñoz, M. De Roman, L. Leija, P.R. Hernandez, M. Del Valle, 2003 *An electronic tongue using potentiometric all-solid-state PVC-membrane sensors for the simultaneous quantification of ammonium and potassium ions in water*, Analytical and Bioanalytical Chemistry, vol 377 pp 248-256

[5] A.K. Deisingh, D.C. Stone and M. Thompson, 2004 *Applications of electronic noses and tongues in food analysis*, Int. J. Food Sci. Technol. Vol 39 pp 587–604.

[6] M. C. Janzen, J. B. Ponder, D. P. Bailey, C. K. Ingison, and K. S. Suslick, 2006 *Colorimetric*

sensor arrays for volatile organic compounds, Anal. Chem. Vol 78 pp 3591–3600.

[7] K. Cantrell, M.M. Erenas, I. Orbe-Paya, L.F. Capitan-Vallvey, 2010 *Use of the Hue Parameter of the Hue, Saturation, Value Color Space As a Quantitative Analytical Parameter for Bitonal Optical Sensors,* Anal.Chem. vol 82 p 531.

[8] T. Mayr, C. Igel, G. Liebsch, I. Klimant, O.S. Wolfbeis, 2003 *Cross-Reactive Metal Ion Sensor Array in a Micro Titer Plate Format*, Anal.Chem. vol 75 p 4389.

[9] A. Martínez-Olmos, S. Capel-Cuevas, N. López-Ruiz, A.J. Palma, I. de Orbe and L.F. Capitán-Vallvey, *Sensor Array-Based Optical Portable Instrument For Determination Of Ph,* Sensors And Actuators B, 2011 doi:10.1016/j.snb.2011.02.052

Competition at the Wireless Sensor Network MAC Layer: Low Power Probing interfering with X-MAC

Sven Zacharias, Thomas Newe

University of Limerick

E-mail: Sven.Zacharias@ul.ie

Abstract. Wireless Sensor Networks (WSNs) combine sensors with computer networks and enable very dense, in-situ and live measurements of data over a large area. Since this emerging technology has the potential to be embedded almost everywhere for numberless applications, interference between different networks can become a serious issue. For most WSNs, it is assumed today that the network medium access is non-competitive. On the basis of X-MAC interfered by Low Power Probing, this paper shows the danger and the effects of different sensor networks communicating on a single wireless channel of the 2.4 GHz band, which is used by the IEEE 802.15.4 standard.

1. Introduction

Wireless Sensor Networks (WSNs) are an emerging technology in the area of sensory and distributed computing. They enable new possibilities in measurements and monitoring by enabling very dense, large deployments of cheap, small hardware. Therefore, phenomena can be monitored live in full detail. A WSN consists of many sensor nodes (also called motes). These are small devices built of sensors, a microcontroller, a wireless communication interface and an energy source. To access the collected data, one or more base stations (also called sinks) act as data collectors, gateway and/or storage devices in the network. A key feature of WSNs is self-configuration without pre-configuring each single sensor node. Generally, sensor networks have to work reliable over a long time without human maintenance. Besides running out of energy, communication problems are one of the main dangers for reliability. In the following, the most common communication standards are presented and the crowd of communicating and potentially interfering devices is shown. Furthermore, a scenario where two WSNs that are operating and jamming each other on the same frequency channel is described and simulated. Finally some basic countermeasures are presented.

2. Wireless Transmitting Technologies on the 2.4 GHz band

The IEEE 802.15.4 standard is designed for low-rate Wireless Personal Area Networks (LR-WPANs) [1]. Contrary to its name this standard is not only used for small single-hop Personal Area Networks, but also for multi-hop sensor networks. Many standards, like ZigBee [2], WirelessHART [3] or 6LoWPAN [4] are based on 802.15.4 on the lower layers. 802.15.4 specifies two layers according to the Open Systems Interconnection (OSI) model: the Physical and the Medium Access Control sub-layer of the Data Link Layer.

2.1. Physical Layer

The Physical Layer of IEEE 802.15.4 supports 27 channels in the three unlicensed Industrial, Scientific and Medical (ISM) frequency bands. The 2.4 GHz band, being the highest frequency band, offers some advantages: it provides the highest data rate and is useable worldwide. Since this band is also used by other applications, the transmitting units are also cheap. At the moment the usage of the 2.4 GHz band seems to be the most common choice for WSNs.

But the 2.4 GHz band has some clear disadvantages: the waves are not very penetrating at these high frequencies and there are lots of other technologies using this band. The most common applications sharing these frequencies are microwave ovens, wireless DECT phones, baby phones and other proprietary wireless devices, as well as harmonics of monitors, Bluetooth devices and Wireless Local Area Networks (W-LAN). A very good overview of possible interference sources and their effects is provided by the technical report of the Jennic Corporation [5]. The following will focus on W-LANs, especially IEEE802.11b, since they are, as shown in the technical report, far more interfering than Bluetooth, which uses Frequency Hopping. Boano et al. [6] analyse the stability of MAC-Layers, namely NULLMAC, X-MAC [7], Low-Power Probing [8], Low-Power Listening and CoReDac [9], against interference in general. They also optimise X-MAC to stand interference better and present a new, more robust, version of X-MAC.

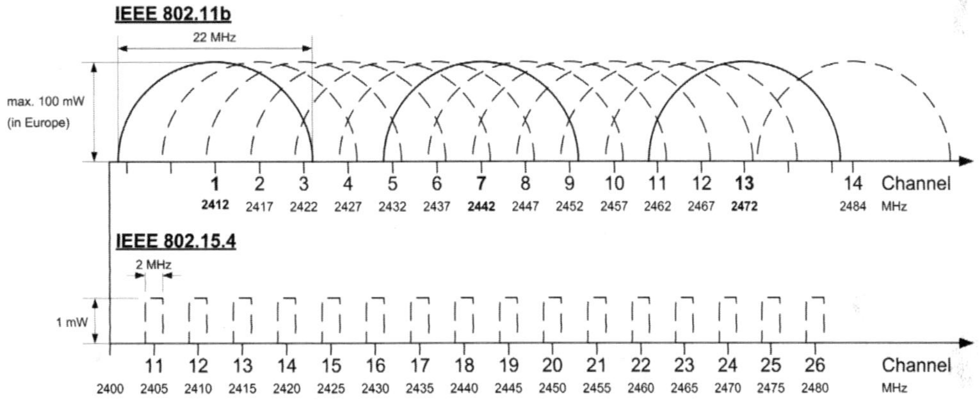

Figure 1. Comparison of the usage of the 2.4 GHz band by IEEE 802.11b and IEEE 802.15.4. Do not scale spectral mask or output power from this drawing. Bold channels are non-overlapping channels (recommended to be used in Europe).

Figure 1 gives an impression of the crowd in the frequency band, which is caused by W-LANs operating according to the IEEE 802.11b standard. The IEEE 802.11b channels are 22 MHz wide and their Channel Centre Frequencies are 5 MHz away from each other, so that they overlap each other. In comparison to the later, the channels of the IEEE 802.15.4 standard are shown. They do not overlap, because they are also spread by 5 MHz, but they are just 2 MHz wide.

A common recommendation in Europe is to use channel 1, 7 and 13 for W-LAN routers in order to avoid interference with other W-LANs. Thus, the IEEE 802.15.4 channels 15, 16, 21 and 22 are not interfered with W-LANs. In the United States, the W-LAN channel recommendation is to use channel 1, 6 and 11, so that the IEEE 802.15.4 channels 15, 20, 25 and 26 would be most suitable for WSNs in this location.

Gnawali et al. [10] show the effect of an interfering W-LAN on the Tutornet testbed used for the evaluation of the Collection Tree Protocol.

It is very likely that in the near future, the frequency bands will be even more used. For WSN deployments, the chosen frequency channel will become more and more important. Since W-LANs are sending quite continuously, they are easy to detect or the used channels may even be known to the creator of a WSN. Because of the size and the ad hoc nature of a WSN, it is not always possible to check the full area for interference. Especially side effects of other WSNs may not be detected in a quick pre-deployment channel test, because WSNs send with very low output energy and sometimes very rarely. The idea of a self-managing or at least out-of the box pre-configured network is crucial for the success of WSNs, so that complex pre-deployment checks are not suitable for most applications. More applications, e.g. Smart Meters for the Smart Grid, will be deployed in near future as off-the-shelf hardware probably having a fixed preselected channel.

With an increasing density of WSNs a new problem can occur: two separate WSNs working on only one 802.15.4 channel. This problem may also occur when the WSN developer uses the default options for his platform or if he ships a preconfigured product for end-users.

2.2. Medium Access Control sub-layer

The Medium Access Control sub-layer of the Data Link Layer joins above the frequency band described in the previous section. As described in the last section there is the chance of two or more networks operating on one channel, and thereby the Medium Access Control of two or more different WSNs will compete to use the channel. There are plenty of MAC protocols for different types of networks and applications. A good overview of MAC protocols for Wireless Sensor Networks is provided by Demirkol et al. [11].

2.2.1. *Common design goals of Medium Access Control protocols.* The most important goal in WSNs is to save energy, since energy is the most restricted resource and if a node runs out of energy it is no longer part of the network. Therefore, it does not only stop sampling, but it also cannot forward messages anymore in a multi-hop network. The main reasons causing energy waste on the MAC layer are Collisions, Idle Listening, Control Packet Overhead and Overhearing as stated by Roy and Sarma [12]. They present a good summary of different MAC protocols and its energy consumptions.

Competition was not taken into account when most of the MAC protocols were designed [11], [12]. The authors review this view under the critical perspective that WSNs will be deployed in huge amounts and as argued earlier the likelihood of competition in WSNs will increase.

3. Simulation

The simulation was designed to investigate the effect of different MAC protocols operating on the same channel as it could happen on the edge of two WSNs. There are plenty of different protocols published. This work studies the behaviour of two asynchronous MAC protocols. These protocols are similar in structure, and the protocols can be tested with equal parameters. Since X-MAC and LPP are implemented in Contiki OS, these versions were used for the experiment.

3.1. X-MAC

X-MAC is a short preamble MAC protocol. It uses an enhanced version of Low Power Listening to save energy. The nodes turn off their radios for most of the time. If a node wants to send, it turns on its radio and sends short preambles (strobes) until it receives an acknowledgement. If it receives an acknowledgement, the message is send.

Non-sending notes wake up after a sleep time for a short listening period to monitor the channel for strobes. Due to this behaviour, the idle listening time is reduced [7].

3.2. Low-Power Probing

Low-Power Probing (LPP) is roughly the inverse approach to X-MAC. Instead of the sender announcing its will to send a message, the receiver is announcing its possibility to receive messages. When using LPP, all nodes are duty cycled and wake up for just a short time. If a node is awake, it sends small packets (probes) to signal being awake and then it listens for a short time. A sending node turns its radio on and listens for the probe of the note that it wants to send to. When receiving it, the message is sent [8].

3.3. Network configuration

X-MAC and LPP were used with their default parameters. The channel is checked by both protocols at a rate of 4 Hz.

X-MAC is parameterised by *on-time*, *off-time*, *strobe-time* and *strobe-wait-time*. These specify how long the receiver listens, how long the radio is turned off, the maximum duration of sending strobes if no acknowledgment is returned and the duration of checking for other strobes to receive messages.

LPP has two parameters to control the duty cycling: *listen-time* and *off-time*. Listen-time states the duration of listening for a probe and off-time is the time duration between two listen periods.

The default Contiki parameters create a roughly comparable energy consumption for X-MAC and LPP.

3.4. Network configuration

Up to five senders report to one base station, all using X-MAC. Each sender sends a short message including a time stamp and counter. This packet can be seen as representative for a measured value and the measurement time. The sender address is added by the lower layer, thus it has not been included on the application layer. Each sender sends a packet every five seconds and this occurs 100 times. In order to test the effect of a different WSN network operating nearby, a second network of two nodes trying to transmit via the LPP protocol every second was simulated also. There was one LPP node in range for interference. Figure 2 shows the simulated scenario. By using the COOJA simulator, it is possible to time the nodes very precisely. So all nodes boot up at the same time, but their sending times were offset by one second each. Hence, there was no interference between nodes of the X-MAC network.

The experiment was executed from two up to six X-MAC senders. Each configuration was simulated without interference and with interference by the second network.

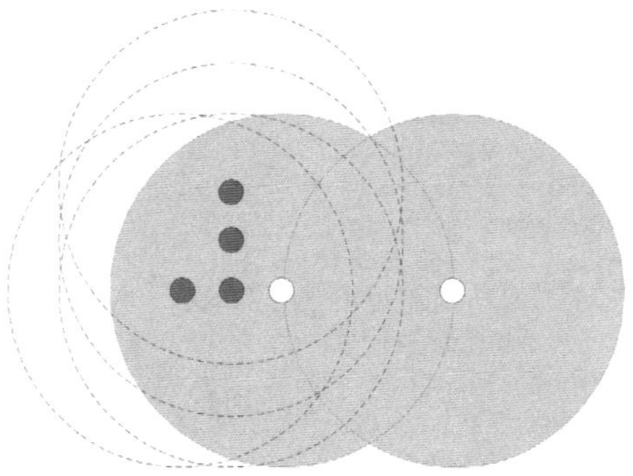

Figure 2. Simulated scenario of an X-MAC WSN interfered by a LPP WSN.

3.5. Results

The results of the simulations have been analysed for two basic features, namely the amount of successfully delivered data packets on the application layer and the estimated energy consumption of the sending nodes. Table 2 shows the number of packets arriving on the application layer. This reveals that the most significant effect on the WSNs is lost packets. In many applications, e.g. industry automation and health care, this problem has to be addressed, since reliable transmission is crucial. Table 2 also reveals that the network configurations with two and three sending nodes are heavily affected. This is due to the fact that the sending intervals are static, thus a bad timing of the two networks (where the timing is similar) can lead to a worst case scenario, in which almost every transmission is blocked. The sending window is in the region of milliseconds, so that these effects can be difficult to investigate in real deployments. In the X-MAC-layer protocol, the errors on sending nodes were logged. These logs can only help to estimate the delivery rate, since not all errors are logged and so the logs are not accurate enough to guarantee delivery.

Table 2. Delivered packets and errors logged by the X-MAC protocol on sending nodes.

Number of sending nodes	Received packets interfered / non-interfered	Logged errors on sending nodes interfered / non-interfered
1	99/100	- / -
2	56/200	10 sending drops / -
3	159/300	144 cyclic redundancy check (CRC) errors / -
4	400/400	- / -
5	500/500	- / -

To compare the energy consumption of the simulated networks, the Contiki-included module on-time counters have been used. The energy is calculated by multiplying the on-times by consumption factors given in Equation 1.

$$\text{estimated electric current [mA]} = \frac{\text{listen} \times 20[\text{mA}] + \text{transmit} \times 17.7[\text{mA}] + \text{cpu} \times 1.8\,[\text{mA}] + \text{lpm} \times 0.545[\text{mA}]}{\text{cpu} + \text{lpm}} \quad (1)$$

The parameter *listen* is the time the radio is in receive mode, *transmit* the time in transmit mode, *cpu* the time the microprocessor has been running and *lpm* the time the microcontroller has been in low power mode. These factors are based on the shell power application included in Contiki. The online energy estimator of Contiki is described by Dunkels et al. [13].

The total energy consumption is not significantly affected by the interference, since no resending was established. The transmission was cancelled often after the sending of the strobes, thus the energy consumption for an aborted sending is quite small. It has to be taken into account that no resending on a higher layer was implemented to understand that the energy consumption has not increased in total. By calculating the energy rate per packet the isolated WSNs achieve values of about 0.03 mA per successfully sent packet. In contrast, the inferred two X-MAC sender network increased to 0.11 mA per successfully sent packet.

4. Conclusion

This work has shown the danger of having two or more WSNs or parts of them operating on a single communication channel. Though by duty cycling the usage of the medium is quite rare, the chances of side-effects exist and have been described here. The interference of different MAC-layer protocols has not been tested to date. The following can be considered as possible solutions to avoid interference:

- Channel Hopping/Spread Spectrum: The frequency channel is changed permanently. This provides relief against single channel interference and increases the protection from eavesdropping. The disadvantage is the additional complexity and the need to keep the nodes synchronous, which can become difficult in big multi-hop or ad hoc networks. Bluetooth is using this technology.

- Random sample/sending intervals: This is a simple method to limit the danger of sending synchronously with another node of the same or of another network.
- Channel pre-deployment check: If the deployment area can be tested, measurements helping to find interference sources and jamming signals can help to design a good working, efficient and reliable WSN.
- Low power sending: Messages should be sent with just the power needed to reach the next hop. This helps to save energy, because multi-hopping consumes less energy than directly transmitting to a more distant node and is less interfering.

Since the amount of deployed WSNs will surely increase in the next few years, interference and cooperation of different systems will become an active research area. This work has shown a particular case and further research is needed. Interference models are not only needed on the Physical layer, but also on the MAC layer.

Acknowledgement
The authors wish to thank the following for their financial support: the Embark Initiative and Intel, who fund this research through the Irish Research Council for Science, Engineering and Technology (IRCSET) postgraduate Research Scholarship Scheme.

References

[1] Institute of Electrical and Electronics Engineers 2003 *Standard 802.15.4-2003*
[2] ZigBee Alliance *www.zigbee.org* (accessed March 2011)
[3] Gerrit Lohmann 2010 *Wireless Introduction* (HART Communication Foundation)
[4] IPv6.com - the source for ipv6 information, training, consulting & hardware *ipv6.com/articles/sensors/IPv6-Sensor-Networks.htm* (accessed March 2011)
[5] Jennic - Technology for a changing world 2008 *Co-existence of IEEE 802.15.4 at 2.4 GHz*
[6] Boano C A, Voigt T, Tsiftes N, Mottola L, Römer K and Zuniga M A 2010 Making Sensornet MAC Protocols Robust Against Interference *7th European Conference on Wireless Sensor Networks* (Coimbra)
[7] Buettner M, Yee G, Anderson E and Han R 2006 X-MAC: a short preamble MAC protocol for duty-cycled wireless sensor networks *Proceedings of the 4th international conference on Embedded networked sensor systems* (New York) pp 307-320
[8] Musaloiu-E. R, Liang, C-J M and Terzis A 2008 Koala: Ultra-Low Power Data Retrieval in Wireless Sensor Networks *Proceedings of the 7th international conference on Information processing in sensor networks* (IEEE Computer Society) pp 421-432
[9] Voigt, T and Österlind, F 2008 CoReDac: Collision-free command-response data collection *13th IEEE Conference on Emerging Technologies and Factory Automation* (Hamburg)
[10] Gnawali O, Fonseca R, Jamieson K, Moss D and Levis P 2009 Collection tree protocol *Proceedings of the 7th ACM Conference on Embedded Networked Sensor Systems* (ACM) pp 1-14
[11] Demirkol I, Ersoy C and Alagoz F 2006 MAC protocols for wireless sensor networks: a survey *Communications Magazine* vol 44 (IEEE) pp 115-121
[12] Roy A and Sarma N 2010 Energy Saving in MAC Layer of Wireless Sensor Networks: a Survey *National Workshop in Design and Analysis of Algorithm*
[13] Dunkels A, Österlind F, Tsiftes N and He Z 2007 Software-based On-line Energy Estimation for Sensor Nodes *Proceedings of the Fourth Workshop on Embedded Networked Sensors* (Cork)

Unobtrusive Wireless Monitoring System for Assisted Living and Improving the Wellbeing of Elderly People

Aidan Browne, Richard Duncliffe, James Spillane, Colin Walsh, Martin Hill, Tom O'Mahony, Fergus O'Reilly.

Department of Electronic Engineering, Cork Institute of Technology, Rossa Avenue, Bishopstown, Cork.

aidan.browne@mycit.ie, duncliffe35@hotmail.com, j.spillane@mycit.ie, colin.walsh@arrisi.com, Martin.Hill@cit.ie, tom.omahony@cit.ie, fergus.oreilly@cit.ie.

Abstract. A novel system to unobtrusively monitor the wellbeing of elderly people based on their activity patterns is presented. The system uses a wireless ZigBee network to monitor the electrical usage in a subject's home and then sends this data to an Apache server via HTTP from a GPRS unit. The data is logged in a MySQL database where pattern analysis is used to identify periods of significant inactivity. When such an event is identified designated contacts are notified by text message. For subjects requiring higher levels of monitoring a portable health monitor can be integrated incorporating a fall detector and panic button to inform of emergency situations.

1. Introduction

People are always concerned about the welfare of their loved ones, especially in their later years and if they are living alone. Thus there exists a need to monitor the well-being of these elderly people while still allowing them to retain their independence. A summary of current technology solutions to monitoring wellbeing is presented in Table 1. An ideal system should be as unobtrusive to the person in question as possible. Such a system is presented here where sensor nodes monitor the electrical usage in a subject's home and transmit the data to a server and database. Pattern recognition algorithms check for periods of inactivity and inform predefined contacts of such events.

.

COMPANY	SOLUTION OFFERED	TECHNOLOGY
GE QuietCare[1]	24/7 detection and warning system	motion detection system
Emergency Response (CARE SERVICE)[2]	alarm button, fall detect, flood detect, temperature, gas and smoke detect, motion	series of sensors, motion detectors
HealthSense[3]	panic alarm pendant, motion, bed and toilet sensors solution, blood pressure, weight scales, pulse oximeters, networked cameras solution	series of sensors

Table 1 Existing Solutions

2. Functionality

For those with good health and living alone, the system developed utilizes low powered sensor nodes to monitor the electrical usage in a subject's home and transmit using wireless protocols to a server and database. Once in the database, pattern recognition algorithms will check for periods of inactivity and if detected, the system will then inform predefined contacts that the subject may be in danger. The novelty of this system is that, once installed, the subject has to do nothing more than their normal daily routine and do not need to interact with the system in any way, putting it out of sight and mind. Care givers or family members can then check the status of the subject by logging in to the secure website provided where graphs of activity can be viewed.

For those with health concerns, a health monitoring device can also be utilized which transmits the vital signs of the subject (pulse rate, body temperature) to the database along with a fall detector and panic button status. This device, while being more obtrusive offers a more complete picture of the health and reasons for inactivity in a person. The panic button gives some control to the user in terms of informing of an emergency while the fall detector would be potentially lifesaving to an individual in their later years.

3. System Architecture

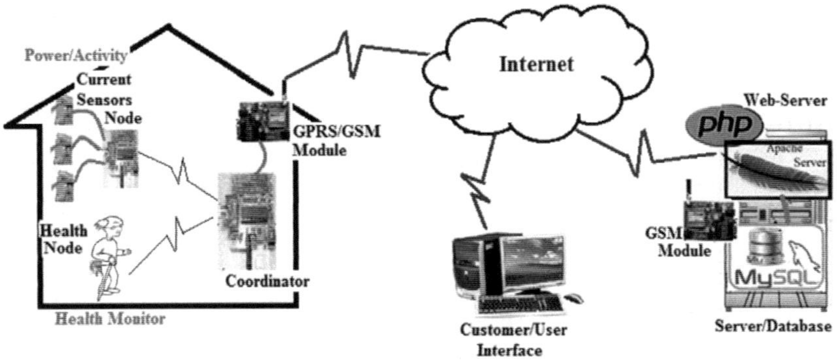

Figure 1 System Architecture

The system architecture as seen in Figure 1 is made up of the domestic Wireless Sensor Network (WSN), web-server and the user interface.

3.1. Domestic WSN

The domestic WSN consists of a ZigBee low power network with a single coordinator node and point to point star network to monitoring nodes. The prototype uses the PICDEMZ development boards [4] and a custom application node has also been developed. The monitoring nodes are designated as power/activity monitors or health monitors as shown in Figure 1. In the power monitor node three current transformers are used to monitor the power present in three targeted household circuits and are sampled every minute. Every ten minutes the stored data is transmitted to the coordinator node via the ZigBee protocol [5][6]. In the health monitoring node temperature and pulse rate are continuously logged and fall events, detected using an accelerometer and panic button events can also send notification of an emergency situation to the coordinator[7][8].

The coordinator receives data every ten minutes from the energy monitor and on an event from the health monitor. The data is sent with a timestamp to a central web-server via a GPRS connection. The coordinator will also store phone numbers for carer's of the elderly person and will notify a care giver if it receives an emergency event notification.

3.2. Web-server

The web-server is an Apache server with activity and health packets received from the domestic WSN via HTTP from a GPRS module. The received data is processed using PHP/Java scripts to extract the required data fields and inserts the data into associated data tables in a MySQL database. The coordinator node ID is used to establish from which customer the data originated. The customer table in the database is polled to extract the customer name and used to derive the table names in which the received data is to be placed.

In In the parser, consecutive readings are compared to identify changes in power consumption. Periods with no change are deemed inactive periods while periods with changes greater than a threshold are tagged as an active period. A column in the database is populated with an active/inactive depending on this comparison. This column is used later when determining the activity patterns of subjects.

3.3. User Interface

The user interface operates by remotely accessing the web-server using PHP/Java scripts to graphically display and report on subject's activity. The user interface provides a user friendly method of efficiently monitoring the activity/inactivity of a large number of remote subjects.

In addition, a Java program continually polls the database extracting all subjects on a system then polls through the database to find their most recent activity. The program checks to see if there has been activity in the morning, and if inactivity is detected after a certain time, a message is sent to their given contact numbers. The program also polls the activity database throughout the day to determine if there has been inactivity. If there has been no activity for a period of time, comparisons are made statistically with a previous number of days for the same time period to determine if the inactivity is unusual. Any unusual events or inactivity will result in a message to the user's carer and /or contact numbers that have been given.

4. System Test

A prototype system has been developed including a full end-to-end data monitoring function. Development efforts have concentrated on:

- optimising the performance of the domestic WSN
- reliable data transfer to the Web-server
- activity/inactivity analysis routines

A power profile of the power monitoring node was measured to evaluate the power consumption and derive the life expectancy of the battery powered node. The power profile is shown in Figure 2. For the timing outlined above the node powered from three 2300mAh AA batteries would yield a node lifetime of 754days which can be seen in the Node Life Expectancy table below (Table 2).

The code used for the energy monitor was deployed and tested on three different platforms, the PICDEMZ development board, a designed node based on the PICDEMZ with reduced peripherals and a PIC18F4620 microcontroller placed on a bread board. The major difference between the three platforms is current drawn in sleep mode, due to varying peripherals attached to the PIC18F4620 microcontroller. The resulting node lifetime, for each platform, can be seen in Table 2.

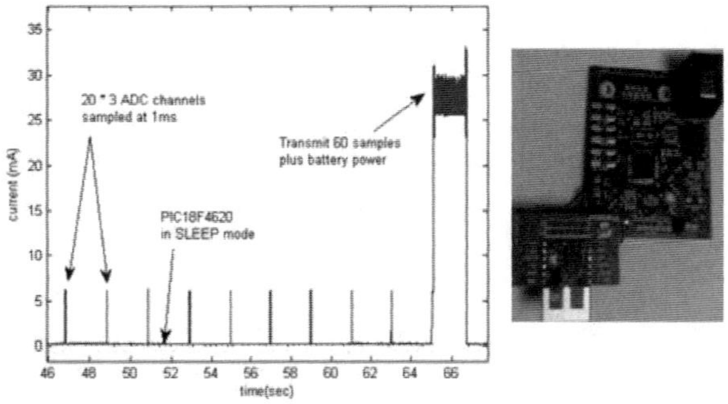

Figure 2 Power Profile of Prototype Power/Activity Monitor

Node Life Expectancy for Three 2300 mAh AA Batteries	
Node	**Life Expectancy (Days)**
PICDEM Z	**235**
Fabricated Node	**169**
Bread-Board (Radio On 1.6 seconds)	**754**

Table 2 Node Life Expectancy

The domestic electrical supply consists of a 50Hz signal at 230VAC; our chosen circuits are sampled using an appropriate rated current transformer for the given circuit depending on the load. As the PIC microcontroller can only read positive values from the sensor via the ADC, a sampling period of 1mS for 20 samples per circuit is used in order to reliably reconstruct the sampled AC signal coming from the current transformer. Figure 3 below shows a graph of a number of samples for a single circuit. The sine waves were produced using a signal generator for validation purposes.

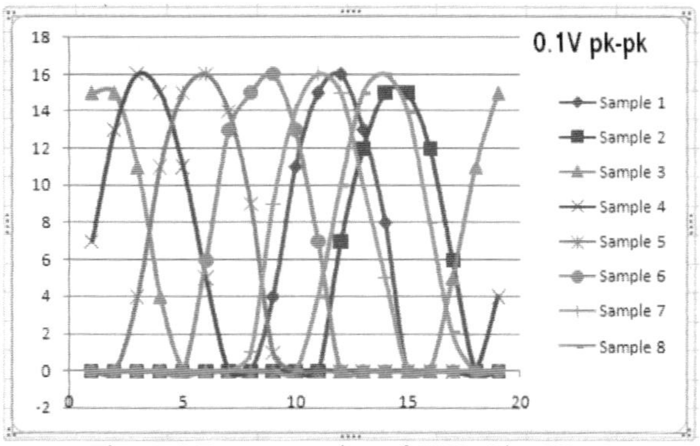

Figure 3 Reconstruction of 50Hz Sine Wave

The peak values obtained are used to determine the power in each circuit. At this sampling period we can reliably determine the peak value to over 96% accuracy. This was tested and proved by varying the peak to peak voltages being input to the ADC on the PIC microcontroller. The expected ADC values for each voltage were calculated and the detected values recorded. The results are plotted below in Figure 4 and a linear response can be observed.

	50Hz Sine Wave	
Volts	Expected	Observed
0.1V pk -pk	15.5	16
0.2V pk-pk	35	35
0.3V pk-pk	46.5	44
0.4V pk-pk	62	63
0.5V pk-pk	77.5	75
0.6V pk-pk	93	90
0.7V pk-pk	108.5	109
0.8V pk-pk	124	128
0.9V pk-pk	139.5	141
1V pk-pk	155	153

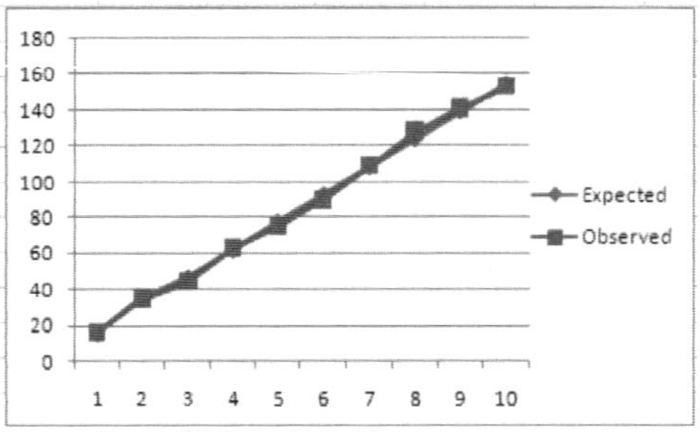

Figure 4 Reconstruction of 50Hz Sine Wave

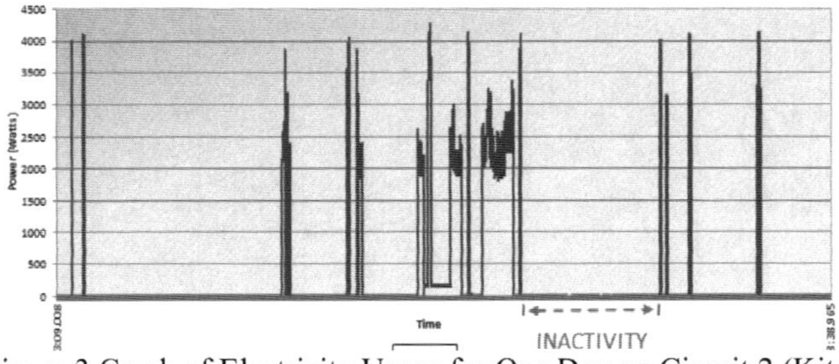

Figure 3 Graph of Electricity Usage for One Day on Circuit 2 (Kitchen)

Figure 3 above graphs one day electricity usage for circuit 2 (kitchen). Inactive periods are easily identified. It is intended that if there is continued inactivity during the day of over 4hours, previous data for the same time period will be analyzed i.e. day-1, day-2...day-7 etc to identify if inactivity is unusual. Various statistical algorithms are in place that analyze the activity count and electricity usage variances and deviations and perform data correlations of previous day's activity to determine and build a pattern model. If unusual behavior is detected a message will be sent to care givers and/or family. Currently field data obtained is limited and sustained testing and increased data could lead to improved and more focused analysis and result sets.

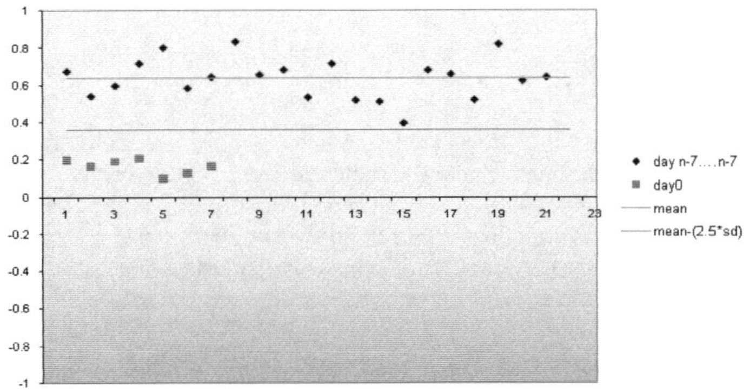

Figure 4 Data Analysis

Messages will also be sent if morning inactivity is detected. Data is also sampled from circuit 1 (shower) and circuit 3 (lighting) to retrieve a more detailed activity outlook.

5. Future Work

The complete system functionality has been established from domestic WSN to storing the data in a database at a central server. Increased Network Stability Testing of the system needs to be carried out to improve and remove any potential weak points. Additional System Security, introducing security measures such as data encryption at the WSN and upgrading from HTTP to HTTPs for transmitting the data to the central server could be implemented to better secure the overall system.

Additional data from deploying the system in a number of household test environments would validate the stability of the entire system and generate data to be used to improve our activity pattern analysis algorithms.

6. Conclusion

A complete system to unobtrusively monitor the wellbeing of elderly people living alone has been presented. The system uses a domestic WSN to monitor electricity usage of the subject's home with this data being regularly transmitted to a remote web-server. The historical electricity data is analysed on a continual basis and when a period of inactivity is detected a message is sent to family and carers. Data has been presented on the domestic WSN performance and from initial trials where inactivity has been identified using a simple activity/inactivity algorithm. A system to notify care-givers when inactivity is identified or in an emergency event is incorporated both at domestic and remote user interface levels.

References

[1] GE Healthcare, QuietCare, system for monitoring elderly people. http://www.gehealthcare.com/usen/telehealth/quietcare/proactive_eldercare_technology.html last retrieved 13/4/2011.

[2] Home Alone Connection, PERS System, personal emergency response system, http://www.homealoneconnection.com/hsm.php last retrieved 13/4/2011.

[3] Healthsense, Technology for better senior living, http://www.healthsense.com/ last retrieved 13/4/2011.

[4] 2004 Microchip Technology Inc. PIC18F2525/2620/4525/4620 Data Sheet, DS39626B.

[5] Möller S., Newe T. and Lochmann S., "Review of Platforms and Security Protocols suitable for Wireless Sensor Networks". The Eighth IEEE Conference on Sensors, 2009.

[6] Huang, H., et al., "An Implementation of Battery-aware Wireless Sensor Network Using ZigBee for Multimedia Service", Proceedings of Intenrnational Conference on Consumer Electronics, 2006, pp. 369-370.

[7] Boqiang Liu Yanyan Zhang Zhongguo Liu, "Wearable Monitoring System With Multiple Physiological Parameters", 5th International Summer School and Symposium on Medical Devices and Biosensors, 2008.

[8] Elaine Lawrence, Karla Felix Navarro, Doan Hoang, Yen Yang Lim, "Data Collection, Correlation and Dissemination of Medical Sensor Information in a WSN", Proceedings of Fifth International Conference on Networking and Services, 2009, pp.402-408.

[9] Tracy Barger, Donald Brown, Majd Alwan, "Health Status Monitoring Through Analysis of Behavioral Patterns", IEEE Transactions on Systems, Man and Cybernetics, Part A: Systems and Humans, 2005, pp. 22-27

Journal of Physics: Conference Series **307** (2011) 012040 doi:10.1088/1742-6596/307/1/012040

A robust controller design method for feedback substitution schemes using genetic algorithms

Mirsha M. Trujillo, Sillas Hadjiloucas and Victor M. Becerra

Cybernetics, School of Systems Engineering, The University of Reading, RG6 6AY, UK

Correspondance email: s.hadjiloucas@reading.ac.uk

Abstract. Controllers for feedback substitution schemes demonstrate a trade-off between noise power gain and normalized response time. Using as an example the design of a controller for a radiometric transduction process subjected to arbitrary noise power gain and robustness constraints, a Pareto-front of optimal controller solutions fulfilling a range of time-domain design objectives can be derived. In this work, we consider designs using a loop shaping design procedure (LSDP). The approach uses linear matrix inequalities to specify a range of objectives and a genetic algorithm (GA) to perform a multi-objective optimization for the controller weights (MOGA). A clonal selection algorithm is used to further provide a directed search of the GA towards the Pareto front. We demonstrate that with the proposed methodology, it is possible to design higher order controllers with superior performance in terms of response time, noise power gain and robustness.

1. Introduction

In its simplest form, a feedback substitution scheme (FSS) comprises a transducer, followed by a filter and controller as well as a second transducer re-converting the energy back to the original domain as illustrated in Fig. 1. $W_r(s)$ is the external signal to be nulled through the first transduction process, $G(s)$ is the transfer function of the transducer that measures the external input, $F(s)$ represents a low pass filter related to a lock-in amplifier involved in the demodulation of the incoming signal, which would be normally modulated above the $1/f$ noise floor, $K(s)$ is the controller to be designed and $W_f(s)$ is the transfer function of the second transducer, which aims to oppose the effect of $W_r(s)$ in the first transduction process $\left(W_f(s) = -W_r(s)\right)$. The output $W_f(s)$ is the combination of the input and noise transfer functions. The feedback signal W_f is generated to counteract the external input. This signal incorporates the low-pass characteristics of the detection process as well as the dynamics of the controller. The entire system operates within certain specifications in terms of response time and overshoot. The main objective in this system is to optimize the two conflicting requirements; response time t_r and noise power gain $\mathcal{G}$. The response time t_r is defined as the time required for the system to produce a signal $W_f = -(W_r - \varepsilon_T)$ where $\pm\varepsilon_T$ is a tolerance level [1].

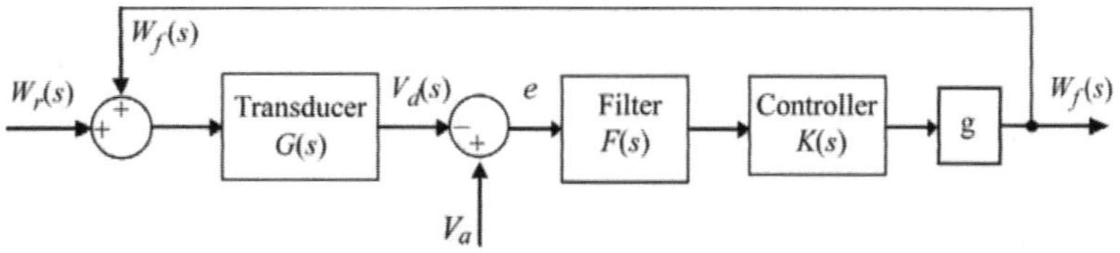

Figure 1. Block diagram for a generic feedback substitution scheme

In the simplest implementation of the system, the transfer function of the transducer is of first order $G=\mu_d/(1+\tau s)$ with time constant τ and gain μ_d. This is a part of the system that may not be changed. The filter $F(s)=\mu_f\left(1+\tau_f s\right)$ and the controller $K(s)=(1+(\tau_i s)^{-1})\mu_i$ have transfer functions that can be changed accordingly (by changing $\tau_i, \mu_i, \tau_f, \mu_f$) to obtain the desired response.

2. Loop shaping design procedure and analysis of robustness

The system described in section 1 can be converted into the system shown in Figure 2a where a Loop Shaping Design Procedure (LSDP) and an H-infinity controller can be synthesized as described by McFarlane and Glover [2]. A co-prime factorization of the shaped plant G_s can also be performed as shown in Figure 2b. The introduction of an explicit co-prime factorization of the system permits the design of a controller with explicit parametrization of robustness. This is of interest as different design procedures such as the one described by Clare and White as well as the new LSDP formulation have different robustness.

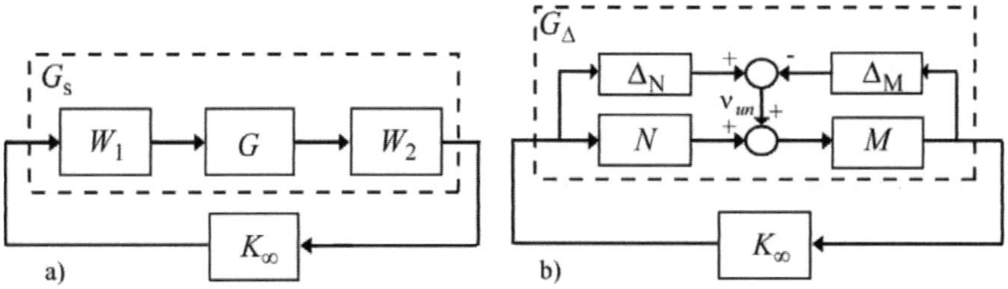

Figure 2. a) LSDP formulation and b) co-prime factorization of the system.

In the original systems shown in Figure 1, the optimization parameters are an overall gain of the system as a single parameter $\mu=\mu_d\mu_f\mu_i$, a normalized time constant of the filter with respect to the detector time constant τ denoted as x and a normalized integration time constant of the controller, y. In the loop shaping design procedure, the design parameters μ, $x=\tau_f/\tau$, and $y=\tau_i/\tau$ are substituted by weighting functions W_1 and W_2 to shape the plant G to achieve the desired response. A normalized co-prime factorization of the shaped plant $G_s=NM^{-1}$ satisfying $MV+NU=1$ is used to model uncertainty as perturbations Δ_N and Δ_M to the co-prime factors of the shaped plant.

The perturbations Δ_N and Δ_M are stable transfer functions with uncertain dynamics and parameters, the only thing known about this transfer functions is that their H^∞ norm is less than a value ε_{sm} which is interpreted as the stability margin of the system. The resulting perturbed plant equation is:

$$G_\Delta=\left\{\left(M+\Delta_M\right)^{-1}\left(N+\Delta_N\right):\left\|\left[\Delta_N \quad \Delta_M\right]\right\|_\infty<\varepsilon_{sm}\right\} \tag{1}$$

From a robust analysis perspective, to maximize the stability margin ε_{sm}, a controller is calculated to minimize the function

$$\gamma:=\left\|\begin{bmatrix} K(1-KG)^{-1}M^{-1} \\ (1-KG)^{-1}M^{-1}\end{bmatrix}\right\|_\infty \leq \frac{1}{\varepsilon_{sm}} \tag{2}$$

where γ is the H_∞ norm of the system in Fig. 2b and $(1-KG)^{-1}$ is the sensitivity function S.

In Figure 3, we consider 3rd order systems composed of a 1st order plant, a 1st order filter and a 1st order controller fulfilling an overshoot requirement of a tolerance of 0.001 using the Clare and White procedure [1]. We plot the normalized response time t_r/τ as a function of the system gain $\mu=\mu_d\mu_f\mu_i$ for different filter time constants $x=\tau_f/\tau$ when the value of y is adjusted to meet the overshoot requirement (tolerance 0.001) and evaluate the resulting system noise power gain and robustness. The resulting systems that demonstrate low robustness margin as well as those with an amplification of noise are shown. Furthermore, the possible systems that result with an unacceptable undershoot are also shown.

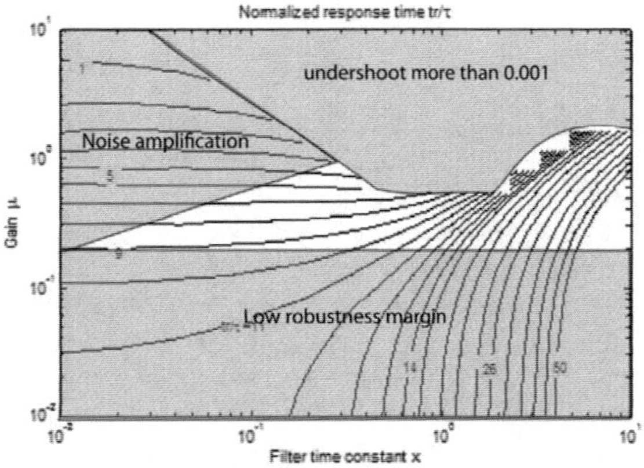

Figure 3. Possible 3rd order systems composed of a 1st order plant, a 1st order filter and a 1st order controller fulfilling an overshoot requirement of a tolerance of 0.001 using the Clare and White procedure. All possible systems that can result by tuning a first-order controller are shown.

Because of the precision of the system's overshoot requirement, a systematic and iterative strategy is needed to design the weighting functions W_1, W_2. The problem specifications are casted out as inequalities [3]. A genetic algorithm is then used to map the surface of all possible H^{∞} controllers that can be designed with the specifications provided through the inequalities defined by the designer [2-8].

First the system requirements are specified using a representation similar to that used in the method of inequalities. These inequalities define a goal attainment problem which is translated into a minimization problem. A Multi-objective Optimization with Genetic Algorithms (MOGA) from MATLAB's Genetic Algorithm and Direct Search toolbox V2.4.1 (R2009a) the function '*gamultiobj*' is used to find the parameters and structure of the weighting functions that lead to an optimum solution.

The design requirements for the system are as follows: The output $W_f(t)$ for a unitary step input $W_r(t)=1$ must have an overshoot $\varepsilon_T = 0.001$ which is specified to a numerical precision of $\varepsilon_T/500$; after the first overshoot, the output $W_f(t)$ must remain within the range $1 \pm \varepsilon_T$; the noise power gain $\mathcal{G}$ in the newly designed system must be minimized for all normalized response times considered. Additional requirements are a zero steady state and a robust stability margin ε_{sm} >0.25. Systems with a normalized response time t_r/τ in the range 1 to 20 are only considered. Using inequalities, the above specifications are casted as follows:

$$\varphi_1 = \left| W_h\left[t, W_r(t)\right] - \frac{\varepsilon_T}{500} \right| \le W_r(t) + \varepsilon_T, \quad t > 0 \tag{3}$$

$$\varphi_2 = \left| W_h\left[t, W_r(t)\right] - \frac{\varepsilon_T}{500} \right| \le W_r(t) - \varepsilon_T, \quad t > t_r \tag{4}$$

$$\varphi_3 = \min \sqrt{\frac{\tau}{4\pi j} \int_{-\infty}^{\infty} |Q(s)|^2 \, ds \,\Big|_{s=j\omega}} \tag{5}$$

$$\phi_4 = \left\| \begin{bmatrix} K(1-KG)^{-1}M^{-1} \\ (1-KG)^{-1}M^{-1} \end{bmatrix} \right\|_{\infty} \le \frac{1}{0.25} \tag{6}$$

A vector of designer objective functions $\Phi = [\varphi_1 \ \ldots \ \varphi_4]^T$ with p_j design parameters and objectives ε_i each, with $\varphi_i(p_j) \le \varepsilon_i$ is constructed with an admissible set composed of all admissible points. The minimization problem is then solved using a MOGA. An auxiliary vector $\boldsymbol{\lambda}$ is required to translate the goal attainment problem into a minimization MOGA problem. Each element of the vector is given by:

$$\lambda_i\left(p_j,\varepsilon_i\right)=\begin{cases}0 & \text{if} \quad \varphi_i\left(p_j\right)\le\varepsilon_i\\ \varphi_i\left(p_j\right)-\varepsilon_i & \text{if} \quad \varphi_i\left(p_j\right)>\varepsilon_i\end{cases} \tag{7}$$

The purpose of this auxiliary vector is to provide the MOGA with a vector of fitness functions where the fitness value is a minimum (zero) if the goal is attained. The difference between the actual value and the goal is returned if the goal is not attained.

We assume the weighting function W_1 to be composed of several lead-lag structures of first or second order. This approach ensures that we can have complex poles and zeros. The coefficients for the lead-lag structures are provided from the design parameters p_j. In order to explore different parameters as well as different structures with the same MOGA with the purpose of finding optimum solutions, the MOGA has to work in a hierarchical structure as proposed in [8]. Here the parameter vector p_j is divided into two different chromosomes (Figure 4). One chromosome corresponds to the coefficients of the weighting function. The other one controls the activation or suppression of elements in the lead-lag structure.

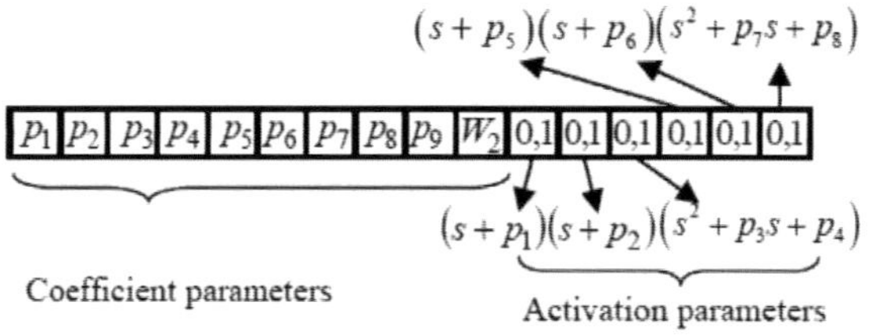

Figure 4. Chromosome separation for hierarchical GA.

The activation parameters p_{11-16}, $p_j\in\left[0 \;\; 1\right]$ change the structure of the lead-lag filters in W_1, whereas the filter coefficients are given from p_{1-9}, $p_j\in\mathbb{R}$. The weighting function W_2 is assumed to have a structure of a simple gain given from p_{10}. In addition, we are including one more integrator with a very small value $(s+\upsilon)^{-1}$ (with $\upsilon=10^{-6}$) to avoid poles in the origin that lead to internal instability when the controller is synthesized. W_1 is calculated from:

$$W_1=p_9\left[\frac{\left(s+p_5\right)\left(s+p_6\right)\left(s^2+p_7s+p_8\right)}{\left(s+\upsilon\right)\left(s+p_1\right)\left(s+p_2\right)\left(s^2+p_3s+p_4\right)}\right] \tag{8}$$

Each parameter that forms part of W_1 and W_2 in the MOGA is initially randomly initiated. For each individual in the population, a controller is synthesized for the system as shown in Fig. 4. That system is simulated using a step response so that from the simulation, the objectives (response time, overshoot and noise power gain) can be measured. Then the MOGA uses these objectives as fitness functions to rank the produced solution according to how close they are to the Pareto-optimal front. The fittest solutions are selected for recombination and mutation. The stopping criterion is based on meeting the overshoot objective while at the same time minimizing the noise power gain. Instead of finding directly the minimum normalized response time, our strategy was to minimize the noise over small bands (±0.15) of set values of response time.

The algorithm starts with random weights which correspond to transfer functions translated into state space and two Riccati equations are solved to finally design the controller. The GA mapped the surface of all possible H^{∞} controllers that can be designed with the specifications provided using the defined inequalities. The proposed methodology permits us to analyze designed plants of up to 7th order assuming an open structure.

As discussed in [2], once the stabilizing controller K_∞ is designed, the implementation of the final controller requires the movement of the two shaping functions at the input and output of the K_∞ to form a new transfer function block $K = W_2 K_\infty W_1$ as shown in Figure 3.

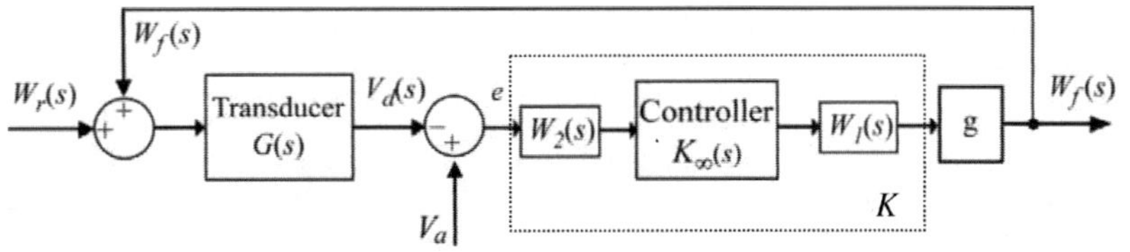

Figure 5. Final LSDP implementation of the system once the K_∞ controller is designed.

3. Comparing results from the two procedures

Figure 6 shows time domain responses for systems designed using the Clare and White procedure as well as the LSDP procedure which uses the MOGA to design the optimal controllers. Both systems have been selected to have the same normalized response time specifications so a comparison could be performed. Both graphs are considering a structured uncertainty by adding a 10% parametric uncertainty to the detector time constant τ as well as to the detector gain μ_d. In terms of Robust Stability both systems can tolerate the modeled uncertainty, The Clare and White system is 76% more sensitive than the LSDP one to variations of μ_d, but 7% less sensitive to variations of τ. In terms of robust performance, the Clare and White system is 5% less sensitive than the LSDP one to variations of μ_d, for variations of τ both systems have the same robust performance. With regard to unstructured uncertainty the new methodology had an improvement of 98% in the robust stability margin as calculated with equation (2) and a reduction of 10.8% in noise power gain using the LSDP MOGA mehtod.

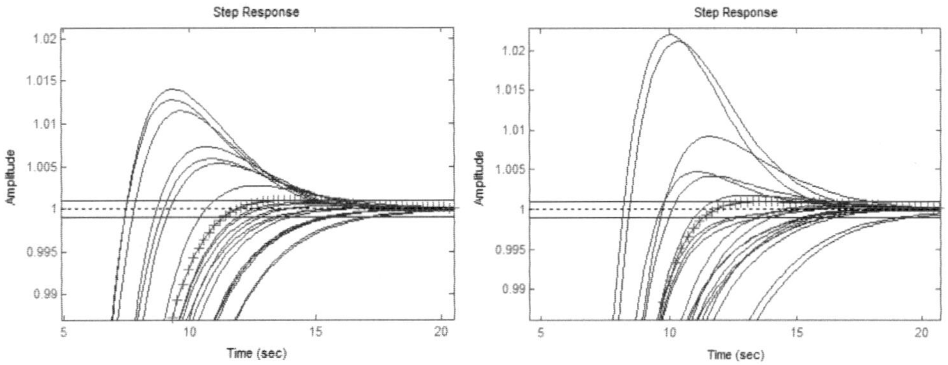

Figure 6. Time domain responses of systems to a step input, with optimal controllers designed using a) the Clare and White method which results in a 3rd order system and b) the LSDP MOGA mehtod which results in systems of higher order. The two horizontal lines close to unity amplitude indicate the allowed overshoot and undershoot tolerance specification for the designed systems and the crossed line indicates the nominal system response time whereas the other lines indicate step responses for systems with parametric perturbation.

4. Conclusion

Optimal feedback substitution schemes having a controller of first or second order can be analytically found satisfying precise overshoot characteristics, while at the same time minimizing response time and noise power gain. The design of higher-order controllers is possible using LSDP and an H∞-

MOGA procedure. In this work we have handled constrains using MOGA converting them into objectives. Because the MOGA has a very limited domain of feasible solutions, we plan to use particle swarm optimization combined with GAs in the future. This is so, because the derived MOGA population often is outside the feasible solution space and another constrain handling algorithm needs to be used to move this population within the feasible solutions space. The current work shows that there is a need to adopt alternative computational intelligence algorithms that will tune the parameters of higher order systems. An advantage of the proposed approach is that parameters W_1, W_2 could also be implemented assuming a non-rational function, so that a fractional order controller could be designed.

The currently developed controller design procedure is applicable to all radiometric detectors governed by first order dynamics and may be extended to 2nd order transducers or higher order transducers which are not currently employed for absolute measurements.

References

[1] J. F. Clare and D. R. White. Response time and noise power gain of electrical substitution radiometers with feedback control. *Applied Optics*. **28**, 15, (1989) 3417 - 3428.

[2] Duncan McFarlane and Keith Glover, "A Loop Shaping Procedure Using H∞ Synthesis", *IEEE Transactions on Automatic Control*, **37** (1992) 759- 769

[3] G. P. Liu, J. B. Yang and J. F. Whidborne, "*Multiobjective Optimisation and Control*", Research Studies Press LTD., (2003).

[4] C. M. Fonseca and P. J. Fleming, "Multiobjective optimal controller design with genetic algorithms", In *Proc. Control 94*, Coventry, England, (1994), pp.745-749

[5] J. F. Whidborne, I. Postlethwaite and D.-W. Gu, "Robust Controller Design Using H∞ Loop-Shaping and the Method of Inequalities", *IEEE Trans. Control Systems Technol.*, **2**, (1994).

[6] Andrew Chipperfield and Peter Fleming, "Multiobjective Gas Turbine Engine Controller Design Using Genetic Algorithms", *IEEE Trans. Industrial Electron.*, **43**, (1996) 583-587.

[7] N. V. Dakev, J. F. Whidborne and A. J. Chipperfield, "H∞ Design of an EMS Control System for a Maglev Vehicle Using Evolutionary Algorithms", *Genetic Algorithms in Engineering Systems: Innovations and Applications*, IEE, 1995, pp. 226-231

[8] K. S. Tang, K.F.Man and D.-W. Gu, "Structured Genetic Algorithm for Robust H∞ Control Systems Design", *IEEE Trans. Industrial Electron.*, **43**, (1996) 575-582.

Determination of Activated Carbon Residual Life using a Microwave Cavity Resonator

A Mason[1], S Wylie, A Shaw and A I Al-Shamma'a
School of Built Environment, Liverpool John Moores University, Liverpool, UK.

A Thomas and H Keele
Dstl, Porton Down, Salisbury, SP4 0JQ, UK.

E-mail: A.Mason1@ljmu.ac.uk

Abstract. This paper presents the continuation of work conducted jointly between Dstl and LJMU. This unique body of work has been, largely, concerned with detecting the residual life of high performance filter materials using electromagnetic (EM) waves within a resonant cavity. Past work has considered both HEPA [1] and ASZM-TEDA[2] activated carbon filter materials. This paper continues the later work, considering the response of ASZM-TEDA activated carbon through the co-ageing of two distinct batches of the material. The paper briefly introduces activated carbon, discusses theory relevant to the work and the methodology used for investigation. A comprehensive set of results is included which seek to validate this technique for determining the residual lifespan of activated carbon.

1. Introduction

Activated carbon (sometimes also referred to as activated charcoal [3]) includes a wide range of processed carbon based materials, and has a microcrystalline structure. Its use in filtration is brought about by the fact that it has a high level of porosity and hence a large surface area where adsorption or chemical reactions may take place.

The preparation of activated carbons involves two main steps [4]: the carbonisation of raw material at temperatures less than 800°C in an inert atmosphere, and the activation of the carbonised product. During the carbonisation process, most of the non-carbon elements (e.g. oxygen, hydrogen and nitrogen) are eliminated as volatile gaseous species by the decomposition of the starting material. The residual elementary carbon atoms group themselves into stacks of flat sheets cross-linked in a random manner; the irregular arrangement of these sheets gives rise to pores. These pores are further enhanced through the activation process, which is usually carried out in an atmosphere of air, CO_2, or steam at 800-900°C. This results in an oxidation of some of the regions within the char in preference to others so that as combustion proceeds, a preferential etching takes place. This results in the development of a large internal surface area which can be as high as $2500m^2/g$ [5], although areas in the region of $500m^2/g$ to $800m^2/g$ are more typical [6].

Activated carbon is available in a number of different forms, including powder, fibres and granules. The type used would usually be application dependant; gaseous applications would normally use the material in the granule form whereas in liquid based applications fibrous forms are often used due to

Published under licence by IOP Publishing Ltd

low hydrodynamic resistance and their ease of moulding to any shape in the adsorption equipment. The ASZM-TEDA product under test in this work is in a granule form, and contains a number of inorganic impregnant. These are Copper, Silver, Zinc, Molybdenum and Triethlyenediamine, and are present in order to improve the effectiveness of the product against warfare gases [4, 7]. Figure 1 and 2 show a sample of the ASZM-TEDA product as viewed by a scanning electron microscope (SEM) at varying levels of magnification.

Activated carbons are excellent and versatile adsorbents, which inevitably leads to a wide number of commercial applications, including gas and water purification, sewage treatment as well as being used in air filtration for clean-rooms and respirators for volatile environments (i.e. gas masks). This later application was the initial driver for this work as issues has been raised in military circles relating to gas masks which utilise activated carbon filters. At present there is no convenient method for determining the residual life of a filter, which may age as a result of adsorbing harmful gases or simply as a result of being stored ready for future use. This has led to a conservative filter change policy which results in significant wastage due to useable filters being discarded.

Since beginning the work, the authors have also considered wider reaching applications of an activated carbon residual life indicator and it would appear that medium to large scale implementations (e.g. clean room and general building filtration) could benefit significantly. In these cases the filter material carries a high cost due to the volumes required, and therefore a technique which could reliably determine residual life could maximise cost effectiveness.

The aim of the research thus far has been to develop and validate an EM resonant cavity sensor for the purpose of residual life indication; the remainder of the paper focuses on this aim.

Figure 1. Activated carbon granules as viewed with a SEM.

Figure 2. A highly magnified view of an activated carbon granule as viewed with a SEM, showing macroscopic surface pores.

2. Microwave Theory

Using microwave cavity resonator techniques with activated carbon began under the assumption that as the product is aged (i.e. through adsorption of material from the air) its relative permittivity (ε_r) must change.

One method which can be used to find the ε_r of a substance is to monitor the effect it has on the resonant frequencies of an electromagnetic (EM) cavity. An EM cavity, consisting of a hollow structure with conducting walls, will resonate when it is excited at an appropriate frequency by a small antenna placed inside the structure. The resonant modes occur when the electric and magnetic fields form standing waves, and so depend upon the internal dimensions of the cavity and the permittivity of any material placed inside. For a cylindrical cavity resonator the fundamental modes are TE_{111} and TM_{010}. TE (Transverse Electric) modes have a magnetic component in the propagation direction and TM (Transverse Magnetic) modes have an electric component in the propagation direction. Each mode will generate a resonant peak with a quality factor (Q), which is inversely proportional to the power dissipated in the cavity for each applied EM oscillation. A high Q indicates a sharp resonant peak that will be more readily analysed and improve the accuracy of the sensor.

The resonant frequency for TE_{nml} and TM_{nml} modes in a cylindrical cavity [8] can be calculated using equation (1).

$$f_{nml} = \frac{c}{2\pi\sqrt{\mu_r \varepsilon_r}}\left[\left(\frac{p_{nm}}{a}\right)^2 + \left(\frac{l\pi}{d}\right)^2\right]^{1/2} \tag{1}$$

Where:

c is the speed of light
μ_r is of the relative permeability
ε_r is the relative permittivity
p_{nm} is the mth root of the of Bessel function of the nth order for TM modes or the mth root of the first derivative of Bessel function of the nth order for TE modes
a is the radius of the cavity
d is the depth of the cavity

All EM modes therefore have the same dependence upon $\sqrt{\varepsilon_r}$, so when the cavity is excited by an appropriate range of frequencies and the resulting spectrum is captured, the resonant peaks corresponding to these modes will shift to lower frequencies as the permittivity is increased.

3. Experimental Setup

3.1. Sample preparation

In order to determine a microwave response to activated carbon relating to residual life, two batches (referred to separately in this document as Batch A and Batch B) of ASZM-TEDA activated carbon granules were artificially aged simultaneously at Dstl facilities.

Briefly, the ageing process included placing the carbon granules in shallow trays at a maximum depth of 2cm and exposing them to an atmosphere of 40°C and 80% relative humidity[1]. The carbon remained in this controlled environment for 7 months, being riffled on a weekly basis to ensure even exposure to the ambient air. Approximately 300g of carbon was removed from the trays each month and divided equally between two sealed containers. The material in one of the containers would then be used to fill a 15ml polypropylene centrifuge tube and sent to LJMU. The material weight varied with age due to water adsorption, although the majority of the weight increase was experienced within the first month as shown in Figure 3. The centrifuge tubes were filled via a "snowflake" filling method [9], which is designed to ensure homogeneity (i.e. minimal air voids and compression zones). This is noteworthy since compression of the carbon granules will inevitably increase EM wave attenuation due to a reduced skin depth [10].

The material in the second sealed container was retained by Dstl and tested for toxic gas breakthrough times. A variety of gases are used in order to test both the adsorptive and catalytic properties of the activated carbon. Figure 4 shows indicative breakthrough times for the co-aged

[1] Further samples have been aged at 22°C but are not discussed in this paper.

samples. The results here are not as linear as one might hope, however they do at least show that the ageing method used appears to be working as intended since the performance of the carbon degrades with time.

3.2. Data acquisition

The apparatus used for testing of the carbon samples is shown in Figure 5. It comprises a Marconi 6200A microwave test set, capable of power magnitude measurements between 20MHz and 20GHz, and a cylindrical microwave cavity. The cavity allows for measurement of reflected (S_{11}) and transmitted (S_{21}) power. Both measurements are taken over the full frequency range of the test set using a bespoke software application which acquires data in defined frequency intervals in order to maximise data resolution. For the purposes of these experiments data was captured at 500MHz intervals, with 1601 data points per interval. Each carbon sample was tested 6 times consecutively and then the average of these measurements taken. In order to minimise the effect of temperature variation over the course of the 8 month test period, the apparatus and samples were kept in an environmental chamber which maintained a temperature of 20°C. It is assumed that once the carbon is placed into the polypropylene tubes little or no further ageing occurs, which has been proven by recapturing the sample spectra on a monthly basis.

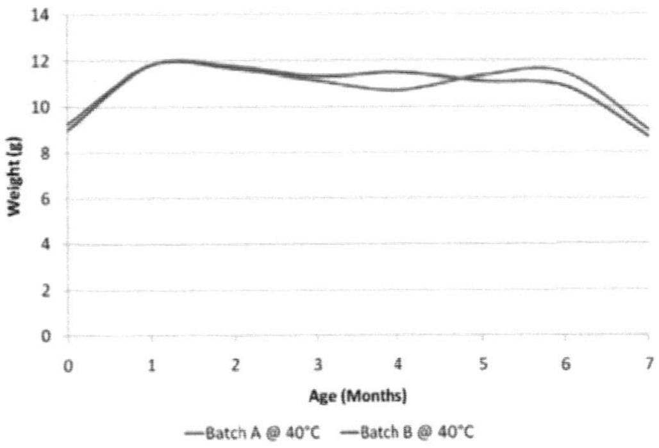

Figure 3. Weight of activated carbon samples based on their age.

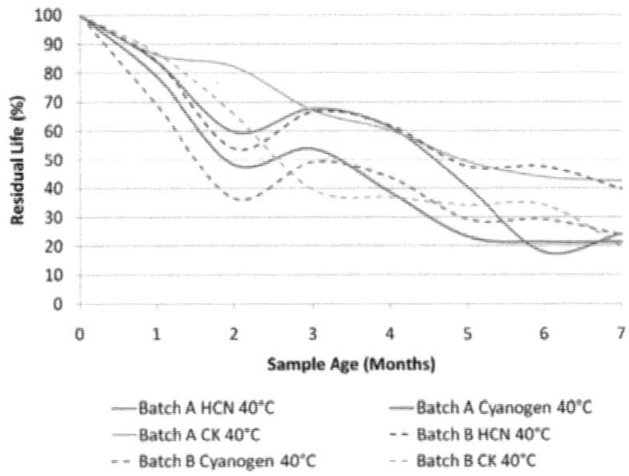

Figure 4. Indicative breakthrough times of the activated carbon samples from both co-aged batches.

Figure 5. Experimental setup for the resonant cavity sensor.

4. Results

Responses shown by materials exposed to microwave radiation in a resonant cavity are often specific to that material. Thus far, the aim of data analysis has been to seek logical patterns in the microwave spectrum recorded during the data acquisition phase. This has been a non trivial process considering the large amount of data per sample that which has been collated. For this purpose a bespoke SQL application has been developed which allows processing of the data in terms of variance calculations, normalisation of data and other methods of statistical analysis.

This statistical analysis is an on-going element of this work. However, searching for patterns in the data which correlate with sample age has proven fruitful. A number of regions, particularly in the S_{21} data, have been discovered which indicate an approximately linear relationship between sample age and some characteristic feature of the acquired spectrum. The pattern search is configured such that it allows for inconsistent non-linearity in order to accommodate potential experimental error, signal noise or other potential anomaly. Two example regions are shown in Figure 6 and 7. The spectrum within each region is shown for batch A and B. Visual inspection of the Figures confirms that the response from each batch is similar, and follows a trend which can be correlated with sample age.

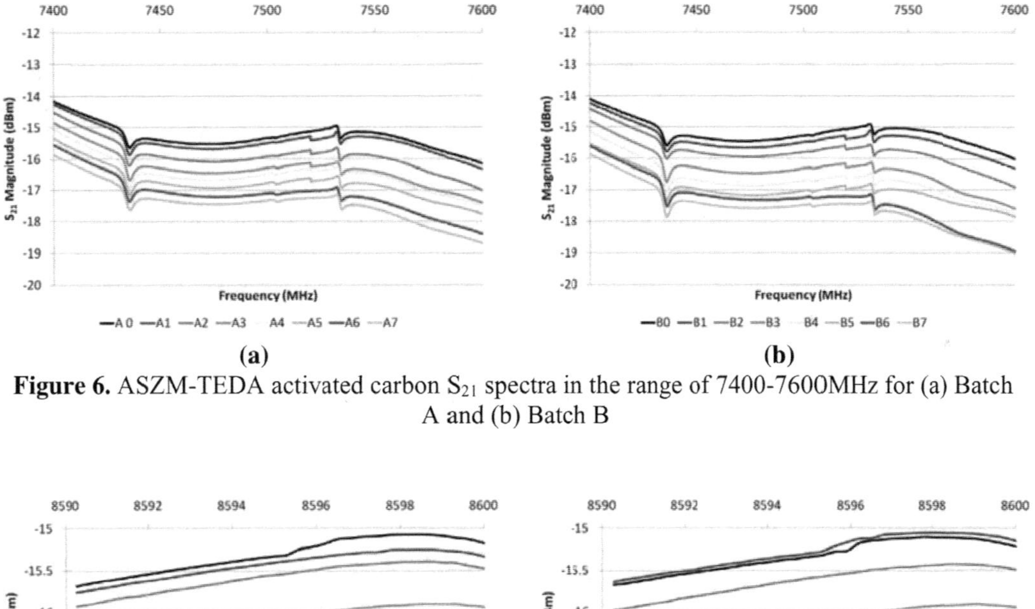

(a) **(b)**

Figure 6. ASZM-TEDA activated carbon S_{21} spectra in the range of 7400-7600MHz for (a) Batch A and (b) Batch B

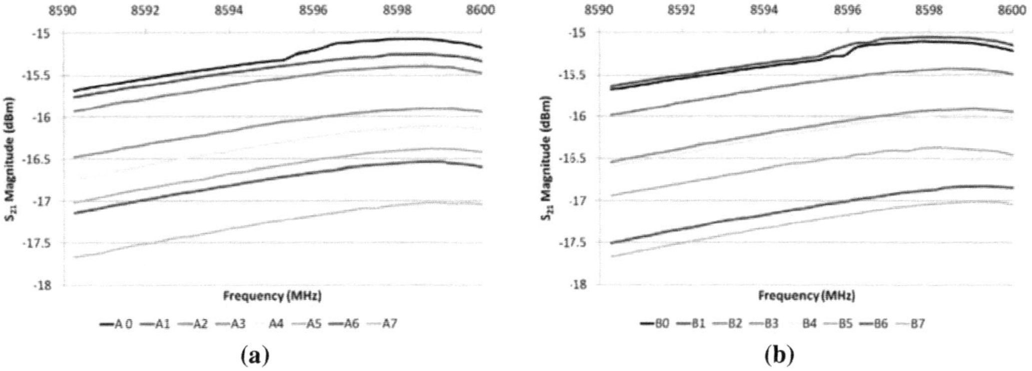

(a) **(b)**

Figure 7. ASZM-TEDA activated carbon S_{21} spectra in the range of 8590-8600MHz for (a) Batch A and (b) Batch B

Initially it was thought that the response shown might simply be due to increased water content in the carbon samples. During the ageing of the samples they are exposed to high humidity, therefore take on moisture from the atmosphere. However, the samples do this relatively quickly; the samples increase in weight significantly between months 0 and 1, but in subsequent months the weight remains relatively stable. This feature is highlighted in Figure 3. Therefore the measured response must be as a result of some other feature of the carbon, which appears to be linked to its ageing. It is known that the carbon is ageing due to the reduced gas breakthrough times shown in Figure 4. Since we know the useful lifetime of activated carbon from gas breakthrough testing, it is possible to correlate this with residual (or useful) remaining life. From the perspective of gas masks, the activated carbon material used in this work would not provide a significant amount of end user protection from harmful gases after 8 months of ageing; it would therefore require discarding.

5. Conclusions and Future Work

This paper describes a technique for determining the residual life of ASZM-TEDA activated carbon material through changes in EM response when placed inside a resonant cylindrical cavity. The paper builds upon previous work in this field [2], and has used a co-ageing method in order to demonstrate cross batch repetition; Figure 6 and 7 show this to be the case. In addition, we know that although the ageing process of activated carbon causes a significant weight gain (in the order of 70%), much of this occurs in the first month of ageing and therefore the sensor response is not simply to water. Ongoing work with the results obtained hope to discover correlations between EM response and carbon ageing at lower frequencies – principle component analysis (PCA) is being investigated as a potential tool in this process.

6. References

[1] A. Mason, *et al.*, "HEPA Filter Material Load Detection Using a Microwave Cavity Sensor," *International Journal on Smart Sensing and Intelligent Systems,* vol. 3, pp. 322-337, 2010.
[2] A. Mason, *et al.*, "Electromagnetic (EM) Cavity Resonance Residual Life Indicator," presented at the 2011 IEEE Symposium on Computers & Informatics, Kuala Lumpur, Malaysia, 2011.
[3] Carbon Resources. (07/11/2010). *Activated Charcoal Information.* Available: http://www.carbonresources.com/activated-charcoal.html
[4] R. C. Bansal and M. Goyal, "Activated carbon and its surface structure," in *Activated Carbon Adsorption*, 1st ed Boca Raton: CRC Press, 2005, pp. 1-9.
[5] H. Marsh and F. Rodríguez-Reinoso, "Porosity in Carbons: Modelling," in *Activated Carbon*, 1st ed: Elsevier, 2006, p. 96.
[6] S. Jhadhav, "Value Added Products from Gasification - Activated Carbon," Indian Institute of Science 2005.
[7] H. Marsh and F. Rodríguez-Reinoso, "Activated Carbon," in *Activated Carbon*, 1st ed: Elsevier, 2006, p. 10.
[8] D. M. Pozar, *Microwave Engineering*, 3rd ed. New York: John Wiley and Sons, 2005.
[9] E. Harris, "Improvements in or relating to the Filling of Granular Materials into Containers," United Kingdom Patent, 1948.
[10] F. Ulaby, "Plane Wave Propagation in Lossy Media," in *Fundamentals of Applied Electromagnetics*, ed: Prentice Hall, 1999, pp. 277-279.

Acknowledgements

The authors would like to thank Paul Gibbons of Liverpool John Moores University for his kind assistance in capturing images of the activated carbon material with a SEM. This has proven both interesting and useful for the research.

Journal of Physics: Conference Series **307** (2011) 012042 doi:10.1088/1742-6596/307/1/012042

Artificial Intelligent Control for a Novel Advanced Microwave Biodiesel Reactor

W. A. Wali[1], K. H. Hassan, J.D. Cullen, A. I. Al-Shamma'a , A.Shaw, S.R. Wylie

Built Environment and Sustainable Technologies Institute (BEST),
School of the Built Environment, Faculty of Technology and Environment
Liverpool John Moores University, Byrom Street, Liverpool L3 3AF, U.K.

E-mail: w.wali@2009.ljmu.ac.uk

Abstract. Biodiesel, an alternative diesel fuel made from a renewable source, is produced by the transesterification of vegetable oil or fat with methanol or ethanol. In order to control and monitor the progress of this chemical reaction with complex and highly nonlinear dynamics, the controller must be able to overcome the challenges due to the difficulty in obtaining a mathematical model, as there are many uncertain factors and disturbances during the actual operation of biodiesel reactors. Classical controllers show significant difficulties when trying to control the system automatically. In this paper we propose a comparison of artificial intelligent controllers, Fuzzy logic and Adaptive Neuro-Fuzzy Inference System(ANFIS) for real time control of a novel advanced biodiesel microwave reactor for biodiesel production from waste cooking oil. Fuzzy logic can incorporate expert human judgment to define the system variables and their relationships which cannot be defined by mathematical relationships. The Neuro-fuzzy system consists of components of a fuzzy system except that computations at each stage are performed by a layer of hidden neurons and the neural network's learning capability is provided to enhance the system knowledge. The controllers are used to automatically and continuously adjust the applied power supplied to the microwave reactor under different perturbations. A Labview based software tool will be presented that is used for measurement and control of the full system, with real time monitoring.

1.Introduction

Biodiesel is becoming an environmentally friendly fuel due to its non-toxic and biodegradable characteristics [1]. With increasing world crude oil prices, the focus of research is to produce biodiesel fuels from very poor quality, waste cooking oil. Fig.1 shows the production of biodiesel from waste oil .

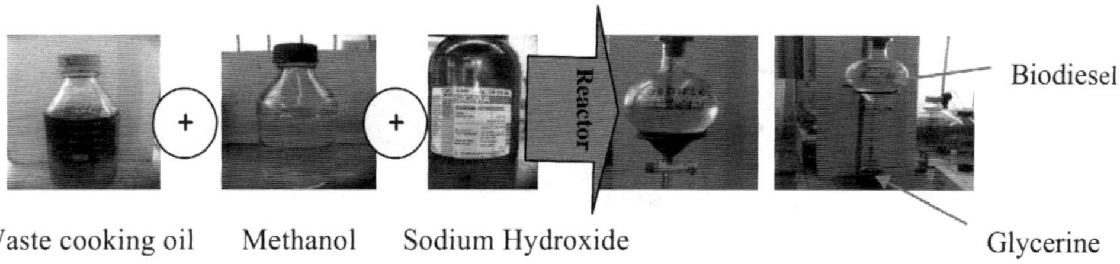

Fig.1. Biodiesel from waste cooking oil.

Energy consumption studies show that the continuous-flow preparation of biodiesel using microwave heating proves to be more energy efficient than the conventional synthesis of biodiesel in large tank reactors[2]. Microwave technology offers the prospect of better conversion and separation processes[3]. The multipurpose microwave reactors developed by Radio Frequency group of Liverpool John Moores University were used to assist the rate of separation of the methanol and oil phases to produce biodiesel from waste cooking oil and animal fats as shown in Fig.2.

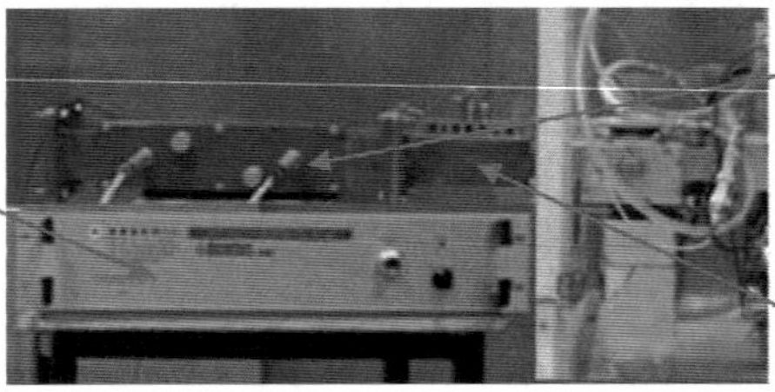

Fig.2 Novel microwave laboratory reactor heating system.

The biodiesel reaction step has the largest effect on the quality and quantity of the final product. An unsuccessful reaction step due to any disturbances or changes in the reaction conditions leads to incomplete conversion of the triglycerides it is therefore advantageous to be able to control and monitor the transesterification reaction until the completion. When dealing with the problem of continuous transesterification reactor control, we need to consider the nonlinear nature of the process, operational, safety limitations imposed on manipulated variables[4], and system is subject to many unexpected disturbances . In this work, the intelligent control methods, Fuzzy logic and Adaptive Neuro-Fuzzy Inference System(ANFIS) were successfully applied to maintain high performance under various operational conditions. Automated the monitored and controlled system was developed using laptop computer and National Instrument's Data Acquisition(DAQ) as shown in Fig.3, with Labview software tool. Labview as shown in Fig.4, stands for **Lab**oratory **V**irtual **I**nstrument **E**ngineering **W**orkbench, and is a graphical programming language which was use to receive the signals from the transducers through the DAQ, process the signals and send the output signals to the equipment to be controlled. The physical quantities of the system to be automated are sensed by the biodiesel system sensors as shown in Fig.5.

Fig.3 Compact DAQ for the biodiesel system

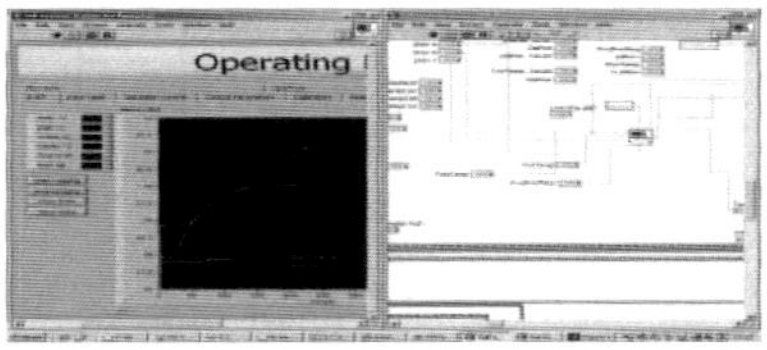

Fig.4 Biodiesel system in Labview

Fig.5 Biodiesel sensors .

2. Experimental system setup

The automated microwave system has been used as a heating source for the reactor to produce biodiesel from waste cooking oil and animal fats. Microwaves at a frequency of 2.45GHz were used to reduce the chemical reaction time, from hours under conventional heating, to minutes for the same volume of waste input. This novel microwave reactor is capable of operating at commercial production rates (kg/hr) instead of laboratory scale (g/day). A Labview-based microwave biodiesel reactor control system as shown in Fig.6 was built in order to implement the Fuzzy logic and Neuro-Fuzzy Inference System(ANFIS) online. All the changes in the control system could be observed in real time and user commands could be accepted during the process.

Fig.6 Biodiesel system.

3. Controllers

3.1. Fuzzy Logic controller design

Fuzzy logic control incorporates human knowledge to describe a system without requiring a mathematical model. A Fuzzy controller is composed of three calculation steps[5]: *Fuzzification:* the function of accepting input values and determining the degree of membership to some pre-selected linguistic terms. *Fuzzy Inference:* consists of determining the relationships based upon If-Then rules used to obtain more weighted outputs. These outputs are also linguistic variables in nature and have a nonlinear relationship. *Defuzzification:* involves converting the output term into a crisp value such that it is compatible with the system. The Fuzzy controller was based on the Mamdani Fuzzy logic Labview Toolkit. The first Fuzzy input represents the error between the measured temperature and the setpoint. The second Fuzzy input represents the change in error. Fig.7 shows the membership functions, (a) for the error and (b) for the change of error over the range of input variable values and linguistically describes the variables universe of discourse. Five membership functions were selected with triangle shape as follows : large positive, small positive, zero, small negative, large negative. The universe of discourse for the error and the change of error were -18 to 18 and -12 to 12 respectively. The left and right half of the triangle membership functions for each linguistic label was chosen to provide a membership overlap with adjacent membership functions.

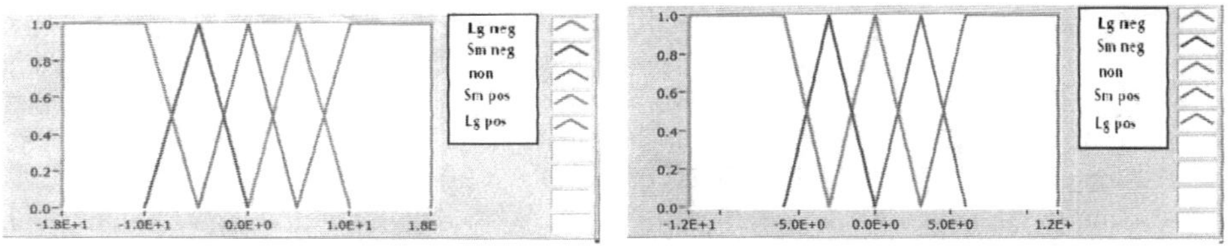

(a) (b)

Fig.7 Fuzzy inputs membership functions for (a) error and(b) change of error.

The Fuzzy logic controller output delivered power which applied to the microwave reactor. Fig.8 shows the membership functions of the output linguistic variables of Fuzzy output.

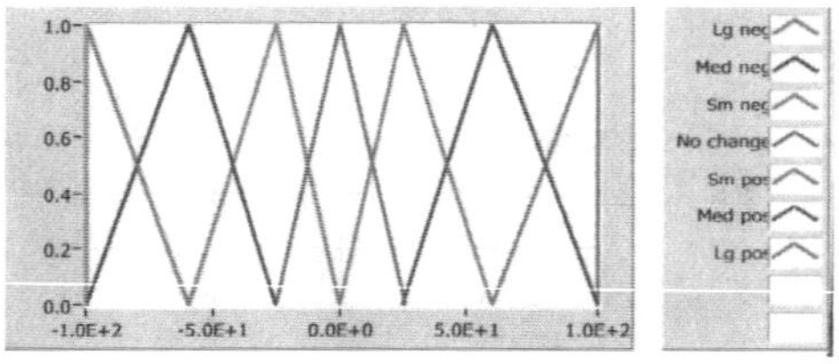

Fig.8 Membership functions of the output linguistic variables .

The output variable was chosen in the interval from -100 to 100 with seven triangle membership functions; large positive, medium positive, small positive, zero, small negative, medium negative, large negative. The rule base of the fuzzy controller decides which of the five membership functions have to fire. The rule evaluation part has 25 fuzzy rules. Table 1 shows the Fuzzy logic control rules developed for the proposed system.

TABLE I
Fuzzy Rules Table For Biodiesel System

e / Δe	Lg neg	Sm neg	Zero	Sm pos	Lg pos
Lg neg	Lg neg	Med neg	Med pos	Med neg	Sm pos
Sm neg	Med neg	Sm neg	Sm pos	Zero	Sm pos
Zero	Sm neg	Sm neg	Zero	Sm pos	Sm pos
Sm pos	Sm neg	Zero	Zero	Sm pos	Med pos
Lg pos	Sm neg	Sm pos	Zero	Med pos	Lg pos

3.2. Adaptive Neuro- Fuzzy Inference System(ANFIS) control.

Adaptive Neuro- Fuzzy Inference Systems (ANFIS) are fuzzy Sugeno put in the framework of adaptive systems to facilitate learning and adaptation. Such frameworks make fuzzy logic more systematic and less relying on expert knowledge [6]. The objective of ANFIS is to optimize the parameters of a given fuzzy inference system by applying a learning procedure using a set of input-output training data. A combination of least square and back- propagation methods are used for training a fuzzy inference system to enhance its performance [7]. In our work , we designed the ANFIS as a first order Sugeno fuzzy model so the consequent part of the fuzzy rules is a linear equation with generalized bell-shaped membership functions of the inputs , which contain three fitting parameters ; centre and half of the width and slope. Two membership functions were chosen on each input, usually this number is determined experimentally. The Initialization of the primes parameters of the ANFIS network are set so that the centers of the membership functions are equally spaced along the range of input variable, where the initial value of consequent parameters are assumed zero. The inverse model control approach was used , in which the controller is the inverse of the plant. In this method a learning task is needed to find the inverse model of the advanced biodiesel microwave system so ANFIS inverse learning for control purpose is performed as shown in Fig.9 . The training data were obtained from real experiments to reflect input-output characteristics of the biodiesel reactor. In the application phase, the obtained ANFIS inverse model was used to generate the control action .

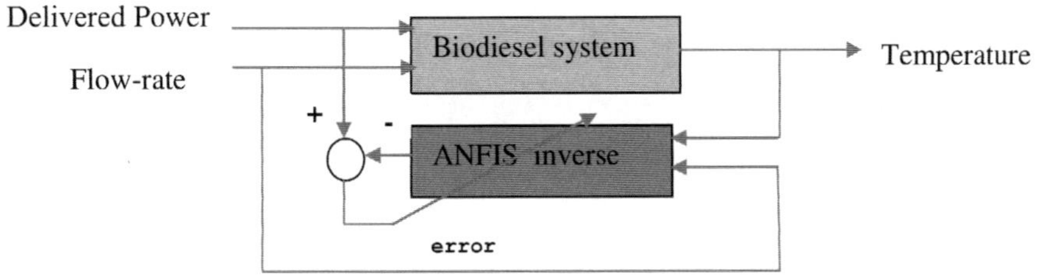

Fig.9 Block diagram for inverse control method (training

4.Experimental results and Conclusion.

The (ANFIS) and Fuzzy logic were implemented online for the advanced biodiesel microwave reactor. All the experimental results are in 5sec control sample intervals and for a 300 ml/min flow rate. The controllers were subjected to a multiple set point tracking beginning from the nominal temperature value of 30°C and changing to 35°C then down to 30°C as shown in Fig.10(a). The ANFIS and Fuzzy controllers successfully tracked the reactor temperature setpoint accurately, but the ANFIS controller had no overshoot and undershoots even in setpoint tracking beginning from the inlet temperature. The ANFIS control response was faster than Fuzzy logic as shown in Fig.10(b). The controllers were then tested under an introduced disturbance in the feed to the flow rate of the reactor. After the start up of the process, the reactor temperature was left regulated by the controllers at 30°C. The disturbance was then introduced when the nominal feed in flow rate was reduced by 20%, it was observed that the (ANFIS) controller was able to bring back the process to its initial setpoint of 30°C with very small effect as shown in Fig.11. For further test under disturbance, the input flow rate was reduced 10% then sudden raised by 30%, the ANFIS controller showed a good ability to disturbance rejection and tracking performance .

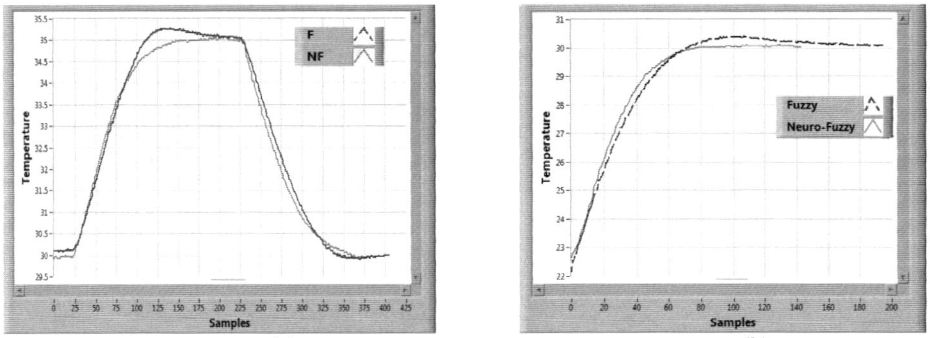

(a) (b)

Fig.10 Neuro- Fuzzy Inference System(ANFIS) and Fuzzy logic controllers (setpoint tracking temperature).

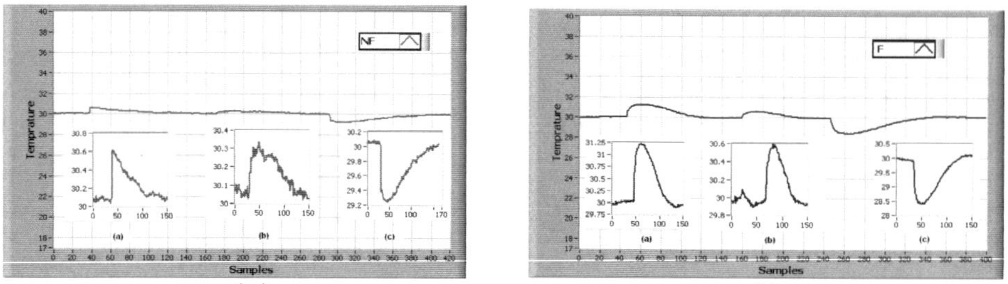

(a) (b)

Fig .11 Disturbance rejection of (a) ANFIS (b) Fuzzy Logic.

6. Conclusion

Intelligent control techniques including Fuzzy logic and ANFIS control were designed and implemented for a novel microwave biodiesel reactor A Labview software tool was used for real time monitoring and control for the full system. The achieved results showed that proposed ANFIS controller is more robust to parameter variations when compared to the Fuzzy controller.

6. References

[1] N. Ellis, F. Guan, T. Chen, C. Poon, "Monitoring Biodiesel Production(transesterification) Using in Situ Viscometer", *Chemical Engineering Journal*, Vol.138, PP.2000-206, 2008.

[2] T. Mchael, Barnard , Nicholas E. Leadbeater ," Microwave Technology and Continuous – Flow Production", *Biodiesel Magazine*, october 2007 Issue.

[3] T.Michael, N.Leadbeater, M.Boucher, L.Stencel, B.Wihite, "Continuous-Flow Preparation of Biodiesel Using Microwave Heating", *Energy&Fuels*, Vol.21, PP.1777-1781,2007.

[4] F. Mjalli, M. Hussain, "Approximate Predictive Versus Self-Tuning Adaptive Contol Strategies of Biodiesel Reactors",*Ind.Eng.Chem.Res.,* Vol.48,PP.11034-11047,2009.

[5] D. Ruan, " Intelligent Systems: Fuzzy Logic Neural Networks and Genetic Algorithms", *Kluwer Academic Publishers*,USA,1997.

[6] K. Assaleh, " Extraction of Fetal Electrocardiogram Using Adaptive Neuro-Fuzzy Inference System", *IEEE Transactions on Biomedical Engineering*, Vol.54, No.1, PP:5968, 2007.

[7] X. K. Zhang, Y, C. Jin, and G. Guo, " ANFIS Applied to a Ship Autopilot Design" *Proceedings of The Fifth International Conference on Machine Learning and Cybernetics, Dalian.* PP: 2233-2236, August, 2006.

Sensors & their Applications XVI IOP Publishing
Journal of Physics: Conference Series **307** (2011) 012043 doi:10.1088/1742-6596/307/1/012043

Real-time non-destructive microwave sensor for nutrient monitoring in wastewater treatment.

N. Al-Dasoqi, A. Mason, R. Alkhaddar, A. Shaw, A Al-Shamma'a.

School of Built Environment, Built Environment and Sustainable Technologies Research Institute (BEST), Liverpool John Moores University, Liverpool, L3 3AF.

E-mail: n.al-dasoqi@2009.ljmu.ac.uk

Abstract. A real-time non intrusive microwave sensor system able to monitor the nutrients found in wastewater has been designed, simulated and implemented. These liquids are continuously flowing through a PTFE pipe and the properties of these liquids gradually degraded in time. Microwaves have the ability to give real-time changes in any material permittivity by means of changing the velocity of the signal, attenuating or reflecting it. The primarily measurements show promising results for future sensor developments which lead to a novel system that can be used in wastewater treatment plants.

1. Introduction

Urban wastewater contains domestic and industrial wastewater with or without rain runoff [1]. In order to maintain a healthy environment and to control diseases this wastewater has to be collected and treated, protecting the environment from the adverse effect of wastewater. The main issue that wastewater treatment works face is how to detect the low concentrations of nutrients before discharging the treated wastewater. In recent years the wastewater treatment industry has suffered from a lack of reliable and efficient tools which support on line measurement of a number of important parameters such as pH, Chemical Oxygen Demand (COD), Biochemical Oxygen Demand (BOD), Chlorine (Cl), Ammonium and phosphorus. All of these parameters give a general idea of the quality of the treated water to be discharged to a water course. This paper presents a novel sensor design that has been used to detect Phosphorous (P). P is one of the key elements necessary for growth of plants and animals; in lake ecosystems it tends to be the growth limiting nutrient and is a backbone of the Kerbs' Cycle (a set of biochemical reactions that occur in the mitochondria) and DNA [2]. The presence of P is often scarce in the well-oxygenated lake waters, unlike nitrogen; phosphate is retained in the soil by a complex system of biological uptake, absorption, and mineralisation. Phosphates are not toxic to people or animals unless they are present in very high levels. Digestive problems could occur from extremely high levels of phosphate [2].

2. The use of microwaves for nutrient detection

Microwaves are electromagnetic waves (EM waves) with wavelengths in the range of (1mm-0.3m) or equivalently, with frequencies between 1and 300 GHz. The EM waves have been successfully used as a sensing element for various industrial applications including water level measurements [3] and material moisture [4].

A microwave resonator can be defined as a section of transmission line with open or shorted ends (ports). Depending on the type of transmission line, the resonators are called, for example, coaxial, microstrip, stripline, slotline, or cavity resonators. When an EM wave signal is incident to one end (or port) of a microwave resonator, some fraction of the signal may be reflected back to that port whilst the remainder exits the other port, therefore being transmitted [5]. Microwave sensors in the form of cavity resonators operate based upon the interaction of the EM waves and the material or object under test. The object changes the velocity of the signal, attenuates or reflects it. The object under test will be placed inside or near to the microwave resonator to determine the transmitted (S_{21}) and reflected (S_{11}) microwave power at discrete frequency intervals. By considering how these parameters vary with the object under test one can accurately determine the composition of the object under test.

Preliminary testing using other microwave cavities has been completed to investigate if this method can be applied to detect P solutions. Initially phosphate (PO_4) was considered using a microwave Vector Network Analyser (VNA) and a cylindrical microwave cavity (see figure 1). The detectable phosphate concentration was high however in this case since 500ppm of Potassium Phosphate (KPO_4) in Deionised Water (DW) was used. Also 500 ppm of four other solutions were introduced which were NaCl, $NaNO_3$, KCl, and KNO3 to test if the microwave sensor is more sensitive to P solution other than others. The DW response was taken as a reference reading for any amplitude shifts (AS) or frequency shifts (FS) that may occur when any of the solutions were introduced to the cavity.

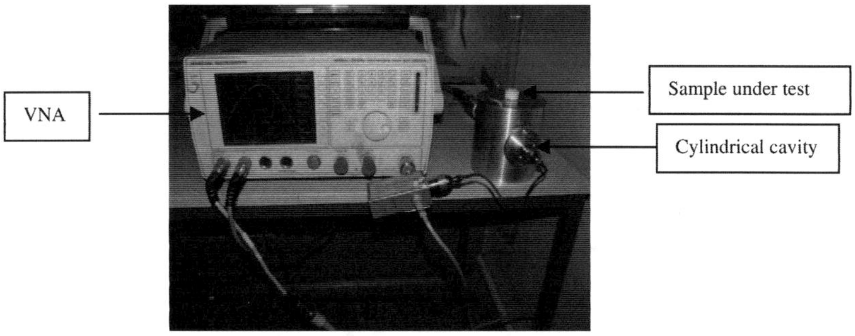

Figure 1. Preliminary cylindrical microwave cavity setup

The transmitted power S_{21} was measured for the six solutions that been introduced to the cavity one after the other. Results in Figure 2 show that the power of the phosphate is 0.7dB less than the other six solutions which have nearly the same power values.

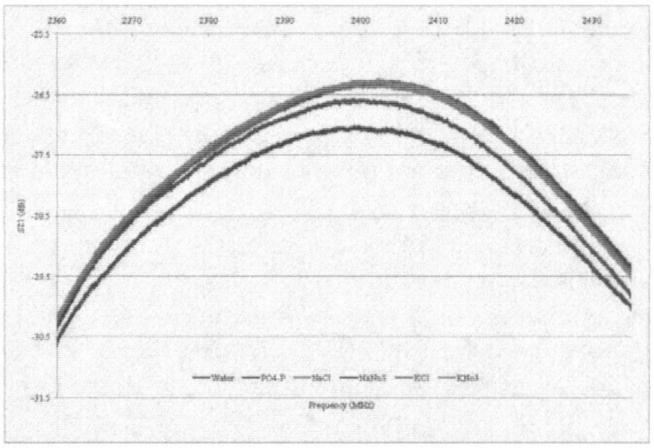

Figure 2. Transmited power response for the cylindrical cavity

3. Microwave sensor design and simulation.

The microwave sensor cavity was designed to work at the TM_{010} (Transverse Magnetic) mode shown in Figure 3. This geometry has been used for material permittivity measurements since 1946 [6] .

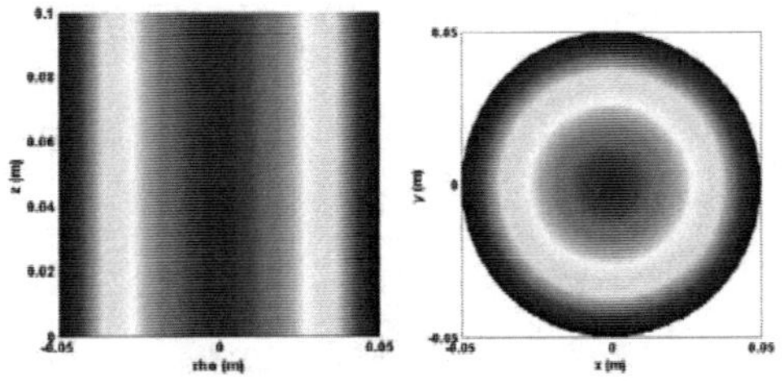

Figure 3. Electric field for TM_{010} mode in Cylindrical Resonant cavity [7]

3.1 Cavity design

The TM_{010} mode is the lowest resonant frequency for a short cylindrical cavity, some of its advantages are that the electric field is maximum in the axes of the cavity, and the resonant frequency is independent of the cavity height (reduction of errors depending on measuring dimensions) [7]; for these reasons a cavity with a PTFE pipe inserted in the middle has been designed as follows:
From a given cylinder radius $a - 37.5mm$, the resonant frequency f_{010} was calculated from (1) as follows:

$$f_{nml} = \frac{c}{2\pi\sqrt{\mu_r \varepsilon_r}} \sqrt{\left(\frac{p'_{nm}}{a}\right)^2 + \left(\frac{l\pi}{d}\right)^2} \qquad (1)$$

$$d < 0.9348a \qquad (2)$$

$$b < \frac{a}{\sqrt{\varepsilon_r}} \qquad (3)$$

Where f_{nml} is f_{010}, p'_{nm} can be determined from [8], d is the height of the cavity and independent of the resonant frequency as $l = 0$, but to get a maximum bandwidth for the next resonant mode the condition in (2) must be applied [7], in this case the next resonant mode is TM_{110}, which is also independent from d . The last measurement to be considered was the maximum radius of a hole b to be made in the resonant cavity for the insertion of the PTFE pipe; this hole must be under the cut off frequency, see equation (3) in order to avoid leakage of the microwave energy outside the cavity, where $\varepsilon_r = 81$ for water. Figure 4 shows the cavity design model.

Table 1. Cylindrical cavity design parameters.

Parameter	Value
a	37.5 mm
b	4mm
d	30mm
f_{010}	1.6326 GHz.

3.2 Cavity simulation
A simulation program called High Frequency Structure Simulator (HFSS) has been used; figure 4 shows the electric field distribution inside the cavity for the first mode (TM_{010}). Where all the design parameters shown in Table 1 were used.

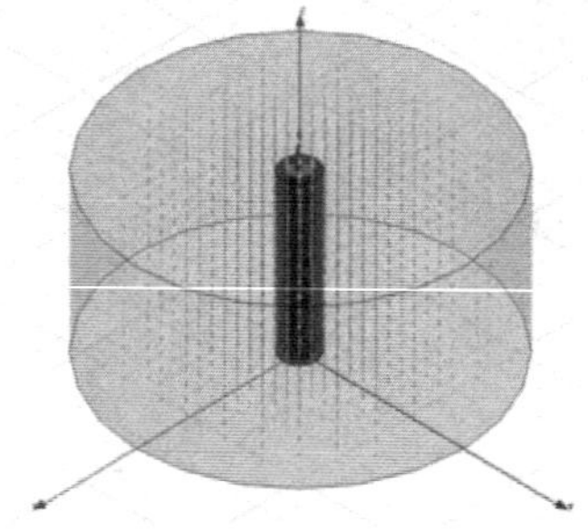

Figure 4. HFSS simulation for the cavity design

4. Microwave sensor cavity system installation
The designed cavity has been installed within a system contains: 1. HPLC Pump which controls the flow rate in the pipe and pumps up the deionised water then the Phosphate solution, 2. VNA which supply the microwave signal to be introduced in the cavity and measure the transmitted power S_{21}, 3. Microwave cavity sensor where all the interaction of the electromagnetic waves and the solution to be measured takes place, 4. Laptop to grab data for more measurements and analysis to be done and 5. Deionised and Phosphate solution bottles, Figure 5.a shows the system setup, and Figure 5.b shows the primarily results obtained from the designed cavity.

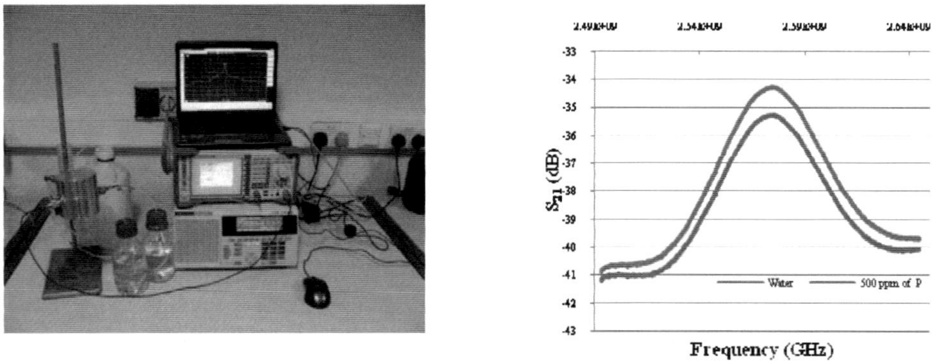

a b

Figure 5. System setup and the primarily results

5. Conclusion and future work
The prototype has been successfully designed, simulated and installed within a system, the primarily measurements showed promising results which would have a significant sensitivity to phosphate as the transmitted power decreases when the phosphate solution introduced into the cavity. This means that the cavity can be used to detect lower concentration of phosphate if more improvements of some parameters (i.e. antenna design) done, Further measurements should be done and more external factors to be controlled such as solution temperatures as the permittivity of water is highly temperature dependent.

6. References

[1] G. Kiely, "Environmental Engineering": McGraw-Hill, 1997.
[2] B. Oram, "Total Phosphorus and Phosphate (Impact on Surface Waters)," W. University, Ed., ed. Dalas, USA: Water Research Center, 2009.
[3] J. D. Boon and J. M. Brubaker, "Acoustic-microwave water level sensor comparisons in an estuarine environment," in *OCEANS 2008*, 2008, pp. 1-5.
[4] T. Lasri*, et al.*, "Microwave sensor for moisture measurements in solid materials," *Microwaves, Antennas and Propagation, IEE Proceedings H,* vol. 138, pp. 481-483, 1991.
[5] P.-N. D. inc. (2009). "S parameters". Available: http://www.microwaves101.com/encyclopedia/sparameters.cfm
[6] F. Horner, "Resonance methods of direct measurments at centimeter wavelengths," vol. 93, pp. 53-68, 1946.
[7] A. J. Cano*, et al.*, "Non-invasive Microwave Sensors for the Monitoring of the state of Liquids Used in the Polyurethane Industry," in *Sensor Technologies and Applications, 2007. SensorComm 2007. International Conference on*, 2007, pp. 56-61.
[8] D. M. Pozar, "Microwave Engineering", 2nd ed. New York: John Wiley & Sons, Inc, 1998.

THEORETIAL AND PRACTICAL NVESTIGATION INTO SUSTAINABLE METAL JOINING PROCESS FOR THE AUTOMOTIVE INDUSTRY

M.A. Al-Jader 1, J.D. Cullen, Andy Shaw and A.I. Al-Shamma'a
Built Environment and Sustainable Technologies Research Institute
Liverpool John Moores University, RF and Microwave Group, Byrom Street, Liverpool, L3 3AF
m.a.al-jader@ljmu.ac.uk

ABSTRACT
Currently there are about 4300 weld points on the average steel vehicle. Errors and problems due to tip damage and wear can cause great losses due to production line downtime. Current industrial monitoring systems check the quality of the nugget after processing 15 cars average once every two weeks. The nuggets are examined off line using a destructive process, which takes approximately 10 days to complete causing a long delay in the production process. In this paper a simulation results using software package, SORPAS, will be presented to determined the sustainability factors in spot welding process including Voltage, Current, Force, Water cooling rates, Material thicknesses and usage. The experimental results of various spot welding processes will be investigated and reported. The correlation of experimental results shows that SORPAS simulations can be used as an off line measurement to reduce factory energy usage. This paper also provides an overview of electrode current selection and its variance over the lifetime of the electrode tip, and describes the proposed analysis system for the selection of welding parameters for the spot welding process, as the electrode tip wears.

Keywords: resistance spot welding, SORPAS, computer simulation, Growth Curve, Expulsion

1. INTRODUCTION

Spot welding is a principle technique used in the automotive industry and has been for many years [1]. The destructive chisel and peel tests are used to examine the weld nuggets created on the production line. The main quality control tests are the destructive chisel test and peel test, which are carried out on welds obtained from the production line of the product [2, 3]. These have been supplemented by non destructive methods [4], The failure rates are detected, examined and evaluated. If there is a substantial increase in failure rates then the whole batch of the product is rejected as faulty. Different metals require different welding force and current, thus the difference in the higher force and current used for a galvanized metal than uncoated steel. It makes a significance change in the energy used for the creation of each spot weld. The following report lay's down a comprehensive view of electrode current selection and its variance over the whole span of the electrode tip. Also describing the advised analysis system for the selection of welding parameters for the spot welding procedure as the electrode tip wears. The unpredictability of the spot welding expulsion at different circumstances will be examined and monitored. Another concern for reducing sparks (splash) caused by resistance spot welding as it can prevent bumps on the bodywork that may cause further complications at later stages. SORPAS is used in order to compare practical tests and theoretical simulations from the practical tests data provided. Throughout the production run, the assurance continuity of nugget size depends on the definitions of optimum welding parameters and the implementation of suitable control for an acceptable spot welding quality. "Weld growth curves" and "weldability lobes" are two accepted methods by the industry to optimize the spot welding process [5]. The technique to form a weld is best exemplified by a weldability lobe which profiles the manufacturing tolerance. Both two and three-dimensional weldability lobes subsist, which are defined in terms of welding time, electrical current and electrode force. Figure 1 shows the spot welding machine operated for the experimental trial, a TECNA 4621 Pedestal Welder [6]. The Top arm is hinged to move down in an arc, while the bottom arm is fixed and contains various sensors used to monitor the welding process in real time [7].

Published under licence by IOP Publishing Ltd

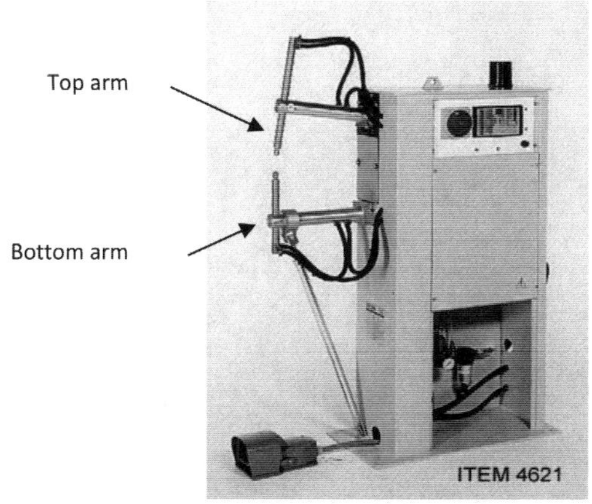

Fig. 1: TECNA 4621 Spot Welding Machine

The weldability lobe can only provide a snapshot of the welding currents range, because as the electrode tips wear the weldability lobe can drift. These two factors are controlled by the interaction between various parameters which control the temperature distribution in the metals during the welding thermal cycle. Galvanised steel typically has narrower lobes and greater electrode wear when compared with uncoated steel [8]. Recently however, a different concern has come to light namely energy consumption within the industry both financial and environmental repercussions [9]. This research greatly improves productivity since damaged tips can be identified and changes during the manufacturing process, rather than performing substandard welds. The quality of the weld is monitored in real time through its electrical properties, i.e. the current, voltage, etc.

2. CURRENT TECHNOLOGIES

In order to determine the efficiency of the existing technology, several experiments were conducted to monitor the step by step degradation in weld quality. The experiments were preformed on coated industrial 0.8mm sheets and the variables were current, voltage, force, water coolant and pressure. To check the quality of the weld destructive testing must be carried out – as shown in Figure 2. Jaguars' common industrial practice is to chisel test all welds on a single car every three months.

The required sensors are placed in relevant positions for most effective detection – as shown in Figure 3. The current sensor (not visible) is a Rogowski (air cored) coil [10]. This sensor gives a voltage output proportional to the current induced through the arms of the spot weld machine. The voltage sensor consists of two leads connected to each electrode. This signal is then passed through a buffer/amplifier circuit into a data logger within a standard computer. The infrared sensor is a photoelectric infrared diode positioned at a distance of 60 mm and at 90° from the electrode tips and also passes through the buffer/amplifier circuit.
Top arm Bottom arm

Fig. 2: Destructive Spot Welding Off Line

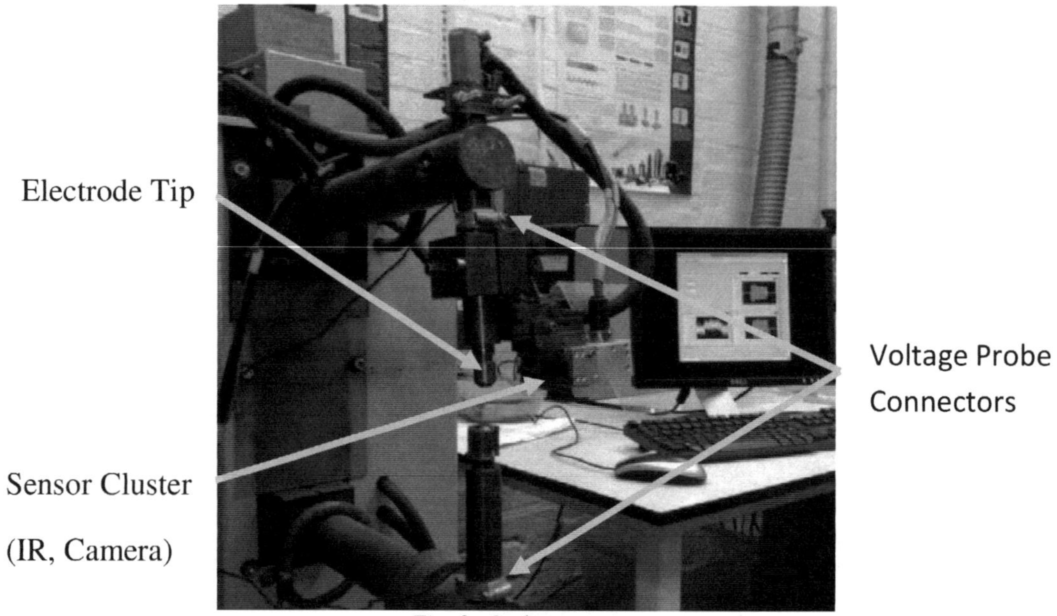

Fig. 3: On line sensors

Through weld ultrasonic testing is also used with the system and is mounted at the top and bottom of the upper and lower electrode arms. The collected data passes through a preconditioning circuit before being fed into a data logger. Typically for 0.8 mm galvanized steel, 0.2 sec (10-50 Hz mains cycles) is used for the weld time, the pre-process data capture is 0.3 sec (15-50 Hz mains cycles) and the post-process capture is 0.5 sec (25-50 Hz mains cycles). This allows the data to be captured before the electrodes are in contact with the metal, during the process and after the electrodes are removed from the metal to provide a complete data set to create the ultrasonic profile.

3. SPOT WELDING SIMULATIONS

SORPAS [11] is professional weld simulation software and has been used by engineers in industry (including automotive, steel making, welding equipment, electronics and other metal processing industries) to support design and evaluation of material weldability combinations, as well as design and selection of electrodes and general optimisation of welding process parameters. It can be used to compare theoretical and practical results – see Figure 4.

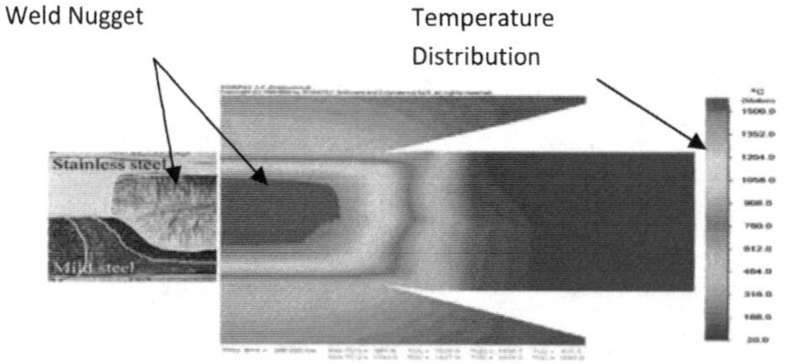

Fig. 4: Comparison of nugget simulated in SORPAS with practical results and the Temperature Distribution

A weld growth curve is constructed by performing several welds at different power settings and taking the average size of the resultant nuggets. The input power will range from settings that produce no welds through to those that cause splash. The range of weld types preformed covers the following criteria: Fig.5 shows the quality of the nugget weld of various welding conditions

- No weld: This occurs when there is insufficient current to melt the parent metal
- Stuck weld: This strictly refers to spot welding galvanised metal. The metal coating has a lower melting point and melts, but the parent metal does not. This results in the metals being stuck together with minimal mechanical strength.
- Undersized weld: This is where a weld created, but upon the destructive testing the nugget is smaller than required size, which according to BS1140 standards [12], is 3.5 times the square root of the thinner parent metal (in mm). This minimum requirement can be overridden by a particular requirement from the manufacturer, such as 4 times the square root of the thickness.
- Acceptable weld: This is the condition where the weld nugget is above the minimum size and below any maximum size (if specified) and does not result in splash (expelled metal).
- Splash weld: This is where some of the molten metal is expelled from the molten nugget, causing the electrodes to collapse in to the metal further, resulting in a thinner weld.

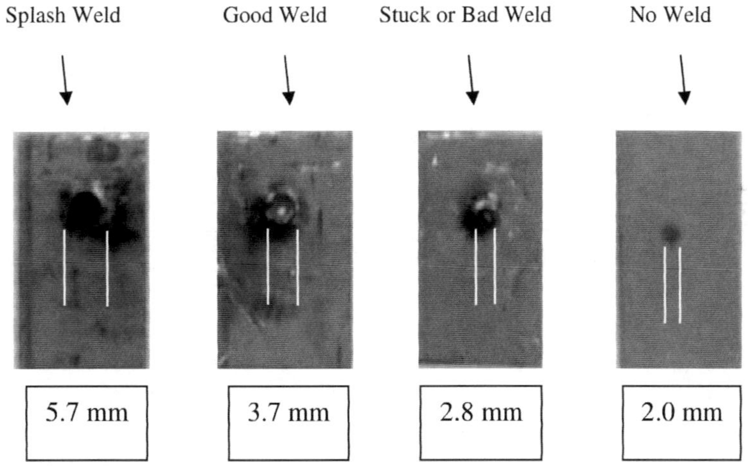

Fig.5: Representation of the quality of the nugget weld of various welding conditions

4. SORPAS SIMULATION

4.1. Weld Growth Curve

The Weld Growth Curve Simulation was carried out using standard coated stainless steel sheet thickness of 0.8mm, 2ltr/min water flow rate, 2.5bar pressure, current ranging from 2 to 11kA, with 0.5kA increments. The red points (see Figure 6) indicate splashes at the interfaces between the sheets, the black points indicate no weld and the green points indicate the welds with a nugget. Figure 7 shows the quality of the nugget weld of various welding conditions.

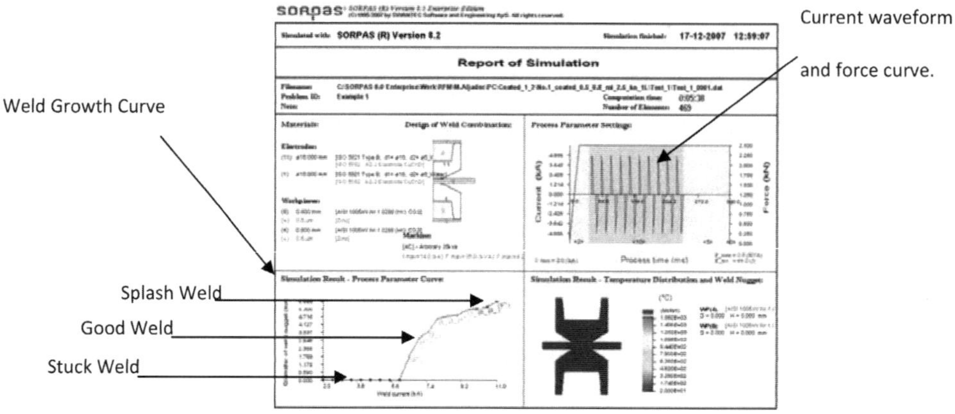

Fig.6: Representation of the welding profile, including materials specifications, current profile, weld growth curve profile

5. THEORETICAL AND PRACTICAL RESULTS

5.1. Weld Growth Curve

Figures 7and 8 show the differences between the theoretical and practical results of the weld growth curves (WGC). The theoretical WGC is resultant from the SORPAS simulation. The Simulation was performed on a typical passenger vehicle body of 1.0mm coated stainless steel sheet. The other welding parameters included 2ltr/min water flow rate, 3.5bar pressure, current ranging from 4 to 9kA, with 0.5kA increments. The black points indicate no weld and the green points indicate the welds with a nugget – as shown in Figure 7. The simulation performed on typical passenger vehicles body of 0.8mm coated stainless steel sheet, with welding parameter include 2ltr/min water flow rate, 3.5bar pressure and current ranging from 4 to 11kA, with 0.5kA increments. The red points indicate splashes at the Temperature Distribution Weld Nugget

interfaces between the metal sheets, the black points indicate no weld and the green points indicate the welds with a nugget – as shown in Figure 8.

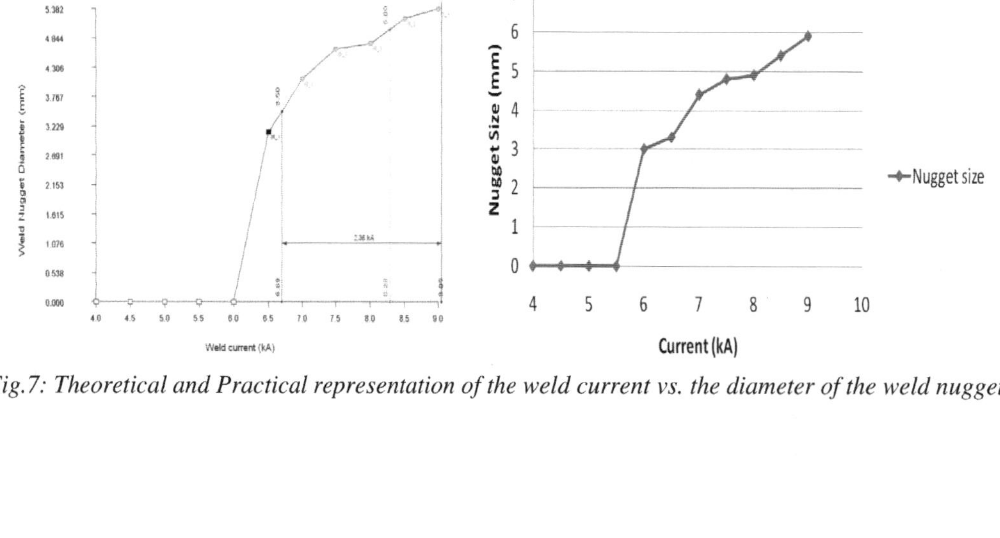

Fig.7: Theoretical and Practical representation of the weld current vs. the diameter of the weld nugget

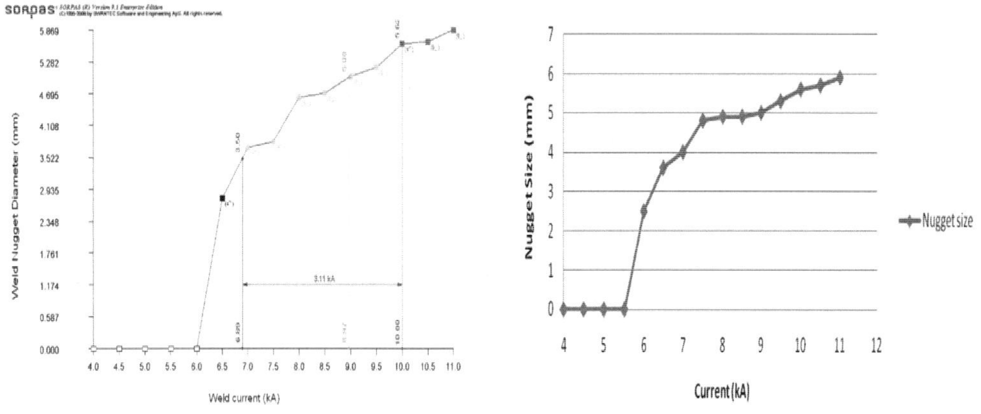

Fig.8: Theoretical and Practical representation of the weld current vs. the diameter of the weld nugget

5.2. Expulsion in resistance spot welding:

Expulsion is a common phenomenon in resistance spot welding and is the ejection of molten metal during welding as shown in Figure 9. It occurs at the interface between the electrode tip and the work piece [14]. This can obviously affect the surface quality and electrode life, but not the strength of the weld it if is limited to the surface. In terms of weld quality, the expulsion is highly inconvenient since it involves loss of liquid metal from the nugget. There are several possible causes of expulsion including; an unexpected gap between the metals, dust particles between the metal surfaces, a greasy metal surface, a reactive chemical compound within the metal or coating, and an unnecessarily high current. Generally expulsion is increasingly expected when greater currents are applied; however, it is still possible to meet the weld nugget requirements in some circumstances.

Fig.9: Practical representation of the Expulsion

5.3. Practical results (expulsion):

As show in Figures 10, 11, and 12 the practical test carried out using standard coated stainless steel thickness of 0.8mm, and 2ltr/min water flow rate. The pressure was varied between 3.5bar, 4.0bar, and 4.5bar, with the current ranging from 4 kA to 9.5kA, at 0.5kA increments. The tests that carried out involve a repetition of 1700 welds per set, where the electrode tips remained the same and replaced for new tips in each set after the 1700 welds.

The three different Figures prove that even though the same specifications are used in each experiment the nugget size produced differs according to the pressure/current combination used.

At Figures 10, 11 and 12, where the pressure was set at 3.5bar, 4.0bar and 4.5bar respectively, The number of nuggets produced that conform with the British Standard nugget size regulation were highest at the 4.0bar followed by 4.5bar and lowest at the 3.5bar pressure. It is concluded that the amount of current required to produce a British Standard nugget at the pressures of 3.5bar and 4.5bar is significantly higher. Therefore not only that the 3.5bar and 4.5bar do not produce a desirable yield, but these pressures are inefficient in terms of the energy consumption.

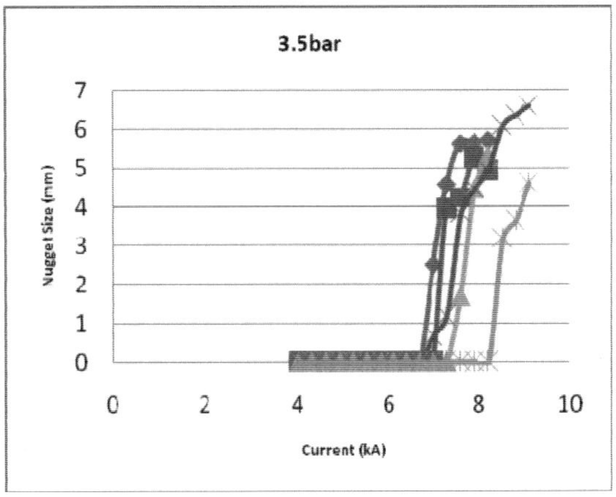

Fig.10: Practical representation of 3.5bar

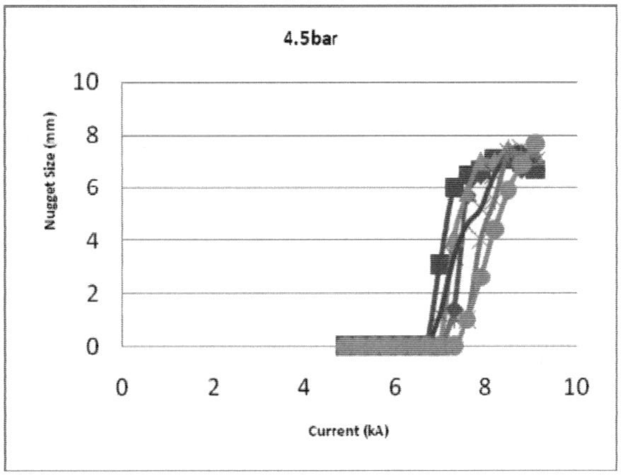

Fig.12: Practical representation of 4.5bar

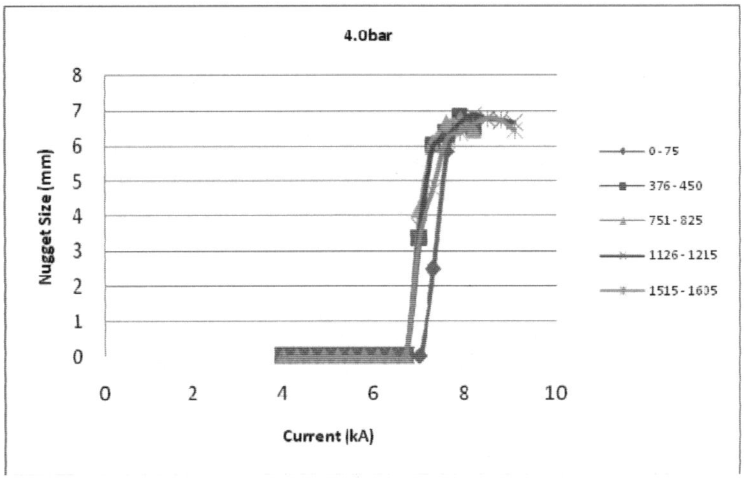

Fig.11: Practical representation of the weld current vs. the diameter of the weld nugget at 4.0bar

6. Conclusion:

The relation between the weld growth curves produced by practical and simulation indicates that although there is an alternation in the results, the correlation is great enough for SORPAS to be adopted in providing the initial parameters for deployment in the industry. The core of this study was on the identification of different spot weld methods using SORPAS. The application of the software was invaluable in the process of spot welding; however its precision varies with the numerous experimental parameters. In the case of the automotive industry, an electrode tip life span can be far greater than recommended (approximately 500 welds) to produce an acceptable weld by reducing the current to the minimum value. Reducing the energy consumption for the welding process is another advantage triggering a positive impact both environmentally and financially, as well as decreasing the likelihood of expulsion. There is ongoing research to enable the process to consume even less energy by minimizing the electrode force as well as control the value of current in order to obtain the optimum condition for good weld. It is envisaged that the introduction of artificial intelligence will endorse for time reduction in weld time, as well as the improvements of monitoring systems that can be used on different spot weld machines. Reduced manufacturing downtime is another outcome of the improved deployment speed.

7. References:

[1] C.F. Larson, Bibliography on resistance welding 1950–1971,Welding Journal 51 (1972).
[2] R.J. Krieger, S.A. Wenk, R.C. McMaster, Non destructive test methods for inspection of welded joints, Welding Journal 33 (1954) 1118–1129.
[3] Nita D 1997 "Resistance Welding Standards" Krupp Camford Internal Report KW002.
[4] Krieger R J, Wenk S A, McMaster R C 1954 "Non destructive test methods for inspection of welded joints" Welding Journal 33.
[5] Gopalakrishnan B, Plummer R W, Iskander W H 2003 "A Comparative Study on Energy Assessment Data from Manufacturing Industry", Proceedings of the 2003 ACEEE (American Council for an Energy Efficient Economy) Summer Study on Energy Efficiency in Industry,ISBN 0-918249-52-X.
[6] Tecna 2003, website, http://www.tecna.net/.
[7] Cullen J D, Athi N, Al-Jader M, Johnson P, Al-Shamma"a A I , Shaw A, El-Rasheed A M A"Multisensor fusion for on line monitoring of the quality of spot welding in automotive industry", Measurementhttp://dx.doi.org/10.1016/j.measurement.2007.01.006.
[8] Liv Williams N T, Parker J D 2004 "Review of resistance spot welding of steel sheets. Part 1.Modelling and control of nugget formation" Inst. Mater., UK ISSN: 0950 6608.
[9] Koenigsberger F, Adair J R 1965 "Welding Technology" 3rd Edition, Macmillan.
[10] D. Ward, Rogowski Coils, Rocoil Limited, 5, Almsford Avenue, Harrogate, North Yorkshire, UK,1965.
[11] SORPAS Version 7.0 User Manual; SWANTEC Software and Engineering ApS; March 2007.
[12] British standards Institution 1993 BS1140:1993 Specification for resistance of spot welding of uncoated and coated low carbon steel.
[13] Williams N T, Parker J D "Review of resistance spot welding of steel sheets. Part 1. Modelling and control of nugget formation" Inst. Mater., UK ISSN: 0950 6608 April 2004.
[14] Hongyan Zhang, Jacek Senkara, Resistance Welding Fundamentals and Applications, ISBN 0-203-49752-X.

Commencement measurements giving fundamental surface tension determinations in tensiometry.

D. Carbery[1], N.D. McMillan[2], D. Morrin[1], M. O'Neill[2], B. O'Rourke[1], P.Pringuet[4], S. Riedel[2], S.R.P. ,Smith[3]

[1] Institute of Technology Carlow, Kilkenny Road, Carlow, Ireland
[2] Drop Technology Ltd, Whitestown Business Park, Tallaght, Dublin 24, Ireland
[3] Physics Centre, University of Essex, Wivenhoe Park, Colchester, Essex CO4 3SQ, UK
[4] Bahrain Polytechnic, BP33349, Isa Town, Kingdom of Bahrain

E-mail: norman.mcmillan@itcarlow.ie

Abstract: This study provides experimental testing of a ray-tracing model of the tensiotrace that explores the measurement potential of a well-defined optical position in the tensiotrace signal known as the 'commencement'. This point is defined as the first measureable optical coupling in the fiber drophead between source and collector fibers for light injected inside a growing drop. Tensiotrace ray-tracing model is briefly introduced. Empirical relationships of commencement measures from a wide-ranging study are presented. A number of conclusions can be drawn from the successful linking of computer predictions to these experimental relationships.

1. Introduction

1.1 Experimental

The instrument arrangement is shown in Figure 1

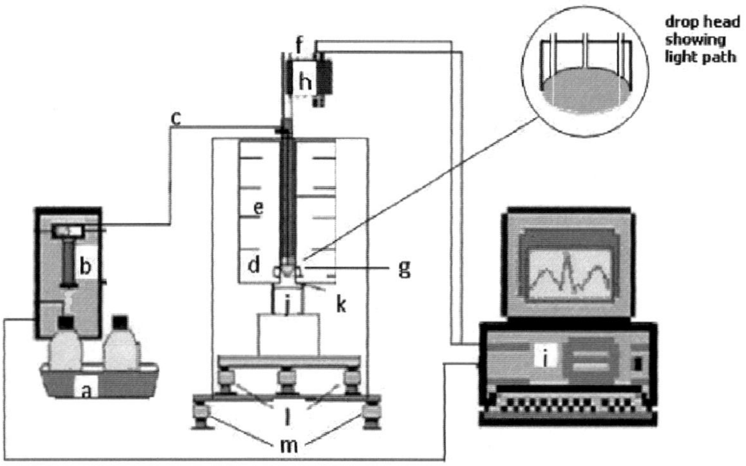

a	Samples for analysis	**h**	Detector
b	Stepper-motor-pump	**i**	Computer
c	Delivery tube	**j**	Chamber to ensure saturation of atmosphere
d	Drophead	**k**	Temperature sensors
e	Temperature control	**l**	Auto-levelling drives
f	Light source	**m**	Anti-vibration balancing feet
g	Optical eyes		

Figure 1 Schematic diagram of the tensiograph instrument.

Journal of Physics: Conference Series **307** (2011) 012045 doi:10.1088/1742-6596/307/1/012045

The liquid **a**, which is to be analysed is pumped via the liquid delivery tube **c** to the drophead **d** using a motor stepper pump **b** and a pendant drop grows at a uniform volume rate (isochoric) in size until it separates from the drophead. Throughout this process, in what is in essence a single wavelength operation, light from a light emitting diode (LED) **f** is injected into the drop through the source-fibre and the reflected signal, picked up by the collector-fibre is transmitted to a photodiode **h**. The amplification of the low photocurrent into an adequate voltage signal is necessary before the signal is passed to an acquisition board [an electronic system based on a programmable interface controller (PIC) and other suitable electronics hardware] incorporated in a personal computer (PC) **i**. The optical eyes **g** sense the drop on the drophead – the initial trigger occurs when the initiation drop falls through the optical eye and triggers the start of the data acquisition.

This indicates when the measurement drop has arrived on the drophead and then the data acquisition continues until this drop falls off the drophead and the second trigger locates precisely the ending of recording of the tensiotrace (a data set recorded over the life cycle of the drop). Two-stage anti-vibration mounts **l** and **m** protect the drop from vibrations. The analogue to digital converter (ADC) of the acquisition board converts the analogue voltage (amplified optoelectronic signal from the detector) into a digital signal, which is then subsequently transferred to the memory of the PC where it is stored for later analysis. The source-detector employs a CCD (charged couple device) or CMOS (complementary metal oxide semi-conductor) detector in conjunction with a spectral source such as fibre deuterium, xenon, halogen or tungsten sources. Such a system provides a spectral array of tensiotraces, one for every measurement wavelength.

1.2 The Tensiotrace

The key concept of the tensiograph is that all the various physical and chemical processes that modulate the coupled light in the drop produce a signal, which is unique in a theoretical sense for every specific combination of properties of the liquid under test (LUT). These processes include partial reflection of the light beam inside the pendant drop, absorption of the light from chromophores in the liquid, scattering of the light from turbid particulate matter and changes in the emission angle from the source fibre. The detected light intensity is amplitude modulated during the life cycle of a drop i.e. the reflecting surface of the inner drop changes continuously. A tensiotrace (Figure 2) is a typical sensogram which is a very recent terminology.

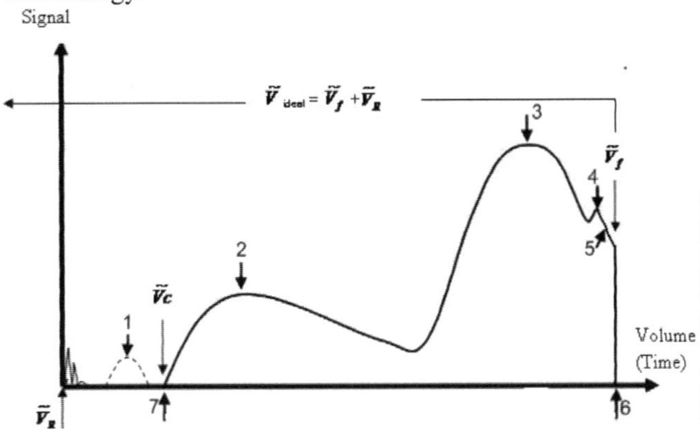

Label	Description
0	Separation vibration
1	First-order peak
2	Rainbow peak
3	Tensiopeak
4	Shoulder peak
5	Separation peak
6	Drop period
7	Rainbow peak commencement

Figure 2 Tensiotrace showing the important trace features associated with drop mechanics, labelled 0 to 7

The first-order peak is only observed in very special drophead designs. It is not seen in the current drophead design shown schematically in Figure 1 - the first order coupling is geometrically not possible because the measurement drop is too flat with the widely spaced fibres. The x-axis can be either volume or time as both are equivalent for constant volume delivery from stepper pumps. Two important measurands are noted: tensiopeak period or time to tensiopeak (t_3) and time of drop period (T_1).

The positioning of the fibres in the drophead has been engineered to optimise the measurement of various physical/chemical properties. The rainbow peak height and position in the tensiotrace is a good measure of refractive index. It gets its name from the fact that in rainbows a measurement of the angle of the bow to that of the observer provides a measurement of the refractive index of the liquid. The height of the tensiopeak provides a good measurand for determining the colour and/or turbidity of the solution in that the peak decreases respectively with the increasing absorptive or scattering power of the liquids. The drop period can be used to give a good measure of the surface tension/density ratio. Other physical measurements have been shown to be possible with the tensiograph, but despite the obvious usefulness of an instrument that provides so many measurements, the principal use of the tensiograph is as a fingerprinting technique. It has been demonstrated, over a considerable period of time that the combination of the complex of physical and chemical properties of the drop ultimately defines the form of the tensiotrace and thus the tensiotrace provides a very sensitive fingerprint technique for liquids, or indeed solid samples dissolved in a solvent.

1.3 The Ray-tracing Model of the Tensiotrace

The computer model is used to investigate theoretical issues regarding the tensiotrace. This RAST ray-tracing model is described in these proceedings by Pringuet et al[1]. We know that there are only numerical solutions to the Laplace-Young equation and it follows that there can only be numerical (computer model) solutions to the analysis of a tensiotrace. The drop shapes in our model are only dependent on the shape factor parameter $\beta_S = \frac{\rho g r_0^2}{2\gamma}$ where ρ is the density of the liquid, γ is the surface tension and r_0 is the drophead radius. It is clear that the tensiotrace is also sensitive to both the refractive index and absorbance (turbidity is not considered here but is a dimensionless quantity that is defined in an analogous way to absorbance). The model takes into account all the physical properties of the drop-under-test (DUT) and is an excellent qualitative tool, but in addition can deliver quantitative relationships such as those below in which the empirical relationships are quantitatively predicted. It will not be necessary to elucidate this RAST model further here.

2 Commencement Measurement

The only existing tensiograph study of remnant drop relationships are those by Gao and Zeng[2].They used the vibration of the remnant drop to investigate the size of this drop. The drop volume is $\overline{V_R}$ in Figure 2.This was a method suggested initially by McMillan, Feeney et al[3] and developed then by McMillan, Carbery et al[4]. The real issue of studying the size of the remnant drop has not been advanced much in these studies that were somewhat limited and largely qualitative. The new work reported here is very different and shows for the first time the real power of tensiography in dealing with these vital issues of defining the volume of a remnant drop and thus experimentally advancing the work of Harkins and Brown[5] and supplementing the recent theoretical contribution by Basaran[6]. The rainbow peak commencement is a very precisely defined point of optical coupling when a measurable amount of light reaches the detector. Figure 3 shows drawings taken from camera recordings of remnant drops of various liquids.

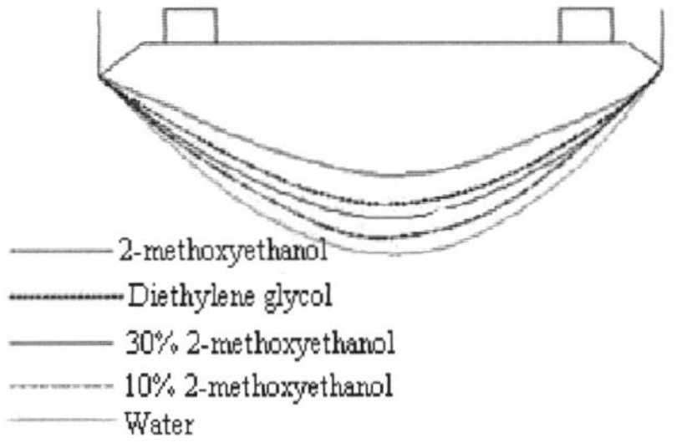

Figure 3. Drawings made from camera images of various remnant drops showing physically the variability in these drops for different liquids

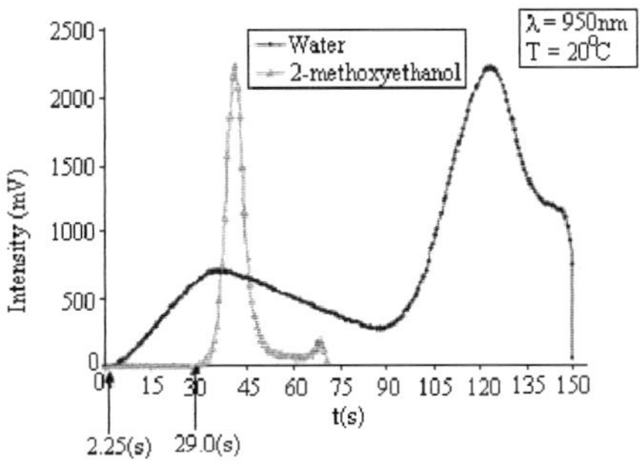

Figure 4. Tensiotraces of water and 2-methoxyethanol showing large variation in the measured commencement position

These experimental studies on the remnant drop underlie all the subsequent discussion in this section as tensiography provides a measurement tool for studying the remnant drop.

Figure 4 shows the experimental results for the very striking variability in the measured rainbow peak commencement for a water and alcohol type drop, illustrated by a typical representative of this liquid type 2-methoxyethanol. The measured times are for water 2250ms and for ethanol 29000ms. These two liquid types are commonly taken as the two extreme liquid types.

It has been shown that it is not possible to accurately determine the rainbow peak commencement of these traces by eye as replicate measurements show too big a standard deviation. However, an accurate method has been used to determine the 'commencement point' of the tensiotrace using specially written software. In brief, the software smoothes the data as can be seen in Figure 5 and then with a set threshold value determines a commencement point as so defined by the threshold. The practical results show that this measurement is both precise and accurate and provides an excellent measurement point.

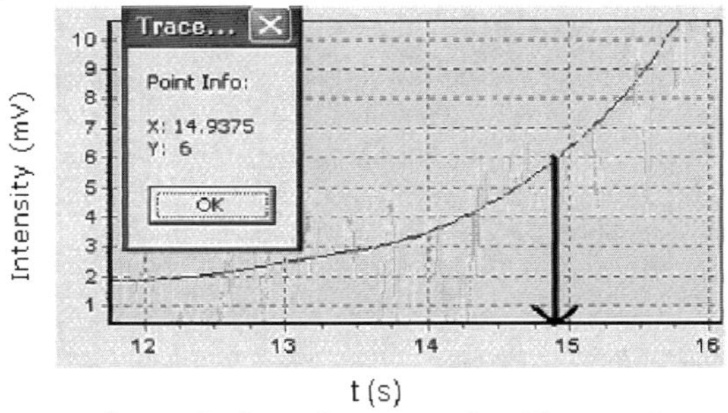

Figure 5. Enhanced trace of a typical tensiotrace using Trace-miner software for the determination of the rainbow peak commencement.

3.Comparison of Experimental and Theoretical/Modeling Results

3.1 Remnant Drop and Rainbow-peak Commencement Studies

Figure 6 shows the modelling results of the series of alcohols for remnant drop volume against the shape factor (β_S). It is clear that the modelling predicts a very linear relationship, which of course underscores the fact that correction factors have well defined mathematical relationships with this shape factor. The RAST model predicts this variation.

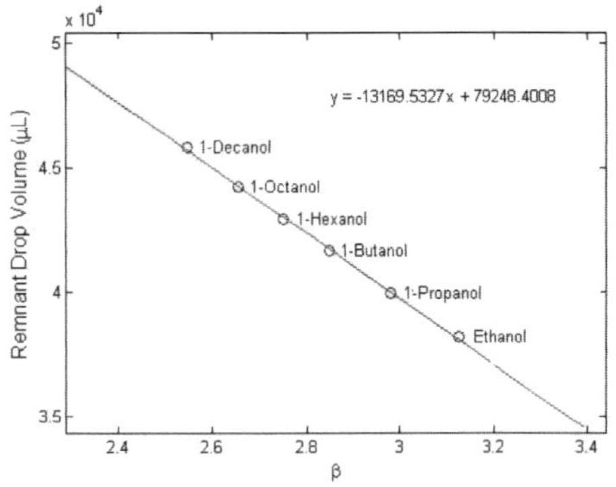

Figure 6 Analysis of remnant drop volume against the shape factor (β_S) on a 8.5mm drophead for a series of alcohols.

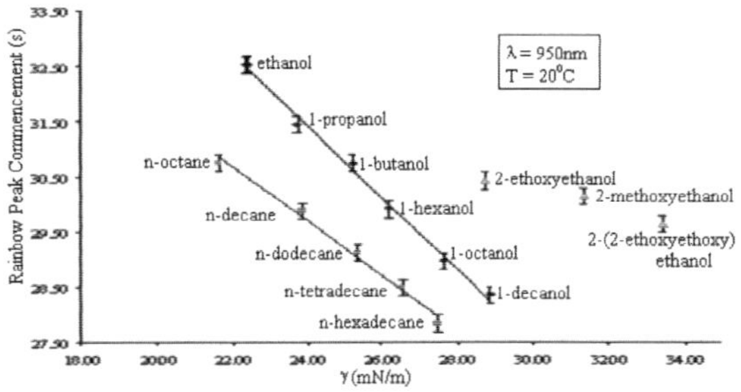

Figure 7. Relationship between rainbow peak commencement and surface tension values for alkane and alcohols.

The graphical plots of a series of rainbow peak experimental results on commencements measured against surface tension (γ) for a series of alkanes and alcohols are shown in Figure 7. It shows that the family of chemicals lie on two different lines. For the branched alcohols no trend line is shown but the four chemicals 1-butanol, 2-ethoxyethanol, 2-methoxyethanol and 2-(2-ethoxyethoxy) ethanol; lie on a second order trend line. The RAST model predicts this variation.

It is found that R_{RP} divided by refractive index (water being used as a standard here) against normalised surface tension or β_S is a first-order relationship. R_{RP} is defined as the ratio of rainbow peak commencement to rainbow peak commencement of water The experimental and modelling results for R_{RP} against normalised surface tension (or β_S) is a straight line for all Micellar surfactants although the ageing of the surface is an issue here and sufficient time must be left in order to bring the surface to an equilibrium situation. A universal relationship appears to exist for all liquids for R_{TP} (Ratio of the tensiopeak period using water as the reference liquid) against normalised drop volume. This is a significant experimental finding for this study[7].

Drop volume of course, because of its historical importance in the classical surface science technique of tensiometry, can be considered as the primary measurand of tensiography. Several other results that show first-order tensiographic relationships exist and Morrin[8] has documented many more empirical experimental findings and indeed even more low order trend-lines.

The results of a recent experimental study show that the "tensiopeak-period" (time to reach tensiopeak) (t_3) accurately predicts the value of the drop period T_1.

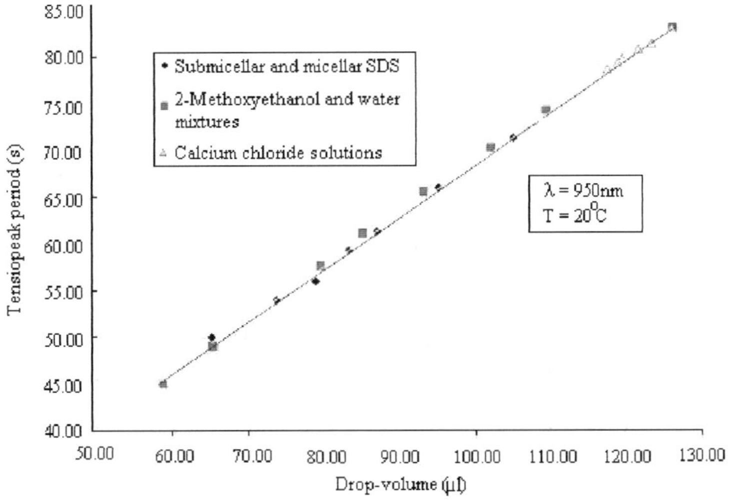

Figure 8. Tensiopeak period variation with drop volume.

Figure 8 shows the experimental linear results for the relationship between tensiopeak period (t3) and drop volume ($\widetilde{V}_f$) for a range of non-surfactant liquids. Drop volume is related to drop period (T_1) as follows:

$$\widetilde{V}_f = qT_1$$

where q is the liquid flow rate.

The significant advantage of using t_3 is that the drop does not separate from the drophead. Further results confirm the relationship between T_1 and t_3 for a wide range of liquids. These experimental studies were all done on a 9mm diameter drophead. Modelling work confirms these results. This relationship has been investigated by McMillan et al[9]. The limitation for the tensiometric method that depends on correction factors is highlighted from the chaotic and obviously unpredictable process revealed in Carbery's camera studies[10]. These chaotic processes mean in practice tensiometric correction factors will not be defined with accuracy.

4. Conclusion

Commencement is extremely well defined point in this optical signal and a reliable experimental method has been devised to obtain a quantitative measure of this point. The study reveals that commencement is a very useful tensiographic measurand. The commencement measures in extensive studies of families of chemicals, surfactants, salts and other liquids have been shown to graphically deliver useful surface measurements.

The tensiotrace is of course an optical signal derived from the evolving dropshape of a pendant drop. It is well-known that pendant dropshapes are universal set of curves that can be represented in a dimensionless form predicted by the dimensionless β_S factor. The tensiotrace is obtained from the RAST ray-tracing model but this trace is generated not only from dropshapes but also refractive index, absorbance and turbidity which are all dimensionless quantities. Some experimental results have been presented demonstrating that tensiographic measurements can directly predict surface tension of the liquid-under-test (LUT). Taken in the whole, these fact suggest that tensiography like profile analysis tensiography (PAT) provides measurements of the LUT or DUT that are based on similar dimensionless methods. This conclusion regarding tensiography is of some real importance as the PAT method is becoming accepted as one of the fundamental method in surface science precisely because of these measurement considerations.

6. References

[1]Pringuet1P, McMillan N.D, Doyle G, Smith S.R.P, ONeill M, Riedel S, RAST Model: simulation of tensiotraces to facilitate drophead engineering, Sensors & their Applications XVI.
[2] Gao, L. and L. Zeng (2003). "A remnant drop method for improving the uncertainty in surface tension measurements of volatile liquids using fibre drop analysis." Measurement Science and Technology **14**(8): 50-54.
[3] McMillan, Feeney
[4] Carbery, D., Riedel, S., McMillan, N. (2001). An experimental investigation into the engineering basis for a new fiber optic small volume drop surface analyser. Sensors and their Applications XI / ISMCR, City University, London, Institute of Physics.
[5] Harkins, W.D. and Brown, F.E., The determination of surface tension (free surface energy) and the weight of falling drops: the surface tension of water and benzene, J. Am. Chem. Soc., 41, 499, 1919.
[6] Yildirim, O., Xu Q. and Basaran, O.A., Physics of Fluids 17, 062107-13, 2005.
[7] McMillan, N., O'Neill, M., Carbery, D., et al. (2009). A new optical method of continuously analysing the surface properties of a single pendant drop while obtaining quality bulk spectral and refractive index measurements of the liquid-under-test. Journal of Physics: Conference Series, Institute of Physics Publishing.
[8] Morrin D, 2008 PhD, Tensiographic Studies, Institute of Technology, Carlow
[9] McMillan N.D, O'Rourke B, Morrin D, Pringuet P, Smith S.R.P, O'Neill M, Hammond J, Riedel S and Carbery D, A new optical method of continuously analysing the surface properties of a single pendant drop while obtaining quality bulk spectral and refractive index measurements of the liquid-under-test, Journal of Physics: Conference Series **178** (2009) 012007 doi:10.1088/1742-6596/178/1/012007

[10] Carbery D, McMillan N.D. et al Sensors & their Applications XVI.

Experimental observations with ultra-speed imaging revealing new complex detail of drop separation processes from a large diameter drophead

D. Carbery[1], N.D. McMillan[2], M. O'Neill[2], S. Riedel[2], D.Kennedy[3] and T.Nicholls[4].
[1] Institute of Technology Carlow, Kilkenny Road, Carlow, Ireland
[2] Drop Technology Ltd., Tallaght Business Park, Whitestown, Tallaght, Dublin 24
[3] School of Engineering, DIT, Bolton Street, Dublin 1
[4]Photron Ltd., Bottom Road, West Wycombe, HP14 4BS, Bucks, United Kingdom

Abstract. An ultraspeed (5000 frames a second) camera has been used to record the drop separating from the photometric instrument the tensiograph. New insights into drop separation processes which have been evaluated in terms of the correction factors used in tensiometric methods for measuring surface tension. An innovative approach to drop separation process drop satellite labelling is proposed and the paper ends with suggestions as to practical improvements to the photometric measurement of what is a very long-established method.

Keywords : drops; separation; Yildirim; Basaran; high-speed; imaging; water; methanol; satellite.

Contact details: des.carbery@itcarlow.ie

1. Drop separation

This study looked at the physical reality of drop separation with some of the highest performance camera system currently available with 5000 frames a second. Details and new analysis of the pendant drop separation process are given. These are the first observations of the drop separation process on such a large 8.5 mm drophead. The study is primarily quantitative but also highlights qualitative issues with regard to measurements related to pendant drops.

The details revealed in the study and the identification of the component drops are consistent with the theoretical work of Yildirim and Basaran et al[i]. When a drop (hanging from a capillary or drophead) increases in size, due to a gradual liquid input, it reaches a stage when part of it separates (primary drop). Other separating parts thereafter are called satellite drops. What remains on the capillary or drophead is termed by these authors the pendant but by others more commonly referred to as the remnant drop. Figure 1 is a schematic of the pendant, satellite and primary drops.

The series of events leading to these separate drops can be termed a drop separation process. During such a process there are many reasons why a drop changes in shape, separates and continues to evolve. The physics of an equilibrium process would force the drop to form into a spherical shape, but is of course acted on by many forces including force of gravity, surface tension, surface waves, fields of flow within the drop and external friction. This study looks at this dynamic non-equilibrium process. Camera images of water and methanol drop separations are presented, commenting on significant stages.

2. Experimental arrangement

Figure 2 shows the experimental arrangement. A quartz drophead 8.5mm diameter was used. The drophead was held vertically and liquid continuously pumped to it through capillary tubing. The pump used was a Hamilton PSD/4 stepper motor driven syringe pump with a 250ul syringe. The pump speed was set at 200 which gave a flow rate of 1.25µl/s. A drop was formed on the drophead and images of the drop were captured just before the time that the drop became unstable and continued to be taken through the drop separation process.

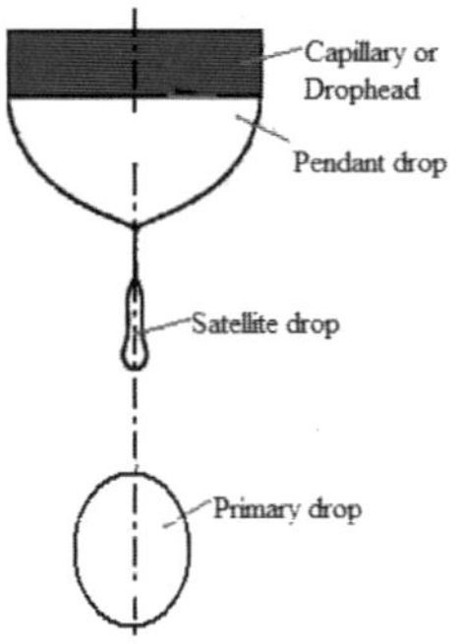

Figure 1 Pendant, satellite and primary drops

A Photron camera (Fastcam SA1.1 model) was used for capturing the images. The maximum resolution of the camera was 1000*1000 pixels and the images were taken at 5000 fps and 5400 fps (frames per second).
The drop liquids were at room temperature (20°C). Table 1 shows the liquid properties[ii] of the Analar methanol and water drops.

Properties	Water	Methanol
Density (kgm^{-3})	998.6	791.1
Surface tension (mNm^{-1})	73.05	24.05
Viscosity (cP)	1.002	0.6

Table 1 Liquid properties of the drops under test

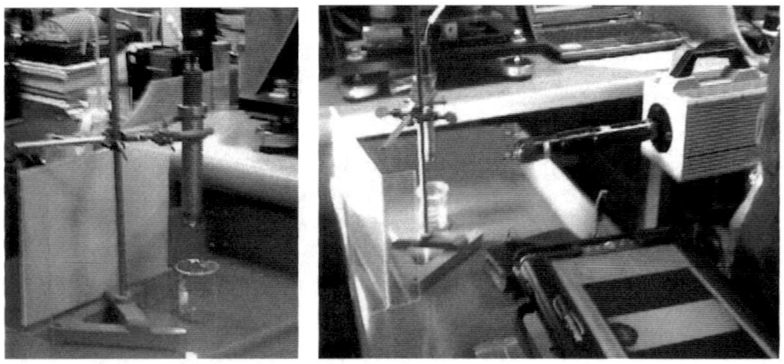

**Figure 2 Drophead with the capillary tubing connected at the top (on left).
Drophead with the extended camera lens and a screen image(on the right)**

The following camera images were taken:
i. Individual camera images of separating water drops
ii. Individual camera images of separating methanol drops
iii. Camera images of the contact edge of the methanol drop at the drophead
The camera speed for the drop images was 5,000 fps.

Journal of Physics: Conference Series **307** (2011) 012046 doi:10.1088/1742-6596/307/1/012046

3. Camera images

The general sequences of water and methanol drops Figure 3 shows images captured during the separating process of a water drop. The frame numbers are given so that the time between images can be calculated on the basis of the camera speed.

At first the full drop is shown, then the obvious necking showing positive and negative radii of curvature of the drop edge. Next, the almost cylindrical midsection reduces to a pinhead before the primary drop separates. At this stage the pendant drop is fully extended before pulling back upwards when the satellite drop separates. As the changing shape satellite falls, a daughter satellite separates upwards and continues until it contacts the pendant drop when it rebounds. A similar observation to the latter was made by Hauser et al[iii] who high-speed motion picture camera (1200 f.p.s) to observe the drop separations from a 4.58mm capillary.

Figure 4 shows images captured during the separating process of a methanol drop. The process is generally similar to the water set but the proportions of size are very different. A significant difference occurs however after the primary drop separates when there are greater number of satellite drops. The general asymmetrical shape of the methanol drop is noted but not explained. The drophead was levelled and rechecked but may not have been exact. (This issue has been addressed in the commercial tensiograph instrument[iv] with a sensitive levelling system together with vibration control.).

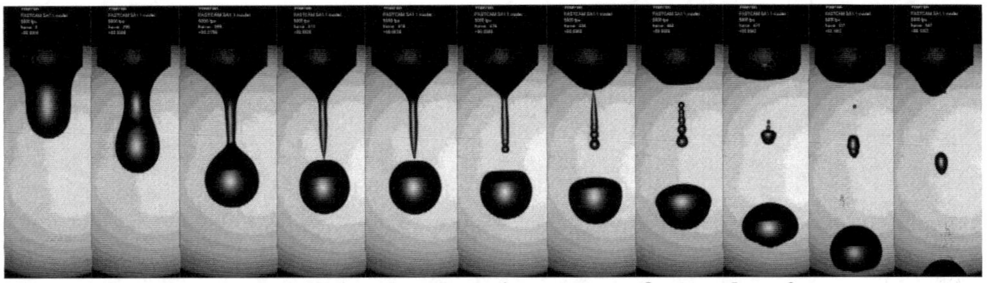

Figure 3 Images captured during the separating process of water drop 2. Frames 0, 299, 399, 413, 414, 424, 434, 444, 471, 501, 541.

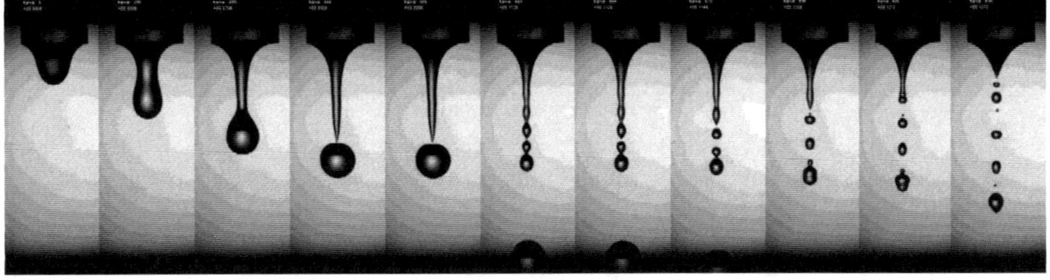

Figure 4 Images captured during the separating process of a methanol drop. Frames 0, 299, 399, 444, 445, 563, 564, 573, 594, 605, 636.

4. Images related to Basaran studies.

There is a direct correspondence between this study and the theoretical description of this author but the detail is too great to deal with here.

Convention to identify drops in a separating process

For clarity in the study of the separating drops, a convention is adopted to identify all drops in this process. The convention is similar to that used by Basaran but is extended further. Each drop is identified by an index d, the pendant drop is designated drop d 1. Figure 5 shows the convention used.

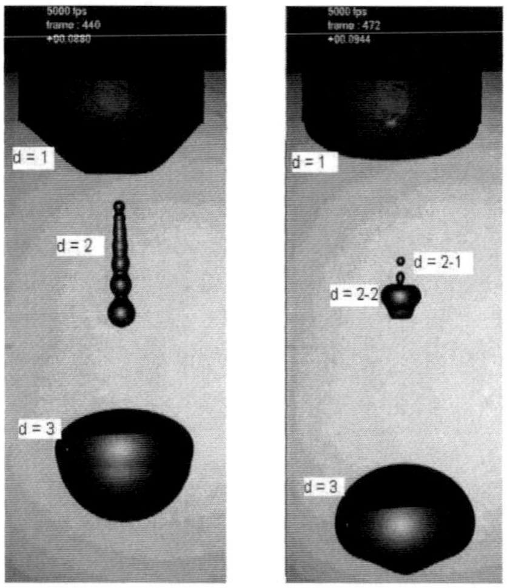

Figure 5 Water drops identified

The first frame in Figure 5 shows how the convention is used for the initial pendant, satellite and primary water drops. However, when a further drop separates from the satellite d 2, the convention is extended as shown in the second frame.

5. Satellite drop

The resulting complex effects on the satellite drop comes from the interplay of forces as the drop oscillates under their action. The lower part falls more slowly than the upper part. Generally, surface tension and gravity forces oppose each other on the lower part while they act together on the upper part. However this is not totally consistent as can be seen from the results that follow. Figure 6 shows the falling speeds of the top and bottom surfaces of the water satellite drop.

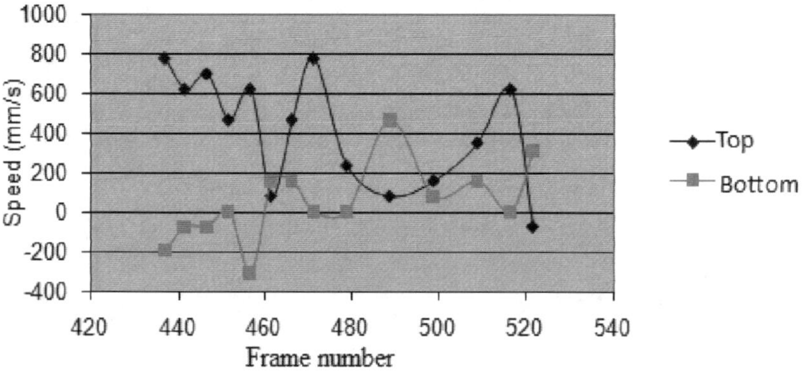

Figure 6 Speed of fall of top and bottom of water satellite drop.

6. Second satellite drop

Figure 7 shows the sequence of images leading to the separation of a second satellite water drop (d2-1) from the main satellite drop (d2), then re-designated (d2-2).

Figure 7 Images of the separation of the satellite drop d2 into d2-1 and d2-2. Frames 405, 406, 407, 410, 412, 413, 416, 417.

This second separated satellite makes two oscillation before successfully leaving the main satellite. The first attempt is shown in Frame 405. The first two images here show what we will term an attempt to separate, while the third and fourth two images show its return to the main satellite. There are possible explanations for this event; one is that the force of attraction increases because of the change of surface shape of the second satellite (becoming more pointed and therefore electrical charge increases). While another explanation is that the main satellite shape is vibrating and its top surface rises upward to contact the separating second satellite.
However, the second attempt at separation is successful and is shown in Frame 412 The second satellite drop separates and is projected vertically upwards. The size of this second satellite (less than 1% of the main satellite) is estimated from the camera image as a sphere of 0.35mm diameter (volume 0.022μl). The gravitational force on it is approx 0.22μN. The separating force needs to exceed this and also must overcome the force of attraction and then requires a further additional force to accelerate the satellite upwards. It can be seen that a third satellite almost separates but is attracted back. During all of this time the main satellite drop is vibrating and being pulled together into a more spherical shape as it continues its fall downwards.

7. Drop separation times

The times at which the drops separate in the process were recorded and one set of results are shown in Table 2.

Separation times of water drops				
	Water drop 1		Water drop 2	
	Frame	Time (ms)	Frame	Time (ms)
Primary drop (d3) separates	350	0	413	0
Satellite drop (d2) separates	373	4.3	435	4.4
Satellite drop (d2-1) separates	412	11.5	471	11.6

Table 2 Separation times of water drops

8. Conclusion

This study set out to investigate the details of the process of separating drop with particular reference to measurements such as the diameter and height of the drop. Many observations were made and recorded, both experimental and theoretical. The full study is presented in Drop Metrology[v]. Some are presented in this paper and others may be the subject of a further paper.

This study presents the ultra-detail of this separation process of large pendant drops.

It shows clearly:

i. a pendant drop adapts to an increase in volume by changing shape, together with a change in contact angle at the drophead,

ii. the drop reaches a maximum volume when it can no longer be supported on the drophead and thereby begins the process of drop separation (This point of instability or "scission" point has not been exactly identified in this particular study. It would be useful in a further study to identify this event.),

iii. the separation of the primary drop,

iv. the separation of the satellite drops,

v. the change in shape of the separate drops when subjected to the various forces,

vi. the similarities and differences between two liquids in the same process.

The measurements have delivered graphical analysis for the first time with the high speed resolution of the camera. Also, an identity system for the separating drops has been established.

The final point to make from this study is the camera images explain in a practical way why the correction factors used in tensiometry are not very precise, the small satellite drop that rebounds from the remnant drop can on other days fuse with the drop. The atmospheric electrical charge will vary daily and the smallest tilt of the apparatus will clearly make a difference. Whether a drop fuses or not will have implications for the volume of the remnant drop and as a consequence the 'correction factor'[vi].

---

[i] Yildirim, O., Q. Xu, et al. (2005). "Analysis of the drop weight method." Physics of Fluids **17**: 062107

[ii] Lide, D., *CRC handbook of chemistry and physics*. 2004: CRC press.

[iii] Hauser, E., H. Edgerton, et al. (1936). "The Application of the High-speed Motion Picture Camera to Research on the Surface Tension of Liquids." The Journal of Physical Chemistry **40**(8): 973-988

[iv] McMillan, N., S. Smith, et al. (2008). "Quantitative drop spectroscopy using the drop analyser." Measurement Science and Technology **19**: 055601.

[v] Carbery, D. (2010), PhD Thesis IT Carlow

[vi] Harkins, W. and F. Brown (1919). "The determination of surface tension (free surface energy), and the weight of falling drops.- The surface tension of water and benzene by the capillary height method." Journal of the American Chemical Society **41**(4): 499-524.

Journal of Physics: Conference Series **307** (2011) 012047 doi:10.1088/1742-6596/307/1/012047

Comparison of the TLDA with the Nanodrop and the reference Qubit system

M O'Neill[1], J McPartlin[2], K Arthure[1], S Riedel[1] and ND McMillan[1]

[1]Drop Technology Ltd., Tallaght Business Park, Whitestown, Dublin 24, Ireland.

[2]Vitamin Research Unit, Department of Clinical Medicine, Sir Patrick Duns Laboratory, CPL, St. James's Hospital, Dublin 8, Ireland

martina.oneill@droptechnology.com

Abstract. The TLDA (transmitted light drop analyser) is a new microvolume UV-visible drop spectrophotometer new to the market. Assays were compared between a TLDA, NanoDrop and the Qubit system which is based on the gold standard for DNA assay, PicoGreen. The evaluation was carried out by the Trinity Biobank in St. James Hospital, Dublin on Buccal swabs and Blood samples. The data is discussed in this paper. The Qubit system is seen as the reference method in most studies as this is believed to provide more accurate results than UV absorbance methods because it distinguishes between DNA, RNA, free nucleotides, and other contaminants. The Qubit system uses fluorescent dyes to measure the concentration of the molecule of interest. The results compare both the spectrophotometric methods against the Qubit fluorescence technique.

1. Introduction

Correct quantitation of small amounts of DNA is very important for a wide variety of molecular biology applications. Recent review of methods is given by Maquat et al[1]. Quantitation of DNA is simple in theory but can be difficult in practice. Three methods of quantitation are assessed in this study – TLDA, NanoDrop and Qubit.

A relevant study has been conducted to demonstrate the value of the Qubit system as the reference system in DNA and protein studies[2]. The Qubit fluorometer is a device from Invitrogen for quantifying proteins, RNA, or DNA. It uses various assays, which contain sensitive dyes that fluoresce in proportion to the amount of protein, RNA or DNA respectively. PicoGreen dye was used in this study. PicoGreen is a fluorescent nucleic acid stain for quantitating double-stranded DNA (dsDNA). It has an excitation maximum at 480 nm (lesser peaks in the short-wave UV range) and an emission peak at 520 nm. When bound to dsDNA, fluorescence enhancement of PicoGreen is exceptionally high and little background occurs since the unbound dye has virtually no fluorescence. The problem of protein contamination of DNA is highlighted in many studies such as that of Wieczorek[3].The first paragraph after a heading is not indented (Bodytext style).

The TLDA[4] and NanoDrop are both UV/Vis spectrophotometric instruments. Drop UV spectrophotometry is one of the most common analytical methods for nucleic acids analysis. NanoDrop has been used in several BioBank studies such as that of a Spanish team led by R. Palmirotta[5]. The instrument has wide applications also for protein assays such as those of Desjardins et al[6]. The two UV instruments have been compared against PicoGreen, seen as the gold standard for DNA analysis. A UV/Vis spectrophotometer measures the intensity of light passing through a sample (I) and compares it to the intensity of light before it passes through the sample (I_o). The ratio of I to I_o is the transmittance. The absorbance, A, is based on the transmittance and is derived using the following formula:

$$A = -\log(\%T) \quad (1)$$

The absorbance of UV light as measured in a spectrophotometer can be combined with the concentration factor to enable calculation of the concentration and purity of dsDNA. One 260nm absorbance unit of double-stranded DNA = 50 µg/mL.

Even after DNA is isolated from an organism, protein frequently remains present in the DNA solution. Protein is tightly bound to DNA and complete removal is not always possible. Both protein and DNA of course absorb UV light, but have different absorbance spectra. Figure 1 below shows the typical UV absorption spectra for both protein (BSA) and DNA.

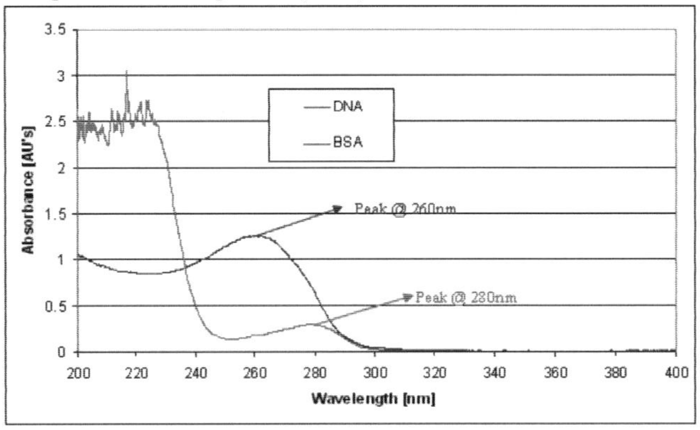

Figure 1. Absorption spectra for BSA and DNA

The peak of light absorption for DNA is at 260 nm, while protein absorbs at 280 nm, mainly due to tryptophan and tyrosine side chains. When a solution contains both protein and DNA, absorbance at 260 nm is primarily due to the DNA present, but a small amount is due to the protein. At 280 nm, the absorbance is predominantly due to the protein present. Thus, the purity of the DNA can be calculated by examining the ratio of the two absorbance values. A ratio of the absorbance values at 260/280 nm of approximately 1.7 to 1.8 predict "clean or pure" DNA, whereas lower values may be indicative of significant protein contamination. However, the 260/280 absorbance ratio is not always an accurate representation of DNA purity. Certain samples may be very difficult to evaluate at 280nm due to interference. Therefore, peptide bonds, which absorb at 228 nm, are often a more constant indicator of the presence of protein in a sample. Thus, absorbance readings measured both at 230 nm and at 280 nm provide a more accurate estimate of proteins or peptides that may be present in nucleic acid samples. The ratio of the 260/230 absorbance values should be >1.5 since nucleic acids have an absorbance minima at 230 nm. Both the TLDA and NanoDrop provide absorbance readings at 230, 260 and 280nm, and examine both the 260/280 and 230/280 ratios for every sample. Buffer salts can contribute significantly to absorbance readings, especially below 260 nm. Thus, it is imperative to blank the spectrophotometer in the same buffer in which your DNA sample is diluted. The Blank used in these experimental trails was TE Buffer.

The procedure used for NanoDrop measurements is well-established; the sample is pipetted directly onto the measurement surface. No cuvette is used. Using a patented retention system, a column is drawn between the ends of two optical fibers to establish the measurement path. The TLDA from Drop Technology uses a patented system which uses surface tension to hold the small volume sample drop in place on the quartz measurement surface. The computer model of the drop spectroscopy system is given by McMillan et al[6]. The theory derived in this study for a pendant drop is relevant to the sessile drop TLDA system. The TLDA sample is a 1.5-3µl sample droplet onto the quartz optical surface that is pipetted onto the drophead rotating instrument loading system. A quartz plinth is necessary as quartz is transparent in the UV. Surface tension alone holds the sample in place for the spectral measurement. With the rotating camber closes the fiber light source is aligned directly over the detector. The light passes through an aligned sessile drop from a source fiber to the detector. The pathlength is determined mathematically from the drop size selected in the software.

2. Experimental

2.1. Measurement Details

DNA and most of the common contaminants found in DNA absorb in the 230nm to 320nm region. Hence an absorbance measurement in this region allows the DNA concentration to be calculated and provides information about the contaminant levels. The most important wavelengths to note are:
230nm: Phenol and guanidium salts absorbs strongly at 230nm. Therefore, high absorbances at this wavelength can be an indication of carry-over of either of these compounds into the test sample.
260nm: DNA absorbs light most strongly at 260nm. The absorbance value at this wavelength can be used to estimate the DNA concentration using the equation Concentration (µg/ml) = (A_{260} reading – A_{320} reading) x 50, which is derived from Beer's Law.
280nm: Tyrosine and tryptophan absorb strongly at this wavelength. Hence, the absorbance at 280nm is used as an indicator of protein contamination.
320nm: The absorbance at 320nm provides a measurement of the turbidity of the sample. It is normally subtracted from the A_{260} value as a background reading for the calculation of DNA concentration. Excessive values here may indicate non-specific contamination.

A good quality DNA sample should have a 260/280 absorbance ratio of 1.7-2.0 and a 260/320 absorbance ratio of greater than 1.5. For an accurate measurement, realistically the absorbance value at 260nm must lie between 0.1 and 1, so dilution of concentrated samples may be required.

For pure DNA, the observed 260/280 nm ratio will be near 1.8. Ratios above this usually indicate the presence of RNA. Pure RNA will have a 260/280 ratio of approximately 2.2-2.3. 260/280 ratios below 1.8 often signal the presence of a contaminating protein or phenol. Alternatively, protein or phenol contamination is indicated by 230/260 ratios greater than 0.5. Where DNA, RNA and protein is present, a 260/280 ratio greater than 1.8 and a 230/260 ratio greater than 0.5 is expected.

The capability of the TLDA to measure DNA was tested with Buccal swabs and Blood samples. The same samples were run on the Nanodrop and Qubit systems. As the TLDA and NanoDrop are both UV systems a direct comparison wanted to be made with these. The findings were directly compared with the standard PicoGreen assay using the Qubit system. Double replicates read within acceptable correlation of variance between samples, 10% allowance.

2.2. Buccal swabs.

The comparison data from the Buccal swabs is shown in Figure 2 below.

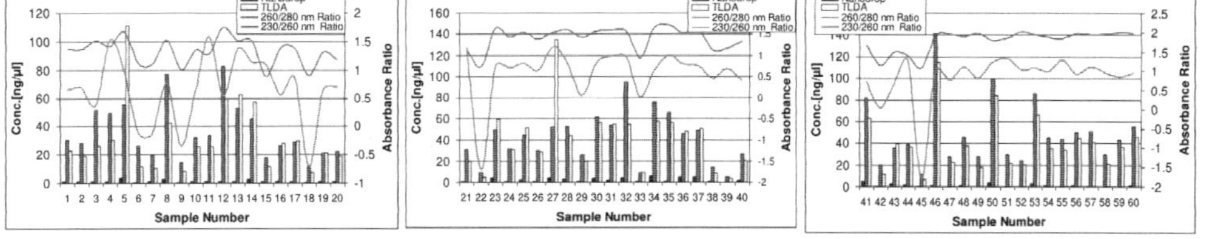

Figure 2. Buccal swabs data measured with the Picogreen Assay, Nanodrop and TLDA. Also shown are the 260/280 and 230/260 absorbance ratios.

Comment:
Most of these samples have a 260/280nm value below 1.8 (average = 1.5) and a 230/260nm value greater than 0.5 (average = 0.68), which signifies the presence of a contaminating protein.
However, the deviation between the concentration results from both the Nanodrop and the TLDA with the Qubit (Picogreen) are shown in Figure 3 below.

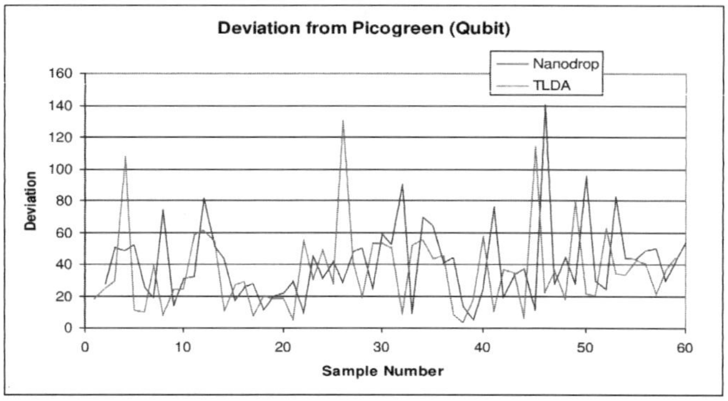

Figure 3. Deviation of the Nanodrop and TLDA concentration results from the Picogreen values.

The average deviation value for NanoDrop was found to be 41 whereas with the TLDA it was 36 ng/µl.

2.3. Blood swabs

The comparison data from the Blood swabs is shown in Figure 4 below.

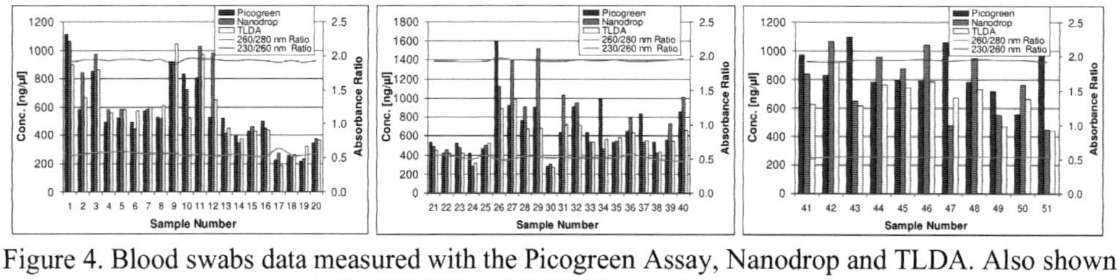

Figure 4. Blood swabs data measured with the Picogreen Assay, Nanodrop and TLDA. Also shown are the 260/280 and 230/260 absorbance ratios.

Comment:
The graphical results above indicate that there was only a small trace or perhaps in some cases no protein or phenol contamination as the 260/280nm value is above 1.8 (average = 1.93) and the 230/260nm value is around 0.5 (average = 0.54). There may also be RNA present as the 260/280 value averages to be 1.93.

However, the deviation between the concentration results from both the Nanodrop and the TLDA with the Qubit (Picogreen) are shown in Figure 5 below.

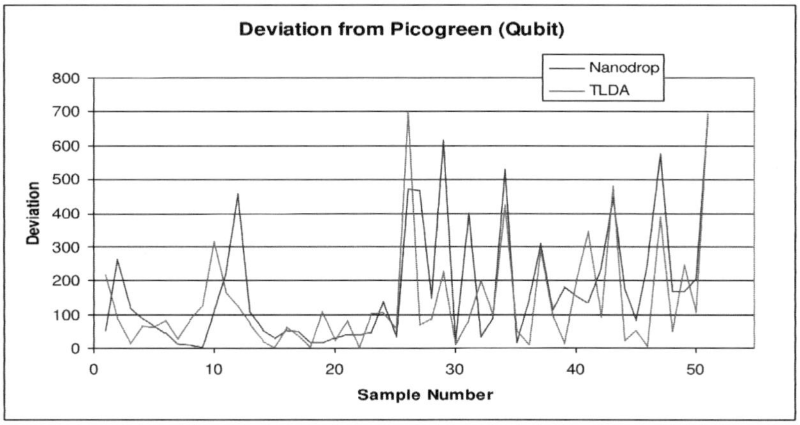

Figure 5. Deviation of the Nanodrop and TLDA concentration results from the Picogreen values for the blood data.

The average deviation value for NanoDrop was found to be 175 whereas the TLDA was slightly lower with a value of 139 ng/µl.

3. Discussion

Due to the dye-based quantification of the Qubit system the dyes are quite specific, allowing you to quantify only either the RNA or DNA. This is a big advantage of the Qubit system over the TLDA or NanoDrop (spectrophotometric instruments) as they calculate the total amount of RNA or DNA into one quantity. The spectrophotometric instruments count anything that absorbs light at 260nm. This could be DNA, RNA, single stranded or double stranded, proteins or contaminant. The Qubit uses PicoGreen technology, which fluoresces only when it binds to double stranded DNA etc., making it a more accurate method of quantification where the molecules of the sample are unknown.

The use of the TLDA and the NanoDrop is faster than setting up the Qubit system, because you have to prepare the dye and buffer for each sample, you have to prepare two standards for each set of samples, and you have to place each sample and standard in its own tube for quantification. The TLDA and the NanoDrop requires less sample volume. For concentrations the Qubit requires 10-20µl of sample in comparison to typically 2µl with the two spectrophotometric instruments. A big disadvantage with the Qubit system is that you need to have a rough estimate of your sample concentration to know which dye to use and to know what amount of sample to use in the quantification, otherwise you will get an out of error range. Also the Qubit system does not tell you the purity of the sample.

Both spectrophotometric methods are simple to use, very sensitive, require a small sample volume and no cuvettes or tubes are required. Also, a major advantage is that no standard curve needs to be generated prior to measurements. As an added benefit, the samples are also recoverable after quantitation.

4. Conclusion

Approximately, 60 buccal swab DNA samples and a similar number of Blood derived DNA samples have been assayed and analysed. It was found that for buccal samples both the spectrophotometric methods gave much higher results than the Qubit standard. This was expected, and is traditionally interpreted as resulting from contamination in the sample (protein, etc) as judged by the 260/280 ratio of all of these samples (1.4-1.6). Much better comparisons can be made between all three assay with DNA derived from blood where this ratios was invariably 1.8-2.0.

Overall, any fluorescence based quantification is not going to be as straightforward as the spectrophotometric methods, but the extra effort produces improved results. The methods are measuring different things. The Qubit (PicoGreen) is measuring the amount of dsDNA in the sample, while the spectrophotometric (TLDA & NanoDrop) instruments is measuring anything in the sample that scatters or absorbs light at the wavelength used for the measurement. The correlation of the results above is not always great. The TLDA and NanoDrop readings are usually higher that the PicoGreen, indicating that the DNA sample may contain a mixture of double and single stranded DNA, contaminants which scatter light, or UV-absorbing materials that are not nucleic acids. Therefore, the main comparison should be made between the TLDA and the NanoDrop.

Regarding practical issues, it was found that the TLDA is more user friendly than NanoDrop. It had about the same 'repeat analysis' incidence (i.e. when duplicates did not agree within 10% for blood samples). It would have been preferable to reverse-pipette each applied sample to obtain better results, but it was not possible as the wastage of samples was not permitted.

References

[1] Maquat L, Colwick S P, Arriano C M and Kaplan N O 2008 *Methods in enzymology* Vol 447 (Academic Press)

[2] Invitrogen, Accurate and sensitive quantitation of nucleic acids, even at low concentrations: Comparison of the Qubit^{IM} quantitation platform with spectrophotometry. March 2010 (http://www.invitrogen.com/site/us/en/home/support/Newsletters-and-Journals/BioProbes-Journal-of-Cell-Biology-Applications/BioProbes-Issues-2010/BioProbes-62/Comparison-of-Qubit-Quantitation-Platform-With-Spectrophotometry.html)

[3] Wieczorek D, Stayer C and Schagat T 2007 Automated DNA Purification from Oragene DNA/Saliva Samples Using the Maxwell 16 System *American Medical Associatio, Manual of Style 10th edition* [Internet] http://www.promega.com/pt/resources/articles/pubhub/enotes/automated-dna-purification-from-oragene-dna-saliva-samples-using-the-maxwell-16-system/

[4] McMillan N D, O'Neill M, Smith S R P, Hammond J, Riedel S, Arthure K and Smith S 2009 Reliable TLDA-microvolume UV spectroscopy with applications in chemistry and biosciences for microlitre analysis and rapid pipette calibration *J. Opt. A: Pure Appl. Opt.* **11** 054016 (11pp) doi:10.1088/1464-4258/11/5/054016

[5] Palmirotta R, Ludovici G, La De Marchis M, Savonarola A, Leone B, Spila A, De Angelis F, Della Morte D, Ferroni P, and Guadagni F April 2011 *Biopreservation and Biobanking*, **9(1)** pp.35-45. doi:10.1089/bio.2010.0027.

[6] Desjardins P, Hansen J B and Allen M 2009 Microvolume Protein Concentration Determination using the NanoDrop 2000c Spectrophotometer Nov 2009 *J. Visualized Experiments* **33** URL: http://www.jove.com/index/Details.stp?ID=1610 DOI: 10.3791/1610

[7] McMillan N D, O'Neill M, Pringuet P, O'Mongain E, O'Rourke B, Riedel S, Neill M, Augousti A, Wüustneck N, Wüstneck R, McMillan D D G, Colin F, Hennerbert P, Pottecher G and Kennedy D 2008 Quantitative drop spectroscopy using the drop analyser: theoretical and experimental approach for microvolume applications of non-turbid solutions *Meas. Sci. Technol.* **19(5)** 055601 18

Application of digital imaging techniques to flare monitoring

Shaun J Rodrigues and Yong Yan
Instrumentation, Control and Embedded System Research Group
School of Engineering and Digital Arts, University of Kent, CT2 7NT, U.K

E-mail: sjr43@kent.ac.uk and y.yan@kent.ac.uk

Abstract. This paper presents a technique for detecting and monitoring flares in harsh industrial environments with the use of an imaging sensor combined with digital image processing. Flare images are captured via an imaging fibre and analysed to detect the flare's presence and region of interest. The flare characteristics are then determined using various image processing algorithms. A prototype system is designed, constructed and evaluated on a purpose built laboratory scale flare test rig. Results indicate that the imaging based technique has potential for the detection, monitoring and analysis of flares amidst various background conditions in the chemical and oil industries for plant safety, pollution prevention and control.

1. Introduction

In oil refineries and chemical processing plants it is a common practice as a part of the plant process to release gases via a stack to burn [1] in the atmosphere, when more fuel gas is produced than the actual requirement and for emergency events due to plant malfunctions [2]. These flares are produced at a considerable height above a stack [3] to elude the possibility of plant area ignition for human safety and to avoid hazards due to heat releases. The flared gases are ignited by a pilot flame, which is located at the root of the flare, so that purge gases and the vent gases can be readily combusted [4]. The pilot flame should be continuously "ON" and has a continuous fuel gas supply to do so [5]. It is mentioned [6] that the chemical process used for flaring is high temperature oxidation that burns hydrocarbons which reacts with atmospheric oxygen to form CO_2 and water. The environmental regulations indicate that the requirement is for smokeless burning [7], whereas the safety requirement is the issue of thermal radiation [6]. The smokeless burning of a flare is considerably obtained by assisting it with steam [7] that causes turbulence but creates some sound levels.

Traditionally flares are monitored for abnormalities by an operator on a 24x7 observation, which is not reliable [8]. Other sensing techniques have been used to monitor flares. These include infrared spectroscopic sensing [9] that is generally very inaccurate due to environmental interferences [3]. A thermocouple approach has also been used but has problems of failures due to corrosive effects of the environment, high maintenance requirement, and limitations of single point measurement. To replace a thermocouple requires an expensive shutdown [10]. CCTV cameras have also been used to monitor flares [11]. Similarly, a colour video camera [2][8] has been used to detect and analyse the characteristics of a flare. Apart from this work, there is little report of the use of imaging sensors for flare monitoring. It is hence recognised that there is a demand for an imaging based flare monitoring system that could be easily viewed from the control room, have least maintenance requirement and could also assist by monitoring the presence of the flare and presenting certain important flare characteristics on a graphical user interface. This paper presents a demonstration system that is designed for flare monitoring through the application of an imaging sensor in conjunction with an imaging fibre.

2. System Description

Figure 1 shows the basic concept of the imaging based flare monitoring system. It consists of an imaging fibre of 6k resolution and an objective lens with a 90 degrees field of view (Figure 2). The objective lens is mounted with a focus directed towards the flare on the platform whilst the imaging sensor, a 640x480 pixel C-mount colour camera, is placed at the ground level. The image signal is sent through USB (or Ethernet) to a microcomputer that incorporates flare analysis and monitoring software. An alternative configuration of the system is to use an embedded system instead of a microcomputer, indicating whether the flare is present and its approximate size, i.e. low, medium or high.

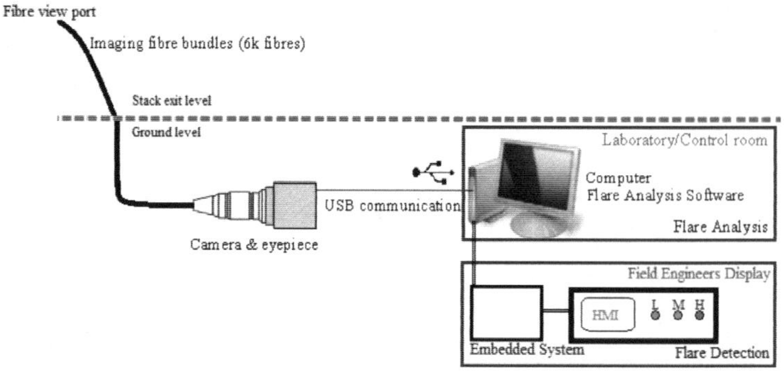

Figure 1: Concept of imaging based flare monitoring

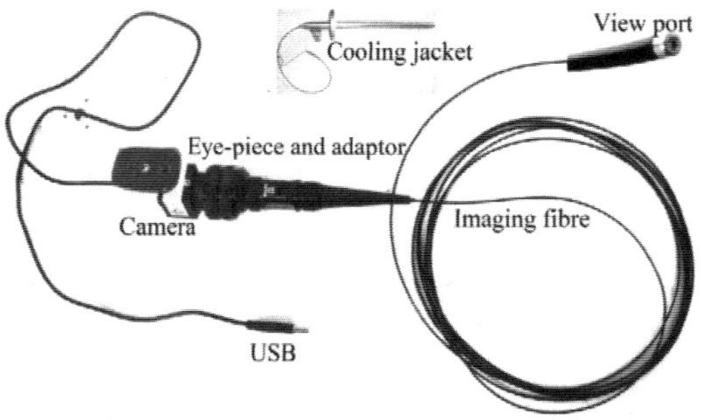

Figure 2: Arrangement of the imaging fibre, view port, imaging sensor and cooling jacket

Since gas is available in abundance, a single layer cooling jacket, as shown in Figure 2, will be deployed to prevent the fibre view port and the front end of the fibre bundles from excessive heat from the flare. The view port of the fibre has a protective window to avoid contamination of the objective lens. The fibre from the platform level to the camera at ground level would have a protective shield.

3. Image Processing Principles

The flare images captured from the imaging system (Section 2) are processed using a novel method for detection and a combination of various existing image processing algorithms to quantify the characteristics and behaviours of the flare.

3.1. Flare detection and visual dynamism

The first flare frame is captured and saved for a fixed time, depending on the camera frame rate, until the next frame is captured. The two frames F_x and F_{x+1}, which represent the first and second frames, respectively, in a video sequence, are subtracted to obtain the difference image wherein the like-pixel regions of no change within the frames F_x and F_{x+1} will result in the R, G and B values of the difference image to '0', i.e.

$$X = \sum_{n=1}^{N}\left[\sum_{x=1}^{C}(F_x - F_{x+1})\right]_n \qquad (1)$$

Where N and C are the number of frames/processing set and the maximum frame rate of the camera, respectively. Similar to frame subtraction, the two frames are also added to obtain the summation image:

$$Y - \sum_{n=1}^{N}\left[\sum_{x=1}^{C}(F]_x + F_{x+1}[\square)\right]_n \qquad (2)$$

Here the flare dominated regions within the respective frames F_x and F_{x+1} will result in the R, G and B values to its maximum, i.e., '255'. The value 'N' in this research is taken to be two frames out of twenty-five frames per second that the camera can capture. The higher the value of N at a higher frame rate, the better the difference image. For every check of the condition that two consecutive frames are captured, the first data set is processed until C=25fps after which the cycle repeats.

The above processing technique in turn relates to the flare background information and the flare (size/shape) identification, respectively. The two processed images (X and Y) are further used to obtain the disparity visualisation (Z) in absolute, which displays the dynamism of the flare by overlapping the two images over each other and assisting in obtaining the region of interest, i.e., the pixel areas cover the flare dynamism in the video images for the two frames, respectively,

$$\mathbf{Z = \|[(X+Y)\ (F_{x+1})_c]\|} \qquad (3)$$

Where $\|[(X+Y)\quad (F_{x+1})_c]\| \geq 0$ and $\|[(X+Y)\quad (F_{x+1})_c]\| = 0$ if and only if $(X+Y) = (F_{x+1})_c$. This is dependent on the camera frame rate 'C', as a higher frame rate would represent a closer disparity visualisation of the two frames in time, thus indicating rapid changes of the flare.

3.2. Flare analysis

Image processing techniques are then used to convert the image frames to gray-scale images (F_{gx} and F_{gx+1}) and binary images (F_{bx} and F_{bx+1}), where F_{gx} and F_{gx+1} will have values between 0 to 255 for gray-scale while F_{bx} and F_{bx+1} will have values of '0' or '1'. A logical 'OR 'operation is then performed for reliable and robust processing of the binary images of the two frames as one of the preliminary steps prior to edge detection and contour identification. This process can be summarised as follows:

$$\mathbf{F_x(R,G,B) \rightarrow F_{gx}(0\ to\ 255) \rightarrow F_{bx}\ (0,1)} \qquad (4)$$

$$\mathbf{F_{x+1}\ (R,G,B) \rightarrow F_{gx+1}\ (0\ to\ 255) \rightarrow F_{bx+1}\ (0,1)} \qquad (5)$$

$$\mathbf{P_b = F_{bx}\ (0,1)\ OR\ F_{bx+1}\ (0,1)} \qquad (6)$$

Where 'P_b' denotes the processed binary image. Edge detection using the Canny algorithm and contour identification is then performed for the purpose of flare analysis. All the algorithms are developed and tested within the Matlab environment.

4. Experimental Results and Discussions

To evaluate the image processing algorithms for flare detection and analysis, experiments were carried out on a purpose built laboratory-scale test rig that can replicate industrial flares. The flare images were captured at a rate of 25 frames per second using the imaging system (Section2). However, the images presented in this section are straight from the camera firmly angled towards the flare. The fuel used was propane. The image processing algorithm was tested under different background conditions, including day and night conditions as well as steam assisted flares and air assisted flares.

4.1. Flare detection and visual dynamism

The two frames F_x and F_{x+1}, as shown in Figures 3(a) and 3(c) were captured from a stream of images by the camera under day and night conditions, respectively. The application of equations 1 and 2 has resulted in the difference image and summed image as shown in Figures 3(b) and 3(d) that are then used for flare detection and determination of the visual dynamism and region of interest.

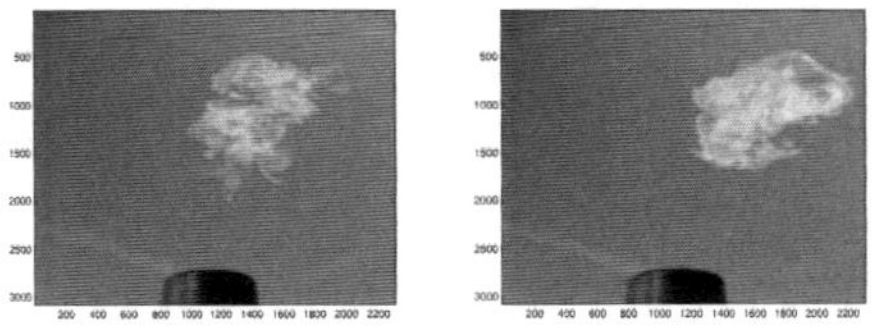

(a) Two consecutive images of a flare in day light

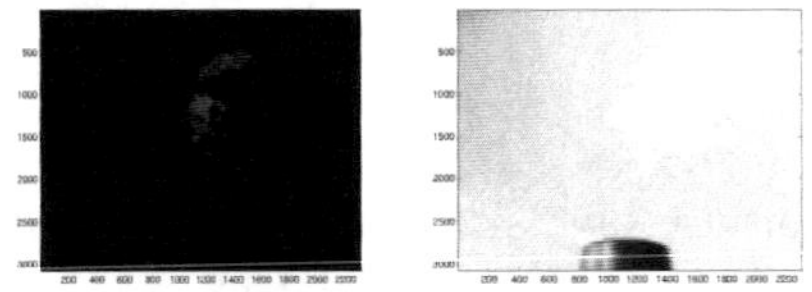

(b) Results of initial processing of the two images in (a)

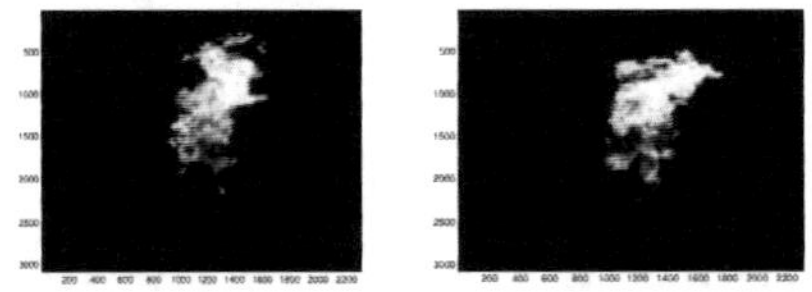

(c) Two consecutive images of a flare at night

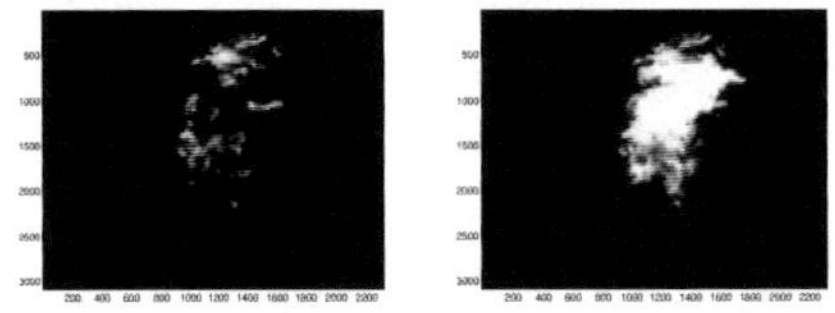

(d) Results of initial processing of the two images in (c)

Figure 3: Pre-processing of flare images

The processed images are significant as frame subtraction indicates that the like-pixel regions of almost no change or very slow change within frames F_x and F_{x+1} that includes the background and a part of the stack will result in the R, G and B values of the difference image to '0' which is displayed as black colour. Similarly, on performing frame addition the pixel regions of the flare dominated regions or the fast moving regions for the two consecutive frames F_x and F_{x+1} will result in the R, G and B values to its maximum, i.e., '255' which is displayed as white colour. The second processed image is also used to determine the size, shape and region of interest of the flare. The algorithm then takes these images and performs the disparity visualisation (Z) calculation as per equation (3). The absolute of the two processed images is used to find the difference from the most recent image, thereby displaying the image that assists in visualizing the dynamism of the flare.

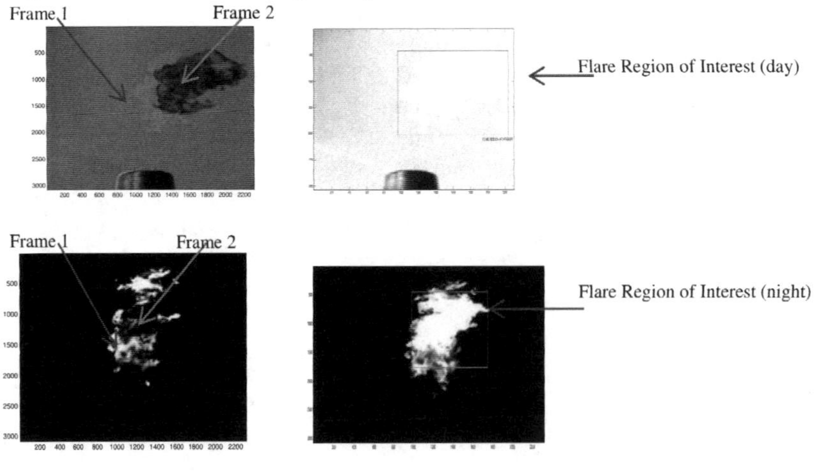

Figure 4: Visual dynamism and region of interest detection under day and night conditions

Figure 4 shows the visual dynamism of the flare that assists in identifying the direction of the flare, the spread within the atmosphere, the flare shape and the energy or rapid changes of the flare. The result confirms that the technique as described is simple and effective for flare detection that also indicates the dynamism, shape and structure of the flare, The region of interest is obtained reliably by finding the max height of the flare which is obtained by locating the lowest and highest vertical pixels with a maximum R, G, B value of '255', respectively, and the width, which is obtained by locating the lowest and highest horizontal pixels with a maximum R, G, B value of '255', respectively. These maximum pixel values are then taken as a reference to draw a rectangle automatically within the software GUI indicating the region of interest of the flare. This technique, however, requires the viewing end of the imaging fibre to be securely installed and any movement of the fibre while capturing two consecutive frames could result in false detection as the algorithm would underperform.

4.2. Flare Analysis

Out of the flare images obtained in time that depends on the camera frame rate, every two successive image frames were converted to gray-scale images as shown in Figure 5. These images illustrate the observations for one set of images captured under different conditions of flaring, both assisted and non-assisted. There was consistency in the observations for a particular condition and to analyse these effects such as missing combustion spots and soot formations for various assisted conditions on the flare that are fairly random due to the highly dynamic nature, the gray-scale images of the two frames were then converted into binary images. When the binary images were observed, changes in the flare characteristics are clearly evident due to fuel addition, air assistance and steam assistance. To obtain accurate and reliable results a logical 'OR' operation was performed on the binary images after which edge-detection using the Canny algorithm and contour identification techniques were applied. It was observed consistently for images due to respective conditions that the non-assisted flare produced a brighter image with missing combustion spots indicated by the contour identification that were observed inside the outer edge of the flare; this could be due to the lower cracking of the fuel (propane) hydrocarbons as there is limited supply of air. The air-assisted flare did not have significant changes, but it is larger in size due to the richer supply of oxygen for cracking down the hydrocarbons, thus the contour indicating lesser spots of incomplete combustion. The steam assisted flare has the least brightness and is a turbulent flare with clean and leaner combustion as observed from the contours and binary images. This result indicates that image processing techniques can be applied to obtain significant information about a flare and suitable for both online or offline flare analysis.

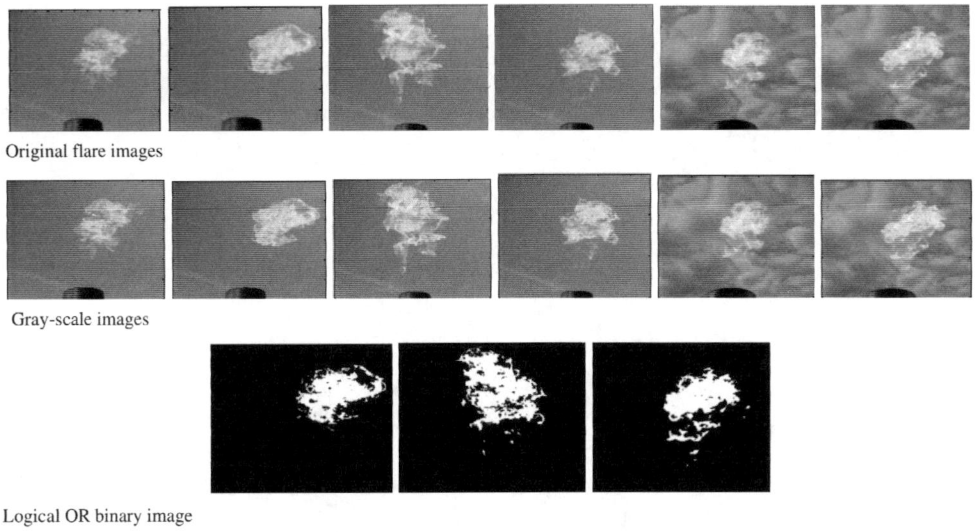

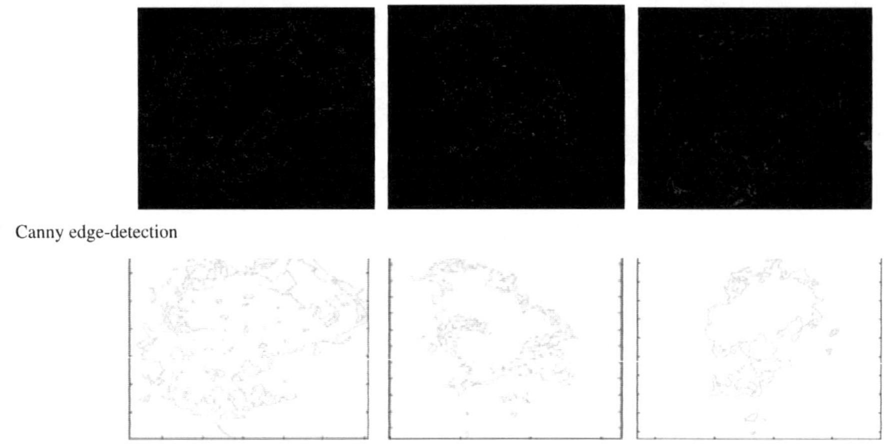

(a) Non-assisted flare (b) Air-assisted flare (c) Steam-assisted flare

Figure 5: Flare analysis using image processing techniques under three different conditions.

5. Conclusion

A flare monitoring system using a colour imaging device and an imaging fibre has been proposed. The novel algorithm for the detection and visual dynamism observation of a flare has been tested successfully for reliability and robustness assessment on a laboratory scale flare test rig under real atmospheric background conditions. Image processing techniques have been applied effectively to obtain relevant information for the purpose of analysing the effects of assistance on the flare. Future work will include the identification of different constituents of the gas flaring and continuous tracking of the type of gas being flared. It is envisaged that the proposed flare monitoring technique will assist the oil, gas and chemical industries to achieve cost-effective maintenance, enhance plant safety and reduce harmful emissions.

Acknowledgements

The authors thank Clive Birch of the Mechanical Workshop at the University of Kent, UK for his excellent support in fabricating the lab-scale flare test rig.

References

[1] H. Okamoto and S. Nakauchi, "Flare monitoring apparatus", US Patent 4,233,596, 1980.

[2] V.G. Venkoparao, R.N. Hota, V.S. Rao, and M.K. Gellaboina, "Flare monitoring for petroleum refineries," 2009 4th IEEE Conference on Industrial Electronics and Applications, 2009, pp. 3022-3027.

[3] R. Dibiano, K. Autenrieth, and G. Hales, "Control of smoke emissions from a flare stack", US Patent 4,505,668, 1985.

[4] W. Waters and J. Zimmer, "Transducer for gas flare pilot flame detection", US Patent 6,164,957, 2000.

[5] "http://www.tornadotech.com/products/combustion/flares/pilot-ignition.html [Accessed 03 February 2011]."

[6] KLM Technology Group, "Flare selection and sizing (engineering design guideline)," Group, 2007.

[7] http://www.exxonmobil.com/NA-English/Files/freddietheflareflier.pdf [Accessed 07 February 2011].

[8] V.S. Rao, V.G. Venkoparao, R.N. Hota, and M.K. Gellaboina, "Flare monitoring," US Patent Pub No. 2009/0046172 A1, 2009.

[9] P.J. Merle and Spellicy L. Robert, "Flare efficiency monitoring by remote infrared sensing: a feasibility demonstration," 1984.

[10] "http://www.williamsonir.com/content10769 [Accessed 18 February 2011]."

[11] "http://iris-systems.com/Watchdog/watchdog.htm [Accessed 23 March 2011]."

Development of an Image Fringe Zero Selection System for Structuring Elements with Stereo Vision Disparity Measurements

Josef E. Grindley, Andrew J. Tickle[1] and Lin Jiang

School of Electrical Engineering, Electronics and Computer Science, University of Liverpool, Liverpool L69 3GJ, United Kingdom

Email: joegrindley@hotmail.com, production_leader@hotmail.com

Abstract. When performing image operations involving Structuring Element (SE) and many transforms it is required that the outside of the image be padded with zeros or ones depending on the operation. This paper details how this can be achieved with simulated hardware using DSP Builder in MATLAB with the intention of migrating the design to HDL (Hardware Description Language) and implemented on an FPGA (Field Programmable Gate Array). The design takes few resources and does not require extra memory to account for the change in size of the output image.

1. Introduction

Image transforms that utilize a Structuring Element (SE) must be applied to an image that has been padded with zeroes or ones outside its fringes, whether zeros or ones are used is dependent on the image transform in use, however in this implementation only zeros are used. When implementing this in DSP, a separate SE consisting of a 5x5 8-bit AND gate array was used prior to the image transform, which was used to apply the packing bits. In this implementation a 5x5 array was used, as the image transform used was 5x5. However, should a 3x3 SE be needed then the data can handle sampling from the center 3x3 of the 5x5 array.

The are several benefits of using an Field Programmable Gate Array (FPGA) over a CPU when performing image processing operations in real time such as pipelining and comparatively lower power consumption, inherently any image processing functions used must be described in HDL. Many equivalent array padding functions are available, for example the function "padarray" in MATLAB performs the same operation in software as the proposed circuit and will be used as a comparison for the results section.

One major advantage of using this implementation is that padding bits are applied around an array without the need to change the array dimensions hence the buffers used do not need to be adjusted in size. While applying the padding bits "on the fly", only one clock delay is needed in the process.

2. Implementation in DSP Builder

The circuit works by using a SE consisting of an array of 8-bit AND gates, as shown in Figure 1, which allows zero to be selected in each appropriate element when the SE is centered at the fringes of

[1] Andrew Tickle is now a Lecturer in Electrical Engineering at Coventry University

Published under licence by IOP Publishing Ltd

the image. This is done by using a bit pattern that characterizes the edges of the fringes, as illustrated in Table 1. As the center of the SE is always selected, the AND gate in the middle of the array is connected to an 8 bit integer "255", although this could be done by connecting the center input to its respective output the AND gate was left in to eliminate synchronization issues and propagation delays. The bit pattern was input into the multiplexers as shown in Figure 2 where its input line to the multiplexer relates the number to the top left of each bit pattern in Table 3. The three numbers to the right are the decimal representation of each shaded section respectively: they are 8-bit making up the 24-bit needed for fringe pattern selection in the AND gate array. These were selected by the pattern select array illustrated in Table 1. The pattern select array was generated in MATLAB and input in parallel with the image. It was the same dimension as the input image and determined by the number to the top left of each section in Table 3.

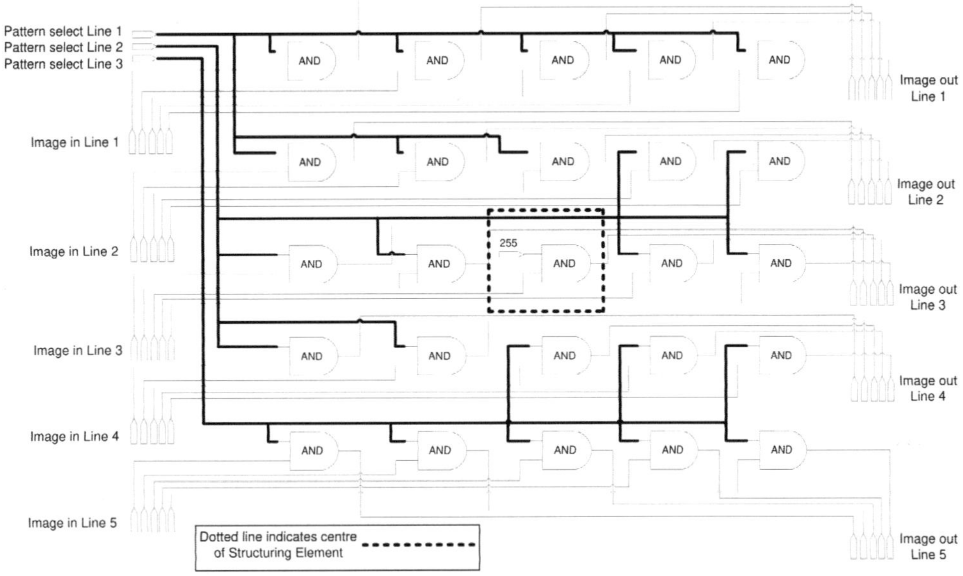

Figure 1 – Illustration of Structuring Element using 5x5 AND gate array

The input to each AND gate from the pattern selection is 1-bit which is connected to all 8 AND gate inputs in parallel. This allows each element in the array to be selected at once using 1-bit as shown in Figure 5 but omitted in Figure 1 to simplify the diagram. The multiplexers shown in Figure 2 were configured as a look up table to select the correct pattern according to the input from the pattern select array. Table 1 shows the pattern select array which is of identical dimensions to the input image and used to select the pattern to be used at the part of the fringe that the AND gate array is centered over, hence applying the padding bits. The pattern select array is 5 bit and requires less memory than the input image. Table 2 shows a binary image for diagram simplicity, the circuit shown in Figure 7 is 8-bit and subsequent testing was done on an 8-bit grayscale image. With the appropriate modification, any bit length could be used. Table 3 shows each possible fringe pattern giving 25 possibilities, each pattern is separated into three 8-bit sections shaded appropriately for illustration. Each sections respective decimal value is in the cells to the right of each pattern; these are labeled L1, L2 and L3 which make up the input constants used by the multiplexers as illustrated in Figure 2. They are indexed and selected by the number to the top left of the pattern which can be seen in the pattern select array.

It should be noted that three multiplexers were used as DSP Builder would not allow multiplexer inputs that have a bus larger than 8 bits wide, hence three of them were needed to select the 24 bits required.

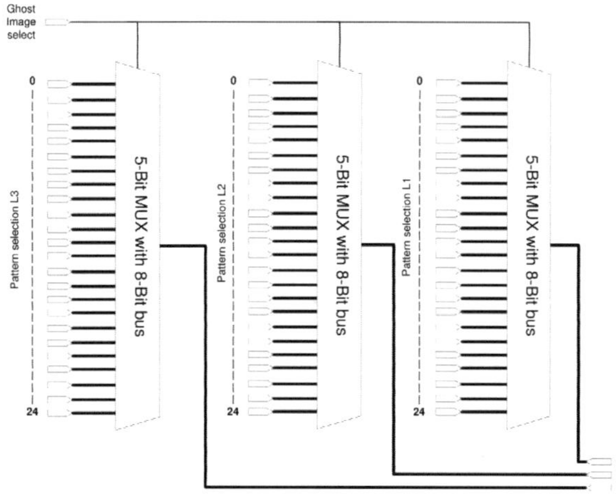

Figure 2 - Illustration of Multiplexer configured to do pattern selection operation

1	2	3	3	3	3	3	3	3	3	4	5
6	7	8	8	8	8	8	8	8	8	9	10
11	12	0	0	0	0	0	0	0	0	13	14
11	12	0	0	0	0	0	0	0	0	13	14
11	12	0	0	0	0	0	0	0	0	13	14
11	12	0	0	0	0	0	0	0	0	13	14
11	12	0	0	0	0	0	0	0	0	13	14
11	12	0	0	0	0	0	0	0	0	13	14
11	12	0	0	0	0	0	0	0	0	13	14
11	12	0	0	0	0	0	0	0	0	13	14
11	12	0	0	0	0	0	0	0	0	13	14
11	12	0	0	0	0	0	0	0	0	13	14
15	16	17	17	17	17	17	17	17	17	18	19
20	21	22	22	22	22	22	22	22	22	23	24

Table 1 - Illustration of pattern select array

0	0	0	0	0	0	0	0	0	0	0	0	0	0	0	0
0	0	0	0	0	0	0	0	0	0	0	0	0	0	0	0
0	0	1	1	1	1	1	1	1	1	1	1	1	1	0	0
0	0	1	1	1	1	1	1	1	1	1	1	1	1	0	0
0	0	1	1	1	1	1	1	1	1	1	1	1	1	0	0
0	0	1	1	1	1	1	1	1	1	1	1	1	1	0	0
0	0	1	1	1	1	1	1	1	1	1	1	1	1	0	0
0	0	1	1	1	1	1	1	1	1	1	1	1	1	0	0
0	0	1	1	1	1	1	1	1	1	1	1	1	1	0	0
0	0	1	1	1	1	1	1	1	1	1	1	1	1	0	0
0	0	1	1	1	1	1	1	1	1	1	1	1	1	0	0
0	0	1	1	1	1	1	1	1	1	1	1	1	1	0	0
0	0	1	1	1	1	1	1	1	1	1	1	1	1	0	0
0	0	1	1	1	1	1	1	1	1	1	1	1	1	0	0
0	0	0	0	0	0	0	0	0	0	0	0	0	0	0	0
0	0	0	0	0	0	0	0	0	0	0	0	0	0	0	0

Table 2 - Illustration of input image with packing bits

3. Result out Images and Discussion

As the output image of Figure 7 shows packing bits that are a line two elements wide have been applied to the fringes of the input image in Figure 6. As the output of the AND gate array is equivalent to a 5x5 structuring element, the outputs were sampled and compared with the image in Figure 7 to validate the results and found to be identical thus demonstrating the design works correctly. When noise is apparent in the image input the noise is propagated to the output. The design does not do any noise filtering but the padding mechanism still works effectively regardless.

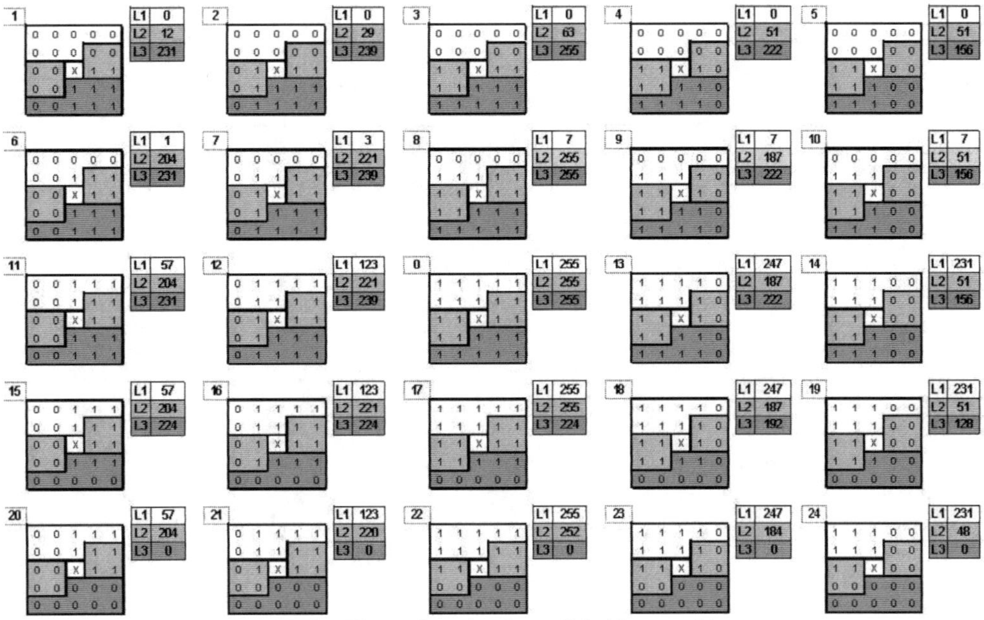

Table 3 – Illustration of each possible fringe position.

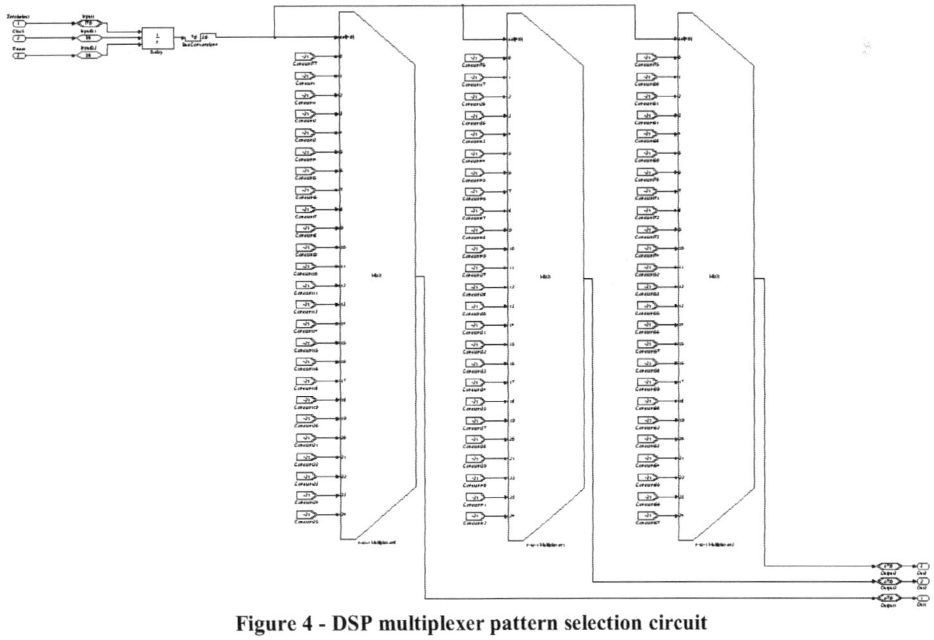

Figure 4 - DSP multiplexer pattern selection circuit

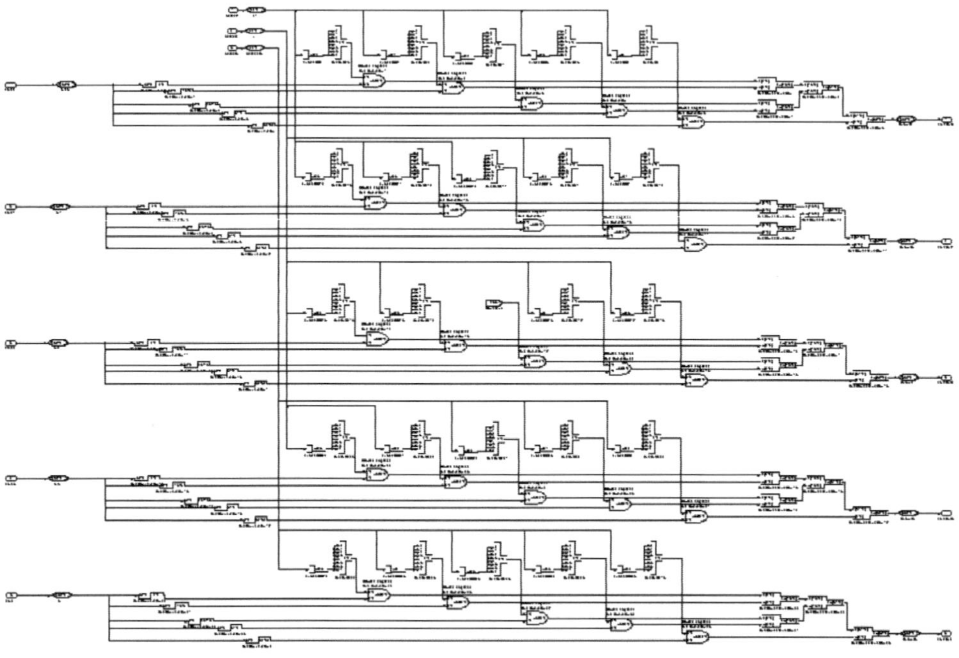

Figure 5 - DSP implementation of 5x5 AND gate array

Input Image

Figure 6 - Input image

Output Image

Figure 7 - Output image

4. Conclusions and Future Work contractors

The results indicate the design is working correctly. The pattern select array was generated by the a MATLAB m file as DSP builder was being used, when implemented on an FPGA the pattern select array must be generated either by a Nios processor or another CPU on the FPGA if the image frame size is variable, if not it could be loaded into memory as a constant when programming the FPGA. The system was originally designed to apply padding bits to the outside of two stereo images then apply the census transform where it was also found to function without any issues, this is intended for use in disparity distance measurements of which work is currently being undertaken for use in mobile robotics.

5. References

[1] Altera 2010 *DSP Builder Reference Manual 9.1 SP1,* www.altera.co.jp/literature/manual/mnl_dsp_builder.pdfAltera 2010 *DSP Builder Reference* (accessed 3 April 2011)

[2] *Manual 9.1 SP1,* www.altera.co.jp/literature/manual/mnl_dsp_builder.pdf (accessed 3 April 2011)

[3] Tickle A J, Smith J S and Wu Q H 2007 *Development of Morphological Operators for Field Programmable Gate Arrays* Proc. IOP Conference: Sensors and their Applications XIV, Liverpool John Moores University, Liverpool, UK

[4] MATLAB reference library, http://www.mathworks.com/help/toolbox/images/ref/padarray.html (accessed 7 March 2011)

[5] Kuhn M , Moser S and Islerl O 2004 *Efficient ASIC Implementation of a Real-Time Depth Mapping Stereo Vision System*, Department of Information Technology and Electrical Engineering, ETH Zurich

[6] Middlebury Computer Vision , vision.middlebury.edu/ (accessed 13 March 2011)

Comparison of Surface Plasmon Resonance and Localized Surface Plasmon Resonance-based optical fibre sensors

J Cao[1, 2], E K Galbraith[1], T Sun[1] and K T V Grattan[1]

[1] School of Engineering and Mathematical Sciences, City University London, Northampton Square, EC1V 0HB, UK
[2] School of Mechatronics Engineering, Harbin Institute of Technology, P.R China

Email: jie.cao.1@city.ac.uk

Abstract. In this work, two of surface plasmon resonance (SPR) and localized surface plasmon resonance (LSPR)-based optical fibre sensors have been successfully developed and cross-compared. With one SPR sensor being coated with a thin layer of gold film and the other gold-nanorods (GNRs), forming a LSPR sensor, both sensors are subjected to various refractive index changes. As a result their sensitivities are measured in the form of resonance wavelength shift as a function of refractive index variation. The results demonstrate that the thin-film coated SPR sensor has much higher sensitivity than that of GNRs coated LSPR sensor but with worse linearity.

1. Introduction

Surface plasmon resonance (SPR) and localized surface plasmon resonance (LSPR) are both powerful tools for label-free biosensing in biochemistry due to their high sensitivity to the refractive index change caused by their interactions with molecules such as DNA or proteins, etc. SPR is known as a phenomenon excited when the frequency of evanescent electromagnetic wave propagating at the metal-dielectric interface is resonant with the oscillation of the surface conduction electrons in metal [1]. LSPR often refers to metallic nanoparticles as it occurs when the frequency of incident photon is resonant with the collective oscillation of conduction electrons in metallic nanoparticles [2].

Traditional SPR sensors are fabricated based on the Kretschmann configuration where a thin noble metal film is coated on a prism [3]. LSPR sensors, on the other hand, are normally fabricated on a chip that noble metal nanoparticles are coated on a substrate such as glass slide [4]. However, most of the conventional SPR and LSPR sensors require bulky and expensive optical equipment and data analysis instrument and, hence, are only can be used in a laboratorial environment. Compared to the traditional SPR and LSPR sensors, fibre-optic based SPR and LSPR sensors have shown advantages, such as requiring small sample volume due to the small sensing tip, using simplified optical design and showing potential to be disposable and the capability for remote sensing. Compared to SPR technology, LSPR has its own advantages by being able to control/optimize the sensitivity of LSPR sensor through the change of the sizes and shapes of nanoparticles [5]. This work is aimed to explore the differences between fibre-optic based SPR and LSPR sensor designs, therefore their corresponding sensing capabilities when they both are subjected to refractive index changes. To date, little comparison was made between SPR and LSPR sensors [6-7] and it is the first time that, to the knowledge of the authors, a cross-comparison is made between SPR and LSPR sensors based on optical fibre.

Therefore, in this work, we studied the characteristics both of fibre-optic based SPR and LSPR sensors by comparing their fabrication and sensitivity to bulk refractive index (RI) change. The SPR sensor was constructed by coating a thin gold film on the unclad core of an optical fibre. To make a comparison with SPR sensor, GNRs were synthesized and immobilized on the same structure of an optical fibre to fabricate LSPR sensor. The sensitivity study was performed by dipping the sensors into solutions with different refractive index, and monitoring the corresponding resonance wavelength shift. The results show that optical fibre based SPR sensor has much higher bulk refractive index sensitivity than LSPR sensor but with worse linearity.

2. Experimental set-up

Prior to the experimental tests, a SPR sensor was constructed by coating a thin gold film on the surface of an optical fibre with its cladding being removed. To make a LSPR sensor, GNRs were first synthesized and then immobilized on the surface of an optical fibre with cladding being removed. Subsequently both sensors were dipped into solutions with different refractive index, their corresponding resonance wavelength shifts were monitored to allow the cross-comparison of the sensor sensitivities observed.

2.1. Preparation of SPR sensor

A piece of 10 cm of multimode (600μm) optical fibre was chosen specifically to allow the easy removal of fibre cladding of about 1cm length at one end. Subsequently the fibre was finely polished at both two end surfaces. To prepare SPR fibre probe, 50 nm thick of gold film was coated on the surface of unclad portion by using a sputter coater. In order to make a reflective optical fibre sensor probe, a thin silver layer was coated onto the end surface of the fibre to allow the reflection of the incident light.

2.2. Preparation of LSPR sensor

2.2.1. Synthesis of GNRs

GNRs were prepared by following the method reported by Nikoobakht et al [8]. Firstly, seed solution was prepared by mixing 5 mL of 0.2 M CTAB solution with 5 mL of 0.5 mM HAuCl4 solution. To the stirred solution, 600 μL of 0.01 M NaBH4 was added, which resulted in the formation of a brownish yellow solution. After additional 2 min vigorous stirring, the seed solution was kept at room temperature (25°C) without disturbing for at least 2 h before use. To prepare gold nanorods solution, 25 mL of 0.2 M CTAB solution was added to 0.5 mL of 0.01M AgNO3 solution and 25 mL of 0.001 M HAuCl4 solution respectively. To the mixed stirred solution, 350 μL of 0.0788 M ascorbic acid solution, which works as a mild reducing agent, was added and changed the color of growth solution from dark yellow to colorless. To grow GNRs, 60 μL of seed solution was added to the growth solution at about 30 °C. The color of the solution was gradually changed within 20 min. The solution was kept at 30 °C for 24 h. Excess amount of CTAB was removed by twice careful centrifugation at 7000 rpm for 20 minutes each time. Finally the GNRs were dispersed in DI water.

2.2.2. Preparation of LSPR sensor

Prior to the immobilization of GNRs onto the surface of an optical fibre, a piece of 10 cm of multimode (600μm) optical fibre was chosen specifically to allow the easy removal of fibre cladding of about 1cm length at one end. Subsequently the fibre was finely polished at both two end surfaces. Prior to the immobilization of GNRs onto the surface of the unclad section of the fibre, the fibre was cleaned by piranha solution (H_2O_2: H_2SO_4; 30%:70%) for 30 minutes. (It should be noted that this solution is extremely dangerous and thus was handled with extreme care). After a thorough rinse with DI water, the optical fibre was placed in an oven set at 110°C for 2 h. The fibre was then immersed in

5% mercapto-propyl-trimethoxysilane (MPTMS) ethanol solution for 5 h before being sonicated in ethanol for 3 minutes. Subsequently it was thoroughly rinsed with ethanol and DI water to remove unbound MPTMS and to be blow-dried using N_2. The mercaptosilane modified optical fibre was incubated in the GNRs solution overnight to enable the immobilization of GNRs onto the fibre surface. In order to make a reflective optical fibre sensor probe, a thin silver layer was coated onto the end surface of the fibre to reflect the incident light back.

2.3. Solutions with different refractive index

In order to test the sensor performance, solutions with different refractive index were prepared, ranging from 1.333 to 1.423, through the use of water and mixed solvents of methanol, ethanol and toluene at different ratios. A refractometer purchased from Fisher Scientific was used to measure the refractive index of testing solutions.

2.4. Experimental set-up

Figure 1 (a) shows the schematic diagram of a sensor system set up to evaluate the sensor performance. As shown in the diagram, light from a white light source is coupled to the sensor probe, SPR probe or LSPR probe, through a 2x1 fibre coupler. The sensor probe is dipped into a solution with a known refractive index and the reflected signal is captured by a mini-spectrometer (Ocean Optics Maya 2000pro), which is connected to the other side of the coupler. Thus the signals obtained could be displayed and monitored by a computer connected to the mini spectrometer. Figure 1(b) shows the photo of the bench-top sensor system setup.

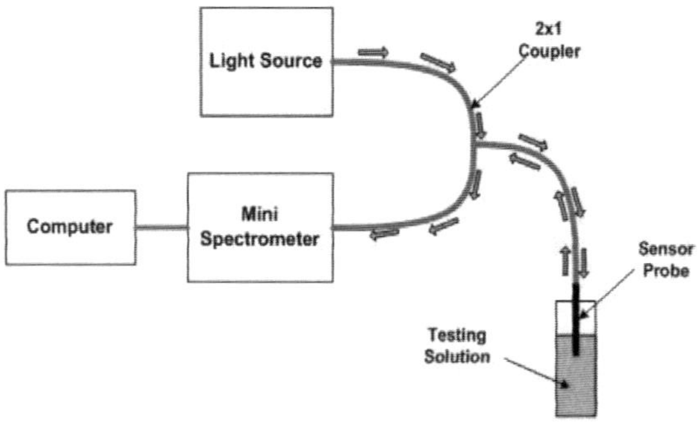

(a)

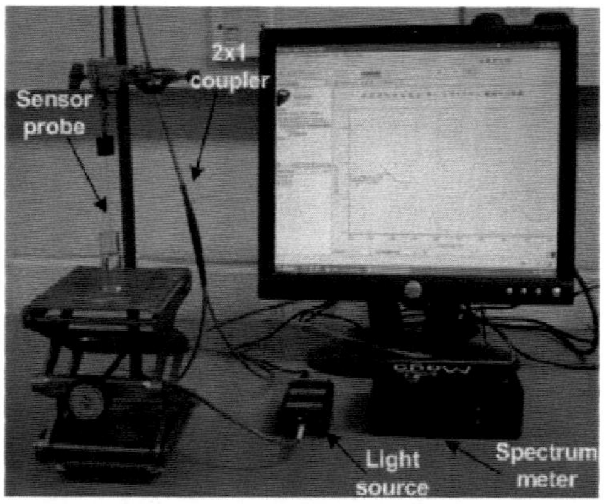

(b)

Figure 1. (a) Schematic diagram of the system. (b) Photo of the experimental setup.

3. Experimental results and discussion

GNRs possess two plasmon resonance bands, with one being the transverse plasmon wavelength (TPW) at about 520 nm and the other the longitudinal plasmon wavelength (LPW) appearing between the visible and near infrared wavelength range and varying with the aspect ratio (length to width) [9]. It is known that the LPW is highly sensitive to refractive index changes and the sensitivity increases with the increase of the aspect ratio of the GNRs while the TPW is insensitive to refractive index change [10]. Therefore, in this work, LPW shift of the GNR-coated sensor was specifically chosen to make a comparison with the SPR wavelength shift of the gold-film coated sensor. The average aspect ratio of the GNRs synthesized for this experiment is about 3.5. The advantage of using LSPR sensor is that it can be fabricated easily through a chemical process rather than using expensive equipment, such as sputter coater or evaporator, as used for the thin-film coated SPR sensor preparation.

Figure 2 shows clearly both the refection mode of the gold-film coated SPR sensor, in Figure 2(a), and absorbance mode of the GNRs-coated LSPR sensor, in Figure 2(b), as a function of refractive index change. The shift of dip in SPR sensor spectra and peak of LPW in LSPR sensor spectra were measured as a function of refractive index change. When the refractive index varies from 1.333 to 1.423, a total wavelength shift of 310 nm was observed for SPR sensor and 70 nm for LSPR sensor. It indicates that the refractive index sensitivity of SPR sensor is about 3444 nm/RIU, which is more than four times higher than that of the LSPR sensor, of which the sensitivity is about 778 nm/RIU. The results obtained form LSPR are similar to what have been reported by other researchers based on a chip structure [7].

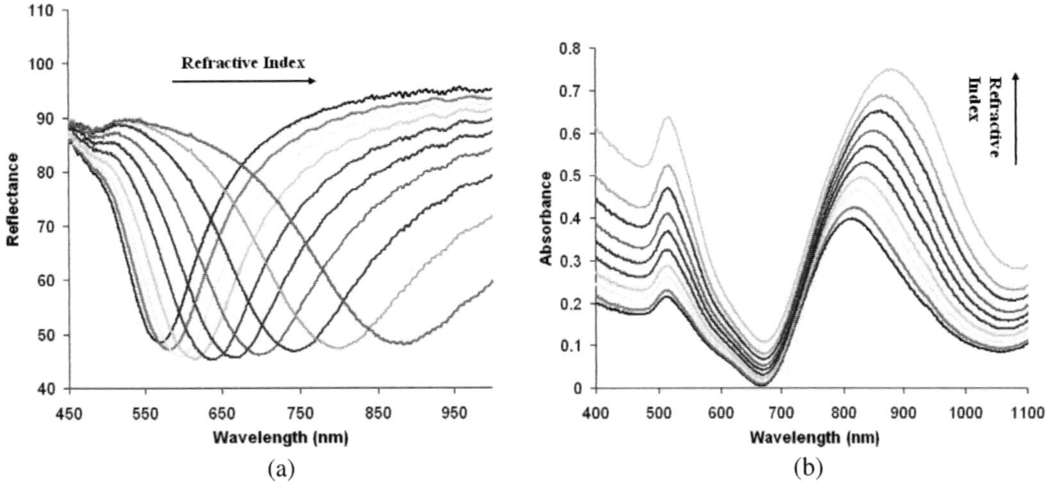

Figure 2. Overlapping spectra of (a) gold film based SPR sensor and (b) gold nanorods based LSPR sensor with increase of refractive index.

Figure 3 summaries the peak/dip wavelength changes of the sensors with the change of refractive index. It is noticeable that the sensitivity of SPR sensor only has a good linearity at a short refractive index range from 1.33 to 1.39. When the refractive index is higher than 1.39, the dip wavelength versus refractive index increases more dramatically, which means the SPR sensor becomes more sensitive. The LSPR sensor, on the other hand, has a much better linearity than the SPR sensor through the whole RI range.

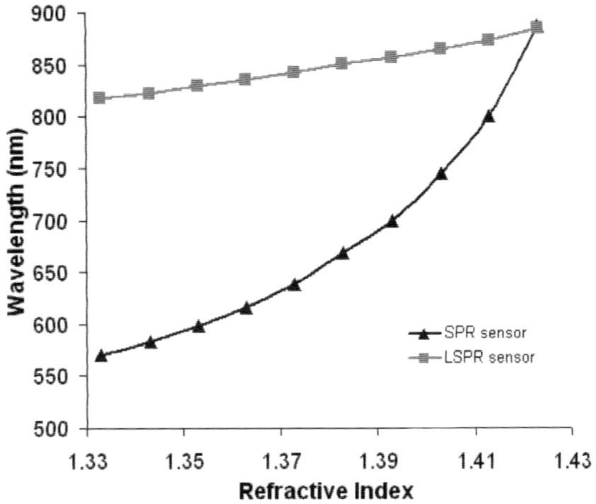

Figure 3. SPR versus LSPR wavelength shift with refractive index change.

4. Conclusion

In this work both gold-film coated SPR sensor and GNRs-coated LSPR sensor have been successfully fabricated, evaluated and cross-compared. When the aspect ratio of the GNRs is around 3.5, it can be concluded that SPR sensor has demonstrated to be more sensitive than that of the LSPR sensor when the solution refractive index changes. However the LSPR sensor has shown several advantages including ease of fabrication and linear performance. In addition, LSPR sensor sensitivity can potentially be tuned through the change of the GNR dimensions, such as aspect ratio, to allow an optimized sensor to be created for some specific sensor applications.

5. Acknowledgment

The authors would like to thank the Engineering and Physical Sciences Research Council (EPSRC) in the UK for the funding support via various schemes. The first author would like to take this opportunity to thank the support from the China Scholarship Council for his two-year scholarship to study abroad. The authors also would like to give special thanks to Prof Shum Ping and Dr Yu Xia from Nanyang Technological University for preparing the SPR sensor.

References

[1] Reather H 1988 *Surface Polaritons on Smooth and Rough Surfaces and on Gratings* (Berlin: Springer-Verlag)
[2] Bohren C F and Huffman D R 1983 *Absorption and Scattering by Small Particles* (New Yourk: Wiley-Interscience)
[3] Kretschm E and Raether H Z 1968 *Naturforsch A: Astrophys. Phys. Phys.Chem.* **12** 2135
[4] Nath N and Chilkoti A 2004 *Anal. Chem.* **76** 5370
[5] Yu C X and Irudayaraj J 2007 Anal. *Chem.* **79** 572
[6] Yonzon C R, Jeoungf E, Zou S L, Schatz G C, Mrksich M and Van Duyne R P 2004 *J. Am. Chem. Soc.* **39** 12669
[7] Svedendahl M, Chen S, Dmitriev A and Kall M 2009 *Nano. Lett.* **9** 4428
[8] Nikoobakht B and El-Sayed M A 2003 *Chem. Mater,* **15** 1957
[9] Sharma V, Park K and Srinivasarao M 2009 *Mater. Sci. and Eng., R* **65** 1
[10] Jain P K, Eustis S and El-Sayed M A 2006 *J. Phys. Chem. B* **110** 18243

Dielectrophoretic Self-Assembly of Au Nanorods for Sensing Applications

A. Pescaglini, U. Emanuele, A. O'Riordan and Daniela Iacopino [1]
Tyndall National Institute, University College Cork, Lee Maltings, Cork, Ireland

E-mail: dcaniela.iacopino@tyndall.ie

Abstract. Anisotropic 1-D metal nanostructures are attractive building block for future opto-electronic nanoscale devices and systems. However, a critical challenge remains the lack of manipulation methods that enable controlled positioning and orientation of metal nanostructures in a fast, reliable and scalable manner. To address this challenge, we explore dielectrophoretic based assembly of discrete gold nanorods and demonstrate site selective assembly and orientation of these rods. The demonstrated optical sensitivity of such large order nanostructures to the local environment opens the way to development of nanoscale sensing devices.

1. Introduction

Colloidal gold nanostructures have received widespread attention in recent years due to both their unusual properties and promising applications.[1-3] Among gold nanostructures, gold nanorods are particularly interesting since their optical properties are tunable throughout the visible and near infrared region of the spectrum as a function of nanostructure size, shape, aggregation state and local environment.[4-6] For nanorods the plasmon band splits in two: the longitudinal plasmon band corresponding to light absorption and scattering along the long axis of the rod, and the transverse plasmon band, corresponding to light absorption and scattering along the short axis of the rod. As result, the light emitted from or scattered off nanords is strongly polarised along the rod length axis, making them particularly attractive for specialised optoelectronic and bio-chemical sensing applications.[7-10] However, the potential of 1-D metal nanostructures for integration into future nanoscale sensing devices and systems will depend on the availability of new scalable manipulation methods that enable controlled positioning and orientation of nanorods in a fast, reliable and scalable manner. One approach, compatible with these requirements, is the exploitation of electric fields to drive the transport, localization and self-assembly of nanostructures at selected areas on a receptor chip substrate. Self-assembly of a variety of structures ranging in size from microns to nanometres including: GaAs based LEDs [11], biological cells [12], semiconducting and metallic nanoparticles [13], DNA molecules [14] carbon nanotubes [15] metallic and semiconductor nanoparticles [16] using electric fields has been widely reported. By contrast, there have been very few reports of electric field assisted assembly of gold nanorods on substrates suitable for optical sensing applications.[17]

Herein, we report on the dielectrophoretic manipulation and self-assembly of discrete gold nanorods suspended in water solutions at fused silica chip substrates. Dielectrophoresis (DEP) is defined as the

[1] To whom any correspondence should be addressed.

Published under licence by IOP Publishing Ltd

motion of uncharged polarisable particles in a non-uniform electric field [18]. The DEP forces arise from the interaction of the electric field with a dipole induced in the particle which exerts a translational motion on the particle. Depending on the permittivity and conductivity of both the particle and supporting medium, and also on the frequency of the applied electric field, particle motion is directed either towards (positive DEP), or away from (negative DEP) regions of high electric field strength. While DEP occurs under both direct current and alternating current (AC) conditions, AC electric fields are most commonly employed to: suppress electrophoresis of solvent molecules, eliminate motion due to particle surface charge and prevent undesired electrochemical interactions at the electrode surface, such as electrolysis, enabling the application of higher drive voltages. In this work, we demonstrate dielectrophoretic-based self-assembly of gold nanorods 14 nm in diameter and aspect ratio of 3. Nanorods suspended in water, were self-assembled on fused silica substrates, following application of an applied voltage of 20 V_{pp} at 1 MHz. Typical assembly times were on the order of ca. 30 minutes. Following assembly, nanorod arrays were characterised using scanning electron microscopy (SEM) and optical microscopy. Electric field simulations were undertaken to elucidate the electric field strengths and dielectrophoretic forces resultant on arrays of nanorods. Finally the dependence of assembled nanorod arrays optical properties from the local environment were investigated by darkfield microscopy.

2. Experimental

2.1. Au nanorod synthesis

Seed Solution. CTAB solution (5 mL, 0.20 M) was mixed with 5.0 mL of 0.00050 M $HAuCl_4$. To the stirred solution 0.60 mL of ice-cold 0.010 M $NaBH_4$ was added, which resulted in the formation of a pale brown solution. Vigorous stirring of the seed solution was continued for 2 min. after the solution was stirred, it was kept at 27 °C.

Growth of Nanorods. CTAB (25 mL, 0.20 M) was added to 1 mL of 0.0010 M of AgNO3 at 27 °C. To this solution, 25 mL of 0.00010 M $HAuCl_4$ was added, and after gentle mixing of the solution 0.35 mL of 0.0788 M ascorbic acid was added. Ascorbic acid as a mild reducing agent changes the growth solution from orange to colorless. The final step was the addition of 60 μL of the seed solution at 27 °C. The color of the solution gradually changed within 10-20 min. the temperature of the growth solution was kept constant at 27 °C until growth process was complete (90-120 min).

2.2. Au Nanorod characterisation:

Scanning electron microscopy (SEM) images of individual Au nanowires were acquired using a field emission SEM (JSM-6700F, JEOL UK Ltd.) operating at beam voltages of 2 kV. Transmission electron microscopy (TEM) images were acquired using a high resolution TEM (2100, JEOL UK Ltd.) operating at beam voltages of 200 kV. Optical micrographs were acquired using a calibrated microscope (Axioskop II, Carl Zeiss Ltd.) equipped with a charge-coupled detector camera (CCD; DEI-750, Optronics) and with darkfield condenser for darkfield imaging

2.3. Fused silica assembly chip fabrication:

Transparent fused silica chip substrates comprised two gold comb electrode structures. The electrodes were arrayed such that the electrode fingers (20 μm width, 1000 μm length and 50 μm pitch) were interdigitated and separated by a gap of 10 μm. In this manner, each electrode finger was electrically isolated from its nearest neighbours. Conducting interconnection tracks connected each comb electrode to a unique contact pad. Electrodes, interconnection tracks and contact pads were fabricated by sputtering Au (200 nm) onto a Cr (10 nm) adhesion layer on a four inch fused substrate. Metallisation was patterned using standard photolithography and lift-off techniques.

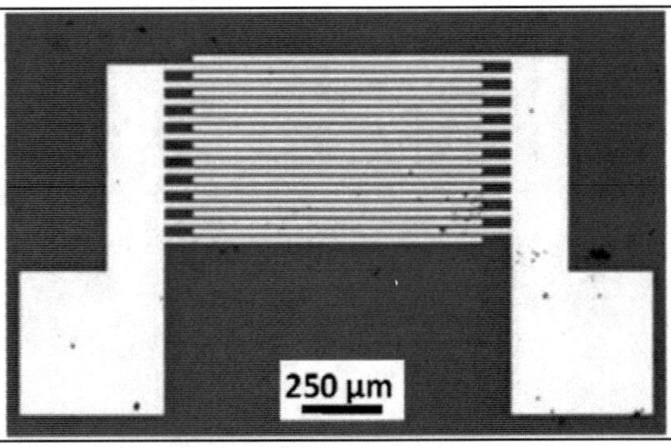

Figure 1: Optical micrograph of comb receptor electrode structure.

2.4. *Optical characterization Darkfield Polarised Spectroscopy*

Brightfield optical images of fabricated electrodes and darkfield optical images of gold nanorod arrays dielectrophoretically self-assembled on electrodes were acquired with an optical microscope (Zeiss axioscop II) equipped with a CCD camera (Optronics DEI-750) using a 100 W halogen lamp as illumination source. Polarised darkfield spectra of linear arrays of gold nanorods were measured with an inverted IX-71 Olympus microscope with an oil immersion darkfield condenser (Olympus U-DCW) and 50X objective. The sample was illuminated with a 100 W halogen lamp. The light collected by the objective was directed into the entrance of slit of a 0.3 monochromator (SP-2356, Acton Research) equipped with a thermoelectrically cooled, back illuminated CCD (Spec10:100B, Princeton Instruments). The experiments were performed by first recording an image of the sample with a mirror in the light path of the spectrometer. Once a suitable linear array was selected, the mirror was switched with a grating (300 groves mm^{-1}) to disperse the scattered light. The acquisition time for the spectra was one minute. Polarized spectra were obtained by placing a polarizer between the illumination source and the condenser.

2.5. *Electric field simulations*

Electric field simulations were undertaken to elucidate local electric field magnitudes present at comb electrode structures. Finite element simulations were performed using a partial equation solver (COMSOL Multiphysic) within the electrostatic limit $\nabla(\varepsilon\nabla\varphi) = 0$. Because the electrode length was much larger than the other dimensions and larger than the gap spacing, the problem could be conveniently reduced to a 2D problem without affect the simulation accuracy. The geometry consisted of two half electrodes 200 nm thick separated by 10 µm on top of 100 µm thick quartz substrate ($\varepsilon = 5\varepsilon_0$) and covered by 100 µm thick water ($\varepsilon = 80\varepsilon_0$) with 20 V_{pp} across the electrodes.

3. Results and Discussion

3.1. *Gold nanorods synthesis and characterization*

Gold nanorods were fabricated according to the seed method described in details by El Sayed et al. [19] This method affords high yield of gold nanorods with an intense longitudinal plasmon resonance peak centered at 765 nm and a smaller transversal Plasmon peak centered at 520 nm (see figure 2a). Figure 2b shows a SEM image of gold nanorods deposited on a Si substrate. The average length of gold nanorods measured from statistical analysis on SEM images (image J, 100 nanorods measured) showed an average nanorod length of 41 nm. The diameter of gold nanorods measured from TEM images was calculated to be 14 nm.

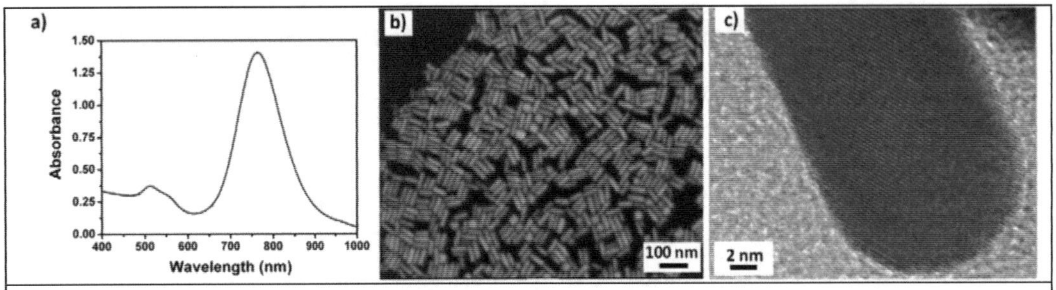

Figure 2: (a) UV-vis spectrum of gold nanorods dispersed in aqueous solution; (b) SEM micrograph of gold nanorods deposited on SiO_2 substrate; (c) high resolution TEM image of a single gold nanorod.

3.2. Gold nanords dielectrophoretic alignment:

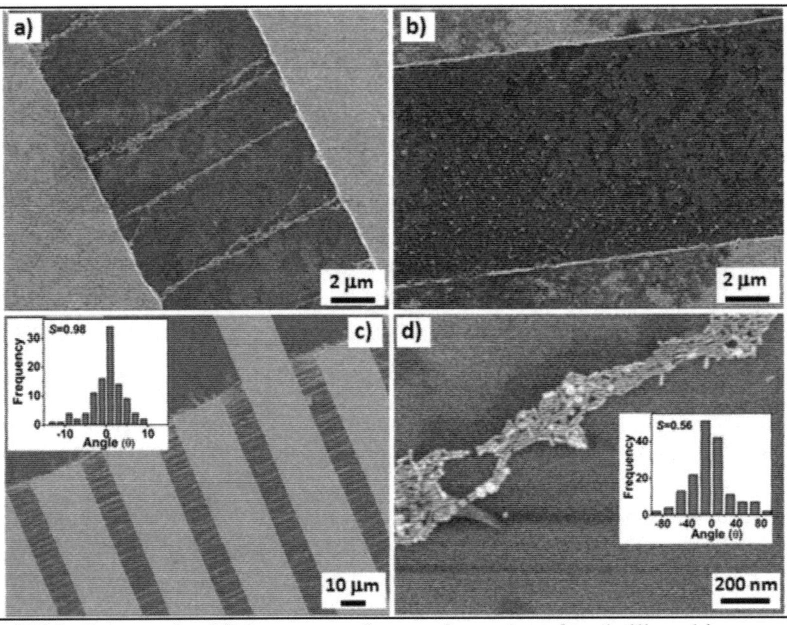

Figure 3: (a) SEM micrograph of linear nanorod array formed on fused silica chips upon application of electric field (20 V_{pp} at 1 MHz); (b) SEM micrograph of random deposited gold nanorods on fused silica chips in absence of applied electric field; (c) Low magnification SEM image showing long range alignment of nanorod arrays aligned normal to the electrodes. Inset: angular distribution plot of the assembled nanorod arrays; (d) High magnification SEM image showing internal nanorod alignement within assembled arrays. Inset: angular distribution plot of nanorods within linear arrays.

Dielectrophoretic self-assembly experiments were undertaken by applying a sinusoidal AC voltage under ambient light conditions to the receptor electrodes at fused silica substrates. The as-prepared nanorod solution (1 ml) was centrifuged three times at 13000 rpm for 5 min and redispersed in 100 µl of DI water (resistivity = 18.2 MOhm cm) in order to minimize the concentration of free ions in solution due to the CTAB excess. A 5 µl drop of the nanorod solution was placed on top of the metal electrodes and left to dry under an applied voltage of 20 V_{pp} at 1 MHz using a function generator (HP8116A Hewlett Packard) via direct probing to contact pads (ca. 30 min). As result of the applied electric field large arrays of gold nanorods aligned perpendicular to the electrodes (and parallel to the

electric field lines) were formed. Figure 3a shows and SEM image of the well separated large nanorod arrays covering the entire electrode area. In contrast, only random gold nanorods between electrodes were observed in absence of applied electric field (Figure 3b). The average diameter and length of formed arrays was estimated to be respectively ca. 100 nm and 10 μm.
The extent of nanorod array alignment on the substrate was quantified by determining an orientational parameter, S, defined as

$$S = \langle 2\cos^2\theta - 1 \rangle$$

Where *è* is the angle between an array and the electric field direction, and brackets,?⟨⟩? denote an ensemble average. An order parameter of S = 0.98, calculated using a sample of greater than 100 separate nanorod arrays, indicated a high degree of alignment (see figure 3c). Closed up examination revealed that gold nanorods within individual arrays were also partially aligned respect to the field direction (see figure 3d). S values between 0.4 and 0.6 were calculated for individual gold nanorods taken from different array lines.

3.3. Dielectrophoresis-mediated process:
In a neutral prolate ellipsoid nanoparticle (nanorod), an electric field E_n parallel to the *n* axis induces a dipole moment given by [20]

$$\boldsymbol{P}_n = V\varepsilon_m K_n^{*}\boldsymbol{E}_n$$

Where *V* is the nanorod volume, *n= L,T* label the Longitudinal and Transversal axis, ε_m the dielectric constant of the medium and K_n^{*} the Clausius-Mossotti factor

$$K_n^{*} = \left[\frac{\varepsilon_p^{*} - \varepsilon_m^{*}}{\varepsilon_p^{*} + A_n(\varepsilon_p^{*} - \varepsilon_m^{*})}\right]$$

in which $\varepsilon^{*}_{\frac{p}{m}} = \varepsilon_{\frac{p}{m}} + i\,{}^{\sigma_{\frac{p}{m}}}\!/_{\omega}$ is the complex dielectric function of the particle and the medium respectively and

$$A_L = \frac{1-e^2}{2e^3}\left[ln\left(\frac{1+e}{1-e}\right) - 2e\right] \qquad A_T = \frac{1-A_L}{2}$$

is the depolarising factor referred to the longitudinal and transversal axis where $e = \sqrt{1-\left({}^{1}/_{R}\right)^2}$ is the eccentricity and *R* the aspect ratio of the particle. When the interaction between the induced dipole and the applied AC electric field is the dominant process in the system, the nanorod experiences a torque Γ which aligned the longitudinal main axis along the field lines

$$\langle\Gamma\rangle = \frac{1}{2}V\,(A_T - A_L)Re(K_T^{*}K_L^{*})\boldsymbol{E}^2\sin(2\theta)$$

With θ the angle between the longitudinal axis and the electric field and a dielectrophoretic force $\mathbf{F_{DEP}}$ with the n axis parallel to the field direction

$$\langle\boldsymbol{F}_{DEP}\rangle = \frac{1}{2}V\varepsilon_m\, Re(K_n^{*})\nabla(\boldsymbol{E}^2)$$

Where ⟨□⟩ indicates the time average. Metal nanorods are particularly interesting to study due to their conductivity and permittivity much higher than that of the medium. The Clausius-Mossotti factor shows a crossover from positive to negative value well above any reasonable frequency excitation, assuring a positive dielectrophoretic force. For gold rods in water negative dielectrophoresis occurs at frequency as large as 10^{18} Hz [21], therefore in the MHz range, the Clausius–Mossotti factor can be approximated with a constant value $K_n \sim A_n^{-1}$ that is only geometry-dependent [22]. Within this approximation, and using an aspect ratio R=3, the effective polarisability along the main axis defined by

$$\alpha_L = V\varepsilon_m K_L$$

Can be estimated as $\alpha_L = 5 \times 10^{-32}$ C m^2/V. Comparing the dipole energy in electric field with the thermal energy of the particles, we can estimate the minimum electric field strength necessary to overcome the Brownian motion

$$E_{min} \cdot P_L \approx kT$$

Using $\alpha_L = 5 \times 10^{-32}$ C m^2/V and T = 300 K we obtain $E_{min} \sim 3 \times 10^5$ V/m. Simulation of the electric field across the electrode gap in figure 4a showed that the electric field reached a maximum value of 10^7 V/m near the electrode edges and never moved below 10^6 V/m inside the gap, so that dielectrophoretic forces are aspected to play a dominant role. This hypothesis was confirmed by the long range self-assembled nanorod arrays obtained experimentally. Formation of linear arrays oriented normal to the electrodes and the alignment of individual nanorods within linear arrays demonstrated that electric field-mediated assembling was the predominant process respect to Brownian motion, electrothermal flow and capillary forces during drying. It is interesting to note that in comparable electric field conditions, gold nanorod alignment has been previously observed in Si/SiO_2 substrates, but no large scale ordering has been reported at the best of our knowledge.[22] We speculate that the quartz substrate and the large gap length-scale played a determinant role in our observed self-assembly process. It is known that in a Si/SiO_2 substrate the conductive Si-doped layer forms an effective back-gate that cannot be completely screened by the SiO_2 layer with a finite thickness. [23] As result, the modified electric field intensity induces positive dielectrophoretic force near the surface preventing the assembly of nanorods around the centre of the gap [22]. The absence of this effect in quartz substrates used in this work constituted an important advantage during the assembly process. Moreover, the large gap size compared with the nanorod size allowed a long scale assembling mediated by nanorod-nanorod interaction that preserved the axial dipole symmetry, resulting in line-like structure formation. Ongoing experiment are focusing on elucidate how this process take place.

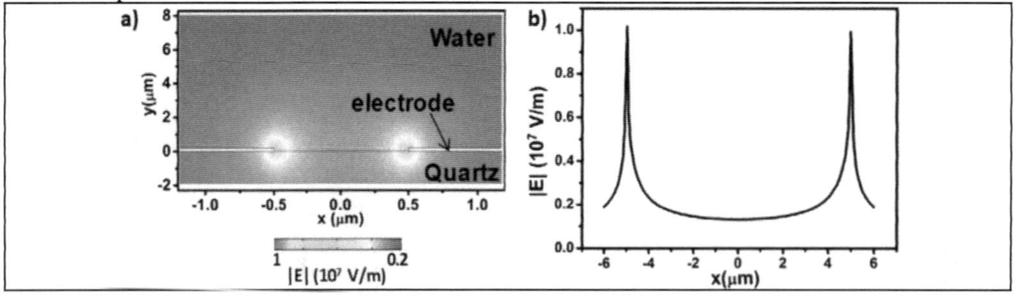

Figure 4: Simulation of the electric field for electrodes spaced 10 μm apart with an applied potential of 20 V_{pp}. (a) Contour plot of a cross section through the electrodes; the color bar indicates the field

strength; (b) electric field obtained from the simulation at a distance of 10 nm above the electrode surface.

3.4. Optical properties of Gold nanorod linear arrays:

The optical characteristics of the dielectrophoretic arrays were investigated by optical microscopy under darkfield illumination. Figure 5a shows a scattering image of the self assembled nanorod arrays aligned normal to the electrodes. Internal alignment of gold nanods within individual linear arrays was also detected by polarized darkfield microscopy. When nanorod arrays were rotated between crossed polarizers, their intensity varied between maximum brightness when the long axis was at 45° and 135° to the analyzing polarizer (Figure 5b) and minimum brightness (extinction) when at 0° and 90° to the analizing polarizer (Figure 5c). This change in optical brightness between crossed polarizers indicated a certain degree of uniaxial alignment of gold nanorods within the wires, in agreement with the calculated S value of 0.56.

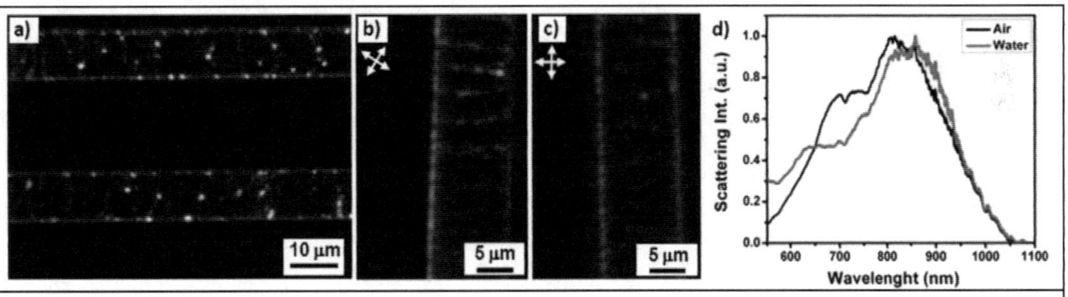

Figure 5: (a) Transmitted darkfield micrograph of gold nanorod arrays dielectrophoretically self-assembled at regions of high field strength showing arrays oriented normal to the electrodes; darkfield polarized images acquired as the sample was rotated in a plane between crossed polarizers, at (b) maximum brightness, (c) extinction. The solid white arrows indicate the orientation of the polarizer and analyzer with respect to the nanorod array direction; (d) Rayleigh scattering spectra of a gold nanorod array in air (1) and water environment (1.33). The Plasmon resonance maximum peak shifts from 813 nm to 836 nm.

It is known that the position of the plasmon absorption peak of metal nanorods depends on the dielectric properties of their surrounding media. In particular, recent studies performed on gold nanorods dispersed in various organic solvents, have shown a linear relationship between the square of the peak position of the longitudinal plasmon mode and the square of medium refractive index. [24] In the present study polarized darkfield spectroscopy was used to measure Rayleigh scattering spectra of individual nanorods arrays in media of various refractive index. A transparent substrate with dielectrophoretically self-assembled nanorod arrays was mounted on an inverted miscroscope equipped with an oil immersion darkfield condenser and was placed in contact with a custom made flow cell. First of all, the scattering spectrum of a selected array was measured in air. Subsequently scattering spectra of the same array in water were also recorded. As shown in Figure 5d a red shift in the longitudinal plasmon peak of the array from 813 nm to 836 nm was observed as the environment changed from air to water, corresponding to a change in refractive index from 1 to 1.33.

4. Conclusions:

In this work, we report the dielectrophoretic self-assembly of discrete gold nanoroed suspended in water at receptor electrodes pre-fabricated at fused silica substrates. The approach is a facile nanorod manipulation method that enables controlled and reliable positioning and orientation of nanorods, resulting in formation of large nanorod linear arrays. Such arrays were aligned parallel to the applied electric field with a high order parameter fit = 0.98. Polarised darkfield microscopy has been used to demonstrate the shift in scattering light observed in individual wires in contact with media of increasing refractive index. As such, the dielectrophoretic method represents a fast, robust and scalable self-assembly approach for manipulation of 1-D metal nanostructures. The sensitivity of such assemblies to the local environment could be potentially important for exploration and development of future sensing devices and systems.

5. Acknowledgments:

This work was supported by the European Union Seventh Framework Programme (Snapsun project).

6. References:

[1] M. A. El-Sayed, *Acc. Chem. Res.*, 2001, 34, 257-264.
[2] D. L. Fedheim and C. A. Jr. Foss, *Metal Nanoparticles: Synthesis, Characterisation and Applications*, Marcel Dekker: New York, 2002.
[3] M. C. Daniel and D. Astruc, *Chem. Rev.,* 2004, 104, 293-346.
[4] M. A. El-Sayed, *Chem. Res.* 2001, 34, 257-264.
[5] Y. Sun and Y. Xia, *Anal. Chem.,* 2002, 20, 5297-5305.
[6] C. Noguez, *J. Phys. Chem. B*, 2003, 111, 3806-3819.
[7] H. Wang, T. B. Huff, D. A. Zweifel, W. He, P. S. Low, A. Wei and J. X. Cheng, Proc, *Natl. Acad.Sci. USA*, 2005, 102, 15752-15756.
[8] C. Yu and J. Irudayarai, *Anal. Chem.* 2007, 79, 572-579.
[9]C. J. Murphy, A. M. Gole, S. E. Hunyadi, J. W. Stone, P. N. Sisco, A. Alkilany, B. E. Kinard and P. Hankins, *Chem. Comm.* 2008, 5, 544-557.
[10] J. N. Anker, W. P. Hall, O. Lyandres, N. C. Shah, J. Zhao and R. P. Van Duyne, *Nat. Mater.* 2008, 7, 442-453.
[11] A. O'Riordan, P. Delaney and G. Redmond, *Nano Lett.* 2004 4(5), 761-765.
[12] H. A. Ohl and I. Hawk, *Science*, 1966, 152(3722):647-&.
[13] N. G. Green and H. Morgan, *J. Phys. D-Appl. Phys.*, 1997 30(11), L41-L4.
[14] C. L. Asbury and G. van den Engh, *Biophysical Journal,* 1998, 74(2), 1024-30.
[15] J. Suehiro, G. B. Zhou,and M. Hara, *J. Phys. D-Appl. Phys.* 2003, 36(21), L109-L14.
[16] A. Bezryadin, C. Dekker and G. Schmid, *App. Phys. Letts.* 1997, 71(9), 1273-5.
[17] B. Ozturk, C. Blackledge and B.N. Flanders *Appl. Phys. Lett.* 2006, 88, 073108-3
[18] H. A. Pohl. *Dielectrophoresis*. New York: Cambridge University Press; 1978.
[19] B. Nikoobakht and M. A. El-Sayed, *Chem. Mater.* 2003, 15, 1957-1962.
[20] H. Morgan and N. G. Green, *E AC Electrokinetics: colloids and nanoparticles*, Reserch Studies Press LTD, 2003.
[21] B. C. Gierhart, D. G. Howitt, S. J. Chen, R.L. Smith, and S.D. Collins, *Langmuir*, 2007, 23, 12450-12456
[22] W. Ahmed, E. S. KooiJ, A. Van Silfhout, and B. Poelsema, *Nano Lett.*, 2009, 9,11,3786-3794
[23] C. W. Marquardt, S. Blatt, F. Hennrich, H.v. Lohneysen, R. Krupke, *Appl. Phys. Lett.*, 2006, 89, 183117
[24] J. Yang, J.C Wu, J. K. Wang and C. C. Chen, Chem. Phys. Lett., 2005, 416, 215-219

Journal of Physics: Conference Series **307** (2011) 012052 doi:10.1088/1742-6596/307/1/012052

Fabrication and Electrochemical Characterization of Micro- and Nanoelectrode Arrays for Sensor Applications

Nur Azura Mohd Said,[a,b] Karen Twomey,[a] Vladimir I. Ogurtsov,[a] Damien W.M. Arrigan,[c] Grégoire Herzog[a]

[a]:Tyndall National Institute, Lee Maltings, University College Cork, Cork, Ireland
[b]:Biotechnology Research Centre, Malaysian Agricultural Research & Development Institute, Serdang, Selangor, Malaysia
[c]: Nanochemistry Research Institute, Department of Chemistry, Curtin University, Perth, Australia
gregoire.herzog@tyndall.ie

Abstract. This paper describes the fabrication of microelectrode arrays, with two different geometries: disc (Designs d1 and d2) and band (Designs b1, b2 and b3) using three critical dimensions (100 nm, 1 μm and 10 μm) leading to 5 different designs, fabricated by the combination of UV photolithographic and e-beam lithographic techniques. Three silicon nitride layer thicknesses (200, 300 and 500 nm) were chosen to determine an optimized transducer design and fabrication process. Cyclic voltammetry characterisation using a simple redox probe ion, ferreocenecarboxylic acid in phosphate buffered saline electrolyte solution, demonstrated steady-state voltammetric curves for d1, d2, b1 and b2. A good agreement between experimental and theoretical data is found for devices d1, d2, b1 and b2. The experimental current for b3, on the other hand, is much lower compared to the calculated one- perhaps due to the overlapping of the diffusion layers of neighbouring microelectrodes in the array.

1. Introduction

Microelectrode arrays are gaining popularity in recent decades, particularly in electroanalytical chemistry. Nevertheless, historically, the greatest drive towards the early development of solid microelectrodes has arose in the 1940s and 1950s when the need to measure oxygen concentrations in living biological organisms became of paramount importance [1] with the first tungsten microelectrode fabricated over 50 years ago by the 1981's Nobel Laureate David H. Hubel[2]. IUPAC has outlined that a microelectrode has dimensions of tens of micrometers or less, down to submicrometer range[3] whilst nanoelectrode is defined as an electrode having dimensions on the order of a few tens of nanometres[4]. Due to their small sizes, microelectrodes offer a series of advantages compared with the conventional macroelectrodes including: (i) enhanced rates of mass-transport, (ii) decreased ohmic drop and (iii) increased signal-to-noise ratio. As a result of these characteristics improved response time, greater sensitivity with limits of detection down to sub-parts per billion and improved response are achieved[5]. The drawbacks of the use of single microelectrodes, however, are that they produced low current output and are more susceptible to mains interference[6,7]. These can be overcome with the use of arrays of microelectrodes, whereby multiple microelectrodes are operated in parallel. According to Arrigan[8], if collections of these electrodes are arranged in an ordered manner with a controlled inter-electrode spacing, they are referred to as arrays; and if the collections are not so ordered and there is no specific control over the inter-electrode spacing, then they are referred to as ensembles. In terms of microelectrode arrays fabrication techniques, Huang et al. (2009)[9] has described mainly five techniques which include: i) assembly techniques (e.g random assemblies of micro-wires[7]); ii) photolithography (thin film fabrication by the deposition of metallic film, namely Ag, Pt, Au and Ir, on Si_3N_4 or silicon oxides layer[5,10,11]) ; iii) screen-printing technique (for mass production and cheaper regular disk arrays[12]); iv) direct electrodeposition; and v) modification of the microelectrode array (molecular modification on the array surface, hydrophobic block, patterned electrodeposition, and pattern-based carbon nanotube/nanofiber microarrays). Apart from that, other techniques reported are the focused ion beam milling for the fabrication of recessed nanoband electrodes[6] and nanopore array electrodes[13], reactive-ion etching[14] and CMOS[15]. We report here the fabrication and the characterisation of electrode arrays, which will be used for the immobilisation of nanosome for the detection of odourants. Two different geometries for the nanotransducers have been selected: disc (Designs d1 and d2) and band (Designs b1, b2 and b3) using three critical dimensions for these structures (100 nm, 1 μm and 10 μm) leading to 5 different designs. Three silicon nitride layer thicknesses (200, 300 and 500 nm) were chosen to determine an optimized transducer design and fabrication process. The summary of the designs used in terms of their geometry and sizes are presented in Table 1.

Table 1 Summary of the five different types of transducers

Design	Geometry	Width/diameter	Length	N electrodes	Surface area/cm^2
d1	Disc	10 μm	-	314	2.466×10^{-4}
d2	Disc	1 μm	-	75	5.89×10^{-4}
b1	Band	10 μm	500 μm	17	8.5×10^{-4}
b2	Band	1 μm	50 μm	9	4.5×10^{-6}
b3	Band	100 nm	5 μm	75	3.75×10^{-7}

2. Materials and methods

2.1. Fabrication

The fabrication of the transducer device is a combination of UV photolithographic and e-beam lithographic methods to pattern electrodes with critical dimensions ranging from 100 nm to 10 microns. A silicon oxide layer is thermally grown on the silicon substrate and alignments marks are patterned on it. Titanium, platinum and gold layers are deposited by e-beam evaporation on the silicon wafer. The 20 nm thick titanium layer acts as an adhesion layer, which will improve the quality of the gold layer deposit. Silicon nitride is then deposited by plasma-enhanced chemical vapour deposition on the surface and will act as a passivation layer. Three silicon nitride layer thicknesses were selected: 200, 300 and 500 nm. Openings of the passivation layer are done in the silicon nitride by plasma etching to form the micro- and nanoelectrode arrays.

2.2 Electrochemistry set-up

The wafers were diced into individual chips (Figure 1). Five different designs of the transducer arrays were designed as described in Table 1. Disc and band arrays are described in Figure1B and 1C.

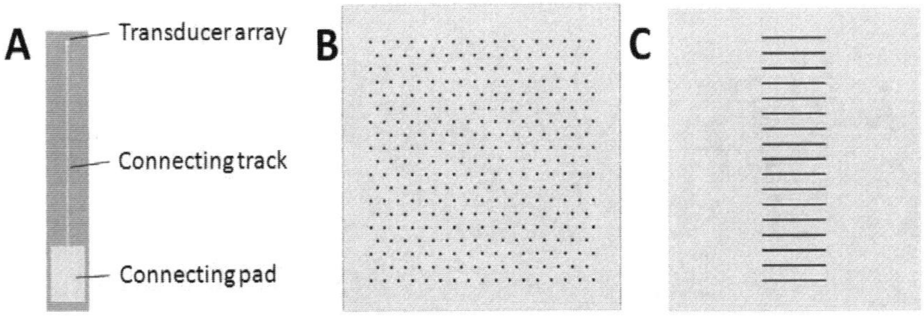

Figure 1: Schematic representation of the fabricated device and of the transducer array of (B) design d1 and (C) b1 layouts.

Prior to use, the nanotransducer devices are treated for 3-5 minutes in air plasma at 100 W, 600 mTorr (Harrick Plasma, New York, USA) to remove residual organic matter that may remain after the fabrication process. and were connected to the potentiostat using a custom made connector kindly provided by Uniscan Limited (Buxton, UK). The electrochemistry study was carried out with CHI660B potentiostat (CH instruments, IJ Cambria, Burry Port, Wales, United Kingdom) in 1 mM solution of ferrocenecarboxylic acid in PBS (0.01 M, pH7.4) using three electrochemical system with with a Ag | AgCl | 3 M KCl commercial reference electrode and a platinum wire commercial electrode. The devices fabricated are tested with cyclic voltammetry (CV) in a three electrode cell configuration at scan rate of 5, 10, 20, 50, 100, 200 and 500 mVs^{-1} over the potential range 0.1 to 0.6 V. All chemicals are obtained from Sigma-Aldrich and purified water with a resistivity of 18 MΩ cm, from an Option R50 system (Veolia Water Systems, Ireland).

3. RESULTS AND DISCUSSIONS

3.1 Visual characterisation of the transducer devices

Scanning electron and optical micrographs of the fabricated transducer devices were taken (Figure 2) and the dimensions were measured (Table 2).

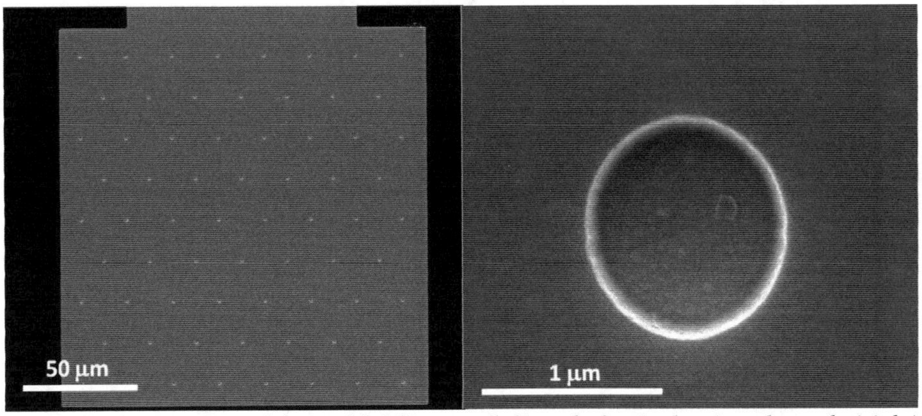

Figure 2: SEM images of design b2 full array (left) and of a single microelectrode (right)

Table 2: Expected and achieved critical dimensions for the transducer devices

		200 nm		300 nm		500 nm	
Design	Critical dimension µm	Dimensions achieved µm	Relative error	Dimensions achieved µm	Relative error	Dimensions achieved µm	Relative error
d1	10	9.31	-0.07	8.89	-0.11	8.89	-0.11
d2	1	1.168	0.17	1.236	0.24	0.964	-0.04
b1	10	9.52	-0.05	10.26	0.03	10.58	0.06
b2	1	1.195	0.19	1.297	0.30	1.03	0.03
b3	0.1	0.250	2.50	0.279	2.79	0.084	-0.16

The larger feature dimensions (d1 and b1) that were realised correspond to the expected with a 10% accuracy regardless of the silicon nitride thickness. However, for the smallest features (d2, b2 and b3), the dimensions achieved vary greatly with the thickness. Especially, it is inherent for silicon nitride thicknesses of 200 and 300 nm that there was no silicon oxide layer between the silicon nitride and the e-beam resist. For the 500 nm thick silicon nitride wafers the dimensions achieved were much closer to the expected dimensions compared to the wafers with thinner silicon nitride layers. The reason for the widening of the opening is due to the lack of selectivity of the etching process for silicon nitride over the e-beam resist.

3.1 Electrochemical characterisation of the transducer devices

The five different designs were characterised by CV and as expected, CV curves for d1, d2, b1 and b2 exhibited steady state shape peak at 100 mV s^{-1} (Figure 3). However CV for b3 is greatly dependant on the scan rate and exhibited peak shape at 10mV s^{-1}. The CV for b3 with different scan rates (not shown here) is similar as reported by Lanyon et al. (2007)[13] using 3 x 3 nanopore array with pore radii of 80 nm and the current obtained also were in the same range as ours (picoA). Berduque et al. (2007)[10] has employed Equation 1.1 and 1.2 for the calculation of theoretical limiting current for both recessed micro-disc electrodes and inlaid micro-band electrodes:

$$i_{\lim} = \left(\frac{4\pi nFCDr^2}{4L + \pi r} \right) N \quad \text{Eq. 1.1}$$

$$i_{\lim} = \frac{2\pi nFCDL}{\ln\left(\frac{64\,Dt}{w^2} \right)} N \quad \text{Eq. 1.2}$$

where D is diffusion coefficient (5.7×10^{-6} cm^2 s^{-1}), F is Faraday constant (96485 $Cmol^{-1}$), A is the surface area, r is the radius, L is the silicon nitride thickness (recess depth), w is the band electrode width, L is the band electrode length, C is the concentration of ferrocene carboxylic acid, t is the time of experiments and n is the number of electrons involved in the reaction.

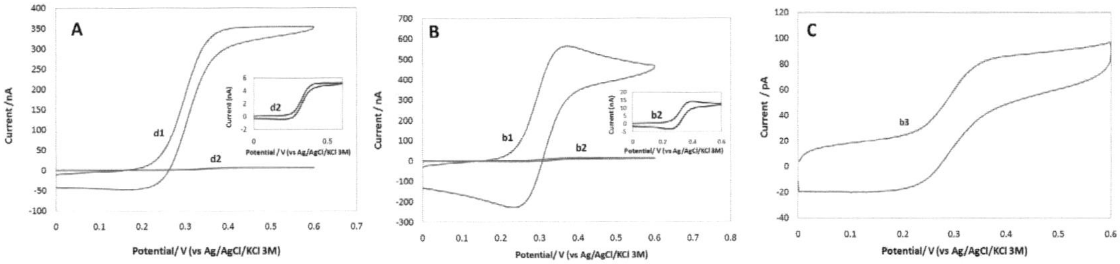

Figure 3: CV for d1 and d2 microdisc array (A); b1 and b2 microband array (B); both at 200 nm recess, scan rate 100mVs^{-1}; CV for microband array b3 at 200 nm recess with scan rate 10 mVs^{-1} (C)

A good agreement between experimental and theoretical data is found for devices d1, d2, b1 and b2, as shown in Table 3. However, the current measured for the nanotransducer device of b3 is 100 times lower than the one expected. Lanyon et al. (2007) also has obtained a lower experimental current compared to the calculated ones for the recessed nanobands used[6]. The decrease of the current might due to the overlapping of the diffusion layers of neighboring microelectrodes in an array hence the behaviour of the array becomes that of a macroelectrode and the voltammograms tail off due to planar diffusion[11,16]. Equation 1.2 also does not take into account the presence of the recess, which for the nanobands in this case, is larger than the critical dimension (width) of the electrode.

Table 3: Comparison between the theoretical current and experimental current for transducers

Design	L / nm	Theoretical current/ nA	Experimental current (nA)
	200	329	344 ± 5
d1	300	321	328 ± 5
	500	306	318 ± 12
	200	5.47	4.9 ± 0.3
d2	300	4.68	3.9 ± 0.6
	500	3.63	3.3 ± 0.7
	200		529 ± 7
b1	300	649	510 ± 28
	500		517 ± 16
	200		13 ± 0.3
b2	300	17	13 ± 2
	500		14 ± 5
	200		0.06 ± 0.002
b3	300	9.4	0.06 ± 0.003
	500		0.07± 0.001

The efficiency of the silicon nitride layer to prevent diffusion of ferrocenecarboxylic acid to diffuse and react at the underlying gold layer was verified by running CV at transducer devices with no opening in the silicon nitride layer. No faradaic current is recorded when there is no opening in the silicon nitride on the control designs (Figure 4).

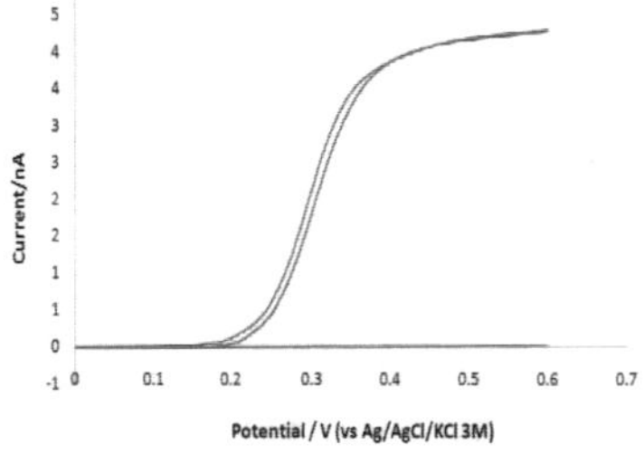

Figure 4: CV at transducer d2 with 314 discs of 1 micron in diameter (blue line) and with no opening in the silicon nitride (red line).

4. CONCLUSIONS

This paper has described the fabrication of microelectrode arrays of microdisk and microband with various recess depth using UV photolithographic and e-beam lithographic techniques. Both microdisk array (d1 and d2) and microband array (b1 and b2) was successfully characterized via visual images and electrochemical study. CV curves for d1, d2, b1 and b2 exhibited steady state shape peak at 100 mV s^{-1} and the experimental limiting current obtained are in the vicinity of the calculated theoretical current. For d3 nanoband, however, the CV shape is more inclined to peak/sigmoidal shape and the CV is greatly dependant with the scan rate. This micro- and nanoelectrode array will be used for the immobilisation of olfactory receptor for the sensing of odourants.

Acknowledgements:

The authors gratefully acknowledge the European Commission (Bioelectronic Olfactory Neuron Device, Grant agreement N° 228685) for support for this work and Malaysian Agricultural Research & Development Institute (MARDI) for the scholarship award.

References

(1) A.M. Bond, K.B. Oldham, C.G. Zoski, *Anal. Chim. Acta*, **1989**, *216*, 177.
(2) D.H. Hubel, *J. Psychiatr. Res.* **1971**, *8*, 301.
(3) K. Štulík, C. Amatore, K. Holub, V. Mareček, W. Kutner, *Pure Appl. Chem.* **2000**, *72*, 1483.
(4) R.G. Compton, G.G. Wildgoose, N.V. Rees, I. Streeter, R. Baron, *Chem. Phys. Lett.* **2008**, *459*, 1.
(5) V. Beni, D.W.M. Arrigan, *Current Anal. Chem.* **2008**, *4*, 229.
(6) Y.H. Lanyon, D.W.M. Arrigan, *Sens.Actuat. B: Chem.* **2007**, *121*, 341.
(7) S. Fletcher, M.D. Horne, *Electrochem. Commun.* **1999**, *1*, 502.
(8) D.W.M. Arrigan, *Analyst*, **2004**, *129*, 1157.
(9) X.J. Huang, A.M. O'Mahony, R G. Compton, *Small*, **2009**, *5*, 776.
(10) A. Berduque, Y.H. Lanyon, V. Beni, G. Herzog, Y.E. Watson, K. Rodgers, F. Stam, J. Alderman, D. W. M. Arrigan, *Talanta*, **2007**, *71*, 1022.
(11) N. Godino, X. Borrise, F.X. Munoz, F.J. del Campo, R.G. Compton, *J. Phys. Chem. C*, **2009**, *113*, 11119.
(12) O. Ordeig, F.J. del Campo, F.X. Munoz, C.E. Banks, R.G. Compton, *Electroanalysis* **2007**, *19*, 1973.
(13) Y.H. Lanyon, G. De Marzi, Y.E. Watson, A.J. Quinn, J.P. Gleeson, G. Redmond, D.W.M. Arrigan, *Anal. Chem.* **2007**, *79*, 3048.
(14) M.E. Sandison, N. Anicet, A. Glidle, J.M. Cooper, *Anal. Chem.* **2002**, *74*, 5717.
(15) R. Lenigk, H. Zhu, T.C. Lo, R. Renneberg, *Fres. J. Anal. Chem.* **1999**, *364*, 66.
(16) T.J. Davies, R.G. Compton, *J. Electroanal. Chem.* **2005**, *585*, 63.

Chromatographic air analyser microsystem for the selective and sensitive detection of atmospheric pollutants.

Jean-Baptiste SANCHEZ[1], Houda LAHLOU[1,2], Yehya MOHSEN[1], Xavier VILANOVA[2], Franck BERGER[1], Xavier CORREIG[2].

[1]Laboratoire de Chimie Physique et Rayonnements, Alain Chambaudet, UMR CEA E4 UFR ST, Université de Franche Comté, 25000 Besançon, France

[2]Departament d'Enginyeria Electronica, Electrica i Automatica, Universitat Rovira i Virgili, Països Catalans 26, 43007, Tarragona, Spain

jbsanche@univ-fcomte.fr

Abstract. The development of industry and automotive trafic produces Volatile Organic Compounds (VOCs) whose toxicity can affect seriously human health and environment. The level of those contaminants in air must be as low as possible. In this context, there is a need for in situ systems that could monitor selectively the concentration of these compounds. The aim of this study is to demonstrate the efficiency of a system build with a pre-concentrator, a chromatographic micro-column and a tin oxide-based gas sensor for the selective and sensitive detection of atmospheric pollutants. In particular, this study is focused on the selective detection of benzene and 1,3 butadiene.

1. Motivation and results

Recently, a huge interest has been devoted to the fabrication of miniaturized systems that can be practically used for in situ analysis of toxic gases. The use of tin dioxide-based thin film sensors seems to be an obvious compromise between the low cost, their small size, their high sensitivity (small change in gas composition causes change in resistance) and fast response to many pollutant gases. Moreover, these detection devices can easily be miniaturized for the development of mobile systems. In spite of these considerations their selectivity still remains unsatisfactory for their further use as a mobile system. Indeed, these gas microsensors can't provide any kind of selectivity when they are operated alone. A chromatographic separation stage prior to the detection stage is a good alternative to provide selective detection towards a wide range of vapour mixtures. Netherless, to obtain a small and mobile device, it is necessary to miniaturized classical gas chromatographic columns [1].

Regarding the necessity to detect very low concentrations of toxic gases (ppm or ppb), a pre-concentration stage is also envisaged in front of the GC micro-column in order to decrease the detection limit of the global system. In particular, the pre-concentration effect allows ppb level detection for a wide range of chemical compounds. Recent research works demonstrated a successful coupling between a pre-concentrator, a gas chromatographic column and a chemical gas sensor (SnO_2)

Journal of Physics: Conference Series **307** (2011) 012053 doi:10.1088/1742-6596/307/1/012053

for the selective and sensitive detection of benzene and 1,3 butadiene [2]. From these results, we are actually focused on the development of a silicon integrated device able to concentrate, separate and detect target pollutants from the air. This micro-fluidic system will integrate together all analytical functions on a single silicon substrate (Cf. Fig.1).

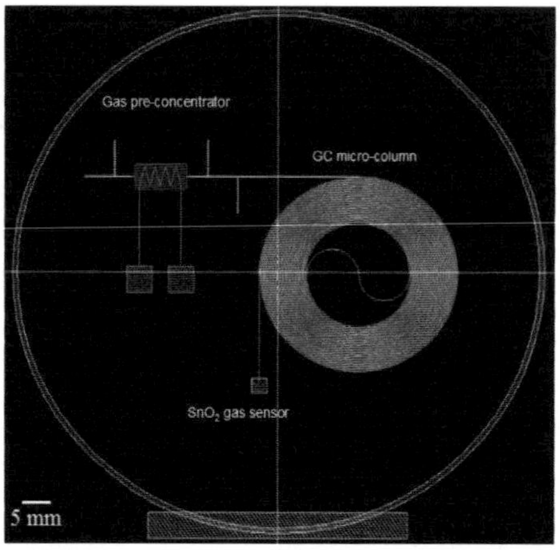

Fig. 1: integrated analytical platform envisaged in this study.

A first step of the study deals with the integration of the pre-concentrator and the GC column. The idea is to propose a common process for the realization of these systems in order to integrate both on the same substrate.

During this talk, a complete description of each discrete element (gas pre-concentrator (Cf. Fig. 2) and GC micro-column (Cf. Fig.3)) will be given.

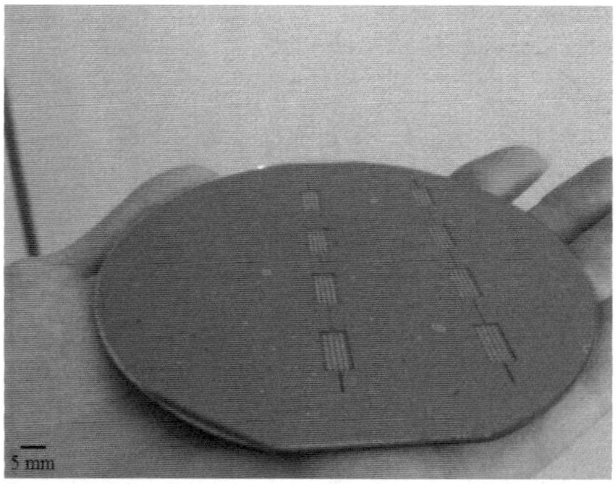

Fig. 2: silicon micro-machined gas pre-concentrators used to concentrate pollutants.

Fig. 3: silicon micro-machined gas chromatographic columns used to separate pollutants.

Then, the performance of these elements will be presented for the detection of benzene and 1,3 butadiene (Cf. Fig. 4).

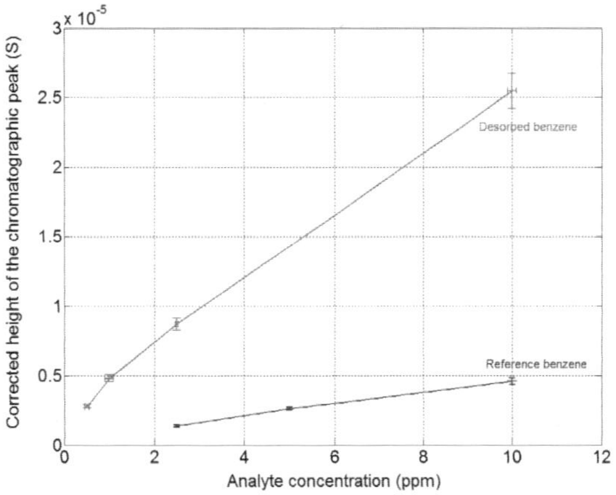

Fig. 4: exemple of the calibration curves for benzene detection with and without the use of the gas pre-concentrator in front of a GC micro-column and chemical gas sensor.

References

[1] J.-B. Sanchez et al., Journal of Sensors, Volume 2010, Article ID 409687, doi : 10.1155/2010/409687

[2] H. Lahlou et al., Sensors and Actuators B: Chemical, (2011), doi : 10.1016/j.snb.2011.02.018.

Journal of Physics: Conference Series **307** (2011) 012054 doi:10.1088/1742-6596/307/1/012054

Development of an ozone high sensitive sensor working at ambient temperature

F. Berger, B. Ghaddab, J. B. Sanchez, C. Mavon

Laboratoire de Chimie-Physique et Rayonnements Alain Chambaudet UMR CEA-E4, UFR ST, La Bouloie, 16 route de Gray, 25030 Besançon Cedex, France.

fberger@univ-fcomte.fr

Abstract. Hybrid SnO_2/SWNTs thin layer were deposited by using sol-gel process. Such sensitive layers showed very high performances for O_3 flow detection at ambient temperature. Limit sensitivity, lower than 21,5 ppb of O_3 in air has been reached by using these hybrid layers. Compared to usefull metal oxide sensors, the main advantage of the use of such hybrid layers, is that these devices enable the detection of O_3 traces at room temperature. The influence of sensor's working temperature is discussed and finally a reactional mechanism for the detection of O_3 is proposed.

1. Motivation and results

Today, the detection of ozone present as pollutant in the atmosphere, is considered as a serious problem. Air quality standard established by the U. S. environmental protection agency (EPA USA) has fixed a limit in ozone concentration of 75 ppb in the atmosphere. For many years, studies regarding metal oxide-based gas sensors showed that these kind of microsensors enable the detection of ozone at a concentration of 50 ppb [1, 2, 3] and show for this reason a real interest for the detection of this pollutant. Nevertheless, their working principle needs the sensor to be heated at high temperature (350°C-500°C), what causes major drawbacks concerning its elaboration (heating device development, thermal stability, high power consumption), and its working principle (thermal degradation of the target compounds to be detected). Concerning precisely the actual development of chemical sensors, recent studies pointed out the advantage of the use of carbon nanotubes. Concerning more precisely the detection of ozone using cabon nanotubes-based sensitive layers, Picozzi et al. [4] and Park et al. [5] obtained respectively a limit in sensivity of 90 ppb and 50 ppb for ozone detection at ambient temperature. An other use of carbon nanotubes concerns the elaboration of metal oxide-based hybrid layers. Recent results pointed out of that such hybrid layers enable the detection of NH_3 [6, 7], NO_2 [8, 9], CO [9], GPL compounds and ethanol [10]. The results presented in this paper concern the development of SnO_2/carbon nanotubes (SWNTs) hybrid layers for ozone detection in traces levels. Marketed SWNTs were dispersed in a tin-based sol and deposited over electrodes with conventional interdigitated design. Thin hybrid layers were deposited from the as-prepared sol by using dip-coating method onto silicon thermally oxidised equipped with interdigitated electrodes. HR-TEM analysis of the sensitives layers enable the determination of the their thickness (Cf. Figure 1) and the characterization of the dispersion of SWNTs into the SnO_2 matrix.

Journal of Physics: Conference Series **307** (2011) 012054 doi:10.1088/1742-6596/307/1/012054

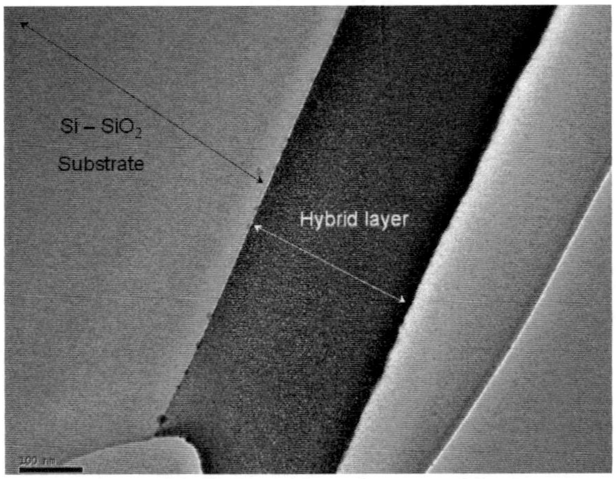

Fig. 1 : HR-TEM cross-section characterization

Then the effect of O_3 onto hybrid layers was investigated. Figure 2 pointed out that a first detection of O_3 led to an irreversible interaction which stabilized the electrical behavior for latter detection.

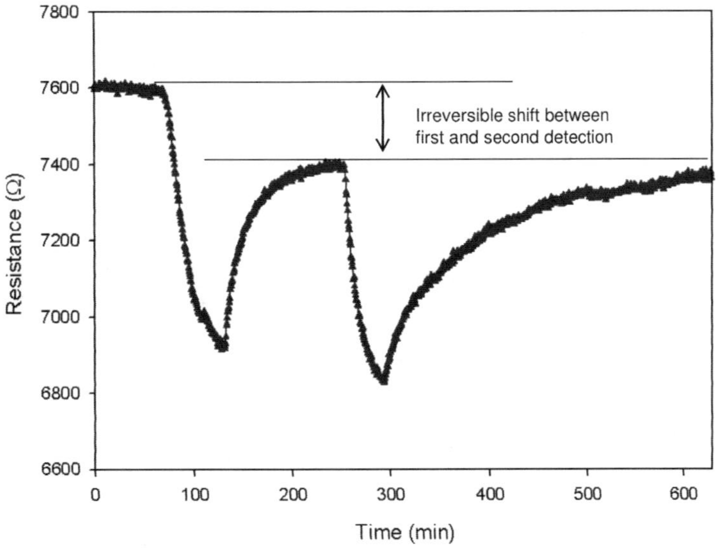

Fig. 2 : Modification of electrical behavior between first and second detection of O_3 (87 ppb).

This phenomenon was explain by an oxidation of the carbon nanotubes located into the SnO_2 matrix. Then the evaluation of the sensors efficiency for O_3 detection was conducted. Results represented in Figure 3 confirm that hybrid layers are very promising devices for the detection of ozone at trace level at ambient temperature.

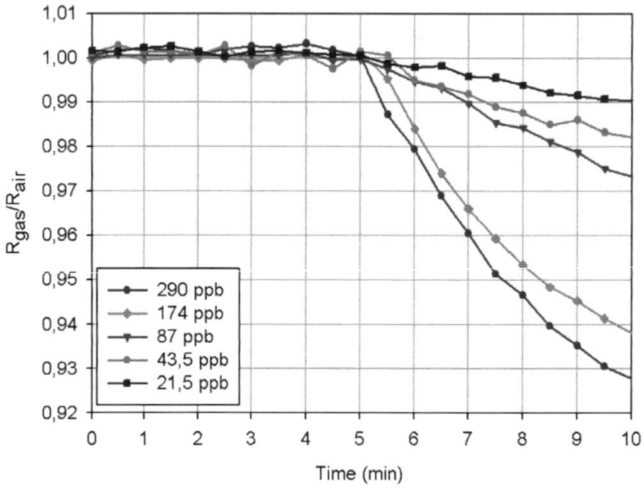

Fig. 3 : Responses of the SWNT's-based sensor to O_3 concentrations (21.5 to 290 ppb) for 5 minutes exposition time.

Finally a rectional mechanism for the detection of ozone using hybrid layers will be proposed, considering the p-n junction formation between SnO_2 and carbon nanotubes.

References

[1] L.Berry, J.Brunet, Sens. Actuators B, 129 (2008) 450.
[2] S.Vallejos et al., Sens.Actuators B, 126 (2007) 573.
[3] K. Aguir, C. Lemire, D. B. B. Lollman, Sens. Actuators B: Chem., 84 (2002) 1.
[4] S. Picozzi et al., J. Vac. Sci. Technol. A 22.4 (2004) 1466.
[5] Y.Park, K.Y.Dong, J.Lee, J.Choi, G.N.Bae, B.K.Ju, Sens. Actuators. B, 140 (2009) 407.
[6] N.D.Hoa, N. V. Quy, Y. S. Cho, D. Kim, Phys. Stat. Sol., 204(6) (2007) 1820.
[7] N. Van Hieu, et al., Sens. Actuators B, 144 (2010) 450.
[8] B.Y.Wei, M.C.Hsu, P.G.Su, H.M.Lin, R.J.Wu, H.J.Lai, Sens. Actuators B, 101, (2004) 81.
[9] E.H. Espinosa et al., Sens. Actuators B, 127 (2007) 137.
[10] Y.L.Liu, H.F.Yang, Y.Yang, Thin solid films, 497 (2006) 355.

Macromolecular sensing at the liquid-liquid interface

Grégoire Herzog,[a] Shane Flynn,[a] Damien W.M. Arrigan[b]

[a]: Tyndall National Institute, Lee Maltings, University College, Cork, Ireland

[b]: Nanochemistry Research Institute, Department of Chemistry, Curtin University, Perth, Australia

gregoire.herzog@tyndall.ie

Abstract. We report here the electrochemical sensing of macromolecules, such as polyLysine dendrimers, at the polarised liquid | liquid interface. Electrochemistry at the liquid | liquid interface is a powerful analytical technique which allows the detection of non-redox active molecules via ion transfer reactions at a polarised water – oil interface. We demonstrate here that different parameters of the polyLysine dendrimers (charge number, molecular weight) have a strong influence on the sensitivity and limit of detection of these macromolecules. This work will help to the development of sensors based on charge transfer at the liquid | liquid interface.

1. Introduction

Electrochemistry at the liquid | liquid interface allows the electrochemical investigation of non-redox active ions, based on their transfer between two immiscible phases. The interface between two immiscible electrolyte solutions (ITIES) is polarised by the means of two reference electrodes and ion transfer can be potentially driven [1]. In recent years, significant advances were made to improve the sensing capabilities of electrochemistry at the ITIES [2]. A wide variety of ions has been detected at the ITIES: K^+, dopamine, peptides, proteins, DNA, dendrimers… Miniaturisation of the liquid | liquid interface has improved the sensitivity and the limits of detection [3]. This improvement comes from increased mass transfer rates at interfaces which have dimensions smaller than the diffusion layer thickness and is analogous to the phenomena observed at micro- and nanoelectrodes [4]. Proteins are biological macromolecules of primary importance as they are the molecular expression of the genetic information. Their nature and function are governed by their complex structure. The presence or absence of a protein, its concentration or alteration of its structure can be indicative of a disease state. Label-free electrochemical detection of proteins has thus attracted a lot of interest in recent years [5]. The electrochemical behaviour of proteins at the liquid | liquid interface is a particularly complex process, involving adsorption of the protein at the interface and the assisted transfer of the anion of the organic electrolyte to the aqueous phase. This process at the liquid | liquid interface has been thoroughly investigated for insulin [6], haemoglobin [2] and lysozyme [7-8]. Electroanalysis at the ITIES can also be indicative of the structure of a protein as changes in the electrochemical behaviour have been investigated in the case of chemical denaturation [9] or enzymatic proteolysis [10]. Dendrimers are a class of synthetic monodisperse and highly branched three-dimensional macromolecules. They are made up of a central core, branched units and terminal groups. By increasing the number of branched units, the dendrimer generation increases. The electrochemical

behaviour of poly(propylenimine) (DAB-AM-*n*) and poly(aminoamine) (PAMAM) dendrimer families at the liquid | liquid interface was investigated in different experimental conditions [11].
We report here the electrochemistry of poly-Lysine dendrimers from generation 2 (G2) up to generation 5 (G5) at the liquid | liquid interface. We have investigated the evolution of the electrochemical signal as a function of the poly-Lysine dendrimer concentration and how the structural characteristics of the dendrimer impact on the analytical sensitivities and the limits of detection. We have compared these values with those reported in the literature for protein detection at the ITIES.
The first paragraph after a heading is not indented (Bodytext style).

Other paragraphs are indented (BodytextIndented style).

2. Materials and methods

2.1. Reagents
The poly-Lysine dendrimers of generation G2, G3, G4 and G5 were purchased from Colcom (Montpellier, France). The core of the dendrimer molecules consists of a linear peptide of 8 lysine groups. For G2, 8 polypeptide units are grafted on the core. The dendrimers characteristics provided by the supplier are gathered in Table 1. All the other chemicals were purchased from Sigma-Aldrich Ltd Ireland. All chemicals were used as received.

Table 1:Dendrimer characteristics

Generation	G2	G3	G4	G5
Molecular weight (g mol^{-1})	8600	22000	65300	172300
Diameter at pH 7 (nm)	4.5	7	11	16
Number of lysine groups	48	123	365	963

2.2 Electrochemical experiments

The experiments were done in a custom made glass cell as shown in Figure 1. Two Ag|AgCl wires serve as reference electrodes (one in the aqueous phase and one in the organic phase) to monitor the interfacial potential difference and the current is measured by two counter Pt mesh electrodes (one in each phase). The organic phase electrolyte was 1,2-dichloroethane (1,2-DCE) containing 10 mM bis-(triphenylphosphoranylidene) ammonium tetrakis(4-chlorophenyl)-borate ((BTPPA$^+$ TPBCl$^-$) and the aqueous phase was 10 mM HCl. The geometric area of the interface was 1.12 cm^2.

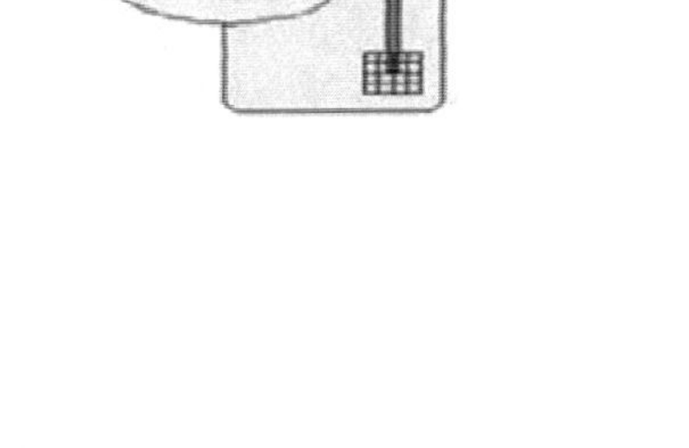

 IOP Publishing
doi:10.1088/1742-6596/307/1/012055

Figure 1: Four electrode electrochemical cell for liquid | liquid experiments. $RE_{org.}$: Reference electrode for the organic phase, $RE_{aq.}$: Reference electrode for the aqueous phase, $CE_{org.}$: Counter electrode for the organic phase, $CE_{aq.}$: Counter electrode for the aqueous phase.

The potential window is limited by the transfer of the background electrolyte salts present in both solutions. As a convention, positive currents are the result of cations transferring from the aqueous to the organic phase or from anions transferring from the organic to the aqueous phase. Similarly, anions transferring from the aqueous to the organic phase and cations transferring from the organic to the aqueous phase result in a negative current. The electrochemical cell used was as described in scheme 1 where x represents the concentration of poly-Lysine dendrimer and N the dendrimer generation.

Scheme 1: Electrochemical cell used

A g	**Ag Cl**	**10 mM BTPPA$^+$ Cl$^-$ in 10 mM LiCl**	**10 mM BTPPA$^+$ TPBCl$^-$ in 1,2 dichloroethane**	**10 mM HCl + x μM GN**	**AgC l**	**A g**

3. Results and discussion

The electrochemical behaviour of polyLysine dendrimers of four generations (G2 to G5) has been investigated at the polarised liquid | liquid interface by cyclic voltammetry. Figure 2 shows the cyclic voltammograms (CV) for increasing concentrations of G2 in the aqueous phase. When the potential window is scanned in the positive direction, a peak is observed at + 0.85 V and on the reverse scan, a peak is observed at + 0.7 V. In our previous work on proteins [2, 6, 8], this pair of peak was attributed to the assisted transfer of $TPBCl^-$, the organic electrolyte anion, to the aqueous phase and to the adsorption onto the aqueous side of the interface, of the protein – $TPBCl^-$ complex. A similar behaviour is observed for the dendrimers. At + 0.45 V, the peak observed is due to the transfer of tetraethylammonium (TEA^+) from the aqueous phase to the organic phase. TEA^+ is a model ion commonly used in electrochemistry at the liquid | liquid interface. Its transfer is inhibited by the presence of dendrimer at the interface. Indeed, the peak current corresponding to the transfer of TEA^+ is decreasing as the dendrimer concentration increases. This was observed for all dendrimer generations. This observation confirms the adsorption occurring at the interface.

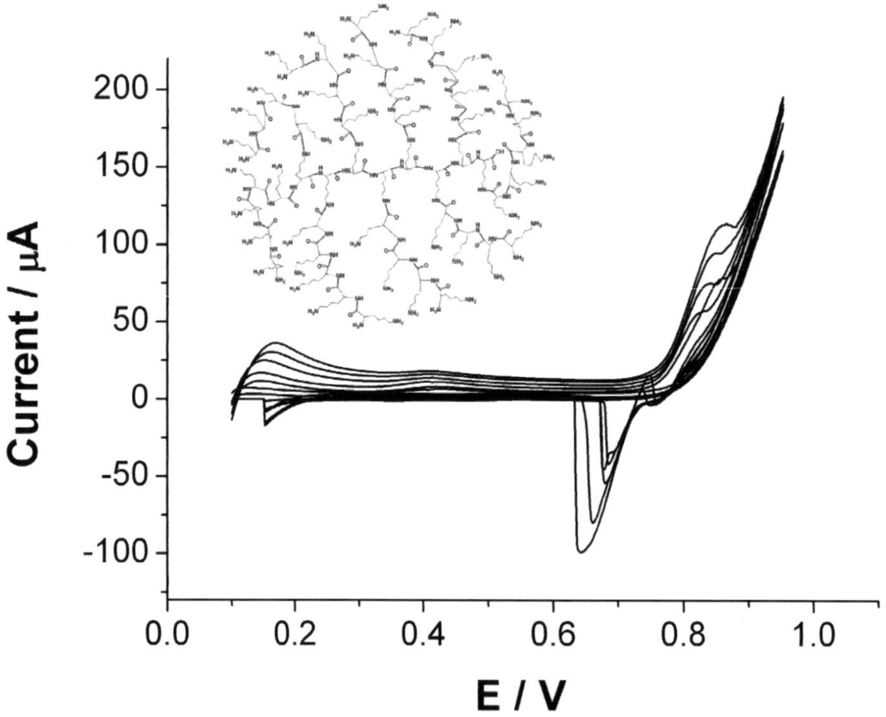

Figure 2: CV of increasing concentration of G2 (0, 11.1, 22, 32.8, 43.4 and 53.9 μM). The concentration of TEA^+ was 146 μM and the scan rate was 5 mV s^{-1}. Inset: Chemical structure of the polyLysine G2 dendrimer.

The peak current increased linearly with the concentration of dendrimers in the aqueous phase, which was observed for each of the dendrimer generation investigated. Indeed, calibration curves of the reverse peak current for the dendrimers G2 to G5 were built. The sensitivity (defined as the slope of the calibration curve i / C, where i is the current value and C the dendrimer concentration) increased with the dendrimer generation (Figure 3). The charge of the dendrimer, z, increased with the generation. Under the present conditions, z is considered equivalent to the number of lysine groups in the macromolecule, as were are at pH 2 and fully protonation of the end lysine groups are expected. Sensitivity values reported in the literature for proteins are much higher than we measured for dendrimers . The large difference indicated that not only charge has an influence on the sensitivity. Indeed, proteins have a tertiary structure with hydrophobic pockets, which play a role in complex formation between the organic electrolyte anions and the proteins adsorbed at the interface [7]. The dendrimers studied are made of lysine, which are strongly hydrophilic.

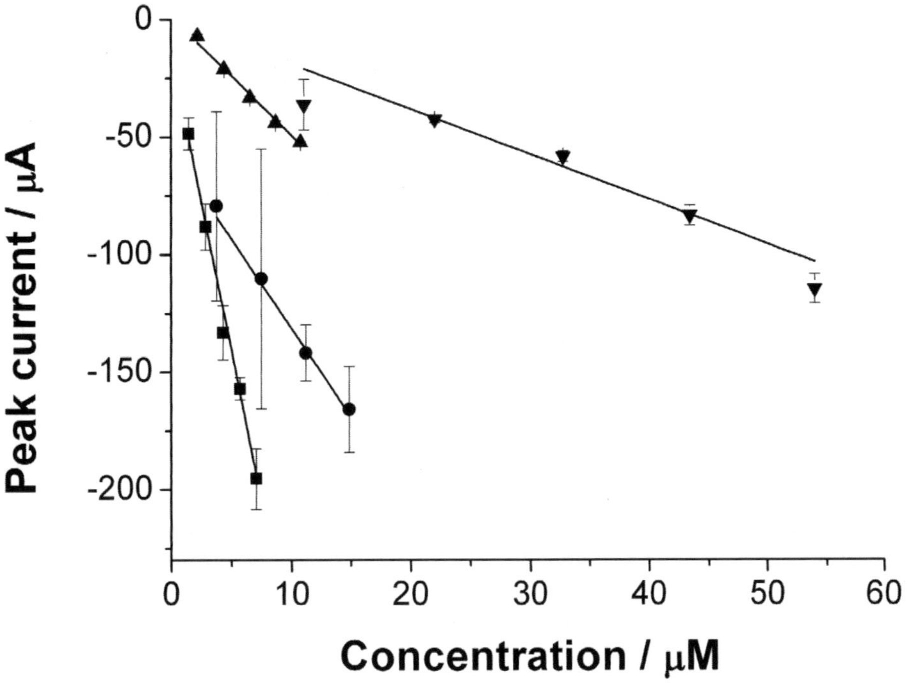

Figure 3: Calibration curves for G2 (▼), G3 (▲), G4 (●) and G5 (■) polyLysine dendrimers.

A limit of detection for the dendrimers G2 to G5 was determined experimentally. As the dendrimer generation increases, the limit of detection decreases to reach a minimum around 1 µM (Figure 4).

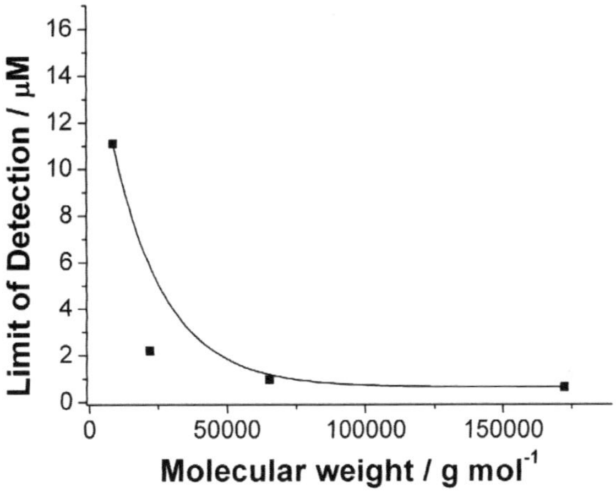

Figure 4: Limit of detection determined experimentally for polyLysine dendrimers G2 to G5 as a function of the dendrimer molecular weight.

The improved limit of detection for the dendrimers of higher generation is consistent with the increase of sensitivity with the dendrimer generation.

4. Conclusions

We report here the detection of polyLysine dendrimers via electrochemical detection at the liquid | liquid interface, A linear relationship is observed between the current measured and the dendrimer concentration. Structural characteristics of polyLysine dendrimers have a strong impact on the analytical performances at the liquid | liquid interface. Sensitivity increases with the charge of the dendrimers but not as fast as for proteins. Dendrimers cannot be used as a model macromolecules to predict the sensitivity and the limit of detection that could be achieved for a protein based on its overall charge. This is due to the tertiary structure of the protein which favours interactions with organic electrolyte anions.

References

[1] H.H. Girault, Electrochemistry at liquid-liquid interfaces, in: A.J. Bard, C.G. Zoski (Eds.) Electroanalytical Chemistry: A Series of Advances, vol. 23, CRC Press, Boca Raton, 2010, pp. 1-104.
[2] G. Herzog, V. Kam, D.W.M. Arrigan, Electrochim. Acta, 53 (2008) 7204-7209.
[3] M.D. Scanlon, D.W.M. Arrigan, Electroanalysis, 23 (2011) 1023-1028.
[4] X.J. Huang, A.M. O'Mahony, R.G. Compton, Small, 5 (2009) 776-788.
[5] G. Herzog, D.W.M. Arrigan, Analyst 132 (2007) 615-632.
[6] F. Kivlehan, Y.H. Lanyon, D.W.M. Arrigan, Langmuir, 24 (2008) 9876-9882.
[7] R.A. Hartvig, M.A. Mendez, M. van de Weert, L. Jorgensen, J. Ostergaard, H.H. Girault, H. Jensen, Anal. Chem., 82 (2010) 7699-7705.
[8] M.D. Scanlon, E. Jennings, D.W.M. Arrigan, Phys. Chem. Chem. Phys., 11 (2009) 2272-2280.
[9] G. Herzog, P. Eichelmann-Daly, D.W.M. Arrigan, Electrochemistry Communications, 12 (2010) 335-337.
[10] G. Herzog, A. Roger, D. Sheehan, D.W.M. Arrigan, Analytical Chemistry, 82 (2010) 258-264.
[11] A. Berduque, M.D. Scanlon, C.J. Collins, D.W.M. Arrigan, Langmuir, 23 (2007) 7356-7364.

Sensors & their Applications XVI IOP Publishing
Journal of Physics: Conference Series **307** (2011) 012056 doi:10.1088/1742-6596/307/1/012056

Non-contact biopotential sensor for remote human detection

A E Mahdi[1] **and L Faggion**[2]

[1]Department of Electronic and Computer Engineering, University of Limerick, Limerick, Ireland.
[2]Joint Research Centre of the European Commission, Institute for the Protection and Safety of the Citizen, Ispra (VA), Italy.

Email: hussain.mahdi@ul.ie; lorenzo.faggion@jrc.ec.europa.eu

Abstract. This paper describes a new low-cost, low-noise displacement current sensor developed for non-contact measurements of human biopotentials and well suited for detection of human presence applications. The sensor employs a simple, improvised transimpedance amplifier that eliminates the need for ultra high values resistors normally needed in current amplifiers required for this type of measurements. The sensor provides an operational bandwidth of 0.5 – 250 Hz, and a noise level of 7.8μV/√Hz at 1 Hz down to 30nV/√Hz at 1 kHz. Reported experimental results demonstrate the sensor's capability in measuring heart related biopotentials within 0.5m off-body distance, and muscle related biopotentials within 10m no obstacles off-body distance, and 5m off-body distance with a concrete wall in between.

1. Introduction

The field of human presence and body remote sensing though walls, rubble or similar obstacles, has received a great deal of interest during the last few years from defence, law and humanitarian agencies around the world. Over the last decade, numerous systems and technologies for remote human detection have been made available. Currently, commonly used technologies include surface penetrating radars/radar vision, radio waves transmitters/receivers, and carbon dioxide and other human waste characteristics based detectors [1-5]. Despite being effective, most available systems are relatively expensive, require prior installation or not easy to move around. They also suffer from a number of practical problems that cause false reading making them prone to evasion [1, 2].

An ideal human presence detection system for above applications should be: a) non-invasive/safe; b) capable of non-contact sensing through solid walls and similar conditions; c) operates on sensing human-related phenomena that are difficult to control by the subjects to be detected; and d) portable or easy to install/operate. One potential means for meeting these requirements is via sensing human biopotentials associated with the activities of many human organs, such as the heart, the brain and the muscles. For example, the heart produces a signal called an electrocardiogram (ECG), the brain produces a signal called an electroencephalogram (EEG), and the activity of muscles, such as contraction and relaxation, produces an electromyogram (EMG) [6]. Owing to their amplitudes and bandwidths, the ECG and the EMG (typical amplitudes/bandwidth of ECG: 0.1-5mV/0.5-100Hz; EMG: 1-10mV/20-500Hz) are relatively easier to measure compared to other biopotentials and, hence, can potentially provide an important means for detection of human presence. However, recording of these signals still relies primarily on galvanic contact of electrode sensors with the skin using Ag/AgCl electrodes in combination with electrolytes [6]. To eliminate the need for direct skin contact, considerable amount of work has been devoted, for example, to using SQIUD magnetometer based systems, which offer superior sensitivity for sensing biopotentials in a non-contact mode up to few centimetres off the body [7]. However, sensor cooling requirements at cryogenic temperatures and associated high set up costs limit the usability of SQUID based sensors. As an alternative to galvanic contact electrodes, capacitive electrodes that do not require direct contact with the body have been demonstrated to record biopotentials, often through several layers of clothes [8, 9]. Advances in modern microelectronics and fabrication techniques have revived interest in this type of capacitive sensing with successful efforts resulting in improving sensor performance and manufacturability. Prance and co-workers [10], for example, developed feedback enhanced and stabilised electrometer based amplifiers that operates on displacement current, have ultra high input impedance and yield very-low noise floor at the operating frequencies of most biopotentials. Further advances in non-contact ECG sensors were reported by Matthews et al [11].

Displacement current is a phenomenon analogous to an ordinary electric current, posited by J.C. Maxwell to explain magnetic fields produced by electric fields around a capacitor. Referring to figure1, capacitive biopotential sensors effectively rely on detecting the displacement current, I_D, that is proportional to the rate of change of the electric field associated with the ECG/EMG signal, V_S. This is effectively achieved by coupling the sensor's amplifier to V_S through a capacitance, C_S, formed by the sensor's metal electrode and the body surface, which typically corresponds to 0.1-10pF. For the low frequency measurements associated with ECG, for example, this weak coupling crucially requires the sensor's input impedance to be in excess of 10^{12} Ω since any finite input resistance would attenuate V_S. In most recently reported capacitive sensors, these and the high gain requirements have been optimally met by using an ultra high input bias resistor, to effectively dump the displacement current, I_D. For example, in the case of the electrical potential probes demonstrated in [10], this as well as improved noise performance has been achieved through the use of an input-bias-stabilisation network employing an ultra high resistor of glass-encapsulated carbon-film type. However, the addition of such high value resistors significantly increases the time constant of the amplifier resulting in a very slow response. It also introduces an extra noise source, due to thermal effect, degrading the sensor noise performance. Plus, resistors with resistances in the range of 10^{12} Ω with small tolerances, high stability and low thermal noise are very expensive and not readily available.

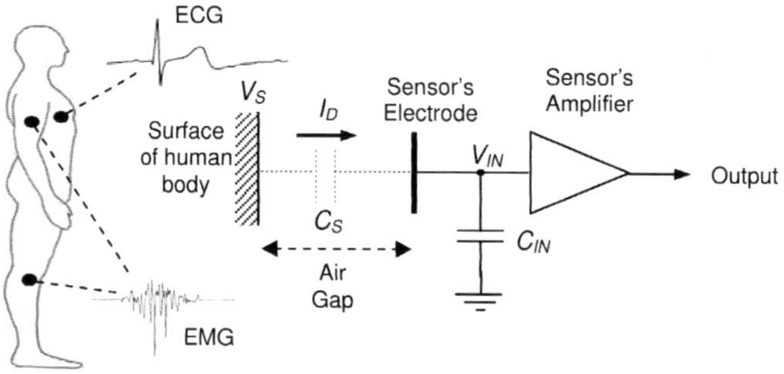

Figure 1. Illustration of the operational concept of a capacitive biopotential sensor.

In this paper we present a new low-cost, low-noise, high sensitivity non-contact sensor for remote detection of human presence via sensing body biopotentials, such as the ECG and EMG. The sensor uses a simple, inexpensive transimpedance amplifier which employs a resistive T-network in its feedback path to achieve high current-to-voltage sensitivity [12]. It operates by feeding the displacement current, I_D, directly into the summing point of the transimpedance amplifier, eliminating the need for an ultra high input bias resistor.

2. System design, construction and characterisation

An outline of the new displacement current sensor is shown in figure 2. The system consists primarily of an electrode forming the sensor head or antenna, an amplification stage, and a filtering stage. Our prime target in this work is to develop a relatively low-cost, portable system for remote detection of human presence. As such, additional consideration was given to issues related to size, power consumption and type of components/materials to be used in developing the system. The electrode in our current prototype is basically an aluminium disc of 5cm in diameter and 0.5 mm in thickness. A 1cm thick lightly charged (statically) dielectric (polystyrene) layer is added to the front of the electrode, as shown, to enhance the sensitivity of the sensor. The amplification stage consists of a transimpedance amplifier followed by a standard voltage follower. This is followed by a bandpass filter formed by cascading an active 1st order low-pass filter, an active 50 Hz notch filter, and a simple RC high-pass filter, giving an effective bandwidth of 0.5Hz – 250Hz. The circuitry forming the amplifier and the filter has been built around four TL082 op-amps from National Semiconductors, and other readily available components. TL082 is a low-power, high input impedance ($10^{12}\Omega$) op-amp with extremely low input noise characteristics (0.01pA/√Hz, 16nV/√Hz). A 5cm×4cm multilayer PCB with an overall thickness of 1.6mm has been developed to mount the circuitry. The PCB also facilitates the circuit guard shown in figure 2. This, as well as a reduction of PCB parasitics, were achieved through a combination of an on-board ground ring encircling the circuitry, and a 4-layer power-ground sandwich layout [13]. No metal case or any other integrated/non-integrated shielding means have been employed in our present prototype of the system.

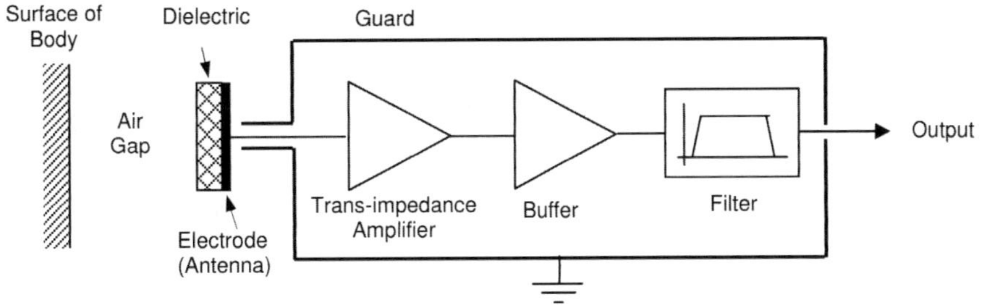

Figure 2. Block diagram of the new displacement current sensor.

Transimpedance amplifiers (TIAs), or current-to-voltage converters, are simple op-amp circuits that are well suited to applications where the current produced by the source is of importance [12, 14]. Figure 3(a) shows the basic configuration of a TIA. As the op-amp tends to maintain its inverting input terminal at ground potential, it forces the input current to flow through the feedback resistor, R_F. Thus, $I_{IN} = I_F$ and $V_O = -I_{IN} R_F$. This results in a current scaling, or signal gain $A_{SIG} = V_O / I_{IN} = R_F$ (V/A). TIAs are usually operated at very high gain and, hence, there is a possibility of closed loop instability. This problem can be eliminated by adding a small capacitor, C_F, in the feedback loop as shown in figure 3(a). In very small current measurements, as the case in our application, the high values required for R_F pose a problem because stable high value resistors are not freely available. In the developd system, this has been avoided by using a resistive T-network, as shown in figure 3(b), whereby the voltage divider formed by R_1 and R_2 increases the effective value of the feedback resistance, and hence A_{SIG} , to R_F $(1+R_2 / R_1)$ [12]. This allows high gain to be achieved using relatively small values resistors. In our current prototype, we achieved a sensitivity of ~10^8 V/A (equivalent to ~160 dB) by setting $R_F = 1\text{M}\Omega$, $R_1 = 1\text{k}\Omega$ and $R_2 = 100\text{k}\Omega$. Figure 3(c) shows the circuit diagram of the actual TIA used in our system, as well as that of the voltage follower and the filter.

The frequency response of the sensor, illustrating its sensitivity/gain, is shown in figure 4(a). Figure 4(b) shows the sensor noise spectral density for non-contact off-body application over a frequency range of 0.01Hz-1kHz. As can be seen, the sensor has a remarkable noise level of 7.8μV/√Hz at 1 Hz down to 30nV/√Hz at 1 kHz. This gain/noise performance is clearly of great importance for a variety of conditions where remote detection of most human biopotentials is required. In particular, it demonstrates very promising potential for our target application which is related to detection of hidden subjects behind walls, inside containers and under rubble.

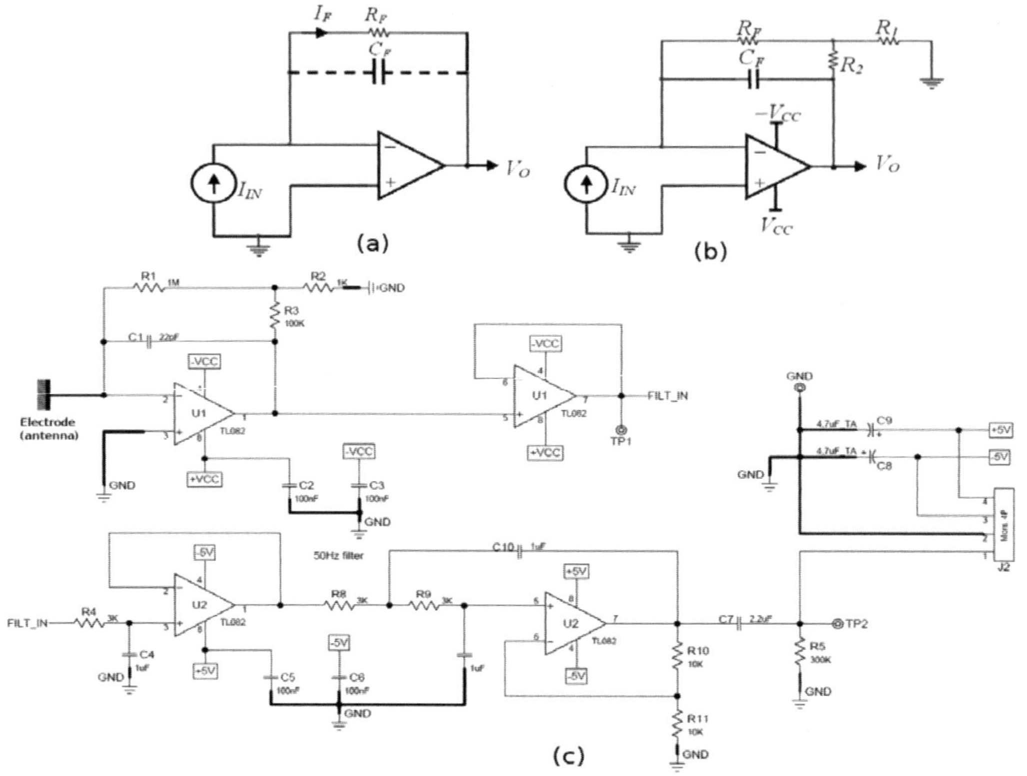

Figure 3. (a) Basic configuration of a TIA (b) A modified TIA circuit with T-network; and (c) circuit diagram of the developed sensor.

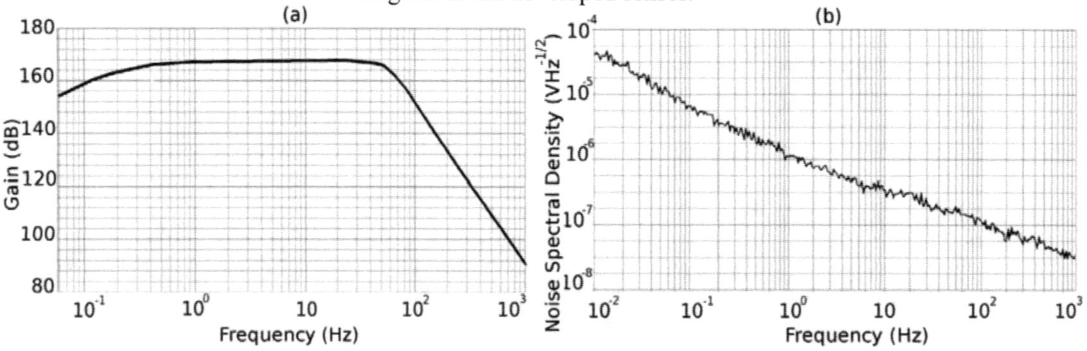

Figure 4. (a) Frequency response and (b) Noise spectral density of the sensor.

3. System performance and experimental results

To demonstrate the performance of developed sensor in remote off-body sensing of human biopotentials, we conducted two experimental tests both in a non-shielded environment. The first test is to demonstrate the capability of the sensor in detecting a human presence based on sensing the subject biopotential signal generated by current flow in the heart. The set-up for this experiment is shown in figure 5(a) which illustrates the relative positioning of the human subject and the sensor electrode. A single sensor is used to record body signals with electrode to body distance, A. It must be stressed here that no electrical connections are made to the body, with the subject sitting down in front of the sensor wearing normal clothing layers and in a normal non-shielded laboratory room. Figure 5(c-f) show the waveforms of measured signals recorded at the output of the sensor, when the electrode is at a distance A = 1, 10, 40 and 50 cm, respectively. We would like to stress that signals shown correspond to the raw sensor output and no additional processing was applied. For comparison, we also show in figure 5(b) the corresponding ECG detected using a standard 3-lead Ramesy ECG1C

Journal of Physics: Conference Series **307** (2011) 012056 doi:10.1088/1742-6596/307/1/012056

electrocardiogram monitor. It is clear that all the waveforms in figure 5(c-f) exhibit the shape of the distinctive periodic PQRST pattern of an ECG signal, displaying near perfect R and S peaks. The patterns displayed do not exactly mimic the conventional on-body ECG and has some time delay with respect to the ECG trace shown in figure 5(b). This feature has also been reported in the measurement of other similar sensors [10], and has been attributed to the multi-polar nature of the dynamic electric fields generated by the cardiac system.

The second experimental test is to demonstrate the sensor's capability in detecting human presence via sensing the subject movement/muscle related biopotentials. The set-up for this test is shown in figure 6(a), which illustrates the positioning of the sensor's electrode relative to a human subject walking/passing in front of the sensor. Again we would like to emphasis that there were no electrical connections to the body, the subject was wearing normal layers of clothing, and the measurements were recorded in a non-shielded environment. Figure 6(b) shows the waveform of measured raw signal at the output of the sensor when the normal distance between the subject and the vertical plane marking the position of the sensor's electrode, denoted *A* in figure 6(a), is 10 m. Figure 6(c) on the other hand shows the waveform of the sensor output when a concrete wall separates the subject from the sensor, with the distances marked *B* and *C* in figure 6(a) being 4 m and 1 m, respectively. For both cases, it is clear that the measured waveforms display the typical shape of an EMG signal normally associated with the activities of the leg and biceps muscles'.

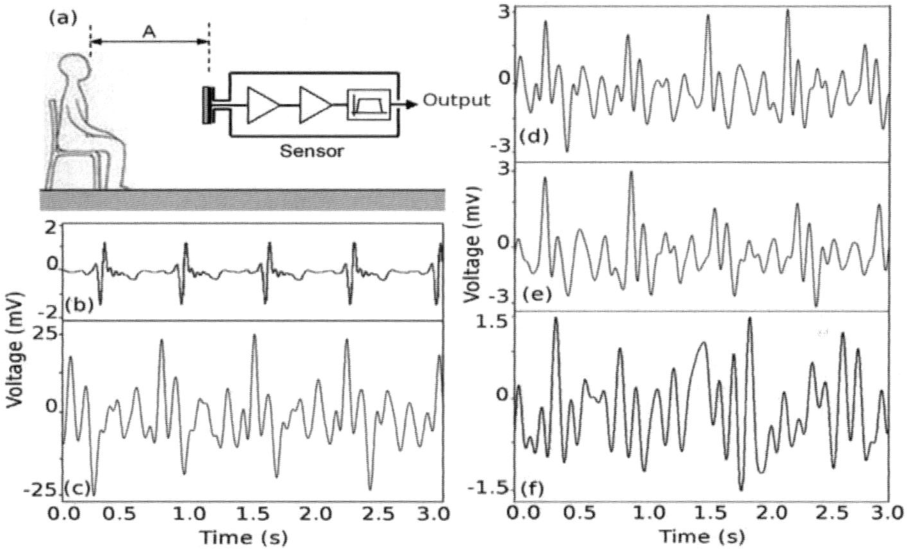

Figure 5. Set-up and measurement results for first experimental test (see text in Section 3).

4. Conclusions

We have described the design, construction and performance of a new low-cost displacement current based sensor for remote detection of human body/presence through sensing heart and muscles related biopotentials. The sensor has been built using readily available inexpensive components, and uses a simple but improvised transimpedance amplifier that employs a T-network utilising relatively low values resistors. Presented data showed that the system offer remarkable noise characteristics. In terms of capability in remotely sensing the biopotentials generated by the human heart and muscle activities, the performance of the sensor is very comparable to recently reported similar sensors which have been developed using modern microelectronics, fabrication and shielding techniques. Currently, our system is undergoing further developments with regards to optimising its noise and sensing performances, as well as integration into a portable scanning system suitable for remote detection of humans hidden behind walls, inside containers, or under rubble.

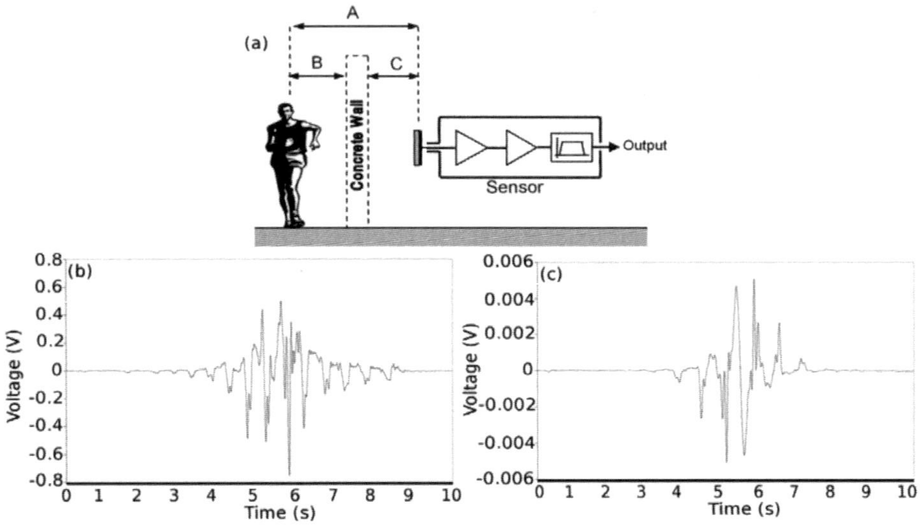

Figure 6. Set-up and measurement results for the second experimental test (see text in Section 3).

References

[1] Dogaru T, Nguyen L and Le C 2007 *Computer models of the human body signature for sensing through the wall radar applications* Public Report ARL-TR-4290, Army Research Laboratory (Adelphi, MD, September 2007).

[2] Brenda A K 2006 Selected systems for the detection of human stowaways in air cargo containers *Proc. IEEE 40th Int. Carnahan Conf. on Security Technology* (Lexington, KY, 16-19 Oct 2006) pp 26-29.

[3] Nishi M, Takahashi, S and Yoshida 2006, Indoor human detection systems using VHF-FM and UHF-TV broadcasting waves *Proc.IEEE Int. Symp. on Personal, Indoor and Mobile Radio Comm.* (Helsinki, Finland, 11-14 Dec. 2006) pp 1-5.

[4] Chi-Wei W and Zi-Yu H 2008 Using the phase change of a reflected microwave to detect a human subject behind a barrier *IEEE Trans. Biomed. Eng.* **55** pp 267-272.

[5] Brown D A 1995 Human occupancy detection *Proc. IEEE 29th Int. Carnahan Conf. on Security Technology* (Surrey, UK, 18-20 Oct 1995) pp166-174.

[6] Thakor NV 1999 *Biopotentials and electrophysiology measurement* in The Measurement, Instrumentation, and Sensors Handbook, ed J G Webster (Boca Raton, FL: CRC Press).

[7] Mahdi A E, Panina L and Mapps D 2003 Some new horizons in magnetic sensing: high-Tc SQUIDs, GMR and GMI materials *Sensors & Actuators A: Phys.* **105** pp 271-285.

[8] Lopez A and Richardson P C 1969 Capacitive electrocardiographic and bioelectric electrodes *IEEE Trans. Biomed. Eng.* **16** pp 299-300.

[9] Lim Y G, Kim K and Park K S 2006 ECG measurement on a chair without conductive contact *IEEE Trans. Biomed. Eng.* **53** pp 956–959.

[10] Prance R J, Beardsmore S T, Watson P, Harland C J and Prance H 2008 Remote detection of human electrophysiological signals using electric potential sensors *Appl. Phys. Lett.* **93** doi:10.1063/1.2964185.

[11] Matthews R, McDonald N J, Fridman I, Hervieux P and Nielsen T 2005 The invisible electrode – zero prep time, ultra low capacitive sensing *Proc.Int. Conf. on Human Computer Interaction* (Las Vegas, NV: 22-27 July 2005) pp 22–27.

[12] Clayton G and Winder S 2003 *Operational Amplifiers* 5th ed. (Oxford, Boston: Newnes).

[13] Semtech 2006 *RF design guidelines: PCB layout and circuit optimization* Appl. Note AN 1200.04 (Semtech: www.semtech.com).

[14] Microchip 2005 *Analog sensor conditioning circuits – an overview* Appl. Note AN990 (Microchip Technology Inc.: www.mirocchip.com).

Self Powered Non-Dispersive Infra-Red CO_2 Gas Sensor

Gas Sensing Solutions Ltd, 60 Grayshill Road, Westfield North Courtyard, Glasgow G68 9HQ

D R Gibson and C MacGregor

des@gassensing.co.uk

Abstract. This paper describes a non-dispersive infra-red CO_2 gas sensor, incorporating a mid-infra-red solid state light source/ detector combination, tuned to match the spectral absorption characteristic of CO_2 gas. Injection moulded optics provide low cost manufacture. Continuous operation power consumption is < 3.5mW and pulsed mode with energy per measurement < 6mJ. Self powered operation using a solar cell is demonstrated together with wireless capability. Performance of two path length variants (20mm and 70mm) is described. The sensor shows invariant temperature output characteristic from -25 to 50°C. Accuracy level is typically ±3% of reading.

1. Introduction

Information concerning carbon dioxide (CO_2) concentration levels is an essential element in air quality monitoring and control. CO_2 is an insidious gas, therefore changes in concentration are difficult for humans to recognize. The gas is safe in low concentrations (typically <1000ppm), however prolonged exposure at moderate levels (>5000ppm) can lead to a range of health related problems such as sick building syndrome [1,2] causing fatigue like symptoms.

Current and incoming legislation requires CO_2 gas monitoring within building environments for optimal control of air quality [3,4]. Moreover, such CO_2 gas monitoring is employed within demand control ventilation systems used for building heating ventilation and air conditioning as a means of reducing energy consumption in addition to control of indoor air quality control.

In industrial environments where process generated CO_2 dominates, for example in breweries, soft drinks, packaging industry, freezer storage etc, the maximum permitted CO_2 concentration according to most standards can be as high as 5,000 ppm during an 8-hour working period. CO2 gas monitoring in this case is required to ensure compliance with industrial health and safety requirements. Such levels are normally not found in home or office environments where people are the main source of carbon dioxide.

Current and emerging building indoor air quality and industrial safety markets are require real-time continuous monitoring of CO2 concentration levels, creating high volume demand for low cost, low power consumption CO2 gas sensors, driven in many instances by legislative requirements. Increasing requirements for wireless networked CO2 sensors is placing emphasis on low power consumption, at levels enabling battery operation and ultimately power scavenging/ self powering technologies.

Currently there are two types of gas sensors commonly available for monitoring CO_2 concentrations in air, i.e., non-dispersive infrared (NDIR) and solid electrolyte sensor types. NDIR sensors have performance advantages in terms of long-term stability, accuracy, and power consumption for CO_2 measurement. Hence, NDIR sensors are the most widely used for the real-time measurement of carbon dioxide. Solid electrolyte types have cost advantages, but the product performance is not good enough to allow their use in any but the least demanding applications.

The NDIR method is an optical method of detecting gases. This relies on the fact that many gases absorb specific wavelengths of infra-red light. It is possible to calculate the gas concentration, by passing light through a defined length and measuring how much light is absorbed at the specific wavelength absorbed by the gas. For CO_2, the commonly used wavelength is 4.26um, a wavelength which is strongly absorbed by CO_2, but not absorbed at all by other commonly found gases or by water vapour.

Current NDIR CO_2 gas sensor utilise an incandescent bulb and pyroelectric light source/ detector combination, with consequent high power consumption – typically 50 to 200mW. This necessitates hard wired installation or short battery life, which is costly and at odds with legislatively driven requirements for multiple wireless capable sensor install in buildings, transport systems and also the need for portable sensors for use in safety and industrial applications.

Moreover, incandescent light sources have additional restrictions for use in applications requiring low power consumption, such as extended stabilisation time (typically > 10 minutes) and limited ability to pulse (typically < 4Hz) for short duration measurement. In addition, output wavelength drift with bulb lifetime necessitates use of complex referencing optics and optical filtering.

In this paper a low power consumption NDIR CO_2 gas sensor based on mid-infrared solid state light emitting diode (LED) and photodiode (PD) is described. LED`s offer the advantage of high source emittance, fast modulation rates, room temperature operation with no burn-in and are more cost effective than laser diodes for deployment in practical gas sensors [5].

Specific performance benefits are described including low power consumption (<3.5mW), rapid stabilisation time (<2s) and ability to pulse for short duration measurements. Moreover, intrinsic wavelength stability of the solid state LED/ PD combination allows use of simplified and low cost injection moulded reflective optics with no requirement for active alignment for sensor assembly.

2. Sensor Configuration

The NDIR sensor described in this paper comprises a narrow bandgap III-V LED light source and photodiode detector. Mid infrared radiation is launched into an optical structure which defines lightpath and chamber into which gas diffuses, causing a reduction in light transmission at the sensing wavelength. Electronics and intrinsic firmware control LED/ PD drive current/voltage, pulsing and signal processing. Output parameter is CO_2 gas concentration, stabilised over the sensor operating temperature range. Sensitivity is determined by optical pathlength.

Sensor constituent parts and assembly are described as follows.

2.1. LED/ PD Combination

Photonics technology has been used [6], producing LED sources and PD detectors operating at mid-infrared wavelengths of interest in gas detection. The technology is based on a pentanary AlGaInAsSb narrow bandgap III-V material combination [7,8].

Spectral characteristics of both LED and PD can be tuned by altering the ratio of the III-V material stoichiometry [8].

Figure 1a illustrates LED and PD configuration. Figure 1b shows actual LED and PD mounted on an electronic bridgeboard ready for assembly onto the optical and electronic chassis.

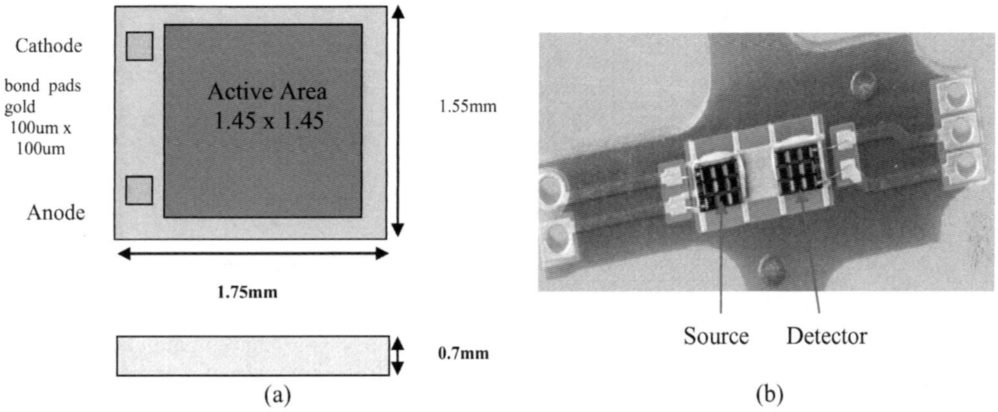

Figure 1 (a) LED and PD detector structure and (b) LED and PD detector mounted on bridgeboard

This technology produces LED and PD`s which are tuned to a relatively narrow bandwidth. Used as a pair the bandwidth of the combined system is narrower still. This narrow bandwidth allows the LED and PD to be used without additional optical filtering, further reducing cost and simplifying the optical design.

LED and PD detector operating voltages are reduced to 2.5V by enhanced impedance matching of the LEDs and PD through segmentation of the LED and PD active areas [9]. Operating voltage at 2.5V is well suited to interface with 5V and 3.3V electronics.

2.2. Performance Modelling, Optics and Signal Processing

Detailed optical modelling of sensor sensitivity and noise has been carried out to identify necessary optical pathlength for CO_2 gas sensor, covering required concentration ranges, accuracies and data processing.

Figure 2 illustrates modelled and measured noise (expressed as percentage of actual CO_2 concentration) for two pathlengths (20 and 70mm). Modelling indicates no one pathlength can provide high accuracy (<±5% of reading) over full range from 200ppm to 100% concentration. The model utilises complete absorption spectrum for CO_2 .

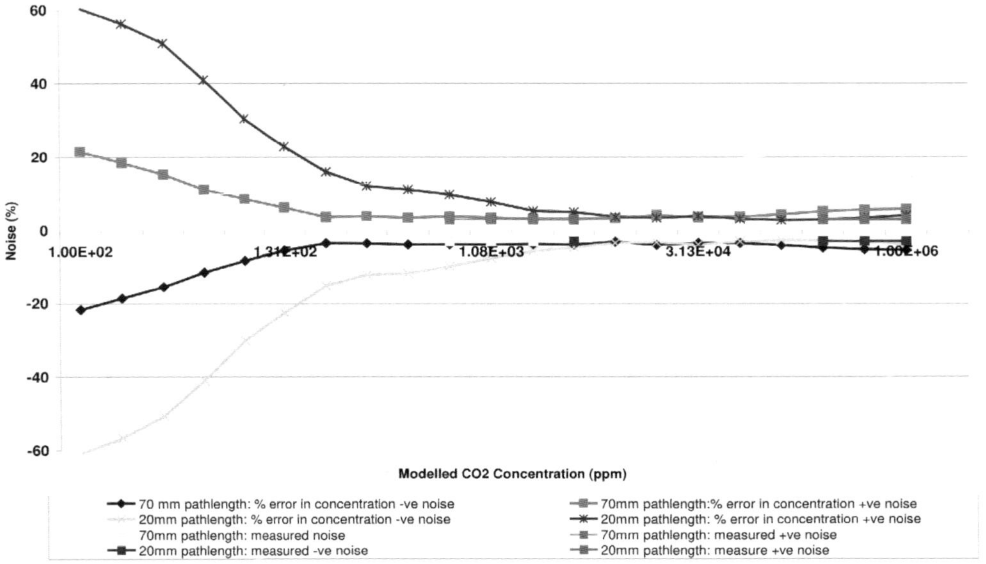

Figure 2 Modelled sensor noise (expressed as a percentage of actual CO2 concentration level) for two pathlengths: 20 and 70mm. Measured noise for the two sensor pathlengths is shown

A minimum of two pathlengths is required to provide accuracy <±5% of reading. 20mm pathlength is suited to concentration levels from 5 to 100% whilst 70mm pathlength provides complementary required accuracy from 200ppm to 5%. Figure 3 shows the optical/electronic assemblies for the 20 and 70mm pathlength sensors, both incorporating low cost, injection moulded reflective optics.

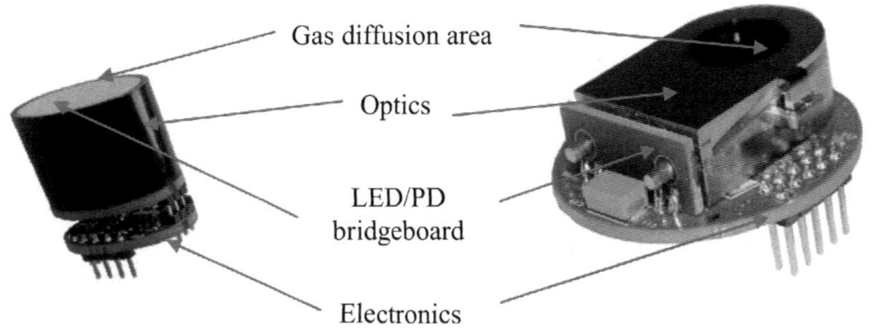

(a) Wide range Pathlength 20mm

(b) Ambient Pathlength 70mm

Figure 3 Assembled (a) wide range (actual size 20mm diameter, 25mm length including electronics) and (b) ambient sensors (note ambient is shown without protective cover. Size with cover 42mm diameter, 15mm thickness including electronics)

Assembly requires no active alignment of LED/PD bridgeboard and optics.

Both sensor configurations can be calibrated from zero to customer specified maximum range. However, wide range is normally supplied in four standard ranges: 0 to 5%, 0 to 20%, 0 to 65% and 0 to 100% whilst ambient is supplied in three standard ranges 0 to 5000ppm, 0 to 1% and 0 to 2%.

Figure 4 shows the signal processing scheme. The Micro Controller Unit (MCU) periodically sends a very fast pulse train to the Infra Red LED (IR LED) via a GAIN stage. The modulated IR light pulses fall on the Photo Diode detector.

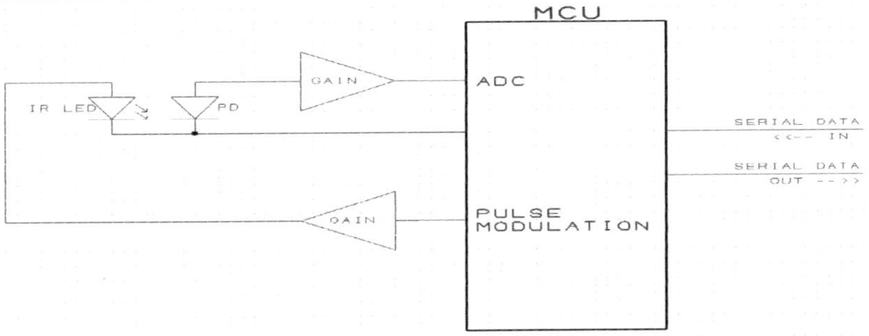

Figure 4 Signal processing configuration

IR light level is detected by the PD and is amplified by a GAIN stage before being captured by the MCU's Analogue to Digital Convertor (ADC) thus making a digital version available for processing. The digital version of the IR light level is processed by the MCU which then provides a serial data output indicating the CO_2 concentration present between the mid-infrared LED and the PD.

3. Performance

Performance of the two pathlength configurations is provided in Table 1.

Parameter	Wide range 20mm pathlength	Ambient 70mm pathlength
Power (continuous)	<3.5mW	
Energy per measurement (pulsed)	6mJ	
Power up stabilisation time	<2s	
Peak current	33mA	
Average current	1.1mA	
Input voltage	3.3V	
Accuracy	±3% (of reading)	
Standard ranges	0 to 5, 20, 65 and 100%	0 to 2000ppm, 1 and 2%
T_{90}	4 secs	30secs
Operating temperature range (°C)	Standard: 0 to 50 °C; extended -25 °C to +50 °C	
Storage temperature (°C)	-30 to +70 °C	
Relative humidity	0 to 95% RH non-condensing	
Dimensions	20mm x 25mm	43mm x 15mm

Table 1 Performance parameters for wide range (20mm) and ambient (70mm) sensors

Figure 5(a) shows temperature invariant performance over six temperatures (-25, 0, 15, 25, 35,55°C):

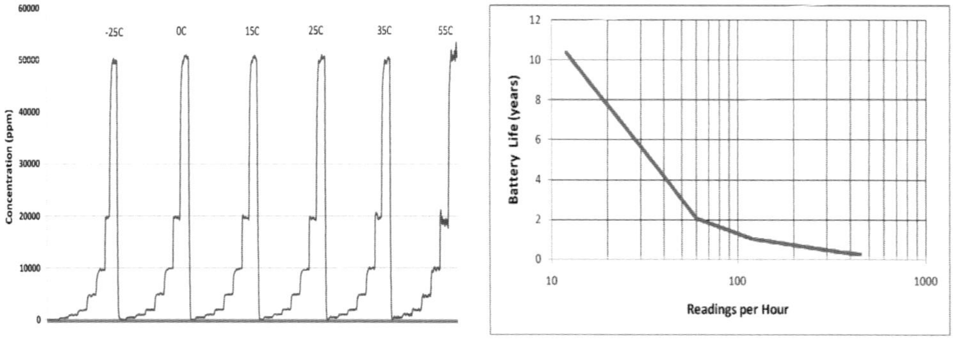

(a) (b)

Figure 5(a) Temperature stability at 500ppm, 1000ppm, 2000ppm, 5000ppm, 1%,2%,5% ove six temperatures and (b) battery life (three AAA batteries) as a function of measurements per hour

Figure 5(b) shows battery life (three AAA batteries) as a function of number of readings per hour, indicating battery life is between five to ten years for typical sampling rates used in building control and industrial safety applications.

Self powered operation with wireless data transmission has also been demonstrated with a 67 x 28mm silicon solar cell with 200lux light level [10]. Self powered operation enables sensor to operate at one reading every 10 minutes.

4. Summary and Conclusions

A novel NDIR CO_2 sensor is demonstrated with less than 3.5mW power consumption in continuous operation and 6mJ/ measurement in pulsed mode. Energy consumption levels in pulsed mode enable self powered operation and wireless data communication using a solar cell.

References

[1] Gupta, S. et al, Building and Environment. 2007, 42, 2797–2809
[2] Jennings, P.R. et al, JAAPA. 2000, 34-6, 39
[3] Ferng, S.F. et al; Journal of Environmental Health. 2002, 65, 14-18
[4] Daisey, J.M. et al; Indoor Air. 2003, 13, 53-64
[5] Smith,S.D.et al, Phil.Trans. R. Soc. Lond. A (2001) **359**, p.621- 634
[6] Krier, A et al, Phys.Stat.Sol.(a) 205, No.1, 129-143 (2008)/ DOI 10.1002/pssa.200776833
[7] Gao,H.H. et al, J.Phys.D, Appl.Phys.**32**, 1768-1772 (1999)
[8] A.Krier et al, J.Phys.D, Appl.Phys.**33**, 101-106 (2000)
[9] Koichiro, U.et al,Phys. stat. sol. (c) 5, No. 9, 2819 – 2821 (2008) / DOI 10 10.1002/pssc.200779264
[10] EnOcean GmbH, Application Note AN313, CO_2 SENSOR - Design concept for solar-powered CO_2 sensor

In-situ monitoring of ammonia gas using an optical fibre based approach

G. Dooly, H. Manap, S. O'Keeffe, E. Lewis

Optical Fibre Sensors Research Centre, Department of Electronic and Computer Engineering, University of Limerick, Ireland.

gerard.dooly@ul.ie

Abstract. An optical fibre sensor for the monitoring of low level atmospheric ammonia concentrations is presented. The measuring technique employed is based on a differential optical absorption approach, rather than a semiconductor based technique which is generally exploited within comparable commercially available products. The sensor described herein demonstrates vast improvements in terms of sensitivity, selectivity and lifespan over ammonia sensors currently available commercially. Extensive laboratory-based experimental tests demonstrate the sensor's ability to monitor concentrations as low as 1ppm without any notable cross-sensitivity issues to atmospheric gases such as nitrogen, oxygen and carbon dioxide. Furthermore, in-situ experimental tests within an agricultural cattle enclosure demonstrate sensor's suitability to environments where low concentration monitoring of ammonia over extended periods of time is necessary.

1. Motivation

In 2008, the environmental protection agency in Ireland estimated that over 91% of all ammonia emissions in Ireland came from the agricultural farming sector as a result of animal manures. In fact it is estimated that 16% of all nitrogen contained within animal waste is released as ammonia gas into the atmosphere [1]. This amounted to 95.3 kilotonnes of ammonia gas from agricultural farming in Ireland being released into the atmosphere in 2008 [1].

After these emissions enter the earth's atmosphere they can lead to many adverse affects including acidification and ground-level ozone. As a result, increasingly stringent European legislation such as the National Emissions Ceilings (NEC) Directive has been established in an effort to reduce the amount of ammonia emissions worldwide [2]. This has led to the demand for a sensitive sensor which for the monitoring of ammonia gas over the last number of years.

In addition, the concentration level of ammonia in an enclosed agricultural environment such as an agricultural cattle enclosure can be of high importance within an agricultural setting. Exposure to excessive levels of ammonia gas within an agricultural environment can lead to a wide range of health issues, both for livestock and agricultural personnel alike. This has also increased the demands for the development of an ammonia sensor. However a sensor capable of operating within the environment selectively and accurately with a relatively long lifespan is of preference.

Journal of Physics: Conference Series **307** (2011) 012058 doi:10.1088/1742-6596/307/1/012058

2. Sensing Technique

The earth's atmosphere comprises primarily of nitrogen, oxygen and carbon dioxide gases. It also contains around 1% of water vapor. This wide composition makes it difficult for sensors operating in such an atmosphere due to inherently challenging cross-sensitivity effects. Commercially available sensors capable of monitoring ammonia are often based on semi-conducting films [3] for detection. Sensors based on this technique can provide a relatively high level of sensitivity; however they can suffer from cross-sensitivity effects and can have a relatively low lifespan. Ammonia sensors have also been reported based on optical absorption within the Infrared region [4]. However due to relatively high absorption strengths for water vapor in this region, these sensors can exhibit cross-sensitivity to humidity.

Recently developed optical devices operating within the ultraviolet region of the spectrum has allowed for the development of low cost optical based sensors within this region. This has allowed for a low cost optical approach based on absorption in the ultraviolet region. This approach can provide an appropriate measurement solution to this issue with minimal cross-sensitivity effects. This technique, often referred to as differential optical absorption spectroscopy, allows for a significantly high selectivity response through the monitoring of absorption lines associated with ammonia gas molecules.

Sensors based on direct absorption spectroscopy utilize the fact that all gas specie absorbs optical light at different characteristic wavelengths. The absorption spectrum for ammonia gas demonstrates significant absorption coefficients within the ultraviolet region at wavelengths between 171nm and 216nm (reported previously by Chen et al) [5]. Through the analysis of similarly reported absorption coefficients for nitrogen, oxygen, carbon dioxide and water vapor (shown in figure 1), it was determined that cross-sensitivity effects within this region were negligible. Experimental tests confirming this relatively high rate of selectivity has been reported in previous published papers [6-7].

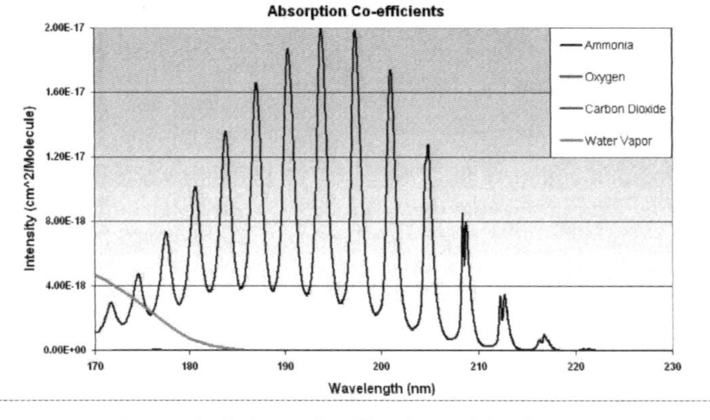

Figure 1. Theoretical absorption line intensities for NH_3, O_2, CO_2 and N_2.

An extrinsic sensor based on an open path configuration is the basis for the sensor setup. The setup consists of a deuterium–halogen lamp (DH-2000 from Ocean Optics) for optical light emission and a high resolution spectrometer (Ocean Optics HR2000) for optical detection. Collimating lenses are located at either end of the gas cell to aid with the propagation of optical light across the open space, where it interacts and absorbs parts of the optical light.

Optical fibres for transmission to the sensing region allow for remote monitoring and online measurements, giving the possibility for data analysis on optical light recordings in real-time. The experimental setup for the sensor is shown in figure 2.

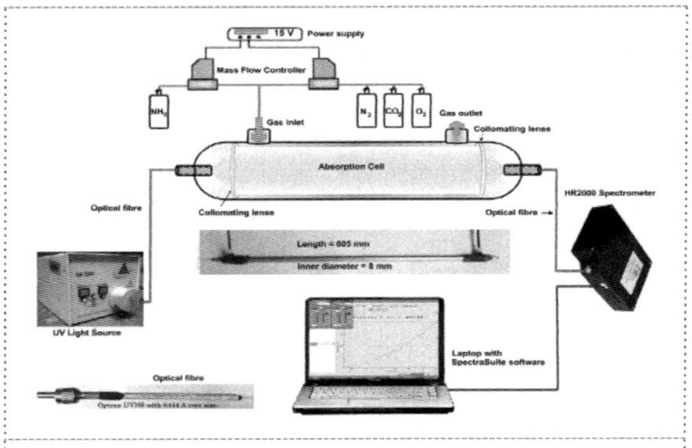

Figure 2. Experimental setup.

A variation of the Beer-Lambert Law, given by equation 1, is used to relate the loss of optical light in the UV region and the concentration of ammonia gas in the gas cell. Where, I: Transmitted intensity; Io: Incident intensity; σ: Absorption cross section for ammonia (cm2/Molecule); NA: Avogadro's constant; L: Path length (cm); T: Temperature (K); P: Pressure (atm).

$$ppm = \frac{-[\ln \frac{I}{I_o}][22.4]}{\sigma \times N_A \times l \times 10^{-9}} \times \frac{T}{273} \times \frac{1}{P} \quad (1)$$

The spectrally resolved data from the spectrometer is interpreted in accordance with the equation 1 using a software based platform and development environment ($\text{LabVIEW}_{\text{TM}}$). The processed data is then displayed to the user in real-time.

3. Laboratory tests

Laboratory based experimental tests were carried out in order to comprehensively quantify the capability of the sensor. A series of mass flow controllers was used to provide a varying concentration of each of the gases during the testing procedure. In addition, a widely used commercially available sensor from Crowcon [4] was used for comparison throughout the testing procedure. Concentration measurement using a commercial sensor is carried out manually. The readings from LCD display from the commercial sensor are noted down for every 60s and the data compared with the optical sensor system, as shown in figure 3.

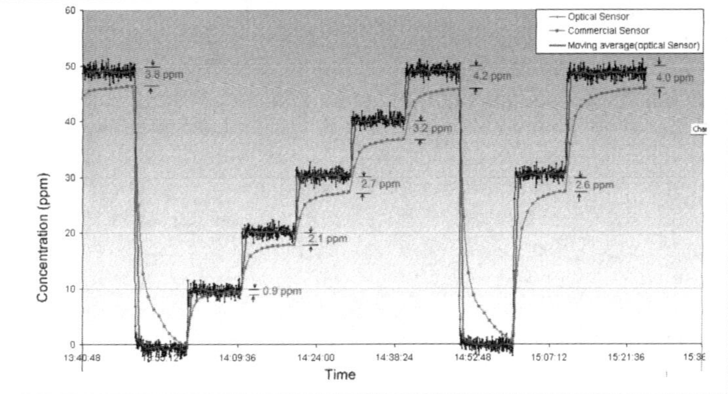

Figure 3. Recorded concentration levels for varying levels of NH_3.

Journal of Physics: Conference Series **307** (2011) 012058 doi:10.1088/1742-6596/307/1/012058

Laboratory based experimental test results demonstrate the sensor's ability to provide accurate measurements compared to the commercial sensor. Calculated ammonia concentrations from the mass flow controllers and the optical sensor are in close agreement with each other, with an averaged error of 1.22% over the whole of the experimental test being recorded. However, the commercial sensor only demonstrates an accurate concentration measurement at 10ppm, above this the error rate increases significantly. In addition it is clear that the response of the optical sensor is significantly better than that of the commercial sensor. Response and recovery times throughout the test were recorded at an average of less than 4s.

4. In-situ tests

In-situ experimental tests were required to assess and fully quantify the performance of the optical sensor within the required environment. These experimental tests were carried out in an agricultural cattle enclosure in the Republic of Ireland, with 22 cattle accommodated across four fenced sections. The experiments were carried out during the peak winter season where ammonia is believed to be existing at a very low concentration [8]. The weather was foggy with some snow and the temperature was recorded at 0°C.

The optical sensor was placed at one corner of the enclosure, approximately 3m from the fenced section where a group of cattle was located. A cylinder of nitrogen gas was used in order to flush the absorption cell prior to the experimental test, while a vacuum pump was used to pump the surrounding air into the cell during the testing procedure. A photograph of the experimental test is shown in figure 4 and the experimental test results are shown in figure 5.

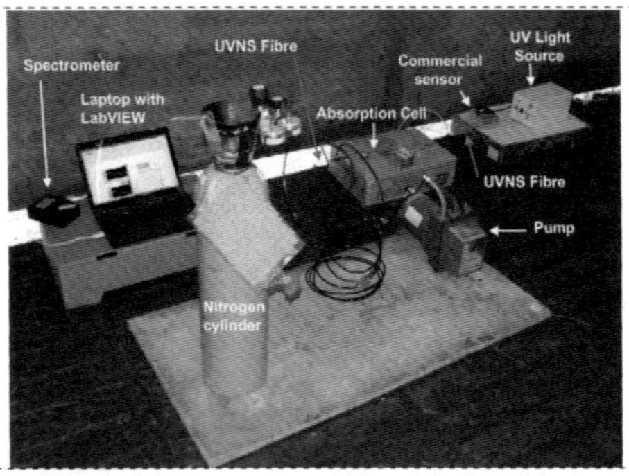

Figure 4. Photograph of experiment setup for in-situ ammonia monitoring in an agricultural cattle enclosure.

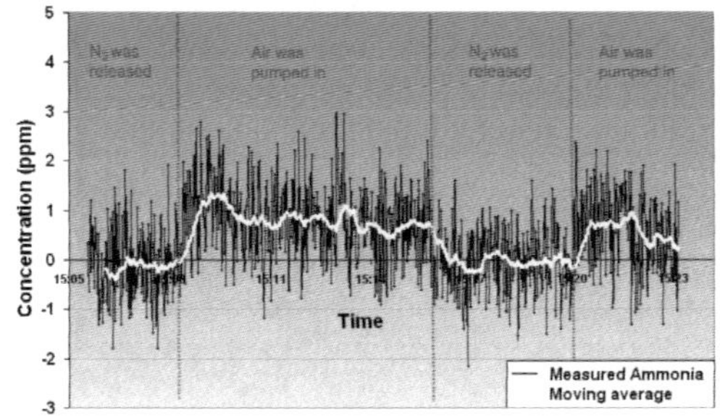

Figure 5. Experimental results from in-situ experimental tests.

To begin the in-situ experimental test, the nitrogen gas was turned off and pump was initiated. Although concentration levels recorded by the optical sensor are significantly low, approximately 1ppm, the sensor is able to accurately determine the presence of ammonia in the surrounding atmosphere. Following a period of approximately 10 minutes, the pump was turned off and the absorption cell is flushed with nitrogen. Once this occurs, the concentration readings from the optical sensor quickly return to zero. Upon the initiation of the pump (and deactivation of nitrogen) for the second time, the sensor once again records a concentration of approximately 1ppm, demonstrating a strong level of repeatability. Although the commercial sensor was operated throughout the testing procedure, it did not record any concentration levels of ammonia, indicating that it was incapable of measuring such low concentrations.

5. Conclusions

A novel optical fibre sensor for the detection of low concentrations of ammonia gas has been described. To achieve a high level of selectivity a technique based on differential optical absorption spectroscopy was developed and relatively low cost components were employed. Experimental tests were carried out using varying concentrations of ammonia as well as other atmospheric gases. Further tests were carried out in-situ within an agricultural cattle enclosure in the republic of Ireland in conjunction with a commercially available ammonia sensor.

Experimental tests were found to be very encouraging, with the lower limit of detection recorded to be at least 1 ppm, response times found to be under 4s and no recordable cross-sensitivity issue noted. These results show a marked improvement over commercial available ammonia sensors in terms of sensitivity, selectivity and response times.

6. Acknowledgements

The authors would like to acknowledge resources made available through the Higher Education Authority, PRTLI Cycle 4 for Environment and Climate Change. Also the authors would like to acknowledge the support of the University of Malaysia, Pahang (UMP) and the Ministry of Higher Education, Malaysia in providing scholarship for research studies.

References

[1] "Trends in NH3 emissions", Environmental Protection Agency, Ireland, http://www.epa.ie
[2] Directive on national emission ceilings for certain atmospheric pollutants, European Parliament and Council Directive 2001/81/EC of 23rd October 2001.More references
[3] A. Galdikas, A. Mironas, V. Strazdiene, A. Setkus, I. Ancutiene and V. Janickis, "Room-temperature-functioning ammonia sensor based on solid-state Cu_xS films" Sensors and Actuators B: Chemical, Volume 67, August 2000, Pp. 76-83
[4] "Tetra: A personal monitor you can trust", Crowcon Detection Instruments, http://www.crowcon.com/
[5] F.Z. Chen, D.L. Judge, C.Y.R. Wu, J. Caldwell, "Low and room temperature photoabsorption cross sections of NH3 in the UV region". Planet. Space Sci. , 1999. 47: p. 261-266.
[6] Manap H., Dooly G., O'Keeffe S., Lewis E., "Cross-sensitivity evaluation for ammonia sensing using absorption spectroscopy in the UV region". Sensors and Actuators B: Chemical. In Press, Corrected Proof.
[7] S O'Keeffe, H Manap, G Dooly, E Lewis, "Real-time Monitoring of Agricultural Ammonia Emissions Based on Optical Fibre Sensing Technology", IEEE Sensors Conference, Waikoloa, Hawaii, November 2010.
[8] N Yamamoto, H Nishiura, T Honjo, Y Ishikawa, K Suzuki, "A long-term study of atmospheric ammonia and particulate ammonium concentrations in Yokohama, Japan", Journal of Atmospheric Environment, 1995. 29(1): p. 97-103.

Integration of amperometric sensors for microchip capillary electrophoresis application

F Dicorato[1], E Moore[1,2], J Glennon[2]
[1] Life Sciences Interface Group,
Tyndall National Institute, Lee Maltings, Dyke Parade, Cork, Ireland
[2] Chemistry Department,
University College Cork, College Road, Cork, Ireland

E-mail: eric.moore@tyndall.ie

Abstract. Capillary electrophoresis is a technique for the separation and analysis of chemical compounds. Techniques adopted from the microchip technology knowledge have led to recent developments of electrophoresis system with integration on microchip. Microchip Capillary Electrophoresis (μCE) systems offer a series of advantages as easy integration for Lab-on-a-chip applications, high performance, portability, speed, minimal solvent and sample requirements. A new technological challenge aims at the development of an economic modular microchip capillary electrophoresis systems using separable and independent units concerning the sensor. In this project we worked on the development of an interchangeable amperometric sensor in order to provide a solution to such electrode passivation and facilitating the use of tailored sensors for specific analyte detection besides. Fluidic chips have been machined from cyclic olefin polymer pallets (Zeonor®) using a micro-injection molding machine.

1. Introduction

Recently particular attention has been given to the development of microchip capillary electrophoresis systems (μCE) because of their fast and efficient separation capabilities [1]. While microchip technology has improved over the last decade, the techniques of detection have lagged behind. Although preferred techniques in the past such as laser-induced fluorescence (LIF) and mass spectroscopy (MS) showed high sensitivity of detection, they often require off-chip control instrumentations limiting the mobility and the portability of the system. Electrochemical detection (EC) seems to be the meeting point between the concept of portability (low volume of the control instrumentation) and the need for increased sensitivity and resolution [2,3]. Furthermore electrochemistry is also highly compatible with integration processes and microfabrication technologies. The electrochemical detection modes used for microchip capillary electrophoresis systems are: amperometric detection, potentiometric detection and conductivity detection. The amperometric detection is generally considered the most powerful method in term of resolution, but it is limited to the electroactive compounds. For this reason, the other two methods are seen as complementary methods for the non-electroactive compounds [4].
A general limitation for the amperometric detection method lies in the fact that the detector electrodes are in direct contact with the solution and may therefore deteriorate due to corrosion processes.
Here we present a new modular approach to the development of a microchip capillary electrophoresis device. Fluidic chips have been designed and developed in our facilities using a micro-injection

molding machine. We have also developed a range of micro-electrodes made of indium tin oxide (ITO), silver and platinum by means of patterning, etching and deposition techniques using the internal fab of the Tyndall National Institute. A three electrodes configuration has been used for sensing: ITO working electrode, AgCl reference electrode and Pt counter electrode. Primary importance has been given on the improvement of an oxidative technique to form an AgCl electrode on-chip. The stability of the sensor has been investigated using cyclic voltammetry and potentiometry with excellent outcomes. The Electroosmotic flow (EOF) has been tested into the Zeonor® using fluorescein dye in 20 mM sodium borate solution. Amperometric detection test of the single sensors showed significant response with homovanilic acid, 4-hydroxy-3-methoxymandelic acid, L-Tryptophan and 3-Indoxyl sulfate.

2. Experimental

2.1. Design

The fluidic chip is composed of two layers in Zeonor®. The design has been ideated in order to have a modular sensor as well as a modular separation channels (bottom layer). In this case bottom layers with different designs for the separation channel (serpentine or squared) could be used with the same top layer. The chip sizes are of 60mm x 15mm x 1mm (LWH) for the top layer and 60mm x 15mm x 2mm (LWH) for the bottom layer. The channels have a squared cross-section with dimensions 50μm x 50μm. The three amperometric electrodes have been evaporated on a pyrex substrate of the following sizes: 28mm x 3mm x 0,5mm (LWH).

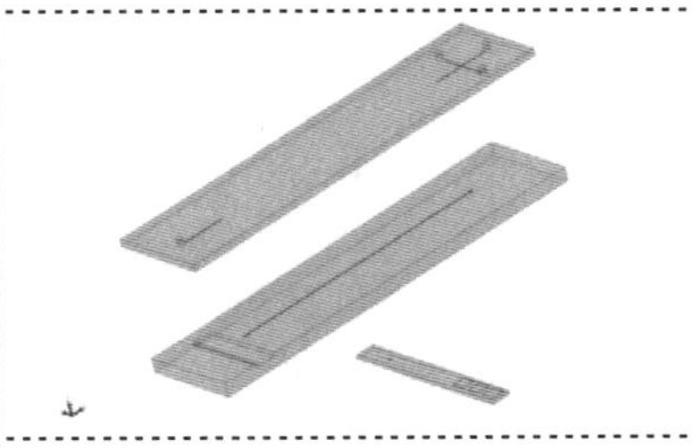

Figure 1. Exploded view of the system in a CAD application

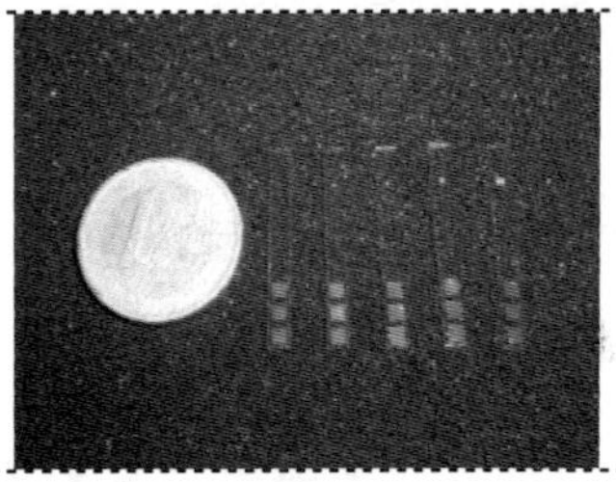

Figure 2. Amperometric sensors in five different designs

2.2. Fabrication of the fluidic chip

The chip layers have been accomplished using the Babyplast® 6/10P microinjection moulding machine available in the Life Sciences Interface lab (Fig. 3 and Fig.4). A solvent vapour bonding technique has been used to attach the mating layers. This technique consists in the exposition of one layer to cyclohexane vapour into a chamber [5]. Over the last years the most popular bonding technique used has been the thermal fusion. In the specific case of microfluidic devices a series of disadvantages have been encountered using this approach. Since the chip is heated slightly above the glass transition temperature of the material, residual internal stress can lead to bulk material reflow with results of deforming small features or clogging the channels. The solvent vapour bonding takes over this issue increasing the polymer chain mobility of the exposed surfaces so facilitating the bonding process [5]. Optimization of parameters as temperature and exposure time have been done running a series of tests.

Journal of Physics: Conference Series **307** (2011) 012059 doi:10.1088/1742-6596/307/1/012059

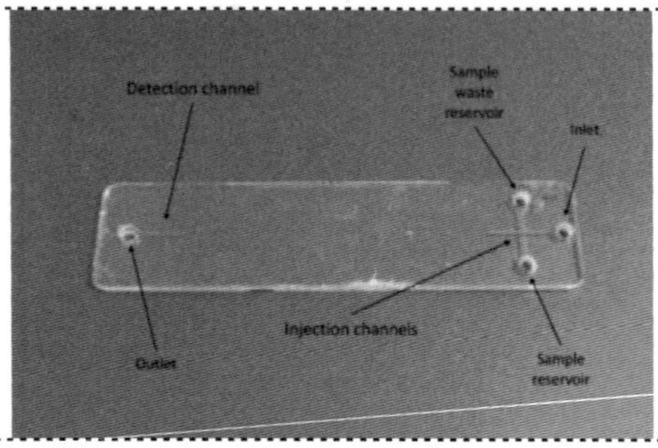

Figure 3. Details of the top layer

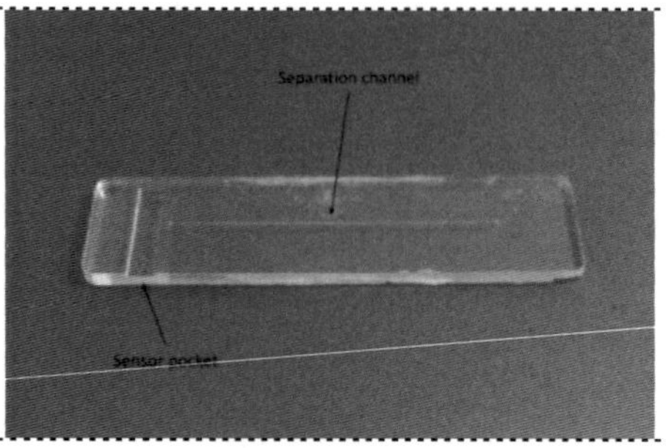

Figure 4. Details of the bottom layer

We found necessary to remove any kind of protrusions which made uneven the surface of the layers probably due to the extraction of the sprue from the mould of the Babyplast® machine. Protrusion were frequently present at the edge of the channels or reservoirs and were removed using a fine sand paper. Cleaning with 2-propanol and deionized water was then necessary to remove any kind of burrs present on the surface. An adequate chamber created with a petri dish and a glass lid was placed on a hot-plate in a range of temperature between 30°C and 40°C, estimated and controlled using a digital thermometer. After the sufficient amount of time to get temperature equilibrium between the solution and the hot plate, the bottom layer was exposed into the chamber for approximately 20 s. The bottom layer was then removed from the vapour and inserted with the top layer in a hand-made aligner in order to facilitate the alignment between the detection and the injection channel with the separation channel. The aligner with the fluidic layers were then pressed in a clamp to evenly distribute the pressure. To improve the bond strength, a treatment with UV light (UV/Ozone Procleaner Plus, Bioforce Nanosciences) for 20 mins was performed. In fact Mair et al. [5] observed a positive shift of about 50% in burst pressure using a UV treatment (13.5 J/cm^2 at 260 nm). Presumably chain scissions are induced by UV light and cross-linking reactions occurs at the bonded interface.

2.3. Fabrication of the amperometric sensor

The amperometric sensor has a three electrodes configuration with an indium tin oxide (ITO) working electrode, a Pt counter electrode and Ag/AgCl reference electrode (Fig.5 and Fig.6). Pads and connections have been done in gold material. Five different designs of the sensor corresponding to five different dimensions of the electrodes have been developed. All of them have been fabricated in the internal fab facility of the Tyndall National Institute. After the deposition of the electrodes, a silicon nitride passivation layer has been grown on the wafer with openings on the electrodes area and pads. This is normally done to avoid such conduction between connections and letting the current flowing just in the electrodes area when dipped in the electrochemical solution. A further treatment with a ferrichloride solution ($FeCl_3$) of the wafer was performed to form an AgCl layer on top of the silver deposited in the fab for the reference electrode [6]. Then a necessary operation of dicing of the wafer was performed. The height of the electrodes is 500 µm for all the five designs to make the alignment easier in the channel 50 µm wide. The widths of the five sensors are reported below in the Table 1.

Figure 5. Sensor representation in a CAD application: pads (red), connections (blue), electrodes (green)

Figure 6. Detail of the electrodes. From the left: Ag/AgCl reference electrode, ITO working electrode, Pt counter electrode

Table 1. Widths in μm of the five different designs of the microelectrodes

	Type 1	Type 2	Type 3	Type 4	Type 5
Reference	5	10	50	100	200
Working	10	20	100	200	400
Counter	10	20	100	200	400

2.4. Integration

The integration of the sensor was carried out using an optical UV glue (NOA 68, Norland Optical Adhesives). The capillary forces drove the glue from the lateral access of the sensor until the sides of the channel. When the glue approached to the walls of the channel, a UV gun permitted the polymerization and hardening of the glue. Even an appropriate interface for the sensor has been built in order to avoid soldering operations of wires onto the pads allowing for easy connection to the electrochemical instrumentation (Fig. 7). The interface is composed of two units: one printed circuit board and a plastic cap designed using a CAD software and developed using a computer numerical control (CNC) machine available in the mechanical workshop of the Tyndall National Institute. The plastic cap was then fixed onto the printed circuit board using plastic screws.

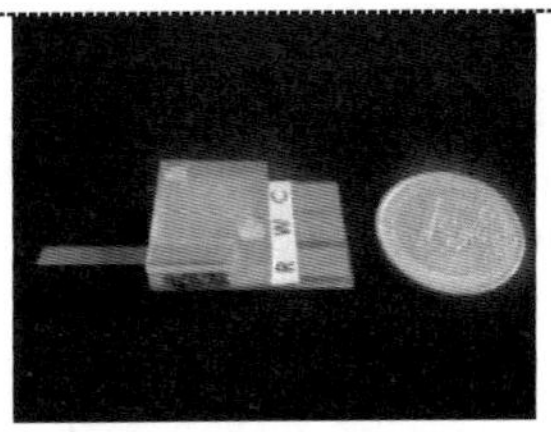

Figure 7. Interface for the sensor.

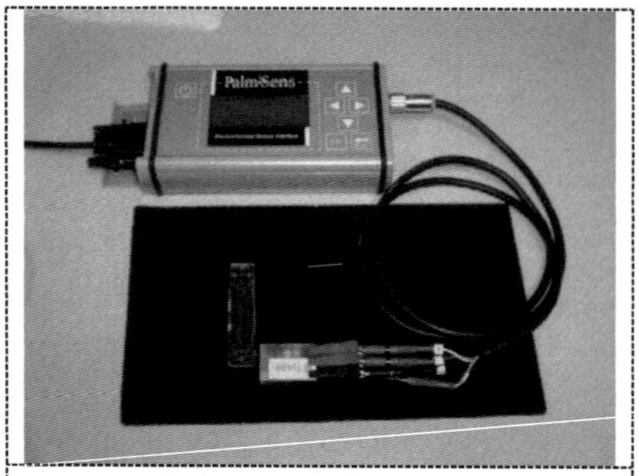

Figure 8. Complete view of the system with Palmsens® connected.

2.5. Electrochemical instrumentation

Amperometric detection, potentiometry and cyclic voltammetry have been carried out using the battery-powered Palmsens® (Palmsens Instruments BV, Houten, Netherlands) (Fig. 8), a portable device which can be connected to a laptop or a pocket PC via cable or Bluetooth connection. The high-voltage power supply used for the electroosmotic flow test is the LabSmith High Voltage Sequencer HVS448 6000D (LabSmith, Livermore, CA, USA).

3. Results and discussion

3.1. Characterization of the electrochemical cell using cyclic voltammetry

All the five design of the sensors have been characterized using cyclic voltammetry in a solution of 15 mM $K_4Fe(CN)_6 \cdot 3H_2O$ in 1 M KNO_3. Five scans at 25mV/s have been delivered for each sensor to check the stability of the AgCl reference electrode and results then compared with a commercial external Ag|AgCl|1 M KCl reference electrode. The measurements were stable for all the five sensor with good and repeatable outcomes as shown in Fig.9. A potential shift has been noticed for the five designs after a comparison with the results obtained using the external commercial Ag|AgCl| 1 M KCl reference electrode. The measurements resulted shifted for a value included in the interval 10 mV – 20 mV probably due to the oxidation technique to form the silver chloride on top of the reference electrode.

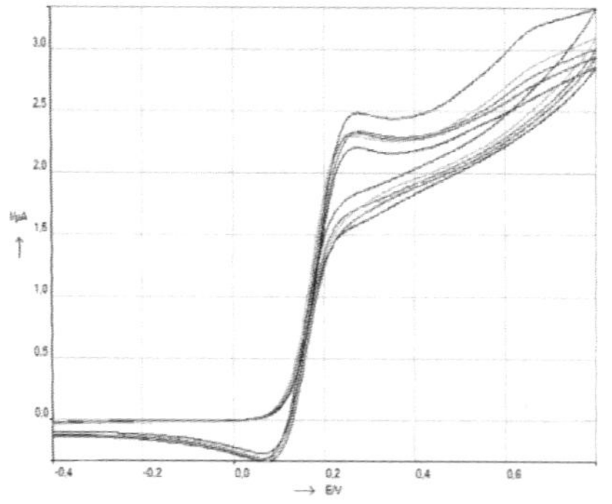

Figure 9. Cyclic voltammetry of the sensor type one in a solution of 15 mM $K_4Fe(CN)_6 \cdot 3H_2O$ in 1 M KNO_3.

3.2. Characterization of the electrochemical cell using amperometric detection

Amperometric detection test have been performed separately for each sensor with the aim of identifying the main compounds which could be detected using an indium tin oxide working electrode. Amperometric detection was started while the analyte sequentially introduced in a 10 mL becker at 50 s intervals each. Analytes at issue were homovanilic acid, 4-hydroxy-3-methoxymandelic acid , L-tryptophan, 3-indoxyl sulfate respectively. The detection potential chosen was 800 mV at a 0.2 s scan rate.

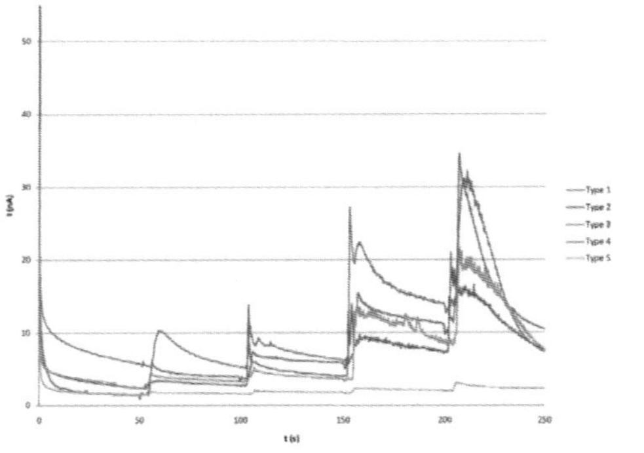

Figure 10. Electropherogram: E = 0.8 V, t_{run} = 250s, scan rate = 0.2, analyte concentration = 25 μM.

3.3. Electroosmotic flow test

Cyclic olefin polymer is a pretty new material used in microfluidic and it is still not clear which is the charge state of the surface. For this reason an electroosmotic flow test was necessary to evaluate the presence of a specific charge on the channel surfaces which could support this kind of flow. The buffer solution used was sodium borate in a concentration of 20 mM. Once the system was filled of the solution and the channel confirmed bubble free, the sample reservoir was loaded with drops of fluorescein. The sequence of injection was then started and pictures taken from an Olympus IX2 optical microscope (Olympus Imaging & Audio Ltd, Southend-on-Sea, Essex, UK) (Fig. 11) with a motorized lens under the chip. Injection sequence is reported in Table 2.

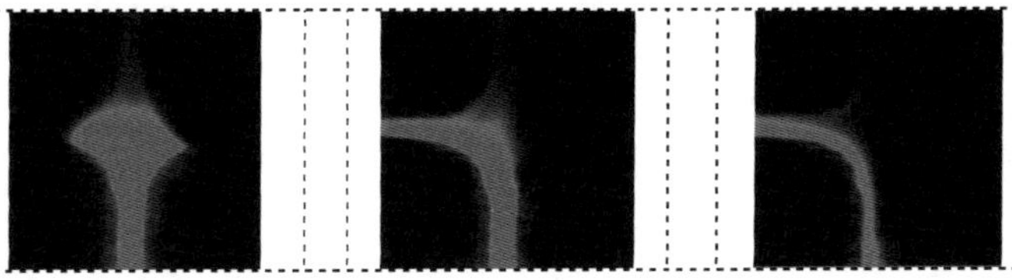

Figure 11. Sequential pictures of the EOF at the intersection (from the left to the right) after 8s, 9s, 10s.

Table 2. Injection sequence for the electroosmotic flow test

	Step A	Step B	Step C
Time (s)	5	1	10
Sample reservoir (V)	100	40	80
Sample waste reservoir (V)	-100	40	80
Inlet (V)	50	200	200
Outlet (V)	50	-200	-200

4. Conclusion

A first attempt of integration of an amperometric sensor using on-column configuration here is done. Zeonor® showed compatibility with the electroosmotic flow and a further step in microfluidic applications has been completed. Amperometric sensors with an on-chip Ag/AgCl reference electrode represent a possible application for microchip capillary electrophoresis systems since the stable and repeteable results obtained. A complete test of the fluidic and the sensors integrated is at the moment under investigation.

5. Acknowledgements

The authors acknowledge the Irish Separation Science Cluster (ISSC) and the National Access Programme (NAP), a programme of the Tyndall National Institute with funding from Science Foundation Ireland (SFI).

6. References

[1] Wang J 2005 *Electroanalysis* **17** 1133
[2] Vandaveer W R, Pasas S, Martin R S and Lunte S 2002 *Electrophoresis* **23** 3667
[3] Wang J 2002 *Talanta* **56** 223
[4] Schöning M J, Jacobs M, Muck A, Knobbe D T, Wang J, Chatrathi M and Spillman S 2005 *Sens Actuators* B **108** 688
[5] Mair D A, Rolandi M, Snauko M, Noroski R, Svec F and Fréchet J M J 2007 *Anal. Chem.* **79** 5097
[6] Kim H R, Kim Y D, Kim K I, Shim J H, Nam H and Kang B K 2004 *Sens Actuators* B **97** 348

An Investigation of Pulse Transit Time as a Non-Invasive Blood Pressure Measurement Method

B M McCarthy, B O'Flynn and A Mathewson
Tyndall National Institute, UCC, Lee Maltings, Prospect Row, Cork, Ireland

E-mail: brian.mccarthy@tyndall.ie

Abstract. The objective of this paper is to examine the Pulse Transit Method (PTT) as a non-invasive means to track Blood Pressure over a short period of time. PTT was measured as the time it takes for an ECG R-wave to propagate to the finger, where it is detected by a photoplethysmograph sensor. The PTT method is ideal for continuous 24-hour Blood Pressure Measurement (BPM) since it is both cuff-less and non-invasive and therefore comfortable and unobtrusive for the patient. Other techniques, such as the oscillometric method, have shown to be accurate and reliable but require a cuff for operation, making them unsuitable for long term monitoring. Although a relatively new technique, the PTT method has shown to be able to accurately track blood pressure changes over short periods of time, after which re-calibration is necessary. The purpose of this study is to determine the accuracy of the method.

1. Introduction

Blood pressure is a major indicator of the health of a person and is one of the five vital signs. However, unlike other vital signs, abnormal blood pressure can be a sign of a long term illness rather than an immediate one and can be used as a predictor of Cardiovascular Disease (CVD). CVD is the leading cause of death worldwide and accounted for 33.54% of all deaths in Ireland in 2009[1].
Blood Pressure is typically measured in a clinical setting using a sphygmomanometer, a stethoscope and a clinician. Although this is considered the golden standard for blood pressure measurement, it has been recognized that there are numerous problems with this method [2], [3],[4]. It is commonly acknowledged that a non-invasive and continuous blood pressure measurement device is preferable to the sphygmomanometer.
Up until recently, there has only been two major non-invasive, continuous blood pressure measurement techniques that have been researched extensively i.e. the oscillometric method and the Penaz method. Both of these methods, however, still use a cuff, which can lead to patient comfort impact and the occlusion of the artery increases the workload of the heart and causes circulatory interference at the measurement site [5].
The Pulse Transit Time (PTT) method is an alternative for blood pressure measurement. The PTT is the time it takes for a blood pressure pulse to travel from one arterial site to another [6]. Typically, the PTT is initiated by the R-wave of the ECG and the finger is used as the second arterial location. The arrival of the pulse at this location is usually detected by the use of a photoplethysmograph sensor. By monitoring the change of the PTT it is possible to track the changes in blood pressure.

The PTT has a number of advantages over the oscillometric and Penaz methods but the most significant advantage is the PTT method does not require an inflatable cuff, making it ideal for a wearable and continuous non-invasive blood pressure measurement device.

2. Background

The relationship between blood pressure and PTT as it is used in this method was first proposed by Chen in 2000 [7]. Chen related blood pressure to a third parameter known as Pulse Wave Velocity (PWV). The following is a summary of this derivation.

PWV is defined as the speed of propagation of a blood pressure pulse. The PWV is dependant of a number of arterial properties, namely the elasticity E, the arterial thickness t, the arterial diameter d and the density of the blood ρ. This relationship is defined by the Moens-Kortwegg equation:

$$PWV = \sqrt{\frac{gtE}{\rho d}} \tag{1}$$

where g is the gravitational constant. For the sake of simplicity, g is usually omitted and the pressure is said to be hydrostatic, since g is constant on earth. This equation is based on the assumption that, over a short period of time, neither the elasticity, the thickness of the walls nor the diameter of the walls changes significantly.

It has been noted that the elastic modulus of the vessel increases exponentially with increasing blood pressure such that:

$$E = E_0 e^{\gamma P} \tag{2}$$

Where E_0 is the elastic modulus at zero pressure, P is the blood pressure (mmHg) and γ is a coefficient ranging from 0.016 to 0.018 ($mmHg^{-1}$).

PTT is related to PWV by:

$$PWV = \frac{K}{PTT} \tag{3}$$

where K is a proportional coefficient, indicating the distance that the pulse has to travel between the two arterial locations.

Using some manipulation [7], blood pressure can be related to PTT directly by:

$$P_e = P_b - \frac{2}{\gamma PTT_b} \Delta PTT \tag{4}$$

Where P_b is the base blood pressure level, PTT_b is the value of the PTT corresponding to the pressure P_b, while ΔPTT is the change in the PTT.

3. Methodology

Data was collected from six volunteers: four male and two female volunteers aged between 24 and 28 years. Each volunteer was tested three times, with half an hour's worth of data being recorded at each sitting i.e. 90 minutes of data was recorded per volunteer.

Blood pressure was simultaneously recorded using an Omron M6 oscillometric blood pressure monitor. This device is recommended by Dabl Educational and was also recommended by a cardiologist. The Omron M6 was worn on the upper left arm and was used to recalibrate the estimated blood pressure at five minute intervals i.e. was used to provide P_b from Equation (4).

Using this method, 540 minutes of data was recorded and 126 blood pressure measurements were recorded.
Electrocardiogram (ECG) data was recorded from a self developed electrocardiogram, the Tyndall Physiological Health Board. This Physiological Health Board uses a three electrode ECG monitor to record the data from each of the patients. A three electrode system was developed since it provided the accuracy that was required for ECG monitoring while not being obtrusive for the patient. The electrodes were placed in a manner which satisfied Einthoven's Triangle [8]. The Physiological Health Board uses a 12-bit ADC (AD7490, Analog Devices) to transmit data to the on-board micro-controller (Atmel Mega 1281) for transmission.
The photoplethysmograph (PPG) data was recorded from another device that was developed within Tyndall. This PPG sensor uses a finger cuff and is placed on the middle finger of the right hand.
Both the ECG and PPG the signals were recorded at a sampling rate of 400Hz using a personal computer based signal acquisition system through LabView and stored in an spreadsheet for post processing.
Pulse Transit Time was defined as the time interval between the R-wave of the ECG and the peak of the PPG signal at the finger site, within the same cardiac cycle. Measurements were taken from 2.00 PM to 4.30PM in a quiet environment at 20°C. The volunteers were asked to refrain from smoking and ingesting caffeine or any other blood pressure altering drugs in the hours prior to the test, as per the recommendations for all blood pressure measurements [4].

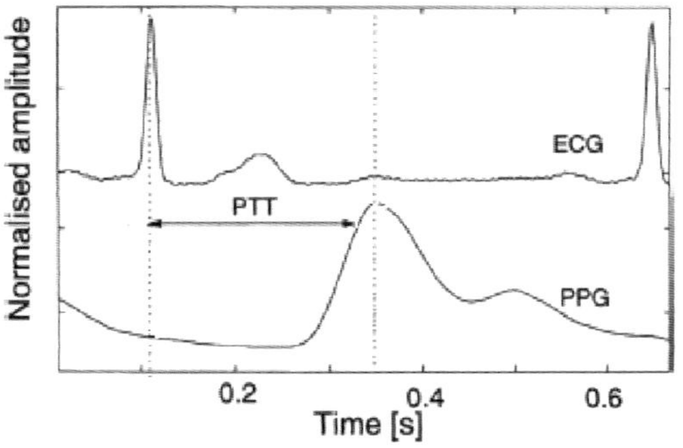

Figure 2. Definition of the Pulse Transit Time

All post-processing of the recorded data was performed using Matlab (MathWorks, Natick, MA, USA) using a self derived algorithm. This algorithm initially filtered all the data to remove excess noise. The filtered data was then analyzed to detect the peaks of the ECG and the peaks of the PPG. From this it is possible to calculate the PTT and ΔPTT. The algorithm then used the data recorded from the Omron M6 blood pressure monitor (P_b) and the ΔPTT to calculate the estimated blood pressure P_e, using Equation (4) .

4. Results

The results from this test described in section 3 can be seen in Table 1. The data is presented in the form of the mean and standard deviation of the error of the estimated Blood Pressure. These results show a good correlation between the blood pressure measured using the clinically validated Omron M6 and the PTT method.
The largest standard deviation is seen from Volunteer 3's results. This can be explained by the fact that this volunteer is both hypotensive and a diabetic. It was noted during testing that this volunteer had poor circulation which adversely affected the signals being recorded from the PPG sensor which was placed on the finger.

Table 1. Summary showing the mean error of the estimated and the standard deviation for each volunteer.

Volunteer	*Estimated Error of Systolic BP*
1	1.81 ± 4.39
2	-0.82 ± 3.69
3	-1.34 ± 7.93
4	-1.44 ± 5.71
5	-2.57 ± 6.23
6	-0.53 ± 3.70

5. Discussions and Conclusions

This study intended to evaluate the PTT method that has been used as an alternative to the use of a sphygmomanometer and other cuff based methods of blood pressure measurement. The results, shown in Table 1, demonstrate a high degree of accuracy between the two methods. The mean difference between all the results for each of the volunteers is less than 3mmHg. However, there is a quite large standard deviation between the PTT and traditional methods. Even though this group also contained a hypotenstive volunteer, the PTT method was able to track this persons blood pressure with reasonable accuracy.

Although volunteer 3 was known to diabetic and hypotensive, this was not the case for volunteers 4 and 5, which also show a large standard deviation. Analysis of the results of all patients have revealed that some of the peaks of the PPG sensor were not correctly identified. It had been previously established, while testing volunteer 3, that poor circulation adversely affects the results but analysis of volunteers 4 and 5 showed a particularly noisy signal. This noisy signal can be attributed to movement of the finger during the recording of the data. Although the filtering that was applied using Matlab was able to eliminate some of this noise, it has become apparent that more filtering of the signals will be required to eliminate noise due to movement.

Overall, this test has shown that it is possible to track blood pressure with reasonable accuracy using the Pulse Transit time. However, this tracking of blood pressure was done using constant re-calibration at 5 minute intervals. It is predicted that extending this calibration interval would mean a deterioration of the results but a further study will be required to establish this level of deterioration..

6. Future Work

Further work will be required to complete the evaluation of the PTT as an alternative to the traditional methods of blood pressure monitoring. The algorithm that have been used to filter the data, to determine the peak detections and to calculate the PTT, will have to be extended to account for motion artefact reduction. This will help to increase the accuracy of the results.

A larger study group will need be tested in the subsequent rounds of testing. It would be preferable that this group will contain a larger amount of patients, both with normal and abnormal blood pressures, to examine the usefulness of the technology for both healthy and ill patients.

7. Acknowledgement

The author would like to thank Intel and the IRCSET Embark Initiative for funding this project, National Access Program for funding research for hardware development, John O'Donoghue in the Computer Science Department in UCC and Tyndall National Institute for providing the facilities and equipment used in this project. Tyndall is part of the SFI funded CSET, CLARITY centre for sensor web Technologies.

References

[1] CSO. (2011, February 2011). *Vital Statistics: Fourth Quarter and Yearly Summary 2009.* Available: cso.ie/releasepublications/documents/vitalstas/2009/vstats_q42009.pdf
[2] R Leca and V Groza, *Instrumentation and Measurement Technology Conference Proceedings, 2008. IMTC 2008. IEEE*, 2008, pp. 309-313.
[3] D Shapiro, *Psychophysiology,* vol. 33, pp. 1-12, 1996.
[4] T G Pickering.,*Circulation,* vol. 111, pp. 697-716, February 8, 2005 2005.
[5] C C Y Poon and Y T Zhang, *Engineering in Medicine and Biology Society, 2005. IEEE-EMBS 2005. 27th Annual International Conference of the*, 2005, pp. 5877-5880.
[6] T Ma and Y T Zhang, *Engineering in Medicine and Biology Society, 2005. IEEE-EMBS 2005. 27th Annual International Conference of the*, 2005, pp. 996-999.
[7] W Chen, *Medical and Biological Engineering and Computing,* vol. 38, pp. 569-574, 2000.
[8] K Berntson and D Lozano, 3rd Edition ed: Cambridge University Press, 2007, pp. 182-210.

Nanomolar Trace Metal Analysis of Copper at Gold Microband Arrays

A Wahl, K Dawson, N Sassiat, A J Quinn and A O'Riordan[1],
Nanotechnology Group, Tyndall National Institute, University College Cork, Lee Maltings, Cork, Ireland

E-mail: alan.oriordan@tyndall.ie

Abstract. This paper describes the fabrication and electrochemical characterization of gold microband electrode arrays designated as a highly sensitive sensor for trace metal detection of copper in drinking water samples. Gold microband electrodes have been routinely fabricated by standard photolithographic methods. Electrochemical characterization were conducted in 0.1 M H_2SO_4 and found to display characteristic gold oxide formation and reduction peaks. The advantages of gold microband electrodes as trace metal sensors over currently used methods have been investigated by employing under potential deposition anodic stripping voltammetry (UPD-ASV) in Cu^{2+} nanomolar concentrations. Linear correlations were observed for increasing Cu^{2+} concentrations from which the concentration of an unknown sample of drinking water was estimated. The results obtained for the estimation of the unknown trace copper concentration in drinking was in good agreement with expected values.

1. Introduction

The development of robust and reliable sensors with lower detection limits is of increasing importance in analytical chemistry, for both biological and chemical sensing applications. To this end, electrochemical based sensors are being increasingly investigated, due to key advantages such as high sensitivity, selectively, low operating voltages, portability, accuracy and low production costs [1]. Further enhancements arise the dimensions of the sensor electrode are scaled down to ultra micro regimes. At ultra-microelectrodes, i.e. with radii less than 25 µm, radial diffusion profiles of analytes begin to dominate causing increased current densities, mass transport, and increased Faradaic to capacitive current ratios [1-4]. In this manner the application of ultramicroelectrodes as sensors allow increased sensitivity in detection of target analytes, for instance in biologically important molecules (hydrogen peroxide, glucose) [2, 5] and environmental analytes (heavy metals) [6, 7].

Toxicity associated with heavy metals is driving the need for detection of trace metals in a variety of media, such as food and drinking water and is as such a concern no longer restricted to environmental assays. Metals such as lead, zinc, cadmium and copper can cause chronic toxicity to children at relatively low doses (e.g. 8 mg/L for copper). A key concern therefore, is the use of copper pipes for the supply of potable water in domestic urban settings. In this regard, the current EU standard maximum concentration of copper in drinking (tap) water is 2 mg/L (2 ppm) [8]. Current approaches to copper detection involve spectroscopic methods with complex wet chemical sample pre-treatment requirements. These approaches are both costly and time consuming and are restricted

[1] To whom any correspondence should be addressed.

to laboratory settings with qualified personnel. It is therefore important to develop new robust copper detection methodologies which may be performed by unskilled personnel at remote locations with high sensitivity. One approach is the application of electrochemical techniques where a voltage is applied to an electrode immersed in an acidified sample of interest for a given time interval. By judicious choice of voltage (versus a reference electrode) copper is selectively platted out of solution onto the working electrode thereby eliminating the requirement for sample pre-treatment such as reaction with compex agents e.g. EDTA (Ethylenediaminetetraacetic acid). While traditional electrochemical analysis of these heavy metals have been conducted at mercury electrodes [9, 10] (due to the fast electrode kinetics), research on detection of metallic analytes at solid-state or noble metal electrodes has achieved increasing popularity, due to increased sensitivity and low limits of detection achievable by anodic stripping voltammetry techniques [9].

At noble metal electrodes, cyclic voltammograms (CVs) exhibit two reduction and two oxidation peaks in the presence of a metallic analyte, one arising from bulk (i.e. multilayer) deposition and oxidation and the other associated with under potential (i.e. single monolayer) deposition and oxidation peak. Deposition of the metallic analytes at an under-potential combined with anodic stripping voltammetry (UPD-ASV) of the monolayer provides a highly sensitive method of analysis for trace concentrations of metals at nanomolar concentrations [9]. In this work, determination of copper employing UPD-ASV at gold microband electrode arrays is presented. UPD involves the pre-concentration of a monolayer of a metallic ion, in our case copper ions, by the application of a lower potential than would be thermodynamically optimal for bulk deposition. This step is then immediately followed by the oxidation of this monolayer via a potential sweep method such as linear sweep voltammetry, from which the amount deposited can be determined. Thus UPD-ASV consists of two consecutive steps relying on key parameters that are the deposition potential, deposition time and stripping potential range of the monolayer. Detection of copper at concentrations as low as 10 nM was easily achievable.

2. Experimental

2.1. Gold Microband Electrode Fabrication:

Gold microband electrodes were fabricated using standard optical lithography, metal deposition and lift-off techniques on silicon substrates comprising a ~300 nm thermally grown silicon dioxide layer. Microband structures and overlaid metal interconnection tracks were patterned on Microposit LOR 10A Lift-off resist ~ 500 nm thick, followed by metal evaporation (Ti/Au 10/200 nm) and lift-off. Silicon Nitride (500 nm) was then blanket deposited onto the devices as passivation layer to prevent unwanted electrochemical reactions occurring at these metal interconnects. Finally trenches were opened in this passivation layer to expose the microband electrode arrays allowing subsequent contact with an electrolytic solution.

2.2. Microband Characterization:

Optical micrographs were acquired using a calibrated microscope (Axioskop II, Carl Zeiss Ltd.) equipped with a charge-coupled detector camera (CCD; DEI-750, Optronics).

2.3. Electroanalysis at Gold Microband Electrodes:

Prior to electrochemical experiments, microband array electrodes were cleaned in pure ethanol (Sigma-Aldrich) followed by thoroughly rinsing with deionized water. All electrochemical studies were performed using a CHI760a Electrochemical Analyzer and Faraday Cage CHI200b (CH Instruments) connected to a PC. Experiments were performed in a three-electrode cell configuration employing an array of microbands as working electrodes, with gold counter and Ag wire reference electrodes (Sigma-Aldrich). Cyclic voltammetry was conducted in 0.1 M sulphuric acid in the voltage range of -0.9V to 1.5V at a scan rate of $5V.s^{-1}$. A range of copper sulfate standards in 0.1 M sulphuric acid were prepared using copper sulphate Cu_2SO_4 (Sigma-Aldrich). All solutions were

Journal of Physics: Conference Series **307** (2011) 012061 doi:10.1088/1742-6596/307/1/012061

prepared with deionized water; 18.2 MΩ cm (ELGA Pure Lab Ultra). All glassware was cleaned and rinsed in DI water prior to all measurements.

3. Results and Discussion:

3.1. Gold Microband Structural characterization:

Gold microband array electrodes were fabricated using photolithography process at silicon wafer substrates bearing a 300 nm layer of thermally grown silicon dioxide, see experimental section. Standard optical lithography was employed to overlay electrical contacts and interconnection tracks onto microband arrays, followed by metal deposition and liftoff. Unwanted electrochemical reactions occurring between metal interconnection tracks and electrochemically active species were prevented by the presence of the silicon nitride layer. A lithographically patterned trench (~30 x 30 μm) was selectively opened in the insulating silicon nitride layer directly above the microband array to allow exclusive exposure of the microband array to the electrochemically active species. Following device fabrication, microband electrodes were characterized using optical microscopy. Figure 1 shows an optical micrograph of a fully fabricated microband array device, with interconnections and silicon nitride passivation. Microband electrodes of width ~2 μm were routinely fabricated.

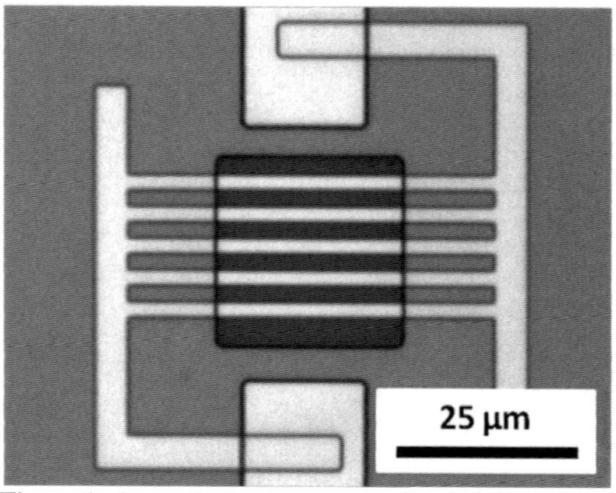

Figure 1: An optical micrograph of a fully integrated and passivated microband electrode array.

3.2. Gold Microband Electrochemical Characterization:

Electrochemical investigations were undertaken employing a potentiostat with a Faraday cage to apply an appropriate potential sweep range in a three-electrode set-up utilizing a microband array working electrode versus Ag wire reference with gold counter electrode.

Preliminary voltammetric experiments were conducted in 0.1 M sulphuric acid to confirm that microband electrodes were fully functional and additionally to clean the metal electrode surface. A typical cyclic voltammogram (CV) recorded in 0.1 M H_2SO_4 for the gold microband electrode is shown in figure 2. Characteristic gold oxide formation and reduction peaks are clearly visible, at 1.02 V and 0.55 V, respectively. A stable reduction peak current at microband array was achieved within a few successive cycles confirming that the electrodes were extremely clean. The low values of current measured (~80 nA) strongly suggests that only the microbands were exposed to the electrolyte and that the passivation layer was functioning. This figure is in agreement with data in literature. [2, 6]

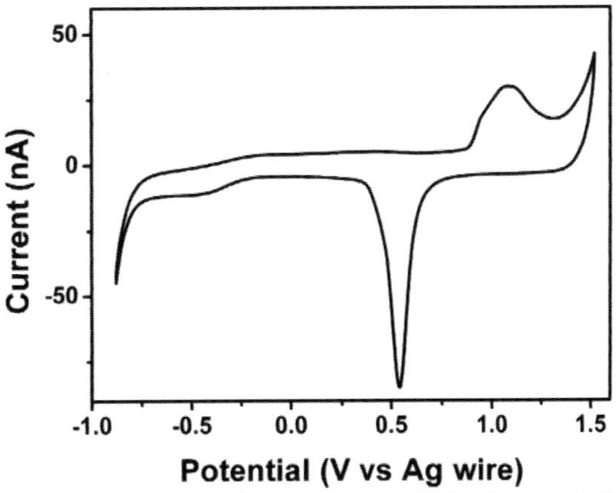

Figure 2: Cyclic voltammogram (CV) of a typical gold microband array in 0.1 M H_2SO_4, from -0.9 to 1.5 V at 5 V s^{-1}.

3.3. *CV of Copper:*

CV was used to determine the potentials of the monolayer formation (i.e. underpotential deposition) and bulk deposition, as well as their stripping (oxidation) potential. Figure 3 illustrates the peaks corresponding to the formation and oxidation of the copper monolayer, respectively at -0.16 V and 0.12 V, within a voltage range of -0.4 V to 1.5 V. Peaks denoting bulk deposition (-0.45 V) and bulk oxidation (-0.09 V) were found to appear when CVs were conducted at more negative potential ranges. The electrochemical deposition of copper monolayer onto the gold microband surface was thus performed by applying -0.16 V for various preconcentration times at the working electrode. Linear Stripping Voltammetry (LSV) was carried out immediately after in 0.1 M sulphuric acid solution.

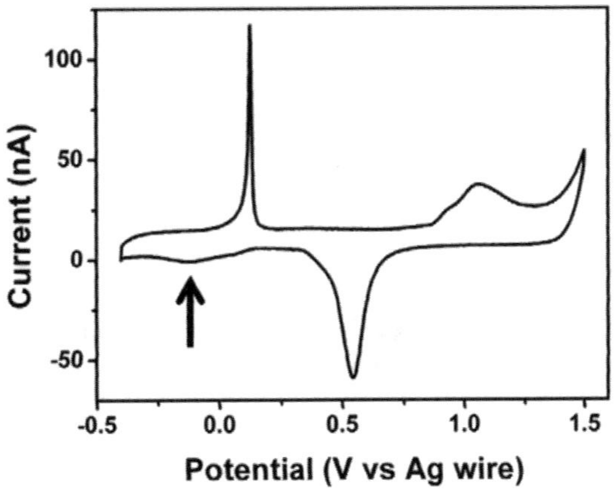

Figure 3: CV of a typical gold microband array in 1 mM $CuSO_4$ in 0.1 M H_2SO_4 from -0.4 to 1.5 V at 5 V s^{-1}, highlighted are the Cu^{2+} UPD reduction peak at -0.16 V (arrow) and stripping peak (0.12 V).

3.4. *Cu^{2+} UPD-ASV Optimization:*

Electrochemical UPD was performed on microband arrays for pre-concentration times of 5, 10, 15, 20 and 30s in a variety of Cu_2SO_4 solutions ranging from 10 nM to 750 nM. LSV was performed immediately after each deposition step, at potentials ranging from -0. 25 V to 0.55 V, at a scan rate of 5V.s^{-1}. Figure 4a, shows an increase in measured electrochemical stripping current with increasing deposition times associated with increased amounts of deposited copper on the microband electrode surface.

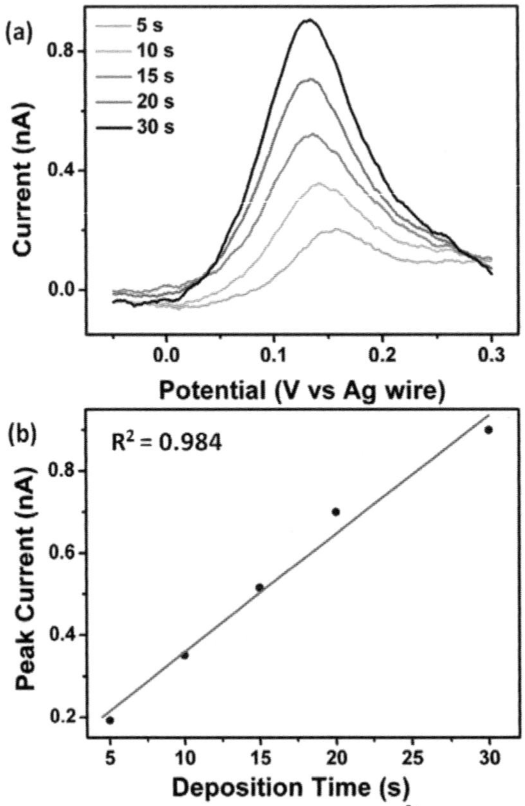

Figure 4: (a) UPD-SV of 10 nM Cu^{2+} in 0.1M H_2SO_4 at a gold microband array electrode. Scan Rate: 5 $V.s^{-1}$. Deposition times: 5, 10, 15, 20 and 30s. (b) Evaluation of the dependence of electrochemical signal vs. deposition time of 10 nM Cu^{2+} in 0.1M H_2SO_4.

To further explore this, the relationship between the increasing signals with respect to the deposition time was analyzed. Figure 4b shows a linear dependence of the copper stripping peak height versus deposition time. This trend was observed for all concentrations of Cu^{2+} examined. The amount of deposited copper on the microband electrode surface was quantified using equation (1):

$$\Gamma = \frac{Q}{(nFA)} \qquad (1)$$

where Γ is the surface coverage (mol/cm^2), Q is the integrated charge of the stripping voltammetric peaks (C), n is the number of electrons exchanged, F is Faraday constant (A/mol) and A is the geometric surface area of the electrode (cm^2). Estimated surface coverage values recorded in 10 nM Cu_2SO_4 at microband electrodes, ranged from 2.29×10^{-10} mol/cm^2 (5 s) to 2.38×10^{-10} mol/cm^2 (30 s) of deposition. To avoid saturation of the electrodes a deposition time of 20 s was selected as a suitable interval for deposition for further measurements.

Journal of Physics: Conference Series **307** (2011) 012061 doi:10.1088/1742-6596/307/1/012061

3.5. Calibration of Cu^{2+} UPD-ASV: Measurement of the sensitivity of the gold microband arrays to UPD-ASV Cu^{2+} was investigated across a range of nanomolar concentrations at the optimized deposition time of 20s. Stripping peak current was observed to increase with respect to concentration, see figure 5a. Analysis of the relationship between the increasing electrochemical signal and copper concentration was found to be linear across the concentration range10 - 750 nM, see figure 5b. From this linear range the sensitivity of the microband electrode arrays was estimated to be 2.8 μA/ cm^2/M

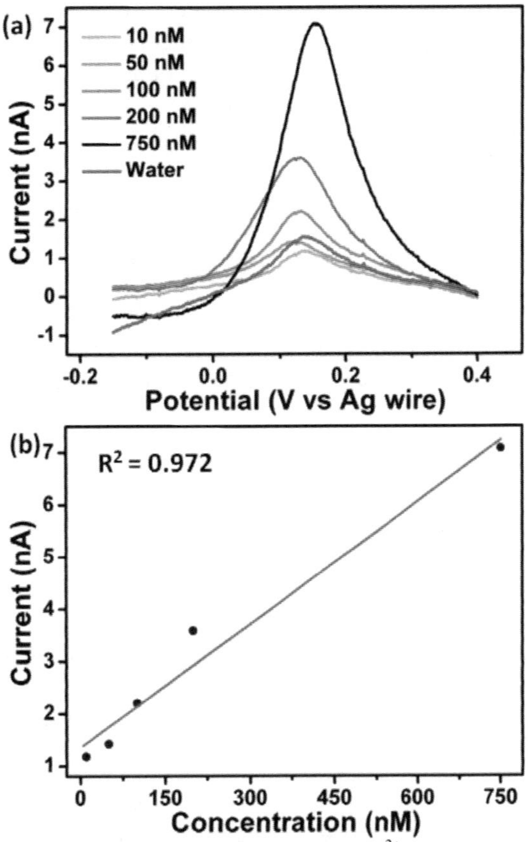

Figure 5: (a) UPD-SV of a series of Cu^{2+} solutions and sample in 0.1M H_2SO_4 at a gold microband array electrode. Scan rate: 5 $V.s^{-1}$. Deposition time: 20s. (b) Evaluation of the dependence of electrochemical signal vs. concentration for a deposition time of 20s.

To investigate application of gold microband arrays for copper detection, determination of copper was undertaken for potable water samples. A domestic sample containing unknown copper concentration was diluted with 0.1 M H_2SO_4 to bring it within the linear range of method. A dilution of 1:100 turned out to be the most suitable sample preparation and UPD-ASV was carried out under conditions described above, the resulting LSV is shown in figure 5a (red line). From the calibration plot obtained for known concentrations of CuSO4, the concentration of copper in the real sample was found to be 1.25 mg/L (ppm), which is lower than the current EU standard of 2 mg/L (ppm) in drinking water (EU Directive 98/83). This work clearly shows the potential of the UPD-ASV technique for the rapid and quantitative detection and measurement of copper ions in real sample matrices.

4. Conclusions:

Gold microband electrode arrays with highly reproducible dimensions were fabricated using photolithography metal deposition and lift-off methods. Cyclic voltammetry in 0.1 M sulphuric acid was employed to confirm the functionality of the microbands as electrochemical devices. Under potential deposition coupled with anodic stripping voltammetry was applied to the microband electrodes as a method of trace metal analysis of copper in 0.1 M H2SO4. The optimized under-potential was assessed to be -0.16 V vs Ag wire and linear sweep voltammetry was used to record the stripping electrochemical response at fast scan rates (5 $V.s^{-1}$). The optimum deposition time at microband electrodes was evaluated to be 20 s, applying this deposition interval to measurements in increasing concentrations of Cu_2SO_4 generated a linear calibration curve across a range of nanomolar concentrations. Detection of copper at concentrations as low as 10 nM was easily achievable. Increasing sensitivity could be simply achieved by increasing deposition time. A sensitivity of 2.8 $\mu A/cm^2/M$ was estimated for a deposition time of 20s. To assess potential real sample applications for these microband electrodes, determination of copper concentration present in tap water was measured using UPD-ASV. From the resulting peak current the concentration of the unknown sample was estimated to be 1.25 mg/L of Cu^{2+}. Future work is now focusing on detection of a variety of other metals including lead, cadmium, selenium and chrome ions in complex matrices.

5. Acknowledgments:

This work was supported by Science Foundation Ireland under the Research Frontiers Programme (SFI/09/RFP/CAP2455).

6. References:

[1] D. W. M. Arrigan, "Nanoelectrodes, nanoelectrode arrays and their applications," *Analyst,* vol. 129, pp. 1157-1165, 2004.
[2] K. Dawson, *et al.*, "Nanofabrication of Robust Nanoelectrodes for Electrochemical Applications," *ECS Transactions,* vol. 28, 2010.
[3] C. G. Zoski, "Ultramicroelectrodes: Design, fabrication, and characterization," *Electroanalysis,* vol. 14, pp. 1041-1051, Sep 2002.
[4] C. G. Zoski and M. Wijesinghe, "Electrochemistry at Ultramicroelectrode Arrays and Nanoelectrode Ensembles of Macro- and Ultramicroelectrode Dimensions," *Israel Journal of Chemistry,* vol. 50, pp. 347-359, Oct 2010.
[5] L. Sanford, *et al.*, "Voltammetric Detection of Hydrogen Peroxide at Carbon Fiber Microelectrodes," *Analytical Chemistry,* vol. 82, pp. 5205-5210, Jun 2010.
[6] Berduque, *et al.*, "Voltammetric characterisation of silicon-based microelectrode arrays and their application to mercury-free stripping voltammetry of copper ions," *Talanta,* vol. 71, pp. 1022-1030, Feb 2007.
[7] J. F. Huang and B. T. Lin, "Application of a nanoporous gold electrode for the sensitive detection of copper via mercury-free anodic stripping voltammetry," *Analyst,* vol. 134, pp. 2306-2313, 2009.
[8] S. C. o. Food, "Opinion of the Scientific Committee on Food on the Tolerable Uppper Intake Level of Copper," H. C. P. Directorate-General, Ed., ed. Brussels, 2003.
[9] G. Herzog and D. W. M. Arrigan, "Electrochemical strategies for the label-free detection of amino acids, peptides and proteins," *Analyst,* vol. 132, pp. 615-632, 2007.
[10] Belmont, *et al.*, "Mercury-plated iridium-based microelectrode arrays for trace metals detection by voltammetry: Optimum conditions and reliability," *Analytica Chimica Acta,* vol. 329, pp. 203-214, Aug 1996.

AUTHOR INDEX

AUTHOR INDEX